Few-Body Systems

Editor in Chief: W. Plessas, Graz

Supplement 12

Few-Body Problems in Physics '99

Proceedings of the 1st Asian-Pacific Conference, Tokyo, Japan, August 23–28, 1999

Edited by
S. Oryu, M. Kamimura, S. Ishikawa

SpringerWienNewYork

Prof. Shinsho Oryu
Department of Physics, Faculty of Science and Technology
Science University of Tokyo, Chiba, Japan

Prof. Masayasu Kamimura
Department of Physics, Faculty of Science
Kyushu University, Fukuoka, Japan

Prof. Souichi Ishikawa
Department of Physics, Faculty of General Education
Hosei University, Tokyo, Japan

Typesetting: Camera-ready by authors
Printing: Druckerei Novographic, A-1230 Wien
Binding: Papyrus, A-1100 Wien

Printed on acid-free and chlorine-free bleached paper
SPIN: 10758524

With 232 Figures

CIP data applied for

ISSN 0177-8811
ISBN 3-211-83503-2 Springer-Verlag Wien New York

Local Organizing Committee

T. Nishikawa (President, SUT)
S. Oryu (Chairman, SUT)
Y. Akaishi (KEK-Tanashi)
R. Hayano (Tokyo)
C. Ishii (SUT)
S. Ishikawa (Scientific secretary, Hosei)
H. Kamada (Bochum)
M. Kamimura (Kyushu)
(Chair of PAC)
Y. Koike (Hosei)
T. Nagae (KEK)
M. Oka (TIT)
K. Sagara (Kyushu)
H. Sakai (Tokyo)
Y. Suzuki (Niigata)
M. Wada (Funabashi)

Programme Advisory Committee

K. Asahi (TIT)
T. Kajino (NAOJ)
T. Katayama (Sapporo)
K. Kato (Hokkaido)
M. Matsuda (Hiroshima)
K. Miyagawa (Okayama)
T. Motoba (Osaka)
T. Ueda (Ehime)
Y. Yano (RIKEN)

Local Advisory Committee

H. Ejiri (RCNP)
K. Ikeda (RIKEN)
H. Kamimura (SUT)
K. Nagamine (KEK/RIKEN)
M. Nagamiya (KEK)
K. Nakai (SUT)
I. Tanihata (RIKEN)
H. Toki (RCNP)
T. Yamazaki (JSPS)
K. Yazaki (Tokyo)

International Advisory Committee

I.R. Afnan (Flinders)
B.L.G. Bakker (Vrije)
C.-G. Bao (Zhongshan)
H.C. Bhang (Seoul)
S. Boffi (Pavia)
V.V. Burov (JINR)
H.W. Fearing (TRIUMF)
A.C. Fonseca (Lisbon)
J.L. Friar (Los Alamos)
B.F. Gibson (Los Alamos)
W. Glöckle (Bochum)
F. Gross (TJNAF)
W.-Y. Pauchy Hwang (Taipei)
Y.E. Kim (Purdue)
T.-S. H. Lee (Argonne)
C.-S. Lee (Chung-Ang)
D.-P. Min (Seoul)
A.N. Mitra (Delhi)
W. Plessas (Graz)
J.-M. Richard (Grenoble)
D.O. Riska (Helsinki)
P.U. Sauer (Hannover)
K.K. Seth (Northwestern)
I. Sick (Basel)
S.A. Sofianos (South Africa)
A.W. Thomas (Adelaide)
W. Tornow (TUNL)
F.G. Wang (Nanjing)
S.N. Yang (Taipei)
Z.-Y. Zhang (IHEP)

Sponsors

Frontier Research Center for Computational Science (FRCCS), Science University of Tokyo

High Energy Accelerator Research Organization (KEK)

The Institute of Physical and Chemical Research (RIKEN)

The Research Center for Nuclear Physics (RCNP), Osaka University

Inoue Foundation for Science

The Ogasawara Foundation for the Promotion of Science & Engineering

Sun Microsystems K.K.

Fuji Zerox

Foreword

The first Asia-Pacific Conference on Few-Body Problems in Physics took place from August 23 to August 28, 1999, at the Noda campus of the Science University of Tokyo in Noda-city and Sawayaka Chiba Kenmin Plaza in Kashiwa-city, a suburb of Tokyo close to the Narita-Tokyo International Airport, with the Frontier Research Center for Computation Sciences (FRCCS) of the Science University of Tokyo as the host institute. The High Energy Accelerator Research Organization (KEK), the Institute of Physical and Chemical Research (RIKEN), the Research Center for Nuclear Physics (RCNP)-Osaka University, the Physical Society of Japan, and the Association of Asia Pacific Physical Societies (AAPPS) supported this conference.

The conference was initiated in the Asia Pacific area as a counterpart to the successful European Conference on Few-Body Problems in Physics (APFB99), in addition to the International Few-Body Conference Series and the Few-Body Gordon Conference series in North America. The Physics of Few-Body Problems covers, as is well known, systems with finite numbers of particles in contrast to many-body systems with very large numbers of particles. Therefore, it covers such wide fields as mesoscopic, atom-molecular, exotic atom, nucleon, hyperon, and quark-gluon physics, plus their applications.

Each of these fields encompasses not only few-body systems but also many-body systems. However, the methods used to approach these different systems sometimes vary, in the quality of the precision required and the approximation techniques utilized. One of the most important aspects to spotlight in few-body physics is that it is based soundly on the fundamental principles of quantum mechanics. The conference, unified by the theme of few-body systems, had as its mission making possible opportunities for the mutual exchange of the technical information among the individual fields plus the creation of new opportunities for (even new fields of) research in physics.

In addition, I would like to stress that this conference was the first such meeting planned in the Asia-Pacific region. Therefore, it had another goal, namely, that APFB99 should encourage young physicists to generate a new wave in this region. It was planned for young scientists to be given an opportunity to experience few-body physics at a higher level, by inviting distinguished few-body physicists in the world to participate and then giving young physicists a chance to present their new results at this conference.

The Asian region is expected to develop rapidly, both scientifically and technologically, in the coming century. Therefore, the aim of the conference was to promote a positive step into the 21st century, based on the theme of "Few-Body Systems".

One of the positive features of this conference was the absence of many parallel sessions; there were only two in the afternoon following the plenary session each morning. Another was the opportunity for intergenerational interactions. APFB99 was organized by a group of few-body physicists from different universities and national institutes in Japan, individuals from diverse fields working on few-body problems such as Hadron Physics, Atom-Molecular Physics, and

Solid State Physics. The core of the organizing committee was, nevertheless, located in the Tokyo area including Noda. Cyberspace communications were very helpful. Furthermore, we were fortunate that one of the editors (S.I.) belonged to FRCCS of the Science University of Tokyo for one year at the Noda Campus. In preparing the scientific program, we largely relied on the advice of the International Advisory Committee. Significant time in the plenary sessions was devoted to non-hadronic physics, and we tried to incorporate new topics at the border of solid state physics (such as nano-physics and micro-clusters) and topics related to few-body problems in some medium.

We also attempted to expand our world dynamically by including talks from neighboring fields for the few-body physics applications such as those which appear inastrophysics and unstable nuclei, together with quantum molecular dynamics as numerical calculation techniques. Contact with the physical world was ensured by the experimental talks in the program.

The Few-Body Problem has a rather long history in Japan, starting from the theory of the nuclear force based upon the meson exchange picture of Dr. Yukawa. Especially in the 1950s were there many nuclear and particle physicists in Japan concerned with few-nucleon systems. The Yamaguchi separable potential and the Fujita-Miyazawa three-body force are still used in our present day discussions. Both distinguished physicists were invited to this conference.

The APFB99 conference, attended by 151 physicists and 12 accompanying person from 17 countries, provided 31 invited talks, 68 oral presentations and 55 posters. Therefore, we kept a five-day program plus a half-day excursion, although that meant the inconvenience of changing the conference site from the Noda campus of the Science University of Tokyo to Sawayaka Chiba Kenmin Plaza in Kashiwa, because the official government holiday on Monday, the 23rd, was included in the conference dates. Nevertheless, we believe that the conference site was proper for an international conference and cost effective for the participants.

World-wide international Conferences in Few-Body Physics were held in Delhi in 1976, Eugene in 1980, Tokyo/Sendai in 1986, Vancouver in 1989, and Adelaide in 1993. However, the Asia-Pacific Conference on Few-Body Problems in Physics was a first attempt at a regional meeting. It was a delight that the international advisory committee decided to make it a series in this form. One of the most impressive aspects of the meeting was the energy of the young scientists in this region, which had a lasting impact on the participants. Dr. Benjamin F. Gibson, for instance, who gave the concluding talk made a surprise award for the best young-scientist presentation. We may hope that the award will be continued in this conference series. We are all anticipating scientific breakthroughs in this field in the next century. They will without doubt come from a new point of view, with new energy, and in a new environment, so that such encouragement of the next generation is the key to the future.

Finally, on behalf of the organizing committee I would like to thank all of the participants who were kind enough to come, and all the international/local advisory committee members who gave us their helpful suggestions. We are also very much obliged to the members of the secretariat Dr. Eizo Uzu, Dr. Shino

Nemoto of SUT, and Dr. Emiko Hiyama of RIKEN, and all the student volunteers of SUT for their devoted service during the conference. The conference was financially supported by the Inoue Foundation for Science, the Ogasawara Foundation for the Promotion of Science & Engineering, the FRCCS of the Science University of Tokyo, and SUN Microsystems K.K. APFB99 could not have been such a success without their kindness.

Noda, January 7, 2000

For the Organizing Committee
Shinsho Oryu

Contents

Atomic and Mesoscopic Systems

Few-Body Problems in Nuclear Astrophysics

Unstable Nuclei and Nuclear Cluster Systems

Hadronic Structure and Quantum Chromodynamics

Relativity

Electromagnetic Interactions

Hypernuclei and YN and YY Interactions

Few-Nucleon Systems

Summary

Few-Body Systems Suppl. 12, 1–6 (2000)

Few-Body Systems

Antiprotonic Helium – A Window into the Antiworld

R.S. Hayano*

Department of Physics, University of Tokyo, 7-3-1 Hongo, Bunkyo-ku, Tokyo 113-0033, Japan

Abstract. The antiprotonic helium, $\overline{p}e^-$ $He^{2+} \equiv \overline{p}He^+$, first discovered at KEK and studied in detail at LEAR, is a unique metastable existence interfacing between matter and antimatter. Our recent high-resolution laser spectroscopy of $\overline{p}He^+$ has reached a precision of 0.5 ppm, and the agreement between our experimental values of transition energies and the calculations has become better than 1 ppm. This agreement in turn sets a severe constraint on the antiproton charge and mass. Future possibilities at the new antiproton decelerator (AD) are also discussed.

1 Introduction

Antiprotonic helium ($\overline{p}He^+$) is an exotic 3-body metastable system consisting of an antiproton, an electron and an alpha particle. We first noticed its existence at KEK when we found that about 3% of antiprotons stopped in liquid helium survived with a lifetime of about 3μs[1]. Subsequently, we measured at LEAR delayed annihilation time spectra (DATS) of antiprotons in various helium media[2], and found that the longevity persists over a wide range of the target helium density (solid to gas).

After an antiproton is captured on a helium atom by replacing one of the two electrons, the $\overline{p}$ usually cascades down by ejecting the remaining electron, and an ionized $\overline{p}He^{++}$ is formed. This ion is quickly quenched by the Stark collision, leading to the prompt annihilation of $\overline{p}$ (see Fig. 1). This is the fate of $\sim 97\%$ of antiprotons stopped in helium. However, if the $\overline{p}$ possesses a large principle quantum number of $n \sim 38$ and a large orbital angular momentum $l \sim n-1$, Auger transitions are much suppressed. Radiative transitions, which tend to follow the $v = n - l - 1 =$ constant propensity rule, are typically $\sim \mu$s per step, and the electron which remains protects the system against quenching collisions. Hence a three-body neutral $\overline{p}He^+$ atom becomes metastable. Due to

* *E-mail address:* hayano@phys.s.u-tokyo.ac.jp

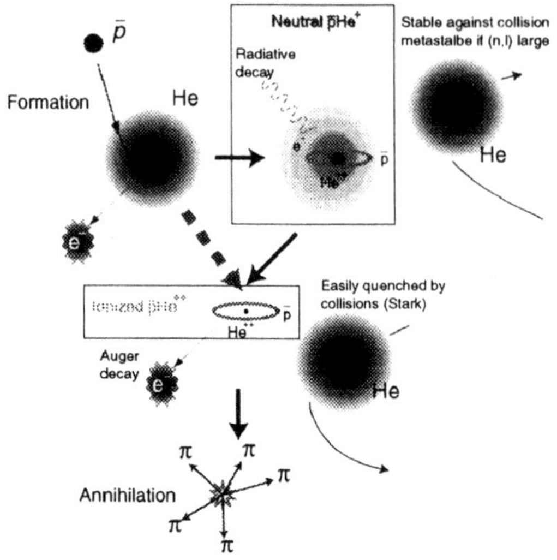

Figure 1. Formation and decay of the metastable $\overline{p}He^+$ atomcule.

its long lifetime, it has become possible to study this atom in detail using a powerful method of high-precision laser spectroscopy.

2 Laser spectroscopy

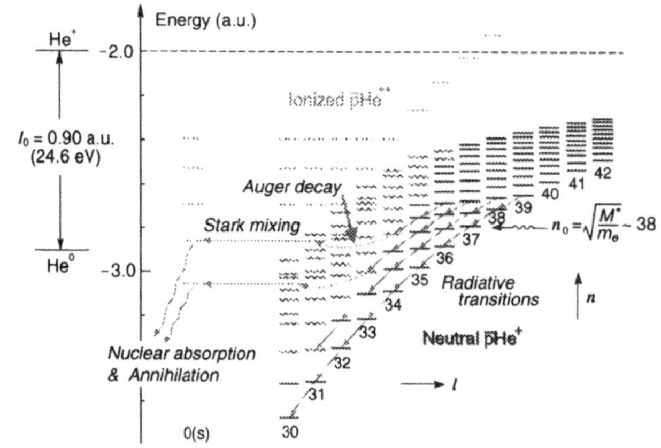

Figure 2. The level scheme of large-(n, l) states of the $\overline{p}He^+$ atomcule. The solid lines indicate radiation-dominated metastable states, while the broken lines are for Auger-dominated short-lived states. Laser-induced resonant annihilation can be observed by 'shooting down' antiprotons from one of the metastable states to a short-lived state.

Fig. 2 shows the level scheme of large-(n,l) states of $\overline{p}He^+$. The solid lines indicate radiation-dominated metastable states, while the broken lines are for Auger-dominated short-lived states. Around $n \sim 38$, where antiprotons are initially captured, the $\Delta n = 1$ level spacing is about 2 eV, hence tunable lasers can be used for high precision studies.

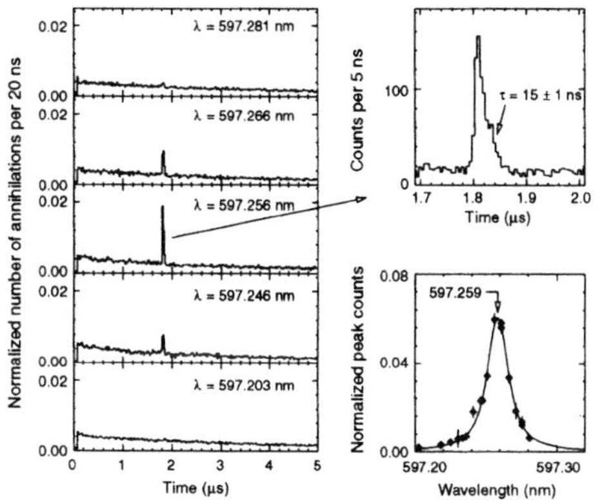

Figure 3. The first observation of laser-induced resonant annihilation in metastable antiprotonic helium atoms. The resonance, observed at 597 nm, was assigned to the $(n,l) = (39,35) \rightarrow (38,34)$ transition. Spikes due to forced annihilation through the resonance transitions are seen at 1.8 μs, when the laser wavelength is adjusted to hit the resonance.

The laser spectroscopy method we developed[3] is to induce a laser-resonant transition between a metastable level and a short-lived level (for example, $(n,l) = (39,35) \rightarrow (38,34)$). We stopped the 200 MeV/c antiprotons from the LEAR storage ring in a cryogenic helium gas target (typically cooled to 5 K at 0.5 bar). For each incoming $\overline{p}$ with no prompt annihilation, we assumed that a metastable $\overline{p}He^+$ was formed in the target, triggered a tunable pulsed dye laser, and measured the annihilation time spectrum. This was repeated until enough ($\sim 10^{5-6}$) events were recorded, when we stepped the laser to the next wavelength.

In 1993, we succeeded for the first time to observe a laser resonant transition at 597 nm ($(n,l) = (39,35) \rightarrow (38,34)$, see Fig. 3)[4], after about a week of scanning around the theoretically predicted wavelengths, which typically had precisions of about 1000 ppm. In 1995, when Korobov achieved a much better precision of 50 ppm[5], our resonance search was greatly accelerated, and we have so far measured the 13 transitions shown in Fig. 4. The agreement between experiment and theory became better than about 5 ppm when relativistic corrections were included[6].

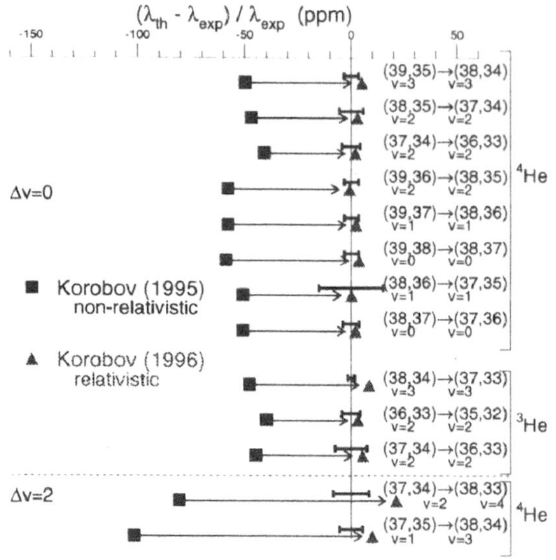

Figure 4. Comparison of the experimental wavelengths of various transitions with Korobov predictions (closed squares without [5] and closed triangles with [6] relativistic corrections). The upper part is for $\Delta v = \Delta(n-l-1) = 0$ intraband transitions and the lower part is for $\Delta v = 2$ interband transitions. The error bars are the experimental ones.

3 Antiproton Rydberg Constant

At this point, we improved the band width of our laser system and performed a more careful scan of the resonance lines. We then found that the resonance vacuum wavelengths for the known transition $(39,35) \rightarrow (38,34)$ depend linearly on the target helium density[7] (theoretical accounts of the density dependence of the energy levels and widths are discussed by Korenman[8] and by Bakalov[9]). Hence, when we compare the experimental values with results of theoretical calculations, we must take the zero-density limit. The wavelength extrapolated to zero-density limit for the $(39,35) \rightarrow (38,34)$ transition was found to be $\lambda_0 = 597.2570 \pm 0.0003$ nm.

This value, with a 0.5-ppm precision, was compared with the result of recent theoretical calculations[6, 10, 11, 12] on the energy of the Coulombic three-body system (see Fig. 5), including relativistic corrections and the Lamb shift. The agreement between our experimental values and the calculations is better than 1×10^{-6}. This excellent agreement in turn provides a precise value of the antiproton Rydberg constant that surpasses the currently known precision of 50 ppm, and sets a severe constraint on the antiproton charge $(-Q_{\overline{p}})$ and the mass $(M_{\overline{p}})$ that both $|Q_p - Q_{\overline{p}}|/e$ and $|M_p - M_{\overline{p}}|/M_p$ be less than 5×10^{-7}, when a more precisely known constraint on the charge-to-mass ratio[13] is combined.

If we assume the CPT invariance between the properties of the proton and

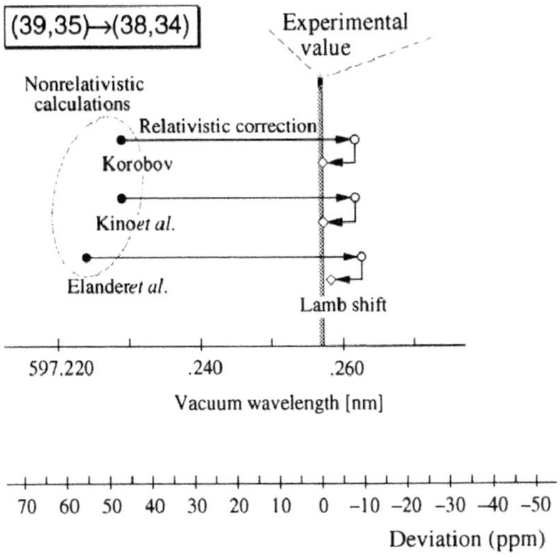

Figure 5. Experimental values of the vacuum wavelengths for the transition $(39,35) \to (38,34)$ compared with recent theoretical values.

the antiproton, as the theoretical calculations do, the agreement is a signature of the excellence of theoretical treatments and calculation techniques of the Coulombic three-body system including QED corrections. On the other hand, if we take the calculation results for granted, this in turn gives a stringent test of the fundamental constants of the antiproton and therefore tests the CPT theorem. We have thus opened a new possibility of determining fundamental constants of the antiproton.

4 ASACUSA at CERN AD

Although LEAR was prematurely shut down at the end of 1996, a new antiproton decelerator (AD) will start its operation in the fall of 1999. One of the first important experiments proposed by the ASACUSA (Atomic Spectroscopy And Collisions Using Slow Antiprotons) collaboration[14] is to precisely measure the hyperfine splitting of antiprotonic helium. Although we have already succeeded to resolve the 1.7 GHz hyperfine splitting in the $(37,35) \to (38,34)$ laser transition with a precision of about 3 %[15], we plan to achieve a much higher precision of $< 10^{-4}$ at AD, using a laser-microwave resonance technique.

We also plan to make an ultra low energy ($E_{\overline{p}} \sim 10 - 1000$eV) antiproton beam using an RFQ linear decelerator and an antiproton catching trap. It will then become possible to make antiprotonic helium and other species of antiprotonic atoms in a near-vacuum condition, perform a much higher precision spectroscopy of these atoms, and to determine antiprotonic charge, mass and magnetic moment with high precision.

Acknowledgement. The author would like to thank CERN PS205 and ASACUSA collaboration members, Drs. T. Yamazaki, J. Eades, E. Widmann, N. Morita, H. Torii and D. Horvath in particular. Special thanks are also due to Drs. V.I. Korobov, G.Ya Korenman, D.D. Bakalov and Y. Kino for stimulating discussions. This work was supported by the Grant-in-Aid of the Japanese Ministry of Education, Science, Sports and Culture.

References

1. M. Iwasaki et al.: Phys. Rev. Lett. **67**, 1246 (1991)
2. T. Yamazaki et al.: Nature **361**, 238 (1993)
3. N. Morita et al.: Nucl. Instr. Meth. A **330**, 439 (1993); H.A. Torii et al.: Nucl. Instr. Meth. A **396**, 257 (1997)
4. N. Morita et al.: Phys. Rev. lett. **72**, 1180 (1994)
5. V.I. Korobov: Phys. Rev. A **54**, R1749 (1996)
6. V.I. Korobov and D.D. Bakalov: Phys. Rev. Lett. **79**, 3379 (1997)
7. H.A. Torii et al.: Phys. Rev. A **59**, 223 (1999)
8. G.Ya Korenman: talk at this conference
9. D.D. Bakalov: private communication
10. N. Elander and E. Yarevsky: Phys. Rev. A **56**, 1855 (1997); **58**, 2256(E) (1998)
11. Y. Kino, M. Kamimura, H. Kubo: in *Proceedings of the XVIIth RCNP International Symposium on Innovative Computational Methods in Nuclear Many-Body Problems, Osaka 1997*; Y. Kino: talk at this conference
12. V.I. Korobov: talk at this conference
13. G. Gabrielse et al.: Phys. Rev. Lett. **82**, 3198 (1999)
14. T. Azuma et al.: ASACUSA proposal, CERN/SPSC 97-19
15. E. Widmann et al.: Phys. Lett. B **404**, 15 (1997)

Few-Body Systems Suppl. 12, 7–14 (2000)

Few-Body Systems

Effective Linear Two-Body Method for Many-Body Problems in Atomic and Nuclear Physics

Y.E. Kim, A.L. Zubarev

Purdue Nuclear and Many-Body Theory Group (PNMBTG)
Department of Physics, Purdue University
West Lafayette, Indiana 47907, U.S.A.

Abstract. We present an equivalent linear two-body method for the many body problem, which is based on an approximate reduction of the many-body Schrödinger equation by the use of a variational principle. The method is applied to several problems in atomic and nuclear physics.

1 Introduction

In this paper we present an approximate method of obtaining the eigenvalue solutions of the system of interacting many (N) bosons using an equivalent two-body method [1] which is based on an approxmiate reduction of the many-body Schrödinger equation by the use of a variational principle. As tests of the accuracy for our method, we have applied the method to the one-dimensional N-body problem with pair-wise contact interactions for which exact solutions are known and have demonstrated that our method is a very good approximation for the case of large N. We have also considered the Bose-Einstein condensation (BEC) of atoms in anisotropic harmonic traps. For both cases, it has been shown that the method gives excellent results for large N [1]. Recently, we have applied our method to estimate nuclear fusion rates for integer spin nuclei confined in an isotropic ion trap [2].

2 Equivalent Linear Two-Body (ELTB) Method

For the N-body system, our method for obtaining the equivalent linear two-body (ELTB) equation consists of two steps.

The first is to give the N-body wave function $\Psi(\mathbf{r}_1, \mathbf{r}_2, ...)$ a particular functional form

$$\Psi(\mathbf{r}_1, \mathbf{r}_2, ...) \approx \tilde{\Psi}(\zeta_1, \zeta_2, \zeta_3), \tag{1}$$

where ζ_1, ζ_2,and ζ_3 are known functions. We limit ζ's to three variables in order to obtain the ELTB equation.

The second step is derivation of an equation for $\tilde{\Psi}(\zeta_1, \zeta_2, \zeta_3)$ by requiring that $\tilde{\Psi}$ must satisfy a variational principle

$$\delta < \tilde{\Psi}|H|\tilde{\Psi} >= 0 \tag{2}$$

with a subsidiary condition $< \tilde{\Psi}|\tilde{\Psi} >= 1$. H is the Hamiltonian.

This leads to a linear two-body equation from which both eigenvalues and eigenfunctions can be obtained. The lowest eigenvalue is an upper bound of the lowest eigenvalue of the original N-body problem.

3 McGuire-Yang N-body Problem

For the one-dimensional N-body problem with the Hamiltonian

$$H = -\frac{1}{2}\sum_{i=1}^{N} d^2/dx_i^2 + c\sum_{i<j}\delta(x_i - x_j),$$

the Schrödinger equation

$$H\Psi = E\Psi$$

is exactly solvable. The bound and scattering states for this system have been found by McGuire [3] and by Yang [4].

For the case $c < 0$, there are bound states [3] for the system of N bosons with the wave function of the following form

$$\Psi = \exp[(c/2)\sum_{i<j} \mid x_i - x_j \mid],$$

and the energy of this bound state is given by

$$E = -c^2 N(N^2 - 1)/24. \qquad \text{(Exact Solution)} \tag{3}$$

To obtain the ELTBE Equation for the McGuire-Yang problem, we seek eigenfunctions Ψ of H in the form of

$$\Psi \approx \tilde{\Psi}(\rho),$$

where $\rho = \sqrt{\frac{1}{N}\sum_{i<j}(x_i - x_j)^2}$.

The ELTB equation for $\tilde{\Psi}(\rho)$ is

$$[-\frac{d^2}{d\rho^2} + \frac{(N-2)}{\rho}\frac{d}{d\rho} + 2V(\rho)]\tilde{\Psi}(\rho) = 2E\tilde{\Psi}(\rho), \tag{4}$$

where

$$V(\rho) = \tilde{g}/\rho, \tag{5}$$

with

$$\tilde{g} = cN(N-1)\frac{\Gamma(N/2-1/2)}{2\sqrt{2\pi}\Gamma(N/2-1)} \tag{6}$$

With the Coulomb like potential $V(\rho) = \tilde{g}/\rho$, Eq. (5), Eq.(4) can be solved analytically.

In the case of large N, we obtain

$$E_0(N) = -\frac{c^2}{8\pi}N^3$$

On the other hand we have for the large N case from Eq. (3), we have

$$E = -\frac{c^2}{24}N^3(1 + O(\frac{1}{N^2})).$$

Therefore, for the McGuire-Yang N-body problem, we have demonstrated that the ELTBE method, Eqs. (4 - 6), is a very good approximation for the case of large N (the relative error for binding energy is about 4.5 %).

4 Anisotropic Trap (Bose-Einstein Condensation)

We consider N identical bosonic atoms confined in a harmonic anisotropic trap with the following Hamiltonian

$$H = -\frac{\hbar^2}{2m}\sum_{i=1}^{N}\Delta_i + \frac{1}{2}\sum_{i=1}^{N}m(\omega_x^2 x_i^2 + \omega_y^2 y_i^2 + \omega_z^2 z_i^2) + \sum_{i<j}V_{int}(\vec{r}_i - \vec{r}_j), \tag{7}$$

For eigenfunction Ψ of H, we assume the solution for Ψ has the following form

$$\Psi(\vec{r}_1, ...\vec{r}_N) \approx \frac{\tilde{\Psi}(x,y,z)}{(xyz)^{(N-1)/2}}, \tag{8}$$

where

$$x^2 = \sum_{i=1}^{N} x_i^2,\ y^2 = \sum_{i=1}^{N} y_i^2,\ z^2 = \sum_{i=1}^{N} z_i^2. \tag{9}$$

We now derive an equation for $\tilde{\Psi}(x,y,z)$ by requiring that $\tilde{\Psi}(x,y,z)$ must satisfy the variational principle (2). This requirement leads to the equation

$$\tilde{H}\tilde{\Psi} = E\tilde{\Psi}, \tag{10}$$

where

$$\begin{aligned}\tilde{H} = &-\frac{\hbar^2}{2m}(\frac{\partial^2}{\partial x^2} + \frac{\partial^2}{\partial y^2} + \frac{\partial^2}{\partial z^2}) + \frac{m}{2}(\omega_x^2 x^2 + \omega_y^2 y^2 + \omega_z^2 z^2) \\ &+\frac{\hbar^2}{2m}\frac{(N-1)(N-3)}{4}(\frac{1}{x^2} + \frac{1}{y^2} + \frac{1}{z^2}) + V(x,y,z),\end{aligned} \tag{11}$$

with

$$V(x,y,z) = \frac{N(N-1)}{2(2\pi)^{3/2}}(\frac{\Gamma(N/2)}{\Gamma((N-1)/2)})^3\frac{1}{xyz}G(x,y,z), \tag{12}$$

and

$$G(x,y,z) = \int_{-\sqrt{2}x}^{\sqrt{2}x} dx' \int_{-\sqrt{2}y}^{\sqrt{2}y} dy' \int_{-\sqrt{2}z}^{\sqrt{2}z} dz' V_{int}(\sqrt{(x')^2+(y')^2+(z')^2})$$

$$\times((1-\frac{(x')^2}{2x^2})(1-\frac{(y')^2}{2y^2})(1-\frac{(z')^2}{2z^2}))^{\frac{N-3}{2}} \tag{13}$$

To the best of our knowledge, Eqs. (12) and (13) have not been discussed in the literature.

Solutions of these equations give for the ground state energy

$$\frac{E}{N\hbar\tilde{\omega}} = \frac{5}{4}\tilde{n}^{\frac{2}{5}} \tag{14}$$

where $\tilde{n} = 2\sqrt{\tilde{\omega}m/(2\pi\hbar)}Na$ (a is the scattering length) and $\tilde{\omega} = (\omega_x\omega_y\omega_z)^{\frac{1}{3}}$. We note that Eq.(14) is the exact ground state solution of Eq.(11) for large N.

For the case of large N we can obtain an essentially exact expression for the ground state energy by neglecting the kinetic energy term in the Ginzburg-Pitaevskii-Gross (GPG) equation [5] (the Thomas-Fermi approximation [6]) as

$$\frac{E_{TF}}{N\hbar\tilde{\omega}} = \frac{5}{7}(\frac{15}{8}\sqrt{\pi})^{\frac{2}{5}}\tilde{n}^{\frac{2}{5}} \tag{15}$$

Comparing Eq. (14) with Eq. (15), we can see that for the case of large N, the ELTBE method is a very good approximation, with a relative error of about 8% for the binding energy.

5 Nuclear Fusion for Bose Nuclei Confined in Ion Traps

We consider N identical charged Bose nuclei confined in an ion trap with the following Hamiltonian

$$H = -\frac{\hbar^2}{2m}\sum_{i=1}^{N}\Delta_i + \frac{1}{2}m\omega^2\sum_{i=1}^{N}r_i^2 + \sum_{i<j}\frac{e^2}{|\mathbf{r}_i - \mathbf{r}_j|} \tag{16}$$

where m is the rest mass of the nucleus.

For ground state wave function Ψ we use the following approximation [1]

$$\Psi(\vec{r}_1,...\vec{r}_N) \approx \tilde{\Psi}(\rho) = \frac{\Phi(\rho)}{\rho^{(3N-1)/2}}, \tag{17}$$

where

$$\rho = \left[\sum_{i=1}^{N} r_i^2\right]^{1/2}. \tag{18}$$

Instead of the above variable ρ we introduce a new quantity $\tilde{\rho}$ defined as

$$\tilde{\rho} = \sqrt{\frac{m\omega}{\hbar}}\rho. \tag{19}$$

Substitution of Eq. (19) into Eq. (16) leads to the following ELTB equation

$$\frac{\hbar\omega}{2}[-\frac{d^2}{d\tilde{\rho}^2} + \tilde{\rho}^2 + \frac{(3N-1)(3N-3)}{4\tilde{\rho}^2} + \frac{\tilde{\gamma}}{\tilde{\rho}}]\Phi = E\Phi, \tag{20}$$

where

$$\tilde{\gamma} = \alpha\sqrt{\frac{mc^2}{\hbar\omega}}\frac{2N\Gamma(3N/2)}{3\sqrt{2\pi}\Gamma(3N/2-3/2)} \tag{21}$$

with $\alpha = e^2/(\hbar c) \approx 1/137$.

The ground state solution of Eq.(20) has the following form

$$\Phi(\tilde{\rho}) = \sum_i c_i \tilde{\rho}^{\frac{3N-1}{2}} e^{-(\tilde{\rho}/\alpha_i)^2/2}. \tag{22}$$

For our case of N Bose nuclei, to account for a short range nature of nuclear forces between two nuclei, we introduce the following Fermi pseudo-potential $V^F(\vec{r})$ [2, 7],

$$ImV^F(\vec{r}) = -\frac{A\hbar}{2}\delta(\vec{r}), \tag{23}$$

where the nuclear rate constant A is given by

$$A = \frac{2Sr_B}{\pi\hbar}. \tag{24}$$

and $r_B = \frac{\hbar^2}{2\mu e^2}$, and S is the S-factor for the nuclear fusion between two nuclei.

6 Fusion Probability and Rates

For N identical Bose nuclei confined in an ion trap, the nucleus-nucleus fusion rate is determined from the trapped ground state wave function Ψ as

$$\tilde{R} = -\frac{2}{\hbar}\frac{\sum_{i<j} < \Psi|ImV^F_{ij}|\Psi >}{< \Psi|\Psi >}, \tag{25}$$

where ImV^F_{ij} is the imaginary part of the Fermi potential, given by Eq.(23).

The substitution of Eq.(17) into Eq.(25) yields

$$\tilde{R} = \frac{AN(N-1)\Gamma(3N/2)}{2(2\pi)^{3/2}\Gamma(3N/2-3/2)}\frac{\int_0^\infty \Phi^2(\rho)\frac{1}{\rho^3}d\rho}{\int_0^\infty \Phi^2(\rho)d\rho}. \tag{26}$$

For large N, the sum in Eq.(22) reduces to one term

$$\Phi(\rho) \approx \tilde{\rho}^{\frac{3N-1}{2}} e^{-(\tilde{\rho}/\alpha_t)^2/2} \tag{27}$$

where $\alpha_t = (\zeta/3)^{1/3}$, and $\zeta \approx \sqrt{\frac{mc^2}{2\pi\hbar\omega}}\alpha N$.

Using our ground state solution, we obtain from Eq.(26)

$$\tilde{R} = \frac{3AN}{4\pi\alpha}\sqrt{\frac{\hbar\omega}{mc^2}}(\frac{m\omega}{\hbar})^{3/2}. \tag{28}$$

We can rewrite Eq.(28) as

$$\tilde{R} = BN\omega^2. \tag{29}$$

where

$$B = \frac{3A}{4\pi\alpha}(\frac{m}{\hbar c}). \tag{30}$$

If the probability of the ground state occupation, Ω, is taken into account, the trap fusion rate is $R_t = \Omega\tilde{R}$. We expect that this probability is very small for a low density static case in the absence of an external potential [8]. However, Ω may not be very small for dynamic cases of Bose nuclei confined in ion traps in equilibrium situations.

Using Eq.(29) we have for the trap fusion rate R_t

$$R_t = B\Omega N\omega^2. \tag{31}$$

For the case of multiple ion traps in a metal with each trap containing N Bose nuclei, we define a trap number density n_t (number of traps per unit volume) as

$$n_t = \frac{N_t}{N} \tag{32}$$

where N_t is the total number of Bose nuclei in traps per unit volume and N is the average number of Bose nuclei in a trap. For this case, the total ion-trap nuclear fusion rate R per unit volume is

$$R = n_t R_t = n_t B\Omega N\omega^2 \tag{33}$$

Both R_t and R do not depend on the Gamow factor in contrast to the conventional case. This is consistent with the conjecture noted by Dirac [9] and Bogolubov [10] that boson creation and annihilation operators can be treated simply as numbers when the ground state occupation number is large. This implies that for large N each charged boson behaves as an independent particle in a common average background potential and the Coulomb interaction between two charged bosons is suppressed.

7 Application to Deuteron-Deuteron Fusion

Using $S = 110keV\text{-}barn$ for deuteron-deuteron fusion, we find from Eq.(24) the nuclear rate constant to be

$$A \approx 1.5 \times 10^{-16} cm^3/sec. \tag{34}$$

and from Eqs.(30) and (34), we have

$$B = 5.2 \times 10^{-22} sec. \tag{35}$$

For an ion trap containing $N \approx 10^6$ deuterons with $\omega \approx MHz (= 10^6 sec^{-1})$ (this may be achievable with the current experimental techniques), we find from Eq.(31),

$$R_t = \Omega \times 5.2 \times 10^{-4} sec^{-1}$$

If Ω can be made larger ($\Omega \approx 10^{-3}$) by cooling the confined deuterons using currently available cooling techniques (laser cooling, collisional cooling, sympathetic cooling, or Sisyphus cooling [11]), we have $R_t \approx 10^{-6} sec^{-1}$ ($\approx$ one per day) which may be a measurable rate for protons and neutrons emitted from the reaction $d(d,p)t$ and $d(d,n)^3He$, respectively.

It is known that metal targets under ion beam bombardment develop surface blisters which may consist of many micro-traps ($\sim \mu m$ size or less). If we assume $\sqrt{\hbar / m\omega} \approx 10^{-4} cm$ (corresponding to $\omega \approx 1.6 \times 10^4 sec^{-1}$) and the trap number density of $n_t = N_t/N$ with $N_t \approx 2.7 \times 10^{-19} cm^{-3}$, we obtain from Eq.(33)

$$R = \Omega \times 10^7 \frac{1}{cm^3 sec} \tag{36}$$

If Ω turns out to be $(10^{-7} \sim 10^{-10})$, we have $R \approx (10^{-3} \sim 1.0) cm^{-3} sec^{-1}$ which may be consistent with anomalous low-energy enhancement of the cross-section for $d(d,p)t$ recently observed by Yuki et al. [12]. Therefore, it is desirable to determine N_t by examining the Pd and Au/Pd/PdO targets used by Yuki et al. [12] in order to obtain more definitive tests of our ion-trap nuclear fusion mechanism. The same mechanism may also provide an explanation and a theoretical justification of the anomalous effect observed for deuteron-deuteron fusion by Arata and Zhang [13].

Acknowledgement

One of the authors (YEK) wishes to thank Professors Ryugo Hayano, Jirohta Kasagi, Shinsho Oryu, Tokushi Shibata, and Toshimitzu Yamazaki for providing useful information and comments.

References

1. Y.E. Kim and A.L. Zubarev: "Equivalent Linear Two-Body Method for Many-Body Problems", PNMBTG 99-2 (May, 1999), to be published in J. of Phys. B: Atomic, Molecular and Optical Physics

2. Y.E. Kim and A.L. Zubarev: "Nuclear Fusion for Bose Nuclei Confined in Ion Traps", PNMBTG 99-3 (August 1999), to be published in Fusion Technology

3. J.B. McGuire: J. Math. Phys. **5**, 622 (1964); J. Math. Phys. **7**, 6 (1966)

4. C.N. Yang: Phys. Rev. Lett. **19**, 1312 (1967); Phys. Rev. **168**, 1920 (1967)
5. L. Ginzburg and L.P. Pitaevskii: Zh. Eksp. Teor. Fiz. **34**, 1240 (1958), [Sov. Phys. JETP 7, 858 (1958)]; E.P. Gross, J. Math. Phys. **4**, 195 (1963)
6. G. Baym and C.J. Pethick: Phys. Rev. Lett. **76**, 6 (1996)
7. E. Fermi: Riverica Sci. **7**, 13 (1936)
8. L.L. Foldy: Phys. Rev. **124**, 649 (1961)
9. P.A.M. Dirac: *The Principles of Quantum Mechanics*, (second edition), Clarendon Press, Oxford 1935, Chapter XI, Section 62
10. N. Bogolubov: J. Phys. **11**, 23 (1947)
11. P.K. Ghosh: "Ion Traps", Clarendon Press (Oxford, 1995). This book contains an extensive review of ion traps including a nearly complete list of references up to 1994.
12. H. Yuki, J. Kasagi, A.G. Gibson, T. Ohtsuki, T. Baba, T. Noda: JETP Letters **68**, 823 (1998)
13. Y. Arata and Y.-C. Zhang: Jpn. J. Appl. Phys. **37**, L1274 (1998); ibid. **38**, L774 (1999)

Few-Body Systems Suppl. 12, 15–24 (2000)

Few-Body Systems

First Principles Description of Shell Structure in Metal Clusters

H. Kamimura[1], O. Sugino[2]

[1] Institute of Physics, Graduate School of Science, Science University of Tokyo, 1-3 Kagurazaka, Shinjuku-ku, Tokyo 162-8601, Japan,

[2] Fundamental Research Laboratories, NEC Corporation, 34 Miyukigaoka, Tsukuba 305-8501, Japan

Abstract. The magic numbers were observed in alkali-metal clusters. In order to explain this striking phenomenon, a spherical jellium model was introduced. This phenomenological model successfully explained the appearance of magic numbers in terms of shell structure, like the shell model in the nuclear theory. For this model Sugino and Kamimura raised questions, asking if a shell structure really exists in small clusters or why the stable shapes of metal clusters are always spherical or spheroidal. Then they performed first-principles calculations for lithium clusters Li_N with $N = 4$ to 36 for obtaining the stable structures of clusters. In this paper their theoretical treatment is reviewed.

1 Introduction

The fragments obtained by successive division of materials are classified into fine particles and clusters, depending on the numbers of constituent atoms. When fragments consist of 10^3 to 10^5 atoms, those are called "fine particles" while those consisting of 10 to 10^3 atoms are called "clusters". Because the size of clusters is very small such as of nanometer-scale, various quantum mechanical phenomena appear. One of them is the appearance of the shell structure in metal clusters, in particular the appearance of "magic numbers" for which clusters are stable. Another interesting and important problem in clusters is whether clusters are different from large molecules or not. In order to investigate these problems, a number of first principles calculations have been performed. Sugino and Kamimura [1, 2, 3] were the first researchers to attack these problems from the first principles by developing a new first principles method. In particular, they tried to clarify a microscopic origin of the magic numbers first observed in alkali-metal clusters, by choosing lithium clusters as an object of study. The gist of their idea is as follows:

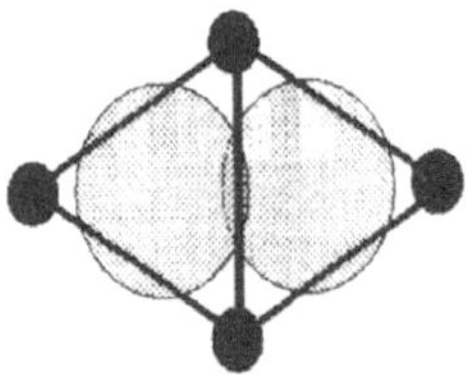

Figure 1. Sketch of "glue bond"

The starting point of Sugino and Kamimura's idea is a set of data for the stability of small lithium-atom clusters. They ask questions why the four-atom and six-atom clusters of lithium are planar structure and why should these unexpected structures be stable. In the first-principles investigation of finding answers to the above questions they have found that the distribution of electron densities in rhombus Li_4 and triangle Li_6 clusters is concentrated about the mid-points surrounded by the interatomic bonds but not at the atoms themselves. The example of Li_4 cluster is schematically shown in Fig. 1. They have called this novel kind of bond among atoms, a "glue bond". Then they have represented the glue bonds as ellipsoids centered on moveable points. By taking account of the electrostatic repulsion between the helium-like cores of the lithium atoms, they have carried through a Hartree-Fock type calculation to determine not only the parameters defining the ellipsoids but also the positions of the nuclei and the points around which the different electrons are centered. In this way they have determined the stable structures of lithium clusters up to the clusters consisting of 36 atoms.

In the present paper we will describe their theoretical treatment and discuss their numerical result, with emphasis of clarifying the microscopic origin of phenomenological shell models in metal clusters on one hand and how many atoms have to be gathered when a metal cluster begins to behave like a solid rather than like a molecule on the other. In the present paper we will first summarize the important experimental results on the appearance of magic numbers in metal clusters. Secondly a brief review will be given on phenomenological theories for the shell structures in metal clusters. Then the first principles description for the origin of magic numbers by Sugino and Kamimura will be made.

2 The Concept of Shell Structure

The concept of shell structure has provided clear explanation for various properties of atoms and nuclei. In our view it is impressible that a shell model theory applies for both atoms and nuclei despite the differences in the forces involved. In the present paper we will show that a similar situation also exists for certain metal clusters. In metal clusters, Knight et al. [4, 5] have found experimentally for the first time that the cluster size dependence on stability, ionization poten-

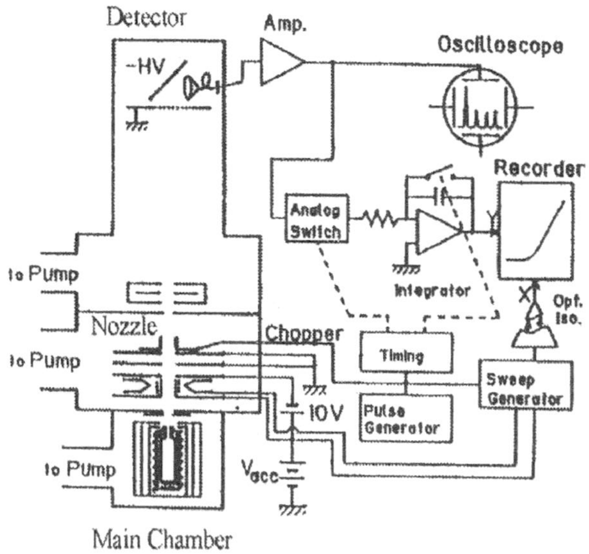

Figure 2. Apparatus sketch

tials, static polarizability, and other properties indicates the existence of magic numbers and thus that of shell structure. These were first observed for sodium and potassium clusters. In this section we describe their experimental findings for the existence of magic numbers in alkali-metal clusters. For this purpose we will first explain the experimental apparatus on how to produce and to detect metal clusters.

Figure 2 shows schematically the apparatus which consists of the cluster beam generator and the time-of-flight (TOP) mass spectrometer. In the main chamber shown in the lower part of the figure, metal vapor from the oven at temperature of about 2300 Kelvin enters the cool condensation chamber. In this chamber metal vapor mixes with a stream of inert gas at pressure of about 1 Torr and temperature of about 77 Kelvin to produce large clusters abundantly. Various size of clusters are formed by expanding the alkali-metal vapor into a vacuum through a nozzle. After passing through a beam-defining collimeter, the cluster beam is ionized continuously by electron impact or laser ionization. The energy of ionizing electrons, E_i, can be varied from 0 to 150 eV. The ionized clusters are extracted and accelerated to 500 eV from an ionization region by a voltage pulse V. The ionized beam is then chopped electrically by an ion deflector. The chopped ion beam is focused to the ion detector at the end of a drift space of 0.95 m long. Since all singly charged ions are accelerated to the same energy by the voltage pulse V, the velocity of a cluster with mass M and charge q is given by

$$v = (2qV/M)^{\frac{1}{2}} \tag{1}$$

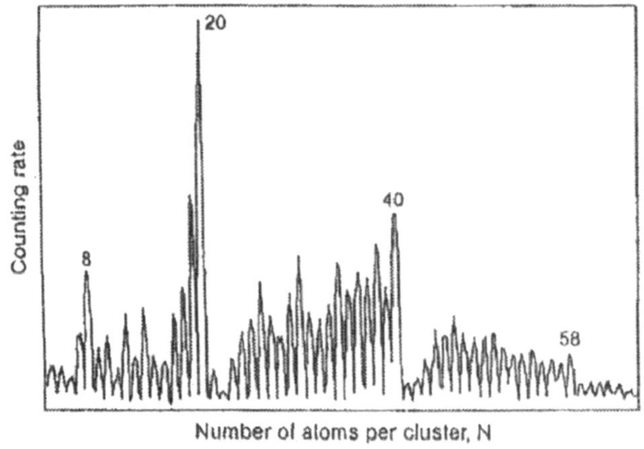

Figure 3. Observed Na cluster abundance spectra. [After Knight et al. [4]]

where v is the velocity of a cluster along the direction of the pulse electric field. This mass-dependent velocity results in distinguishable flight times for different clusters, i.e., it is clear from eq. (1) that a cluster of light mass arrives at the detector shown in the top of Fig. 2 at earlier time than that of heavier mass. One example for abundance patterns observed in the time of flight-mass-spectrum experiments is shown in Fig. 3, where the observed sodium clusters abundance spectrum is shown as a function of the number of atoms per cluster N [5]. Remarkable features of this result is the appearance of "magic numbers" in N such as 8, 20, 40, 58, etc. for which the abundance is strikingly large.

3 Spherical Jellium Model

In order to explain the experimental results shown in Fig. 3, Knight and his coworkers [4, 5] developed a "spherical shell model". In this section we describe briefly the spherical shell model. As to the spherical shell model an excellent review article by de Heer, Knight, Chou and Cohen [6] has been published in Solid State Physics vol.40. Here we follow their description.

In the jellium model, the effect of ionic cores is simulated by a uniformly positively charged background. Here a cluster is assumed spherical. We call it a jellium sphere. The radius of the jellium sphere with Z total valence electrons is given by

$$R = Z^{1/3} r_s, \tag{2}$$

where $2r_s$ corresponds to the average distance between valence electrons. The positive density n_+ is equal to the electron density n_- for a neutral cluster, and they are given by

$$n_+ = n_- = (4\pi r_s^3/3)^{-1}. \tag{3}$$

In the spherical jellium model the electrons in a cluster are treated as the electron gas contained in a sphere of radius R. The electronic structure of this model is calculated by the density functional formalism [7]. In this formalism the total energy of an interacting electron gas is expressed as a functional of the electron density n_- only. The total energy is expressed as the sum of the kinetic energy of electrons, the exchange-correlation energy, coulomb interactions between electrons, and the electrostatic potential generated by the positively charged sphere.

By using the variational method, one obtains a set of Schrödinger-like equations, which are often called the Kohn-Sham equation [8]. In this treatment the many-body effects in the system are replaced by an effective one-electron potential. This significantly simplifies the problem. Here an electron moves in the average potential created by the other electrons. The eigenstates are labeled by the principal quantum number n and the angular momentum quantum number l, where l = 0, 1, 2, 3, 4, 5, ... The states with l = 0, 1, 2, 3, 4, 5, ... are often denoted by s, p, d, f, g, h ..., respectively. Once the wavefunctions and eigenvalues are computed, the total energy of an N-atom cluster, $E(N)$, can be calculated for each N. We further define a quantity

$$\Delta_2(N) = E(N+1) + E(N-1) - 2E(N). \tag{4}$$

This quantity is related to the second derivative of the energy with respect to N and represents the relative binding energy change for cluster with N atoms compared to clusters with $N+1$ and $N-1$. Thus $\Delta_2(N)$ represents one of the measures for relative stability.

4 Magic Numbers

When an energy level is just filled by the electrons in a cluster of N atoms and the next available level is separated from this filled level by a perceivable energy gap, the total cluster energy will have a jump from $E(N)$ to $E(N+1)$. This gives rise to a peak in $\Delta_2(N)$. A peak in $\Delta_2(N)$ indicates that a cluster with N atoms is relatively stable. The stability suggests that this cluster should have a larger abundance in the mass spectrum than a cluster with $N+1$ or $N-1$ atoms. This N is called "magic number". Figure 4 shows the calculated total energy per atom by the spherical jellium model for Na with $r_s = 3.93$ a.u.[9]. In this figure we can see small kinks appearing at N = 8, 18, 20, 34, 40, 58, 92,..., where the total energy changes abruptly, and the energy per atom approaches the bulk limit of sodium (Na) metal at around −2.2 eV when N goes to infinity. The numbers at kinks correspond to the magic numbers. When we compare this theoretical result with the experimental one shown in Fig. 3, we can see that the spherical jellium model is successful in explaining the appearance of the prominent features in the mass spectra but it can not explain many small features in the observed mass spetra. In order to explain these small features, Clemenger introduced the effect of axially symmetric deformation into the spherical jellium model [10]. Clemenger's model resembles the spheroidal

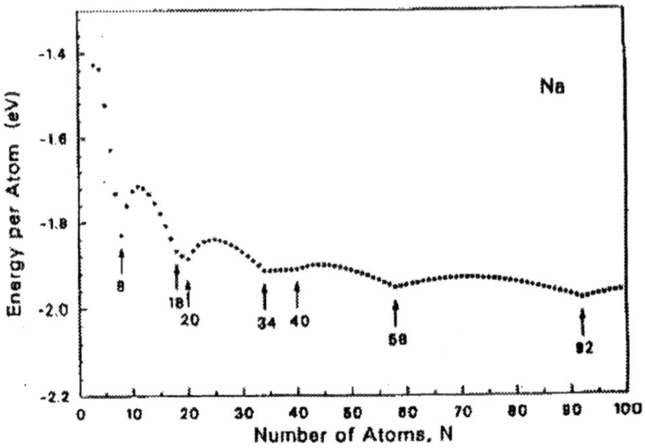

Figure 4. The calculated total energy per atom in the spherical jellium model for sodium. [After de Heer et al. [6]]

model for nuclei by Nilsson[11]. The distortion from a spherical shape causes the energy-level splitting of a system with spherical symmetry which we call "subshells". Thus the observed many small features in the mass spectra can be explained by the existence of subshells.

Ekardt [12] calculated the electron density, effective one-electron potential, energy eigenvalues for closed-shell clusters of sodium with $N = 8$, 58, and 198 as a function of a distance from the center of the jellium sphere. His calculated results for $N = 8$ and 58 are shown in Figs. 5 (a) and (b), respectively. From these figures one can see that the electron charge density in the $N = 8$ cluster varies significantly inside the jellium sphere whose radius R is 8 atomic unit (a.u.) while variation becomes smaller for $N = 58$ whose sphere radius R is 15.48 a.u. When N approaches 198, the electron density becomes nearly constant inside the jellium sphere. This means that the original shell model using the Woods-Saxon type potential may not be a good approximation, but needs a modification for a cluster less than $N = 50$, though it may hold for clusters with N being larger than 100. Further, when geometrical shape of metal clusters are concerned, the jellium approximation, which smears out the atomic potentials, cannot be simply justified for small clusters. These are the demerits of jellium sphere and sheroidal models.

5 First-Prinsiples Study of Small Metal Clusters - Sugino-Kamimura's Theory

In order to remedy the above-mentioned demerits in the phenomenological jellium models and further to investigate the problem of how a cluster of atoms changes from a molecule to a solid with increasing the number of atoms, Sugino

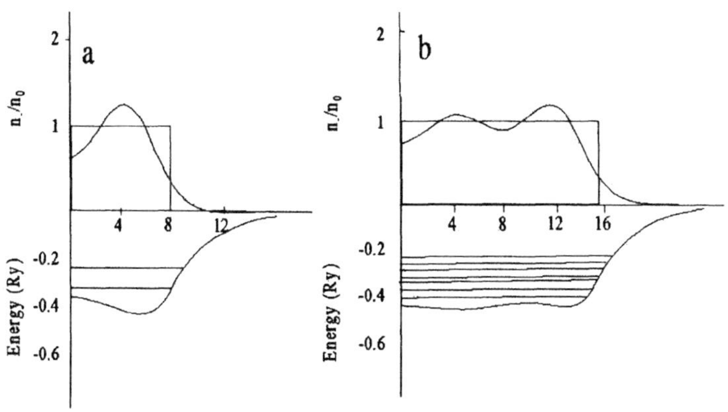

Figure 5. The calculated electron charge density (upper curve), effective potential (lower curve), and energy eigenvalues for closed-shell cluster of sodium: (a) $N = 8$ and (b) $N = 58$. [After Ekardt[12]]

and Kamimura proposed a new theoretical method to calculate the stability of small clusters from the first principles standpoint [1, 2, 3]. In the following we describe their theoretical treatment and present interesting numerical results.

As mentioned above, the sperical potential can not be adopted any more for small clusters with N being less than 50. Instead, the geometry of each nucleus has to be explicitly taken into account. Transition from such molecular cluster to the solid-like one was a subject to the first-principles calculations. Sugino and Kaminura [1, 2, 3] performed *ab initio* Hartree-Fock (HF) calculations for lithium clusters, Li_N, with N=4, 6, 8, 12, and 14. They found that the geometry consists of triangles or tetradedra as shown in Fig. 6. The one-electron HF orbital was found to have a characteristic peculiar to those clusters: When a unitary transformation [13] is applied, the orbitals have a roughly spherical shape and are localized in the triangles or the tetrahedra. The shape of the localized orbital is common to all the investigated clusters and it has approximately the same value for the self-energy, which is defined as the sum of the kinetic energy and self-interaction terms. The HF total energy can be regrouped into the self-energy and the rest that consist of the electrostatic energies between the ions, between the localized orbitals, and among the ions and the localized orbitals. It is therefore concluded that the bonding mechanism is typified by electrostatic energy of the localized orbital. From this result Sugino and Kamimura called the localized orbital "glue bond". Using the glue bond picture, relative stability of clusters can be explained as follows: Li_8, for example, consists of eight ions and four glue bonds. Note that restricted HF is adopted here and two electrons can be accommodated in one orbital. The ions and the glue bonds can be packed similarly as shown in Fig. 6. It can be shown that, for Li_8, a magic cluster, the ions and the glue bonds can be packed more closely compared with other clusters, resulting in the larger binding energy.

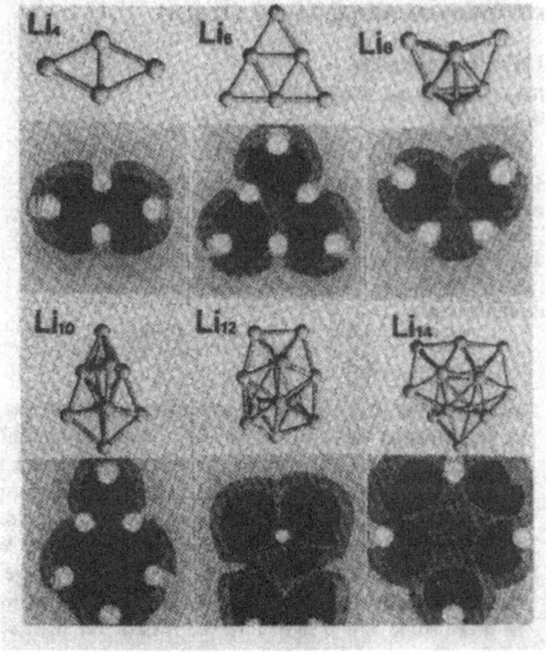

Figure 6. Optimized geometry and electron charge density in Li_4, Li_6 and Li_8. [After Sugino and Kamimura [2]]

Though applying the HF calculations to larger clusters are difficult, the existence of the glue bond allows the HF calculation easier. Sugino and Kamimura [2] represented each glue bond using an ellipsoidal Gaussian function,

$$\phi(\mathbf{r}) = \left(\frac{8 \det \alpha}{\pi^3}\right)^{1/4} \exp\left[-\sum_{i,j}^{x,y,z} \alpha_{ij}(\mathbf{r}_i - \mathbf{R}_i)(\mathbf{r}_j - \mathbf{R}_j)\right] , \quad (5)$$

where R_i and α_{ij} represent the center of the Gaussian function and the Gaussian exponent, respectively, and are determined variationally to minimize the total energy. Such HF calculation is called Floating Ellipsoidal Gaussian Orbital (FEGO) method because not only the ions but also the Gaussian functions are fully relaxed by letting them float in the course of the variational calculation. Since the number of basis function can be reduced to the number of glue bonds in FEGO, computational time can be greatly reduced and thus it can be applied to larger clusters. Using the computed total energy for Li_N, $E(N)$, the energy required to be dissociated into Li_{N-2} and Li_2 ,that is one of the quantities to measure the cluster stability, $E(N) - E(N-2) - E(2)$, was calculated. The result is shown in the upper part of Fig. 7. When compared with the corresponding value obtained by the observed mass spectra, namely $-\ln(I(N)^2/I(N-1)I(N+1))$ which are shown in lowert part of Fig. 7, good

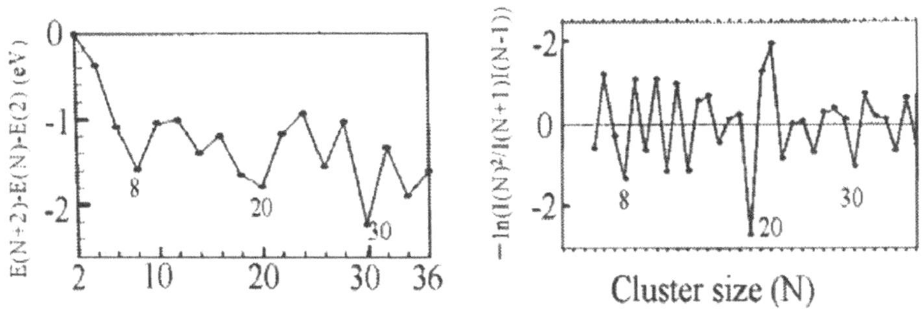

Figure 7. The calculated (left part) and observed (right part) dissociation energy. [The upper part after Sugino and Kamimura [2] and the lower part after de Heer et. al. [6]]

agreement was found between theory and the experiment. Besides, the principal moment of inertia, which characterizes the anisotropy of the cluster shape, was also calculated and showed reasonable agreement with experiment [3]. It is interesting to point out here that, although the assumptions made in constructing the jellium model are not good approximations in these small clusters, the total energies and the anisotropy of the cluster shape obtained by the jellium model coincide with those calculated by the HF.

When studying the computed geometry of the lithium cluster, there is a striking difference between the smaller clusters Li_N with $N < 26$ and the larger clusters. The difference can be seen in the distribution function of the Li-Li distance and the bond angle distribution function, whose calculated results are shown in the upper and lower parts of Fig. 8, respectively. It can be seen from this figure that, for smaller clusters there are Li-Li distances which are located between 7 and 9 bohr, but such distance does not exist for larger clusters. This indicates that the larger clusters have much more ordered structure when the size of the cluster exceeds 26. Above 26 the distribution shows only minor dependence on size. These results suggest that there is a kind of transition at the size of 26. Note that the obtained distribution for larger clusters is not very similar to that of body-centered-cubic(bcc) or the face-centered-cubic(fcc) bulk metal. This indicates that the bulky cluster would appear when the size is even larger.

6 Conclusion.

In this paper we have shown that, although the assumptions made in constructing the jellium model is not very obvious for small clusters with the number of constituent atoms being less than 50, the total energies and the anisotropy of the cluster shape obtained by the jellium model coincide with those calculated by Sugino-Kamimura's Hartree-Fock theory. Their first-principles calculations have shown that the magic numbers correspond to the local minima in the total

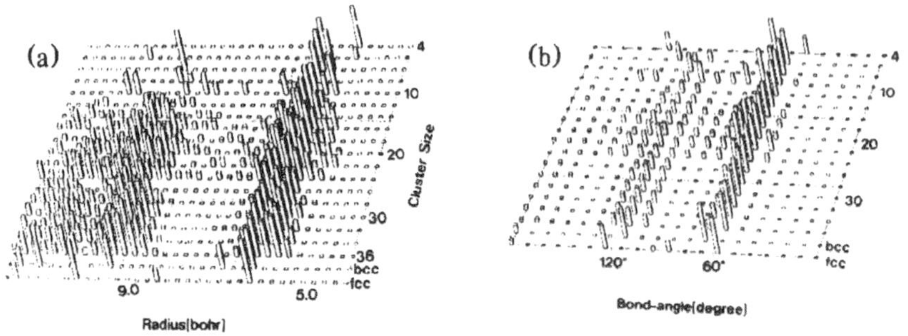

Figure 8. Calculated radial distribution function (left part) and bond-angle distribution function for neutral Li_N clusters with $N = 4$ to 36. After Sugino and Kamimura [2]]

energy calculated as a function of the number of atoms per cluster.

References

1. O. Sugino and H. Kamimura: Mater. Sci. Eng. **B3**, 443 (1989)
2. O. Sugino and H. Kamimura: Phys. Rev. Lett. **65**, 2696 (1990) ; See also Nature **348**, 387 (1990)
3. O. Sugino and H. Kamimura: Z. Phys. D - Atoms, Molecules and Clusters **19**, 67 (1991)
4. W.D. Knight, K. Clemenger, W.A. de Heer, W.A. Saunders, M. Chou, M.L. Cohen: Phys. Rev. Lett. **52**, 2141 (1984)
5. W.D. Knight, W.A. de Heer, K. Clemenger, W.A. Saunders: Solid St. Commun. **53**, 445 (1985)
6. W.A. de Heer, W.D. Knight, M.Y. Chou, M.L. Cohen: Solid State Physics vol.40, 93, eds. H. Ehrenreich and D. Turnbull (Academic Press, San Diego, 1987)
7. P. Hohenberg and W. Kohn: Phys. Rev. **136**, B864 (1964)
8. W. Kohn and L.J. Sham: Phys. Rev. **140**, A1133 (1965)
9. M.Y. Chou, A. Cleland, M.L. Cohen, Solid St. Commun. **52**, 645 (1984)
10. K. Clemenger: Phys. Rev. **B32**, 1359 (1985)
11. S.G. Nilsson: Mat. Fys. Medd. K. Dan. Vidensk. Selsk. **29**, No.16 (1955)
12. W. Ekardt: Phys. Rev. **B29**, 1558 (1984)
13. C. Edmiston and K. Ruedenberg, Rev. Mod. Phys. **35**, 457 (1963)

Few-Body Systems Suppl. 12, 25–32 (2000)

Few-Body Systems

Resonances in Three- and Four-Body Atomic Systems involving Electrons and Positrons

Y.K. Ho

Institute of Atomic and Molecular Sciences, Academia Sinica,
P. 0. Box 23-166, Taipei, Taiwan, Republic of China

Abstract. The recent advances in theoretical studies of atomic resonances in three- and four-body systems by using the method of complex-coordinate rotation are reviewed. Resonances in $(e^-e^+e^-)$, and the $1s2s^2\ {}^2S^e$ resonance in e^--He scattering are discussed. Other applications of the method include doubly excited states of H^- under the influence of external electric fields, as well as photoionization of He, with and without the presence of external electric fields.

1 Introduction

In this talk, I will describe to you some recent activities on the applications of the method of complex-coordinate rotation in the area of atomic collision processes. The method has been most successful in the determination of resonance energy and width for a resonant state. The method requires bound-state-type (so-called L^2-type) wave functions, and results can be precisely determined since usually the bound-bound integrals can be accurately performed. The method was developed into a powerful computational technique in the 1980's following the earlier mathematical development of the ABC theorem [1] for dilatation analytic potentials. Coulomb potential and Yukawa-type potentials belong to this kind of operators.

2 Method of Complex-coordinate Rotation

Let us consider the scattering of a positron (or electron) by a target atom consisting of N particles. Figure 1 shows the energy spectra for the (N+1) particle system before and after the complex transformation. In the figure there are three distinct parts of the spectrum which are relevant to the present discussion: (1) Bound state poles (if they exist) of the (N+1) particle system are located on the negative side of the real axis in the complex E plane; (2) A series of branch cuts extend from the elastic and all inelastic thresholds (for the N, N-1, ..., ets. particle systems) to the right-hand side infinity ($+\infty$) of the energy plane; (3) Resonant poles are located near the cuts and are hidden in higher sheets of the Riemann surfaces.

In the method of complex-coordinate rotation, all the inter-particle radial coordinates r_{ij} are transformed into $r_{i,j} \rightarrow r_{i,j}e^{i\theta}$, where θ is real and positive. Under such a transformation, the spectrum of the (now so-called rotated) Hamiltonian is transformed to the following (see Fig. 1): (1) The bound state poles remain unchanged under the transformation; (2) The cuts are now rotated downward, making an angle of 2θ with the real axis; (3) The resonant poles are "exposed" by the cuts once the "rotational angle" θ is greater than 1/2 arg (E_{res}), where E_{res} is the complex resonance energy, i.e. $E_{res} = E_r - i\Gamma/2 = |E|e^{-i\beta}$, with β the phase factor of the complex energy, and E_r and Γ the usual resonance position and width, respectively. The above coordinate transformation has a great computational advantage for systems with Coulomb interaction. The kinetic part of the Hamiltonian would scale as $\exp(-2i\theta)$, and the (Coulomb) potential part of the Hamiltonian would simply scale as $\exp(-i\theta)$. Under such a transformation, one just calculates the kinetic and potential matrix elements separately, and then scales them according to the above scheme. A complex eigenvalue problem will then be solved, and the eigenvalues are obtained by diagonalizing the analytically continued Hamiltonian $E =< \Phi|H(\theta)|\Phi > / < \Phi|\Phi >$, wherein the eigenvalues and eigenvectors are complex, while the basis functions Φ are real. Discrete resonance complex eigenvalues have the form of $E = E_r - i\Gamma/2$, where E_r gives the resonance position and Γ the resonance width. When we diagonalize the complex Hamiltonian, the whole spectrum of complex eigenvalues would be solved. The complex eigenvalues that represent various parts of the energy spectrum of the (rotated) Hamiltonian are then examined. Some of the eigenvalues represent the bound and resonance states, and some represent the cuts. The complex eigenvalues that represent the rotated cuts have a strong dependence of θ, and that for a resonance is stationary with respect to the changes of θ, once θ is sufficiently large to discover such a resonance. For a complex resonance energy with $\partial E/\partial\theta = 0$, the complex virial theorem $2 < \Phi|T|\Phi >= - < \Phi|V|\Phi >$ is satisfied. The development of the method was documented in the earlier reviews [2, 3].

3 Resonance States in Ps$^-$, ($e^-e^+e^-$)

Over the years, we have applied the method of complex-coordinate rotation to investigate doubly-excited resonances in three-body atomic systems such as H$^-$ [4], He and its iso-electronic squence [5], and positronium negative ion, Ps$^-$ [6], as well as resonances in e$^+$-H scattering [7]. Recently [8], we calculated the high-angular-momentum states of doubly-excited Ps$^-$ by using the properly

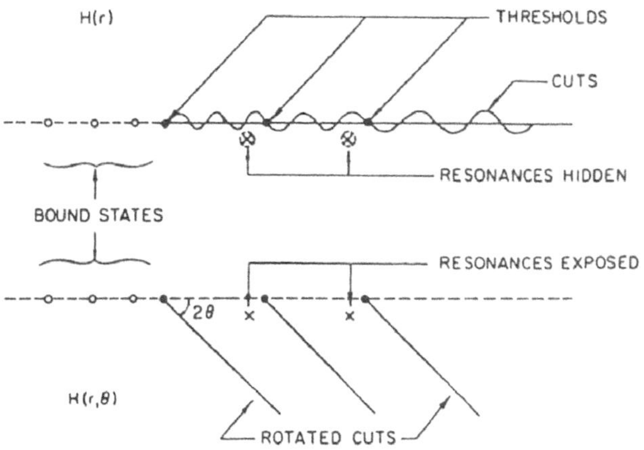

Figure 1. Energy spectra of the Hamiltonian before and after the complex transformation of $r \to re^{i\theta}$

symmetrized products of the Slater-type orbitals,

$$\Phi = A \sum_{\ell a, \ell b} \sum_{ij} C_{a_i, b_j} \eta_{a_i}(r_1) \eta_{b_j}(r_2) Y^{LM}_{\ell a, \ell b}(1,2) S(\sigma_1, \sigma_2), \tag{1}$$

where $\eta_{a_i}(r) = r^{n a_i} \exp(-\xi_{a_i} r)$, and $Y^{LM}_{\ell a, \ell b}$ and S are the eigenfunctions of total angular and spin momentum, respectively. The Hamiltonian in rydberg units is $H = T + V$, with $T = -2\nabla_1^2 - 2\nabla_2^2 - 2\nabla_1 \cdot \nabla_2$, and $V = -2/r_1 - 2/r_2 + 2/r_{12}$. Our results show that a doubly-excited Ps^-, like H^-, behaves very much like an XYX tri-atomic molecule (see Fig. 2). It undergoes rotational and bending vibrational motions [9]. Fig. 3 shows the rotational and vibrational character of the doubly-excited states associated with the Ps(N=5) threshold. Each state is denoted by a set of quantum numbers $^{2s+1}L^{\pi}$[K,T], where K and T are approximately "good" quantum numbers that describe the angular correlation between the two electrons [9].

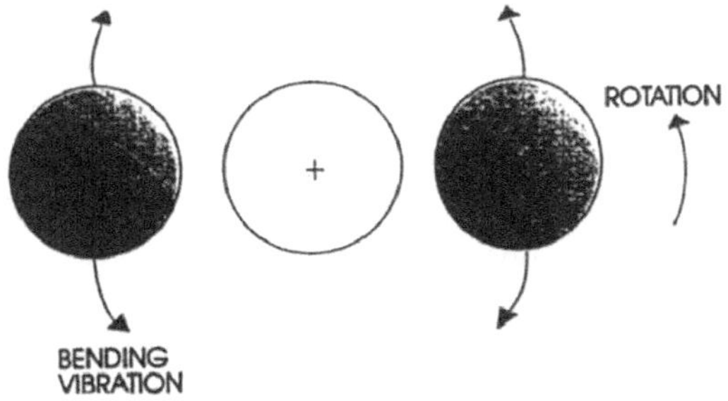

Figure 2. Doubly-excited Ps^- behaves like an XYX tri-atomic molecule [9].

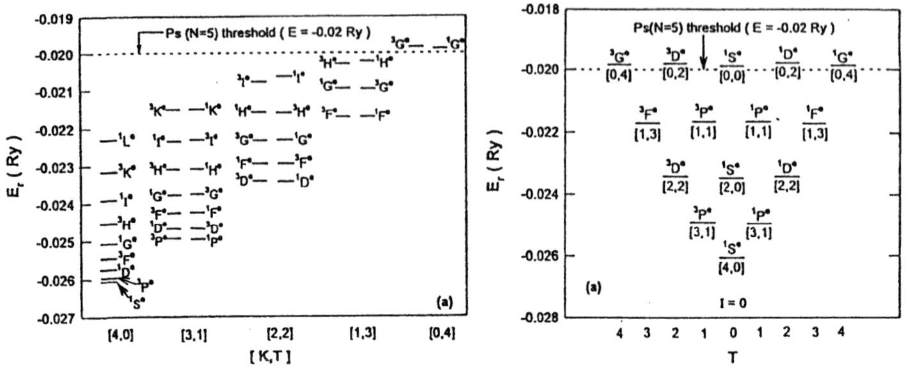

Figure 3. Doubly-excited states of Ps^- associated with the Ps(N=5) threshold showing the rotational (figure on the left) and vibrational (figure on the right) character similar to that of an XYX molecule [8].

4 The $1s2s^2$ ^2S Resonance State of He^-

The $1s2s^2$ ^2S resonance in e^--He scattering was first observed by Schulz in 1963 [10]. In fact, such a resonance was the first resonance observed in electron-atom scattering. Since then a great deal of experimental and theoretical investigations on this state has appeared in the literature, as were documented in a recent review [11]. From the experimental side, the most accurate measurements on the resonance energy are those of Cvejanovic et al. [12], and Brunt et al. [13], who reported energies of 19.367 ± 0.009 eV and 19.366 ± 0.005 eV, respectively. As for the width, the high-resolution measurement of Kennerly et al [14] yielded a value of 11.0 ± 0.5 meV. Earlier theoretical results have a dependence on the target helium wave functions [15]. In a recent work [15], we reported a benchmark calculation for the He^- $1s2s^2$ ^2S state using the method of complex-coordinate rotation and with fully-correlated Hylleraas functions. Correlation effects between all the four charged particles are included when all the six interparticle coordinates are used explicitly in the basis function, i.e.,

$$\phi_i = r_1^{j_1} r_2^{j_2} r_3^{j_3} r_{12}^{j_{12}} r_{23}^{j_{23}} r_{31}^{j_{31}} e^{-\alpha r_1 - \beta r_2 - \gamma r_3} Y_{(\ell_1 \ell_2)\ell_{12},\ell_3}^{LM}(\hat{r}_1, \hat{r}_2, \hat{r}_3), \tag{2}$$

with $Y_{(\ell_1 \ell_2)\ell_{12},\ell_3}^{LM}$, being a vector-coupled product of spherical harmonics for the three electrons to form a state of total angular momentum L. Our resonance parameters of $E_r = 19.36888 \pm 0.00011$ eV and $\Gamma = 11.03 \pm 0.22$ meV agree quite well with the above benchmark experiments.

5 Electric-field Effects on H^-

We have also applied the method of complex-coordinate rotation to investigate the electric-field effects on doubly-excited states of H^-. With the external electric field kept in the z-axis, a dipole operator $\bar{F} \cdot \bar{r}$ is added to the atomic Hamil-

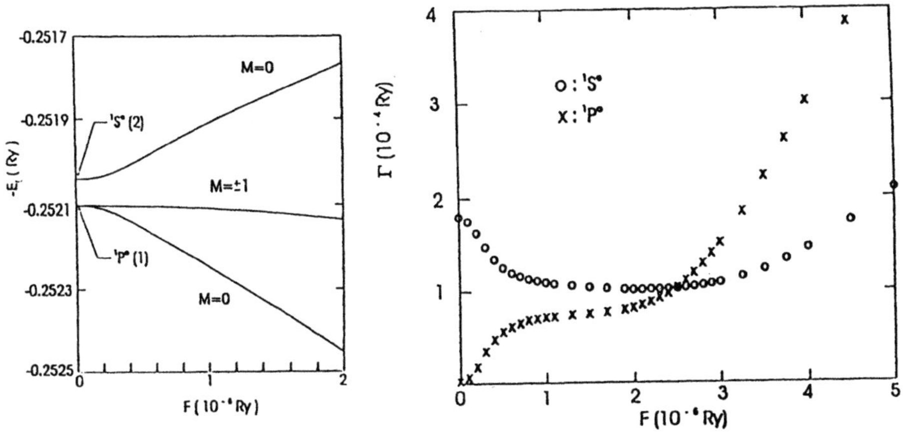

Figure 4. Stark effect for the nearly-degenerate doubly-excited N=2 ^{1}P^o(1) and ^{1}S^e(2) states of H$^-$. The figure on the left is for the energies, and that on the right is for the widths [16].

tonian. Under complex-coordinate rotation, such an operator is simply scaled as $\bar{F}\cdot\bar{r}e^{i\theta}$. For the component $M_L = 0$, states with $\Delta L = \pm 1$ (for example, ^{1}S^e, ^{1}P^o, ^{1}D^e, ^{1}F^o, etc.), and with opposite parity, are coupled together by the external electric field. Fig. 4 shows our results for the field effects on the energies of the Feshbach ^{1}P^o(1) and ^{1}S^e(2) states below the H(N=2) threshold. These two states has an energy separation of about 0.8 meV in the field-free case. It demonstrates the classic Stark effect for the two nearly-degenerate states. Our results [16] agree with the experimental observations [17]. Fig. 4 also shows the electric-field effect on the widths of the ^{1}P^o(1) and ^{1}S^e(2) states [16]. It is seen that in the low-field strength region, the widths for these two states tend to average each other out, a kind of sharing intensities for autoionization. At the high-field region, the one-electron tunneling effect starts to dominate the autoionization processes [16]. In general, a potential barrier is formed by a combined atomic potential and the external electric field. Such a potential barrier is decreased in thickness for increasing field strengths. The electron would hence take a shorter time to tunnel out, and resulting in a broader width. In addition to the field effects on the N=2 resonances, we have also carried out investigations for the electric-field effects on other H$^-$ states [18]. The method of complex-coordinate rotation was also applied to investigate magnetic-field effects on doubly-excited H$^-$ and He [19], as well as the combined electric- and magnetic-field effects (in parallel direction) on doubly-excited H$^-$ [20].

6 Photoionization of He

Recently, we have applied the method of complex-coordinate rotation to calculate photoionization cross sections of ground state He, with and without the presence of external electric fields [21]. The photoionization cross section $\sigma(E)$

from the initial state $|\Psi_0>$ with energy E_0 can be expressed in the dipole-length approximation as (in a.u.) [22]

$$\sigma^L(E) = 4\pi\alpha\ g(E)\mathrm{Im}\sum_{\nu} \frac{<\overline{\Psi_\nu(\theta)}|R(\theta)D|\Psi_0>^2}{E_\nu(\theta) - E}, \qquad (3)$$

where α is the fine structure constant and $\Delta E = E - E_0$ is the transition energy, $D^L = \bar{\varepsilon}\cdot\bar{r}$ is the dipole-length operator for polarization $\bar{\varepsilon}$, and $R(\Theta)$ is the complex rotation operator given by [22]. In Eq (3), $|\Psi_\nu(\theta)>$ is a normalizied eigenstate of $H(\theta)$, where $<\overline{\Psi_\nu(\theta)}|$ denotes its transposed value. Fig. 5 shows the field effects on the $(2{,}6)^1\mathrm{P}^o$ doubly-excited state of He, and the interaction between the $^1\mathrm{P}^o$ and a nearby field-induced $^1\mathrm{D}^e$ state. Since the ground state of He is $^1\mathrm{S}^e$, only the $^1\mathrm{P}^o$ is observable (with one photon) without the presence of external electric field. When an external field is turned on, a nearby $^1\mathrm{D}^e$ state could be induced and also become observable in a one-photon process. The external field would shift both the $^1\mathrm{P}^o$ and $^1\mathrm{D}^e$ states downward until they almost meet each other (avoided-crossing). The two states will then repel each other when the field strength is further increased. Also, when the field is turned on, the intersity of the $^1\mathrm{P}^o$ state is decreased, while that of the $^1\mathrm{D}^e$ state is increased. Eventually, when the field strength is sufficently strong, both states are quenched by the external field. Our predicted phenomenon is awaited for experiments with great interest, as the required field strengths are available in today's laboratories, and photoabsorption experiments for He can be performed in synchrotron radiation facilities around the globe.

7 Conclusion

In summary, this talk describes an overview of the recent applications of the method of complex-coordinate rotation in the area of atomic collision processes. The method is a powerful computational tool to investigate few-body atomic resonances, as the resonance positions and widths can accurately be determined by using bound-state type wave functions, and as such, many benchmark calculations were carried out using this method. The method has also been successful in the investigations of effects on doubly excited H^- and He under the influence of external electric, magnetic, and combined electric and magnetic fields. It is also successful to extend the method to calculate photoionization of ground state helium, and to investigate the resonance profiles, with and without the presence of external electric fields.

Acknowlegdement

The works were supported by the National Research Council. The author would like to thank Dr. A. K. Bhatia, Dr. Z. C. Yan, Dr. I. A. Ivanov, and Dr. T. K. Fang for the collaborative works mentioned here.

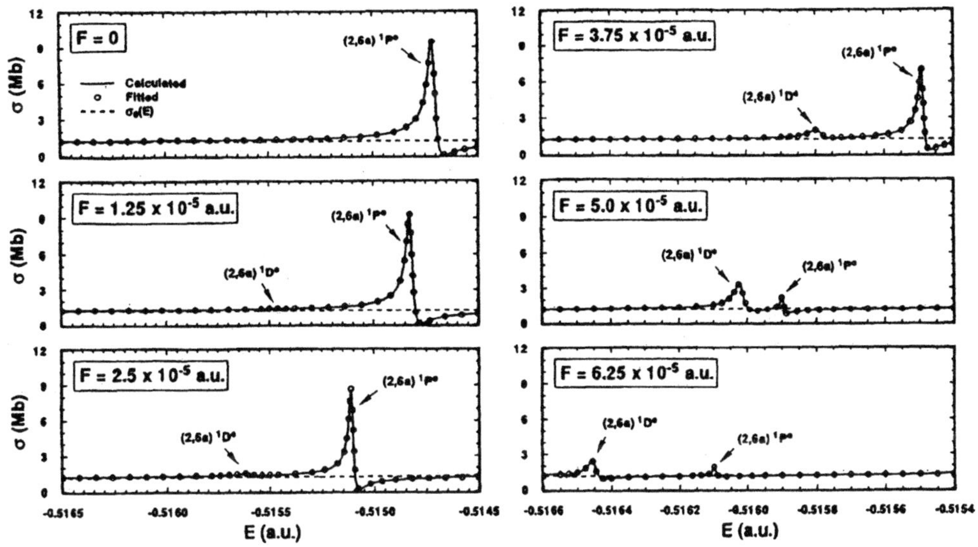

Figure 5. Photoionization of ground state He in the vicinity of the (2, 6a)^{1}P^o and a nearby field-induced (2, 6a)^{1}D^e state, for different applied electric fields [21].

References

1. J. Aguilar and J.M. Combes: Commun. Math. Phys. **22**, 313 (1971); B. Balslev and J.M. Combes: Math. Phys. **22**, 280 (1971); B. Simon: Commun. Math. Phys. **27**, 1 (1972)

2. Y.K. Ho: Phys. Rept. **99**, 1 (1983)

3. W.P. Reinhardt: Ann. Rev. Phys. Chem. **33**, 223 (1982); B.R. Junker: Adv. At. Mol. Phys. **18**, 208 (1982)

4. Y.K. Ho: Phys. Rev. **A49**, 3659 (1994); **50**, 4877 (1994); Y.K. Ho and J. Callaway: Phys. Rev. **A34**, 130 (1986); A.K. Bhatia and Y.K. Ho: Phys. Rev. **A50**, 4886 (1994)

5. Y.K. Ho: J. Phys. **B12**, 387 (1979); Y.K. Ho and A.K. Bhatia: J. Phys. **B30**, 3597 (1997); A.K. Bhatia and Y.K. Ho: J. Phys. **B31**, 3307 (1998)

6. A.K. Bhatia and Y.K. Ho: Phys. Rev. **A42**, 1119 (1990); ibid **A48**, 264 (1993); Y.K. Ho and A.K. Bhatia: Phys. Rev. **A44**, 2890 (1991); ibid **A45**, 6268 (1992); ibid **A50**, 2155 (1994)

7. Y.K. Ho: J. Phys. **B23**, L 419 (1990); Y.K. Ho: Chin. J. Phys. **35**, 97 (1997); Y.K. Ho and C.H. Greene: Phys. Rev. **A35**, 3169 (1987)

8. I. Ivanov and Y.K. Ho: Phys. Rev. **A60**, 1015 (1999); and to be published

9. M.E. Keliman and D.R. Herrick: Phys. Rev. **A28**, 2523 (1983); **A35**, 4051 (1981); C.D. Lin: Adv. At. Mol. Phys. **22**, 77 (1983)

10. G.J. Schulz: Phys. Rev. Lett **10**, 104 (1963)

11. S.J. Buckman and C.W. Clark: Rev. Mod. Phys. **66**, 539 (1994)

12. S. Cevjanovic, J. Corner, F.H. Read: J. Phys. **B7**, 468 (1974)

13. J.N.H. Brunt, G.C. King, F.H. Read: J. Phys. **B10**, 1289 (1977)

14. R.B. Kennerly, R.J. Van Brunt, A.C. Gallagher: Phys. Rev. **A23**, 2430 (1981)

15. Y.K. Ho and Z. C. Yan: Phys. Rev. **A59**, R 2559 (1999)

16. Y.K. Ho: Phys. Rev. **A52**, 375 (1995)

17. H.C. Bryant et al.: Phys. Rev. **A27**, 2889 (1983)

18. Y.K. Ho: Phys. Rev. **A52**, 2062 (1998); **A57**, 2235 (1998)

19. Y.K. Ho: Phys. Letts. **A230**, 190 (1997); J. Phys. **B31**, 1191 (1998)

20. Y.K. Ho: Phys. Rev. **A56**, 2630 (1997)

21. T.K. Fang and Y. K. Ho: Phys. Rev. **A60**, 2145 (1999)

22. T.N. Rescigno and V. McKoy: Phys. Rev. **A12**, 522 (1975); A. Buchleitner, B. Gremaud, D. Delande: J. Phys. **B27**, 2663 (1994)

Few-Body Systems Suppl. 12, 33–39 (2000)

Few-Body Systems

Antiprotonic Helium "Atomcule": Relativistic and QED Effects

V.I. Korobov

Joint Institute for Nuclear Research, 141980, Dubna, Russia

Abstract. In this talk we present the recent theoretical advances in studying metastable states of the antiprotonic helium atoms $He^+\bar{p}$. The α^2 (au) and α^4 relativistic corrections for the bound electron and α^2 corrections for the heavy nuclei as well as the one-loop and two-loops QED corrections such as the electron self-energy and vacuum polarization up to the order of $\alpha^5 \ln \alpha$ would be discussed.

1 Introduction

Now it seems proved [1] that it is possible to calculate nonrelativistic energies for the antiprotonic helium $He^+\bar{p}$ metastable states to the precision of $\sim 10^{-10}$ au or even higher. That exceeds the accuracy of the most precise experimental measurement [2, 3]. It would be shown here that the main uncertainty in the theoretical energies lays in higher order relativistic and QED contributions.

Antiprotonic helium is a system comprised of three particles: helium nucleus, electron, and antiproton. It inherits features both of atomic and molecular nature. In atomic picture electron occupies the $1s$ orbital while antiproton settles into orbitals with $n \approx \sqrt{M/M_e} \approx 38$, where M is the reduced mass of $\bar{p}$. For antiproton to form metastable state requires that the angular momentum l of the antiprotonic orbital be close to $n-1$ (see Fig. 1). In this case $He^+\bar{p}$ atom become stable against (internal) Auger transitions: to remove the ≈ 25 eV binding energy of the electron requires an $\bar{p}$ transition with $\Delta l = l - l' \geq 3$.

$$[He^+\bar{p}]_{(n,l)} \to [He^{++}\bar{p}]_{(n',l')} + e^-, \tag{1}$$

On the other hand, electron is about 40 times faster than antiproton and the molecular "adiabatic" picture essentially helps to simplify numerical treatment of this system. Since the whole system is of atomic size and $(\alpha - \bar{p})$ interaction has high centrifugal barrier, the strong interaction plays no role.

The main decay channel for the metastable states is radiative transitions to lower states that defines an extremely long life-time for the individual state

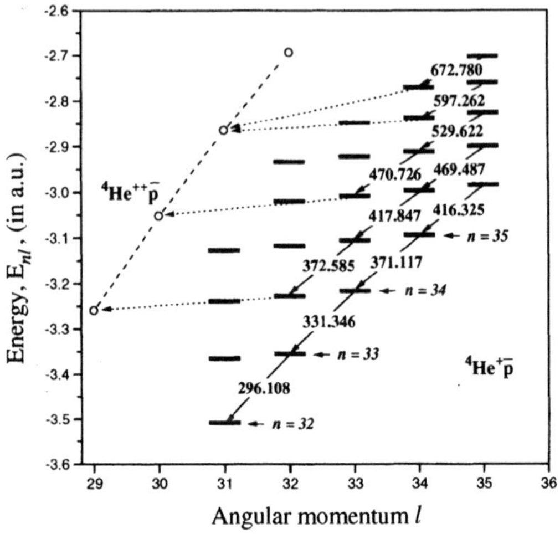

Figure 1. Schematic diagram of energy levels for the metastable states of $^4\mathrm{He}^+\bar{p}$.

of about $1 - 2$ microseconds. For comparison, the life-time of the 2^3P state in helium (see Table 1), where precise measurements of the fine structure are available, is two orders of magnitude shorter.

$\mathrm{H}(2P)$	$\mathrm{He}(2^3P)$	$\mathrm{He}^+\bar{p}$
1.6×10^{-9} s	3.3×10^{-8} s	1.7×10^{-6} s

Table 1. Comparison of lifetimes of $2P$–hydrogen, 2^3P–helium and antiprotonic helium metastable states.

In what follows we discuss briefly the variational method for calculating the metastable states of the antiprotonic helium. And the rest of the talk will be devoted to relativistic and QED corrections to the transition energies.

2 Variational calculation of nonrelativistic energies

The peculiarities of the system under consideration essential for nonrelativistic calculation are a) the high total angular momentum ($L \sim 35$) and b) the molecular "adiabatic" behaviour.

These features make reasonable to perform separation of rotational variables by using the molecular expansion based on the Wigner D-functions

$$\Psi_M^{L\lambda}(\mathbf{R},\mathbf{r}) = \sum_{m=0}^{L} \mathcal{D}_{Mm}^{L\lambda}(\Phi,\Theta,\varphi) F_m^{L\lambda}(R,r,\theta), \tag{2}$$

In the case of semi-adiabatic three-body system the above method converges quickly with respect to $m_{max} \geq L$, where m_{max} is the number of components kept in expansion (1). If m_{max} is smaller than the multipolarity Δl of possible Auger transitions (see Eq. 1), the Hamiltonian projected onto this subspace has a purely discrete spectrum.

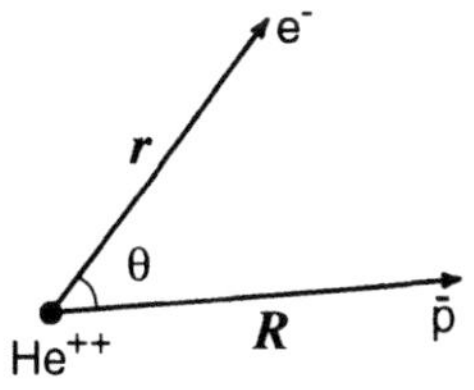

Figure 2. Coordinate system for the variational wave function.

In actual calculations the atomic expansion

$$\Psi_M^{L\lambda}(\mathbf{R},\mathbf{r}) = \sum_{l_1+l_2=L} R^{l_1} r^{l_2} \{Y_{l_1} \otimes Y_{l_2}\}_{LM} G_{l_1 l_2}^{L\lambda}(R,r,\theta), \tag{3}$$

has been used with the coordinate system shown in Fig. 2. It can be proven that the subspace

$$\mathcal{H}_m = \sum_{m'=0}^{m} \mathcal{D}_{Mm'}^{L\lambda}(\Phi,\Theta,\varphi) F_{m'}^{L\lambda}(R,r,\theta),$$

coincides with the subspace spanned by the functions

$$\overline{\mathcal{H}}_m = \sum_{l_2=0}^{m} R^{l_1} r^{l_2} \{Y_{l_1} \otimes Y_{l_2}\}_{LM} G_{l_1 l_2}^{L\lambda}(R,r,\theta),$$

and we can retain in expansion (3) the components with $l_2 \leq m_{max}$.

Final expression for the variational expansion is

$$\begin{aligned} \Psi_M^{L\lambda}(\mathbf{R},\mathbf{r}) &= \sum_{l_2=0}^{l_{max}} R^{l_1} r^{l_2} \{Y_{l_1} \otimes Y_{l_2}\}_{LM} \times G_{l_1 l_2}^{L\lambda}(R,r,\theta), \\ & \qquad\qquad l_1 + l_2 = L \\ G_{l_1 l_2}^{L\lambda}(R,r,\theta) &= \sum_{i=1}^{\infty} [C_i \cos(\nu_i R) + D_i \sin(\nu_i R)] \times e^{-\alpha_i R - \beta_i r - \gamma_i |\mathbf{R}-\mathbf{r}|}, \end{aligned} \tag{4}$$

and the convergence of the method is demonstrated in Table 2. To be honest, we should say that for this example we intentionally choose the state with the Auger width smaller than 10^{-14} a.u. We didn't find a remedy yet how to treat in an accurate way the states with the large Auger width (some advances can be found in [4]). The difficulty arise from the fact that the scattering state wave function has an essentially non-zero density in the region placed far apart from the region where the quasi–bound state is localized.

N	E_{nr}
1500	$-\mathbf{2.98402095449}416$
1700	$-\mathbf{2.984020954497}04$
1900	$-\mathbf{2.9840209544972}0$
2100	$-\mathbf{2.98402095449725}$
2300	$-\mathbf{2.98402095449725}$

Table 2. Convergence of the variational expansion with the number of basis functions. $^4\mathrm{He}^+\bar{p}$ $(36, 35)$.

3 Relativistic corrections

The simplest and the most common way to evaluate relativistic corrections is to calculate the lowest order corrections which are given by the expectation values of the Breit interaction ($\hbar = c = 1$)

$$\begin{aligned} H_{rel} &= -\alpha^2 \sum_i \frac{\boldsymbol{\nabla}_i^4}{8m_i^3} + U_{rel}, \\ U_{rel} &= -\alpha^2 \sum_{i>j} 4\pi\delta(r_{ij}) z_i z_j \left(\frac{c_i}{8m_i^2} + \frac{c_j}{8m_j^2} \right) \\ &\quad -\alpha^2 \sum_{i>j} \frac{z_i z_j}{2m_i m_j} \left(\frac{\mathbf{p}_i \mathbf{p}_j}{r_{ij}} + \frac{\mathbf{r}_{ij}(\mathbf{r}_{ij}\mathbf{p}_i)\mathbf{p}_j}{r_{ij}^3} \right) \\ &\quad -\alpha^2 \sum_{i>j} \frac{z_i z_j \left([\mathbf{r}_{ij} \times \mathbf{p}_j] d_i \mathbf{s}_i - [\mathbf{r}_{ij} \times \mathbf{p}_i] d_j \mathbf{s}_j \right)}{m_i m_j r_{ij}^3} \\ &\quad -\alpha^2 \sum_{j \neq i} \frac{z_i z_j [\mathbf{r}_{ij} \times \mathbf{p}_j] e_j \mathbf{s}_j}{2m_j^2 r_{ij}^3} \\ &\quad +\alpha^2 \sum_{i>j} \left\{ \frac{\boldsymbol{\mu}_i \boldsymbol{\mu}_j}{r_{ij}^3} - 3\frac{(\boldsymbol{\mu}_i \mathbf{r}_{ij})(\boldsymbol{\mu}_j \mathbf{r}_{ij})}{r_{ij}^5} - \frac{8\pi}{3} \boldsymbol{\mu}_i \boldsymbol{\mu}_j \delta(r_{ij}) \right\}. \end{aligned} \tag{5}$$

Here c_i, d_i, and e_i are constants that depend on spin and formfactors of an i-th particle, $\boldsymbol{\mu}_i = (z_i d_i \mathbf{s}_i)/m_i$ is a magnetic moment. However, this approach is limited by the accuracy of about $\alpha^4 \approx 0.3 \cdot 10^{-8}$, which is insufficient for our final goal.

To get relativistic corrections of higher order in α we can consider generalization of the Grotch–Yennie equation [5] to the three-body case. For this equation the free propagator is taken in a form

$$G(E; \mathbf{p}_{\mathrm{He}}, \mathbf{p}_{\bar{p}}, \mathbf{p}_e) = \frac{1}{E - (\boldsymbol{\alpha}_e \mathbf{p}_e + \beta_e m_e) - \mathbf{p}_{\mathrm{He}}^2/2m_{\mathrm{He}} - \mathbf{p}_{\bar{p}}^2/2m_{\bar{p}}} \cdot \frac{1+\gamma_0^{\bar{p}}}{2}$$

Velocity of heavy particles is several orders of magnitude smaller ($v_h^2 \ll v_e^2$), thus the α^2 order relativistic corrections for heavy particles would be enough for the theoretical consideration.

$$\left(\alpha\frac{v_{\bar{p}}}{v_e}\right)^4 \approx (\alpha/40)^4 \approx 1.1\cdot 10^{-15}.$$

On the other hand, inclusion of the first order relativistic correction to the wave function provides the accuracy of $\alpha^6 \approx 1.5\cdot 10^{-11}$. The numerical implementation of the Grotch–Yennie approach is in progress now.

4 Radiative corrections. Asymptotic expansion in terms of α

The displacement of energy due to the one–loop QED corrections for the bound electron is [6, 7]

$$\begin{aligned}\Delta E_{LS}(\text{one-loop}) &= \frac{4Z_i\alpha^3}{3}\langle n|\delta(\mathbf{r}_i)|n\rangle 2\text{Ry}\left\{\left[\ln\frac{1}{\alpha^2}-\ln\frac{k_0(n)}{\text{Ry}}+\frac{11}{24}\right]\right.\\ &+(Z_i\alpha)3\pi\left(\frac{139}{128}-\frac{1}{2}\ln 2\right)-\frac{3}{4}(Z_i\alpha)^2\ln^2\frac{1}{(Z_i\alpha)^2}\\ &\left.+\left[-\frac{1}{5}+(Z_i\alpha)\pi\frac{5}{64}\right]\right\}-\frac{\alpha}{2\pi}\left[i\,\mu_0\left\langle\beta\boldsymbol{\alpha}\cdot\mathbf{E}\right\rangle_{nn}\right]2\text{Ry}+O(\alpha^5\ln\alpha).\end{aligned} \tag{6}$$

Here $\ln k_0(n)/\text{Ry}$ is the nonrelativistic Bethe logarithm [8]

$$\ln\frac{k_0}{\text{Ry}} = \frac{\sum_n |\langle 0|\mathbf{p}|n\rangle|^2 (E_n-E_0)\ln\{|E_n-E_0|/(1\ \text{Ry})\}}{\sum_n |\langle 0|\mathbf{p}|n\rangle|^2 (E_n-E_0)}.$$

The Bethe logarithm is the only quantity in Eq. (6) that depends on the whole wave function and requires significant computational efforts. Recently, a method similar to Schwartz's one (see [9]) has been elaborated that is applicable to both one– and two–electron systems of an arbitrary angular momentum [10]. The final accuracy of expansion (6) is $\alpha^5\ln\alpha \approx 1.0\cdot 10^{-10}$.

Two–loop contribution [11]

$$\begin{aligned}\Delta E_{LS}(\text{two-loop}) &= \frac{Z_i\alpha^4}{\pi}\left[-\frac{6131}{1296}-\frac{49\pi^2}{108}+2\pi^2\ln 2-3\zeta(3)\right]\langle n|\delta(\mathbf{r}_i)|n\rangle 2\text{Ry}\\ &-\left(\frac{\alpha}{\pi}\right)^2\left[\frac{197}{144}+\frac{\pi^2}{12}-\frac{\pi^2}{2}\ln 2+\frac{3}{4}\zeta(3)\right]\left[i\mu_0\left\langle\beta\boldsymbol{\alpha}\cdot\mathbf{E}\right\rangle_{nn}\right]2\text{Ry}+O(\alpha^5)\end{aligned}$$

to the level of $O(\alpha^5)$ depends only on the delta function expectation values and the anomalous magnetic moment of electron

$$\mu_0\left\{\frac{\alpha}{\pi}\cdot\frac{1}{2}+\left(\frac{\alpha}{\pi}\right)^2\left[\frac{197}{144}+\frac{\pi^2}{12}-\frac{\pi^2}{2}\ln 2+\frac{3}{4}\zeta(3)\right]\right\},$$

the latter is usually included in calculation on the stage when Eq. (5) is evaluated.

To complete the whole consideration the finite size corrections should be included that can be done in a form [12]

$$\Delta E_{\text{nuc}} = \sum \frac{2\pi Z_i (R_i/a_0)^2}{3} \langle \delta(\mathbf{r}_i) \rangle,$$

where R is the root-mean-square radius of the nuclear charge distribution. The RMS radius for the helium nucleus and antiproton is, respectively,

$$R(^4\text{He}) = 1.673(1)\text{ fm}, \qquad R(\bar{p}) \stackrel{?}{=} R(p) = 0.862(12)\text{ fm}.$$

We assume in the latter case the equality of the proton and antiproton RMS radius.

5 Results

To present the results of our calculations we take a $(39, 35) \to (38, 34)$ transition, which is the most precisely measured experimentally [3]. The experimental value is

$$E_{\text{exp}} = 7628768(4) \cdot 10^{-8}\text{ a.u.}$$

At present accurate values of the Bethe logarithm for the antiprotonic helium states become available [13]

$$\ln[k_0(39,35)/\text{Ry}] = 4.4327(2), \qquad \ln[k_0(38,34)/\text{Ry}] = 4.4426(5).$$

Results of the theoretical calculation are summarized in Table 3.

E_{nr}	=	7629125.41(10)	$\cdot 10^{-8}$
E_{rc}	=	−419.30(5)	$\cdot 10^{-8}$
E_{se}	=	61.56	$\cdot 10^{-8}$
E_{vp}	=	−1.86	$\cdot 10^{-8}$
E_{2l}	=	0.01	$\cdot 10^{-8}$
E_{nuc}	=	0.04	$\cdot 10^{-8}$
E_{tot}	=	7628765.86(15)	$\cdot 10^{-8}$

Table 3. Summary of contributions to the transition energy (in a.u.). Here E_{nr} is the nonrelativistic transition energy, E_{rc} — relativistic corrections, E_{se} — one–loop self–energy correction, E_{vp} — vacuum polarization, E_{2l} — two–loop corrections, E_{nuc} — nuclear finite size correction.

As is seen from this Table the main uncertainty comes from the Auger width of the daughter $(38, 34)$ state. At this level of accuracy we need to take into account the hyperfine structure of the levels due to the "electron spin–antiproton orbit" interaction [14] that results in a splitting in the transition lines

$$E(L + \tfrac{1}{2}) = 7628769.5(2) \cdot 10^{-8}\text{ a.u.}$$

$$E(L - \tfrac{1}{2}) = 7628761.8(2) \cdot 10^{-8}\text{ a.u.}$$

6 Conclusions

1. Theoretical uncertainty reaches the level close to the uncertainty in the antiproton (proton) mass value. Further improvement both in theory and experiment could bring us a new value for the antiproton mass, which would be more accurate than for the proton!
2. Radiative "recoil" correction for the bound antiproton should be considered. It is expected that it should be small but not negligible and should result mainly from exchange photons of small energies.
3. It is hard but feasible to reach in a not-so-far future the relative accuracy of 10^{-10} for the transition energies.

The author would like to express his deep thanks to Prof. T. Yamazaki and Prof. R.S. Hayano for the numerous discussions and warm hospitality during his (author's) stay at the Tokyo university. This work has been partially supported by INTAS Grant No. 97-11032, which is gratefully acknowledged.

References

1. V.I. Korobov, D. Bakalov, H.J. Monkhorst: Phys. Rev. A **59**, R919 (1999)
2. R.S. Hayano: talk at this conference
3. H.A. Torii, R.S. Hayano, M. Hori, T. Ishikawa, N. Morita, M. Kumakura, I. Sugai, T. Yamazaki, B. Ketzer, F.J. Hartmann, T. von Egidy, R. Pohl, C. Maierl, D. Horváth, J. Eades, E. Widmann: Phys. Rev. A **59**, 223 (1999) and references therein
4. V.I. Korobov and I. Shimamura: Phys. Rev. A **56**, 4587 (1997)
5. H. Grotch and D.R. Yennie: Rev. Mod. Phys. **41**, 350 (1969)
6. R. Karplus, A. Klein, J. Schwinger: Phys. Rev. **86**, 288 (1952)
7. J.R. Sapirstein and D.R. Yennie: in *Quantum Electrodinamics* ed. by T. Kinoshita (Singapore: World Scientific)
8. H. Bethe and E. Salpeter: *Quantum Mechanics of One– and Two–Electron Atoms*, Springer-Verlag, 1957
9. C. Schwartz: Phys. Rev. **123**, 1700 (1961)
10. V.I. Korobov and S.V. Korobov: Phys. Rev. A **59**, 3394 (1999)
11. B.E. Lautrup, A. Peterman, E. de Rafael: Phys. Reports **3C**, 196 (1972)
12. G.W.F. Drake and W.C. Martin: Can. J. Phys. **76**, 679 (1998)
13. V.I. Korobov: in preparation
14. D.D. Bakalov and V.I. Korobov: Phys. Rev. A **57**, 1662 (1998)

Few-Body Systems Suppl. 12, 40–44 (2000)

Few-Body Systems

High-Precision Calculation of Antiprotonic Helium Atomcules and Antiproton Mass

Y. Kino[1,*], M. Kamimura[2], H. Kudo[1]

[1] Department of Chemistry, Tohoku University, Sendai 980-8578, Japan

[2] Department of Physics, Kyushu University, Fukuoka 812-8581, Japan

Abstract. We showed a new possibility to determine the basic constants concerned with antiproton by high-accuracy calculations of the transition wavelength of antiprotonic helium atoms ($\bar{p}He^+$, *atomcule* in short). The Coulomb three-body calculation having large interacting angular momenta of $J{\sim}35$ was carried out by a non-adiabatic coupled rearrangement channel method. Relativistic and QED corrections on the energies were calculated up to the second order perturbation. Excellent agreement with experimental data was obtained. With this agreement the antiproton mass (charge) were determined with the precision of 10^{-7}. The observed transition wavelengths are systematically red-shifted from the calculated value. These shifts couldn't be explained by a binary collision with simple attraction between the atomcule and medium He.

1 Introduction

The Atomcule ($\bar{p}He^+$) is an exotic three-body metastable system consisting of an antiproton ($\bar{p}$), a helium nucleus (He^{2+}) and an electron (e^-). The Atomcule is a compound word of *atom* and *molecule* showing the atom/molecule duality. The transition wavelengths were measured with precise laser spectroscopy (see [1], further references therein). Observed wavelengths of the transitions are given with seven-digit numbers [1]. Seven-digit accuracy is very challenging to the few-body theory, because nine significant figures in the eigenenergies of the three-body Hamiltonian is required in the calculation of the transition wavelength managing large interacting angular momenta of 30–40. Recent precise laser spectroscopy and calculations [2-7] opened a new possibility to investigate basic properties of antiprotonic systems.

The metastable states of interest have a large angular momentum (>30) between the He^{2+} and the $\bar{p}$; consequently, the $\bar{p}$ can avoid nuclear absorption. The e^- is mainly in the 1s state of the He^+ and partially polarized due

* *E-mail address:* kino@mail.cc.tohoku.ac.jp

to a correlation between the e^- and $\bar{p}$ [8]. On the other hand, the effective potential between He^{2+} and $\bar{p}$ resembles the Morse potential which indicates a molecular picture of the atomcule, because the potential consists of attractive Coulomb potential and repulsive centrifugal potential for a high angular momentum state [9]. To calculate the energy levels of this exotic three-body system with high-accuracy taking the unique behavior mentioned above into account, we employ a non-adiabatic coupled rearrangement channel method [10] which has successfully been applied to various few-body systems (see [11], further references therein).

2 Three-body Calculation

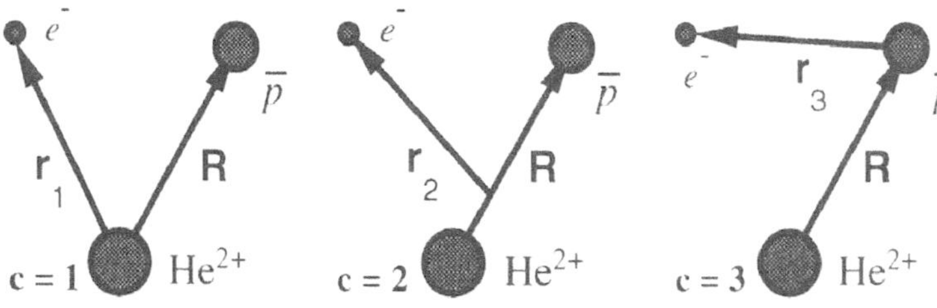

Figure 1. Three rearrangement channels.

In Fig. 1, the three channels are introduced. Channel c=1 with $\mathbf{r}_1$ and $\mathbf{R}$ is suited for describing the atomic picture in which e^- and $\bar{p}$ move around the He^{2+}. Channel c=2 with $\mathbf{r}_2$ and $\mathbf{R}$ is for describing the diatomic molecular picture in which e^- moves around the massive two nuclei. Channel c=3 with $\mathbf{r}_3$ and $\mathbf{R}$ is for describing a strong correlation between e^- and $\bar{p}$. The total wavefunction with the total angular momentum J and its z-component M is described with a sum of the amplitudes of three rearrangement channels, $\Psi_{vJM} = \Phi^{(1)}_{vJM}(\mathbf{r}_1, \mathbf{R}) + \Phi^{(2)}_{vJM}(\mathbf{r}_2, \mathbf{R}) + \Phi^{(3)}_{vJM}(\mathbf{r}_3, \mathbf{R})$, where v is the vibrational quantum number. The amplitude of each channel is expanded in terms of Gaussian basis functions [10]. The Gaussian ranges are taken to be geometrical progression which is suited for describing both short-range correlations and long-range tail behavior [10, 12]. Since the mass of electron is much smaller than nuclei, we didn't use Jacobian coordinate to avoid the complexity of rearrangement of the high-angular momentum. But we treat exactly the mass polarization term of the kinetic-energy operator. Eigenenergies and wavefunctions are solved with the Rayleigh-Ritz variational principle.

With the use of the first-order perturbation theory we calculated the relativistic and QED corrections, $\Delta E^{(1)}_{Jv} = \langle \Psi_{vJM} | \Delta V | \Psi_{vJM} \rangle$, where ΔV consists of relativistic corrections for three particles, Lamb shift and finite size effects of two nucleus [6]. The second order perturbation energies were calculated with the metastable states and *pseudo* states which approximately described the three-body continuum states as closed channels [13]. Both metastable and *pseudo* states are simultaneously obtained in the three-body calculation, and are labeled with quantities v'. Hence the second order perturbation energies are given by,

$$\Delta E^{(2)}_{Jv} = \sum_{v'} \frac{|\langle \Psi_{vJM}|\Delta V|\Psi_{v'JM}\rangle|^2}{E_{Jv} - E_{Jv'}}. \quad (1)$$

The main contribution to $E^{(2)}_{Jv}$ is $v' = v \pm 1$ states in the sum of Eq. 1. The second order perturbation energies were about 10^{-10} a.u. and did not affect the transition wavelength with seven-digit accuracy.

3 Results and Discussions

Table 1. Calculated and observed transition wavelengths which are in unit of nm. λ_C: pure Coulomb calculation. $\hat{\lambda}_C$: with relativistic and QED corrections. λ_E: without He density correction. $\hat{\lambda}_E$: with He density correction. Δx is the uncertainty of $\bar{p}$ mass.

(J_i,v_i)–(J_f,v_f)	λ_C	$\hat{\lambda}_E$	λ_E	$\hat{\lambda}_E$ [1]	Δx
(35,3)–(34,3)	597.2290	597.2573	597.259(2)	597.2570(3)	1×10^{-7}
(34,2)–(33,2)	470.7048	470.7220	470.724(2)	470.7220(6)	3×10^{-7}

The calculated energies agree well with the latest calculation [7]. In Table 1, the calculated and observed transition wavelengths are listed. The relativistic and QED collection reduced the discrepancy between the calculated and observed values by about 40 ppm. The data with He density correction $\hat{\lambda}_E$ [1] are well reproduced within the experimental error. The results show high-accuracy of our method for the Coulomb three-body systems with large angular momenta.

$\hat{\lambda}_e$ were obtained by the extrapolation to the zero density limit. While the data without He density correction λ_e systematically red-shifted from our calculation. (Note that the other observed transition wavelength without the correction also red-shifted [6].) The shifts may be explained by the binary collision between an atomcule and a surrounding He. In Ref. [14] Korenman tried to calculate the collisional shift and the broadening of the transition wavelength with the use of a van der Waals potential having soft repulsion at the small distance and a straight-line trajectory approximation. He claimed that the red-shift were due to the long-range attractive part of the van der Waals potential, but he used simple hydrogen-like wavefunctions of atomcules with effective charges of $\bar{p}$ and e^-. We tried to test the Ref. [14] using our accurate wavefunctions. Using the second order perturbation theory the van der Waals potential between $\bar{p}He^+$ and He are obtained by a dipole approximation and a second order perturbation theory. $V_{vJ}(\rho) = -\alpha_{He}\langle \Psi_{vJM}|\boldsymbol{D}^2|\Psi_{vJM}\rangle/\rho^6 \equiv -C_{vJ}/\rho^6$, where ρ is the distance between $\bar{p}$ and He and α_{He}=1.383 a.u. is the polarizability of He. Here, $\boldsymbol{D} = 2\boldsymbol{r}_{He^{2+}} - \boldsymbol{r}_{\bar{p}} - \boldsymbol{r}_{e^-} = -\boldsymbol{r}_1 - \boldsymbol{R}$, is a dipole operator of $\bar{p}He^+$, in which

$\boldsymbol{r}_{He^{2+}}$, $\boldsymbol{r}_{\bar{p}}$ and $\boldsymbol{r}_{e^-}$ are the position vector of He^{2+}, $\bar{p}$ and e^-, respectively. In Table 2, the strength of the attraction potential C_{Jv} are listed. For comparison the values of Ref. [14] were $C_{35,3}$=2.78, $C_{34,3}$=2.68, $C_{34,2}$=2.68 and $C_{33,2}$=2.60. The discrepancies of C_{if} between ours and Ref. [14] are due to the accuracy of the wavefunction of the atomcule. According to Eq. (8) of Ref. [14] the red-shift (Δ<0) corresponds to $C_{if}=C_{J_i v_i} - C_{J_f v_f}>0$ and the sign of the shift changes in the case of $C_{if}<0$. From Table 2, the values C_{if} are, however, negative for all of the favored transition $\Delta v = v_f - v_i = 0$, $J_f - J_i = -1$. We therefore point out that the red-shift is due to not the long-range attractive potential but an intermediate-range potential which consists of attraction and repulsion. Further calculations which include the short-range repulsive part having the atomcule state dependence in the van der Waals potential are in progress.

Table 2. The strength of the attractive potential C_{Jv} (a.u.). †: $\bar{p}^3He^+$.

$J\backslash v$	0	1	2	3	$J\backslash v$	0	1	2	3
33	1.808	1.728	1.678		32	1.807	1.723	1.673	
34	1.730	1.669	1.622	1.602	33†	1.723	1.748	1.620	1.603
35†	1.661	1.612	1.585	1.584	34†	1.653	1.607	1.582	1.587

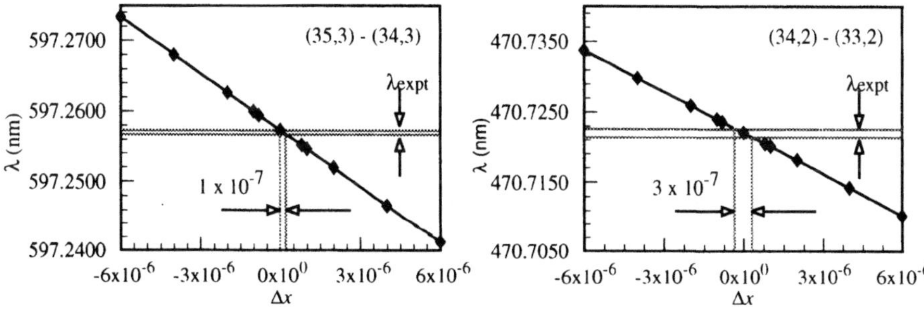

Figure 2. The calculated wavelengths as a function of Δx. Closed diamonds shows the calculated values and solid line is the guide for the eyes.

We assumed that the $\bar{p}$ mass was equal to that of p. The measured charge-to-mass ratio of $\bar{p}$ is equal to the p with 9×10^{-11} uncertainty [15], but the $\bar{p}$ mass and charge were not measured individually with such a high accuracy. The $\bar{p}$ mass was estimated within an uncertainty of 2×10^{-5} [16] using an X-ray measurement of antiprotonic atoms [17]. The transition wavelengths (energies) show a different dependence of mass and charge from the charge-to-mass ratio. To estimate the uncertainty of $\bar{p}$ mass from the experimental data, we recalculated the wavelength using the $\bar{p}$ mass scaled with $1+\Delta x$ against the p mass, $m_{\bar{p}} = (1+\Delta x)m_p$. Here, the $\bar{p}$ charge was scaled simultaneously to keep the charge-to-mass ratio, $e_{\bar{p}} = (1 + \Delta x)e_p$. The value Δx was estimated using an uncertainty of the experimental wavelengths $\Delta\lambda_{\mathrm{expt}}$.

$$\Delta x = \frac{\Delta m_{\bar{p}}}{m_p} = \Delta\lambda_{\mathrm{expt}} \left(\frac{d\lambda}{dx}\right)^{-1}_{\mathrm{cal}} \tag{2}$$

where $(d\lambda/dx)_{\mathrm{cal}}$ was obtained from the slope of Fig. 2. The estimated Δx are listed in Table 1. This work reduced uncertainty of the $\bar{p}$ mass (or equivalently charge) by two orders of magnitude. And this procedure will have a new possibility to determine the basic constant concerned with $\bar{p}$.

Acknowledgement.

We are grateful to Prof. Dr. G.Ya. Korenman, Dr. V.I. Korobov and Dr. H.A. Torii for fruitful discussions. We also wish to thank professor Y.K. Ho for helpful advice. This work is financially supported by the Grant-in-Aid for Scientific Research, Ministry of Education and Culture. YK wish to acknowledge Inoue Foundation for Science for their generous financial support. The computations were partially carried out on the RIKEN supercomputer VPP700/128.

References

1. H.A. Torii et al.: Phys. Rev. **A59**, 223 (1999)
2. V.I. Korobov: Phys. Rev. **A54**, R1749 (1996)
3. V.I. Korobov and D.D. Bakalov: Phys. Rev. Lett. **79**, 3379 (1997)
4. N. Elander and E. Yarevsky: Phys. Rev. **A56**, 1855 (1997)
5. Y. Kino et al.: Nucl. Phys. **A631**, 649 (1998)
6. Y. Kino et al.: Hyp. Interact. **119**, 201 (1999)
7. V.I. Korobov et al.: Phys. Rev. **A59**, R919 (1999).
8. T. Yamazaki and K. Ohtsuki: Phys. Rev. **A45**, 7782 (1992)
9. I. Shimamura: Phys. Rev. **A46**, 3776 (1992)
10. M. Kamimura: Phys. Rev. **A38**, 621 (1988)
11. E. Hiyama et al.: Phys. Rev. **C59**, 2351 (1999)
12. Y. Kino et al.: Phys. Rev. **A52**, 870 (1995)
13. Y. Kino and M. Kamimura: Hyp. Interact. **82**, 45 (1993)
14. G.Ya. Korenman: Hyp. Interact. **119**, 227 (1999)
15. G. Gabrielse et al.: Phys. Rev. Lett. **82**, 3198 (1999)
16. R.J. Hughes and B.I. Deutch: Phys. Rev. Lett. **69**, 578 (1992)
17. P. Roberson et al.: Phys. Rev. **C16**, 1945 (1977)

Few-Body Systems Suppl. 12, 45–49 (2000)

Few-Body Systems

Coupling of Two- and Three-Particle Channels in Slow Antiproton - Atom Collisions

G.Ya. Korenman

Institute of Nuclear Physics, Moscow State University, Moscow 119899, Russia

Abstract. Semiclassical approximation for the problem of coupled 2- and 3-particle channels in $\bar{p} - A$ collisions is developed beginning from the quantum equations. Closed expressions for S-matrix elements are obtained. The approximation avoids inherent difficulties and uncertainties of *postulated* semiclassical approaches, and allows to consider the main quantum effects.

1 Introduction

Collisions of slow antiprotons with atoms attract considerable interest in the recent years. Existing calculations of $\bar{p} - A$ collisions at low velocity for the few-electron atoms (see reviews [1, 2]) give a satisfactory overall picture of antiprotonic atom formation in media. However, the recent experimental data on laser-selected metastable states of antiprotonic helium (see [3] and the references therein) call for more detailed theoretical results on the primary populations of individual levels, especially nearly to the boundary of the population area. New unique data on $\bar{p}$ - atom collisions could be expected from future experiments within the ASACUSA project [4]. Therefore it is necessary to improve consideration of both the elementary processes

$$\bar{p} + A \rightarrow \bar{p} + A, \tag{1}$$

$$\bar{p} + A \rightarrow \bar{p} + A^{+} + e, \tag{2}$$

$$\bar{p} + A \rightarrow (\bar{p}A^{+})^{*} + e, \tag{3}$$

and the kinetics of the processes. Theory of $\bar{p} - A$ collisions at low energy is complicated due to a strong coupling of the channels (1) - (3) at all energies up to $E \rightarrow 0$. Most of the theoretical papers on the problem are restricted by classical and *postulated* semiclassical approaches that lead to some inherent difficulties [2] and neglect any quantum effects for heavy particles. Quantum approaches were used for $\bar{p}$ + H collisions, but they turn out too complicated even for $\bar{p}$ + He collisions. We develop the semiclassical approximation for the problem of coupled 2- and 3-particle channels, beginning from the quantum equations. Closed expressions for S-matrix of the processes (1) - (3) are obtained. This approach, named as Coupled Channels Semiclassical Approximation (CCSA), avoids any inherent difficulties and allows to consider quantum effects, including interference and classically forbidden transitions.

2 Quantum and Quasiclassical Coupled-Channels Equations

Total wave function of the $(\bar{p}+A)$ system can be presented as an expansion in terms of the bound and continuum wave functions of the atomic electrons,

$$\Psi^{JM}(\xi,R)=R^{-1}\sum_j \psi_j^{JM}(\xi)u_j(R)+R^{-1}\sum_\alpha \int d\epsilon\, \psi_{\alpha\epsilon}^{JM}(\xi)u_{\alpha\epsilon}(R), \qquad (4)$$

where $\xi=(\{\mathbf{r_e}\},\Omega)$ is the aggregate of electronic variables and angular variables of the heavy particles, $\psi_j^{JM}(\xi)$ and $\psi_{\alpha\epsilon}^{JM}(\xi)$ are electronic bound state and continuum wave functions joined with angular wave functions of the heavy particles to form the states with quantum numbers J, M of total angular momentum. We will restrict the basis by the ground state of atom A and the single electronic continuum $A^+(\text{g.s.})+e$. (For brevity, we use the atomic basis, but a generalization for the adiabatic basis can be done easily.) Neglecting by the transitions $(\epsilon\to\epsilon')$ we obtain the quantum equations for the coupled channels

$$u_0''(R)+p_0^2(R)\,u_0(R) = \int d\epsilon\, W_{0\epsilon}(R)\,u_\epsilon(R), \qquad (5)$$

$$u_\epsilon''(R)+p_\epsilon^2(R)\,u_\epsilon(R) = W_{\epsilon 0}(R)\,u_0(R) \qquad (6)$$

with the boundary conditions at $R\to\infty$

$$u_0(R)\sim \exp[-\mathrm{i}(k_0R-J\pi/2+\delta_{0J})]-\bar{S}_{00}\exp[\mathrm{i}(k_0R-J\pi/2+\delta_{0J})], \qquad (7)$$

$$u_\epsilon(R)\sim \begin{cases} -\bar{S}_{\epsilon 0}\exp[\mathrm{i}(k_\epsilon R-J\pi/2+\delta_{\epsilon J}+\eta\ln 2k_\epsilon R)] & \text{at } \epsilon<E-I_0,\\ N(E,\epsilon)(\kappa R)^{M/\kappa}\exp(-\kappa R) & \text{at } \epsilon>E-I_0, \end{cases} \qquad (8)$$

where $p_0^2(R)=2m[E-V_0(R)]-J(J+1)/R^2$, $W_{0\epsilon}(R)=2mV_{0\epsilon}(R)$, $p_\epsilon^2(R)=2m[E-I_0-\epsilon-V_1(R)]-J(J+1)/R^2$, $V_0(R)=\langle\psi_0|V(1,\ldots,Z)|\psi_0\rangle$, $V_1(R)=\langle\phi_0(A^+)|V(1,\ldots,Z-1)|\phi_0(A^+)\rangle$, m, E and I_0 are $\bar{p}-A$ reduced mass, initial kinetic energy and atomic ionization potential, respectively, $k_0^2=2mE$, $k_\epsilon^2=2m(E-I_0-\epsilon)$, $\eta=m/k_\epsilon$, $\kappa^2=2m(I_0+\epsilon-E)$. The coefficients $\bar{S}_{\mu\nu}$ together with the channel phase shifts $\delta_{\nu J}$ define S-matrix, $S_{\mu\nu}^J(E)=\exp[\mathrm{i}(\delta_\mu+\delta_\nu)]\,\bar{S}_{\mu\nu}(E)$.

We seek quasiclassical radial functions $u_\nu(R)$ in the form [5]

$$u_\nu(R)=\{A_\nu(R)\exp[-\mathrm{i}\alpha_\nu(R)]-B_\nu(R)\exp[\mathrm{i}\alpha_\nu(R)]\}\,/\sqrt{p_\nu(R)}, \qquad (9)$$

with the additional condition

$$(A_\nu/\sqrt{p_\nu})'\exp(-\mathrm{i}\alpha_\nu)-(B_\nu/\sqrt{p_\nu})'\exp(\mathrm{i}\alpha_\nu)=0, \qquad (10)$$

where $\alpha_\nu(R) = \int_{R_\nu}^{R} p_\nu(R')dR' + \pi/4$, $R_\nu = R_\nu(E,J)$ is a classical turning point in the channel ν. Substitution of Eqs. (9)-(10) into (5)-(6) gives the system of the equations for A_ν, B_ν. Neglect of the terms $\sim \exp[\pm \mathrm{i}(\alpha_\nu + \alpha_\mu)]$ leads to the pairs of quasiclassical equations for (A_0, A_ϵ) and (B_0, B_ϵ), which can be transformed into integral equations

$$A_0(R) = 1 - \mathrm{i}\int d\epsilon \int_R^\infty dR' W_{0\epsilon}(R') \exp[\mathrm{i}(\alpha_0 - \alpha_\epsilon)] A_\epsilon(R'), \quad (11)$$

$$A_\epsilon(R) = C(\epsilon) + \mathrm{i}\int_{R_\epsilon}^{R} dR' W_{\epsilon 0}(R') \exp[\mathrm{i}(\alpha_\epsilon - \alpha_0)] A_0(R'), \quad (12)$$

$$B_0(R) = \bar{S}_{00}(E) + \mathrm{i}\int d\epsilon \int_R^\infty dR' W^*_{0\epsilon}(R') \exp[\mathrm{i}(\alpha_\epsilon - \alpha_0)] B_\epsilon(R'), \quad (13)$$

$$B_\epsilon(R) = C(\epsilon) - \mathrm{i}\int_{R_\epsilon}^{R} dR' W^*_{\epsilon 0}(R') \exp[\mathrm{i}(\alpha_0 - \alpha_\epsilon)] B_0(R'), \quad (14)$$

These equations incorporate the boundary conditions for A_0 and B_0 at infinity. Another boundary condition for the 2-particle channel $A_0(R_0) = B_0(R_0)$ couples solutions of the pairs of Eqs. (11)-(12) and (13)-(14). Boundary conditions for the channel of electronic continuum depend on the value of the channel kinetic energy $\mathcal{E} = E - I_0 - \epsilon$: at $\mathcal{E} > 0$ $A_\epsilon(\infty) = 0$, $B_\epsilon(\infty) = \bar{S}_{\epsilon 0}(E)$, $A_\epsilon(R_\epsilon) = B_\epsilon(R_\epsilon) \equiv C(\epsilon)$; at $\mathcal{E} < 0$, when $\bar{p}$ moves between two turning points $R_1(\epsilon)$ and $R_2(\epsilon)$, $A_\epsilon(R_1) = B_\epsilon(R_1) \equiv C(\epsilon)$, $A_\epsilon(R_2) = D(\epsilon)\exp[\mathrm{i}\Delta(\mathcal{E})]$, $B_\epsilon(R_2) = -D(\epsilon)\exp[-\mathrm{i}\Delta(\mathcal{E})]$, where

$$\Delta(\mathcal{E}) = \int_{R_1}^{R_2} dR \left\{2m[\mathcal{E} - V_1(R)] - (J+1/2)^2/R^2\right\}^{1/2}. \quad (15)$$

3 Semiclassical Approximation for Classically Allowed Transitions

In atomic basis the $\bar{p} - A$ potential $V_0(R)$ crosses the continuum boundary $V_1(R) + I_0$ at some real critical point R_a. It allows to use the stationary phase approximation for the integrals over R in the right side of Eqs. (11) - (14). We suppose that the difference of the terms $\epsilon_s(R) = V_0(R) - V_1(R) - I_0$ is a monotonic function of R, positive at $R < R_a$ and negative at $R > R_a$. In this case, the condition for stationary phase $[\alpha_0(R) - \alpha_\epsilon(R)]' = 0$ gives the single transition point $R = R_s(\epsilon)$ and the single-valued inverse function $\epsilon = \epsilon_s(R)$ that has a sense of energy of electron ejected during the classically allowed transition at the point R. The integrals over R in the right side of Eqs. (11) - (14) are reduced to the values of $A_\nu(R_s)$ or $B_\nu(R_s)$ multiplied by a stepwise function $\theta(R - R_s)$ or $\theta(R_s - R)$ and the factor $w(\epsilon) = -\mathrm{i}\{2\pi/[v(R_s)|\epsilon_s'(R_s)|]\}^{1/2}\exp[\mathrm{i}\beta(\epsilon)]V_{\epsilon 0}(R_s)$ or $w^*(\epsilon)$, where $v(R) = p_0(R)/m$, $\beta(\epsilon) = \alpha_0(R_s) - \alpha_\epsilon(R_s) - (\pi/4)\mathrm{Sign}[\epsilon'(R_s)]$. Then tedious but straightforward transformations allow to express the components $A_\nu(R)$ and $B_\nu(R)$ in terms of the function $C(\epsilon)$. For $\epsilon < E - I_0$ this function can be simply found using the boundary condition at the point R_ϵ. But for $\epsilon > E - I_0$ the conditions at two turning points $R_1(\epsilon)$ and $R_2(\epsilon)$ lead to coupled integral equations for $C(\epsilon)$ and $D(\epsilon)$, which are solved neglecting by the terms $\sim \exp[\pm 2i\beta(\epsilon)]$. The final result for the components of total wave function within the stationary phase approximation has a form

$$\begin{aligned}
A_0(R) &= \theta(R-R_a)+\theta(R_a-R)\exp[-\nu(0,\epsilon_s(R))/2], &(16)\\
B_0(R) &= \bar{S}_{00}(E)\{\theta(R-R_a)+\theta(R_a-R)\exp[\nu(0,\epsilon_s(R))/2]\}, &(17)\\
A_\epsilon(R) &= \mathrm{i}\exp[-\nu(0,\epsilon)/2]\theta(\epsilon_m-\epsilon)\{-w^*(\epsilon)\theta(R_s-R)+\theta(\epsilon+I_0-E)\cdot\\
&\quad [w^*(\epsilon)+w(\epsilon)\exp[-\nu(\epsilon,\epsilon_m)]]\,[1+\mathrm{i}\tan\Delta(\mathcal{E})]/2\}, &(18)\\
B_\epsilon(R) &= \mathrm{i}\exp[-\nu(0,\epsilon)/2]\theta(\epsilon_m-\epsilon)\{-[w^*(\epsilon)+w(\epsilon)\exp[-\nu(\epsilon,\epsilon_m)]\cdot\\
&\quad \theta(R-R_s(\epsilon))]+\theta(\epsilon+I_0-E)\,[w^*(\epsilon)+w(\epsilon)\exp[-\nu(\epsilon,\epsilon_m)]]\cdot\\
&\quad [1+\mathrm{i}\tan\Delta(\mathcal{E})]/2\}, &(19)
\end{aligned}$$

where $\epsilon_m=\epsilon_s(R_0(E,J))$ is the maximum possible energy of ejected electron,

$$\begin{aligned}
\nu(\epsilon,\epsilon') &= \int_\epsilon^{\epsilon'}\gamma(\epsilon'')d\epsilon''=\int_{R_s(\epsilon)}^{R_s(\epsilon')}\Gamma(R)\,dR/v(R),\\
\gamma(\epsilon) &= |w(\epsilon)|^2,\\
\Gamma(R) &= 2\pi\int d\epsilon|V_{\epsilon 0}(R)|^2\delta(\epsilon-\epsilon_s(R))=2\pi|V_{\epsilon 0}(R)|^2\big|_{\epsilon=\epsilon_s(R)}.
\end{aligned}$$

For the elements of S-matrix we obtain within the same approximations

$$\begin{aligned}
S_{00}^J(E) &= \exp\left[2\mathrm{i}\delta_{0J}(E)-\nu(0,E-I_0)-\sum_k\omega_k\gamma(\epsilon_k)\right], &(20)\\
S_{\epsilon 0}^J(E) &= -\mathrm{i}\exp\{-\nu(0,\epsilon)/2+\mathrm{i}\,[\delta_0^J(E)+\delta_\epsilon^J(E)]\}\cdot\\
&\quad \{w^*(\epsilon)+w(\epsilon)\exp[-\nu(\epsilon,\epsilon_m)]\}. &(21)
\end{aligned}$$

The sum over k in Eq. (20) has the same origin as the singularities of the functions $A_\epsilon(R)$ and $B_\epsilon(R)$ at $\Delta(\mathcal{E})=\pi(k+1/2)$. Remembering Eq. (15) it is easy to see that the singularity condition is just the Bohr-Sommerfeld quantization rule for the potential $V_0(R)$, and k is the radial quantum number of the $\bar{p}$ bound state. In other words, the solution (16) - (19) has poles in ϵ at $\epsilon_k=E-I_0-\mathcal{E}_k$ that correspond to atomic capture (3) of $\bar{p}$ to the states with energy $\mathcal{E}_k$. At $\mathcal{E}\to\mathcal{E}_k$ we have

$$\begin{aligned}
\Delta(\mathcal{E}) &\to \pi(k+1/2)+(\mathcal{E}-\mathcal{E}_k+\mathrm{i}\gamma)\pi/\omega_k,\\
\tan\Delta(\mathcal{E}) &\to -\frac{\omega_k}{\pi(\mathcal{E}-\mathcal{E}_k+\mathrm{i}\gamma)},\\
\omega_k &= \pi\left[m\int_{R_1}^{R_2}dR/p_k(R)\right]^{-1},
\end{aligned}$$

where $\gamma\to+0$ provides for a correct behaviour of the wave function at $R\to\infty$ in the $\bar{p}$ capture channel. So, in spite of that the basis doesn't include explicitly rearrangement states, the solution (16) - (19) for the total wave function and the S-matrix of elastic scattering (20) and ionization (21) incorporate the process (3). Explicit amplitude of atomic capture can also be obtained, but it requires rather lengthy transformations and will be published elsewhere. One more important property of the obtained S-matrix is that Eq. (21) includes two terms accounted for the interference of the transitions at two branches of a classical trajectory.

4 Conclusion

We have found total semiclassical solution of quantum equations of coupled two-particle and three-particle channels. S-matrix for elastic scattering and classically allowed (Frank-Condon) transitions into 3-particle channels is obtained in terms of the matrix of potentials with account for atomic capture and interference effects. Total wave function allows to consider also amplitude of atomic capture as well as classically forbidden transitions.

Acknowledgement. I would like to thank Prof. R. Hayano for a kind hospitality at University of Tokyo where a part of the work was fulfilled, and Drs. V.P. Popov and Kh. Tsookhuu for useful discussions and a check of some formulae.

References

1. J.S. Cohen: In: *Electromagnetic Cascade and Chemistry of Exotic Atoms*, p.1. New York: Plenum Press 1990

2. G.Ya. Korenman: Hyperfine Interactions, **101/102**, 81 (1996).

3. M.Hori et al.: Phys. Rev. **A57** 1698 (1998); **A58** 1613 (1998)

4. ASACUSA Collaboration (T.Azuma et al.): A proposal of experiments "Atomic Spectroscopy And Collisions Using Slow Antiprotons". Preprint CERN/SPSC 97-19/P-307 CERN 1997

5. E.E. Nikitin and S.Ya. Umanski: *Theory of Slow Atomic Collisions.* Berlin: Springer 1984

Few-Body Systems Suppl. 12, 50–59 (2000)

Few-Body Systems

Few-Body States in Fermi-Systems and Condensation Phenomena

P. Schuck[1,*], M. Beyer[2], G. Röpke[2], W. Schadow[3], A. Schnell[4]

[1] Institut de Science Nucléaires, Université Joseph Fourier, CNRS-IN2P3 53, Avenue des Martyrs, F-38026 Grenoble Cedex, France

[2] Fachbereich Physik, Universität Rostock, D-18051 Rostock, Germany

[3] TRIUMF, 4004 Wesbrook Mall, Vancouver, B.C., V6T 2A3, Canada

[4] Dept. of Physics, University of Washington, 22901, U.S.A.

Abstract. Residual interactions in many particle systems lead to strong correlations. A multitude of spectacular phenomena in many particle systems are connected to correlation effects in such systems, e.g. pairing, superconductivity, superfluidity, Bose-Einstein condensation etc. Here we focus on few-body bound states in a many-body surrounding.

1 Introduction

Correlated many particle systems such as, e.g. nuclear matter or strongly coupled plasmas, have a complicated dynamical behavior. Only few areas of the density-temperature phase diagram used to characterize the state of the system in thermal equilibrium may be described in the approximation of noninteracting quasiparticles. The dynamics of the quasiparticle is determined by the mean field of the other particles and therefore some prominent features like self-energy corrections and Pauli blocking (Bose enhancement) are sufficiently taken into account [1]. However due to sizable residual interactions many interesting and exciting phenomena, such as clustering, formation of condensates, and phase transitions occur. The description of the dynamics of such systems is particularly difficult since the quasiparticle approach reaches its limits.

A large number of these phenomena such as, e.g. two-particle bound state formation, can already be accounted for by explicitly introducing two-body correlations into the formalism. This may be achieved in the frame work of the Green function method [1] and leads to effective two-body equations that include medium effects in a consistent way.

* *E-mail address:* schuck@isnhp4.in2p3.fr

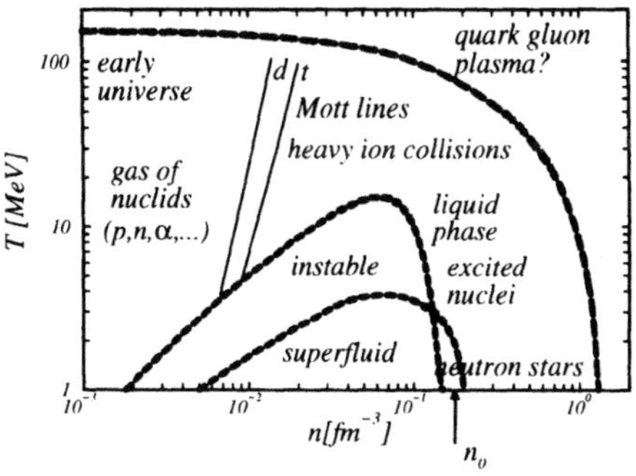

Figure 1. Density temperature phase diagram of nuclear matter. n_0 denotes nuclear matter density $n_0 = 0.17\ \text{fm}^{-3}$.

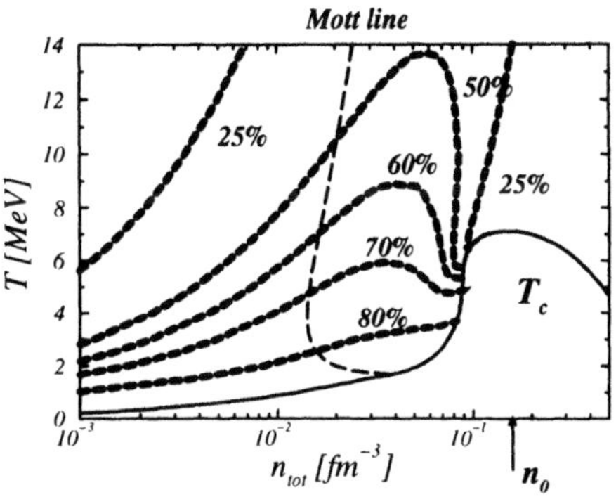

Figure 2. Nuclear matter phase diagram. The the amount of correlated density is given in per cent of the total density [2]. T_c denotes the critical temperature.

The density temperature planes of nuclear matter are shown in Figs. 1 and 2. The phase diagram of nuclear matter turns out to be very rich. In particular the superfluid phase reflecting strong pairing is relevant for the structure of neutron stars [3]. At lower densities bound states occur. This part of the phase diagram may be accessed in the laboratory through heavy ion collisions at intermediate energies. The description of these phenomena in the framework spelled out here constitues a wide and exciting area for an application of few-body methods.

Besides the pairing of nucleons, correlations with more particles are also important. Based on the suitably modified AGS (Alt-Grassberger-Sandhas [5]) formalism the reaction cross section for break-up ($Nd \to NNN$), relevant e.g. for the formation of deuterons in a heavy-ion reaction, has been given in [6]. As a first application the life time of deuteron fluctuations in nuclear matter was calculated in [7]. In [8] we have given an exploratory calculation of the four-body condensate in strongly coupled fermionic systems.

2 Theory

The formalism to derive few-body Green functions within the Dyson equation approach or cluster mean field approximation at finite temperatures and densities has been given elsewhere [4]. Here we give some of the basic results and focus on bound states.

Let the Hamiltonian of the system be given by

$$H(t) = \sum_1 H_0(1) a_1^\dagger(t) a_1(t) + \frac{1}{2} \sum_{121'2'} V_2(12,1'2')\, a_1^\dagger(t) a_2^\dagger(t) a_{2'}(t) a_{1'}(t) \quad (1)$$

where $a_1(t)$ etc. denotes the Heisenberg operator of the particle with quantum numbers s_1, k_1, etc. for spin, momentum etc. In the equal time formalism we may introduce creation and annihilation operators $A_\alpha(t)$, where $A_\alpha = a_1, a_1^\dagger a_2, a_1 a_2 a_3$, etc. for one-particle, hole-particle, three-particle operators. We describe the propagation of these clusters via the Green's function at $T = 0$. However, generalization to $T \neq 0$ is straight forward using the Matsubara technique [1]. The definition of the Green's function reads

$$\mathcal{G}_{\alpha\beta}^{t-t'} = -i\langle T A_\alpha(t) A_\beta^\dagger(t')\rangle, \tag{2}$$

where averaging is due to the density operator ρ_0. For an open system in thermodynamical equilibrium the extremum condition for the entropy leads to the following expression for the density operator of a grand canonical quantum ensemble,

$$\rho_0 = \frac{e^{-\beta(H-\mu N)}}{\mathrm{Tr}\{e^{-\beta(H-\mu N)}\}}. \tag{3}$$

The temperature ($1/\beta = T$) and the chemical potential μ are the corresponding Lagrange parameters. The Dyson equation is

$$i\partial_t \mathcal{G}_{\alpha\beta}^{t-t'} = \delta(t-t')\langle[A_\alpha, A_\beta^\dagger]_\pm\rangle + \sum_\gamma \int d\bar{t}\, \mathcal{M}_{\alpha\gamma}^{t-\bar{t}}\, \mathcal{G}_{\gamma\beta}^{\bar{t}-t'}. \tag{4}$$

We use $\mathcal{N}_{\alpha\beta} = \langle[A_\alpha, A_\beta^\dagger]_\pm\rangle$ in the following, where the upper (lower) sign reflect Fermi (Bose) type operators. The mass matrix is given by [4]

$$\mathcal{M}_{\alpha\gamma}^{t-\bar{t}} = \underbrace{\delta(t-\bar{t})\mathcal{M}_{\alpha\gamma,0}^{t}}_{\text{cluster-mean-field}} + \underbrace{\mathcal{M}_{\alpha\gamma,\mathrm{irr}}^{t-\bar{t}}}_{\text{retardation}} \tag{5}$$

$$(\mathcal{M}^t\mathcal{N})_{\alpha\beta,0} = \langle[[A_\alpha, H](t), A_\beta^\dagger(t)]\rangle \tag{6}$$

$$(\mathcal{M}^{t-t'}\mathcal{N})_{\alpha\beta,\mathrm{irr}} = -i\langle T[A_\alpha, H](t), [H, A_\beta^\dagger](t')\rangle_{\mathrm{irr}}. \tag{7}$$

For $A_\alpha = a_1$, i.e. the one-particle case, neglecting the term $\mathcal{M}_{\mathrm{irr}}$ (representing all retarded irreducible diagrams) leads to the self-energy shift in mean field (e.g. Hartree-Fock) approximation. For a given potential this constitutes a selfconsistent problem to determine μ and β, to be solved by iteration

$$f_1 \equiv f(\varepsilon_1) = \frac{1}{e^{\beta(\varepsilon_1-\mu)}+1}, \qquad \varepsilon_1 = \frac{k_1^2}{2m} + \sum_2 V_2(12) f_2. \tag{8}$$

For $T \to 0$ this function approaches a step function at the respective Fermi momentum.

3 Two-Body Correlations, Bound States and Cooper Pairing

Nuclear matter cannot be considered as a system of independent quasi particles but for a large part it is correlated and, e.g. in the superfluid phase as

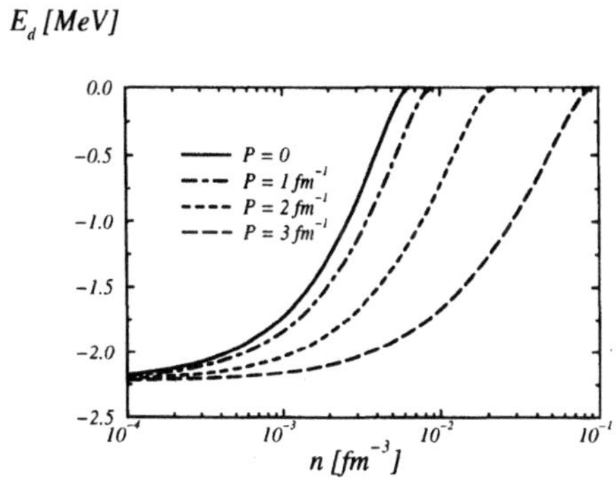

Figure 3. The deuteron binding energy as a function of the nuclear density for $T = 10$ MeV. P denotes the relative momentum between the deuteron and the medium.

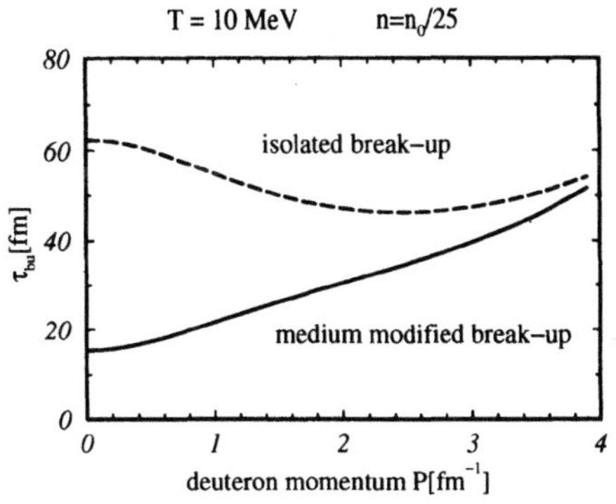

Figure 4. Deuteron life time as a function of the deuteron momentum at finite temperature ($T = 10$ MeV) and density $n = n_0/25$ [10].

demonstrated in Fig. 2 [2]. The basis to treat correlated densities is provided by a generalization of the Beth-Uhlenbeck approach [9]. The nuclear density $n = n(\mu, T)$ is given by

$$n = n_{\text{free}} + n_{\text{corr}}, \qquad n_{\text{corr}} = 2n_2 + 3n_3 + \ldots, \tag{9}$$

where $n_{2,3}$ denotes the two, three-particle correlations, present as bound/scattering states. In first iteration these correlations may be treated on the basis of residual interactions between the quasi particles. The exact two particle equations to be solved are known as Bethe-Goldstone or Feynman-Galitski equations depending on some details [1]. In ladder approximation the two-body equation reads

$$(E - \varepsilon_1 - \varepsilon_2)\, \Psi_2(E) = (\bar{f}_1 \bar{f}_2 - f_1 f_2)\, V_2\, \Psi_2(E), \tag{10}$$

where $\bar{f} = 1 - f$. The normalization is given by

$$\sum_{12} \tilde{\Psi}^{*\nu}_{12} (f_1 f_2 - \bar{f}_1 \bar{f}_2) \tilde{\Psi}^{\nu'}_{12} = \delta_{\nu\nu'}, \tag{11}$$

where the dual wave functions $\tilde{\Psi}^\nu$ are defined by $\Psi^\nu = (1 - f_1 - f_2)\tilde{\Psi}^\nu$. Introducing the two-body t-matrix in a standard fashion both bound and scattering states have been solved [9]. In first approximation V_2 of Eq. (10) corresponds to the bare force. However, at higher densities the force may be substantially screened.

As an example for an in-medium bound state the deuteron binding energy is shown in Fig. 3. The Mott density is defined through the condition $E_d = 0$. Note that the Fermi functions f_1 for particle 1 etc. depend on the relative momentum P between the deuteron and the medium.

The deuteron in the medium also has a finite life time. To second order Born approximation the life time can be extracted from the square of the graphs of Fig. 5. One sees that the width $\Gamma_d^{\downarrow}$ is given by the incoherent decay

Figure 5. Decay of deuteron in nuclear matter induced by surrounding nucleons.

of the deuteron into 3p-1h states. At zero temperature one evaluates the 3p-1h density of states

$$g^E_{3\mathrm{p}-1\mathrm{h}} = \mathrm{Tr}_1\mathrm{Tr}_2\mathrm{Tr}_3\mathrm{Tr}_4\{\theta(H_1-\varepsilon_F)\theta(H_2-\varepsilon_F)\theta(H_3-\varepsilon_F)\theta(\varepsilon_F-H_4)\} \times\delta(E-H_1-H_2-H_3+H_4) \quad (12)$$

$$\propto (E-2\varepsilon_F)^3+\ldots \quad (13)$$

Since the normalization of the deuteron wave function brings a factor $(E-2\varepsilon_F)^{-1}$ we have $\Gamma_d^{\downarrow}\propto(E-2\varepsilon_F)^2$. At finite temperatures $T\ll\varepsilon_F$ we get as in Fermi liquid theory for zero sound a T^2 behavior for $\Gamma_d^{\downarrow}$. In a more elaborated theory one has to consider rescattering of the incoming deuteron with the nucleon(s) of the medium. Summing up all the respective diagrams leads to a generic three-body in-medium scattering matrix, that has been derived in [6, 7]. As a consequence the reaction rate depends strongly on the surrounding medium. The numerical result for the life-time of the deuteron is shown in Fig. 4, where we in addition compare the use of the isolated to the in-medium break-up cross section at $T=10$ MeV and a nuclear density of $n=0.007$ fm$^{-3}\simeq n_0/25$ [7, 10]. The medium dependent elementary cross section leads to shorter fluctuation times. As expected the difference is vanishing for larger momenta. In turn this effect leads to faster chemical relaxation times that is e.g. relevant for the notion of freeze out in heavy ion collisions.

The appearance of a sharp pole at $E=2\mu$ in the respective Green function related to Eq. (10) corresponds to the Thouless criterion for the onset of superfluidity for np Cooper pairs. Depending on the channels considered, it allows us to determine the critical temperatures T_s^c or T_t^c for the respective isospin singlet and triplet channels, as a function of the chemical potential. This is shown in Fig. 2 and later in Fig. 9 for T_t^c. Using the canonical relation $\Delta_{k_F}=1.76T_c$ one realizes that the np gap has a maximum value of $\simeq 10$ MeV. This is more than 3 times larger than the nn gap! Such high values may have an influence in expanding nuclear matter as produced in central heavy ion collisions.

An interesting extension of Eq. (10) is to write this equation down for the case where we have a condensate of pairs, i.e. a superfluid state. For pairs at

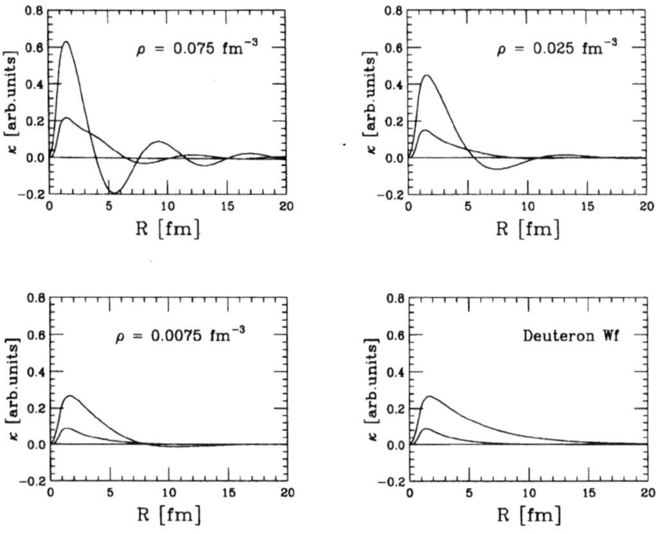

Figure 6. The deuteron (Cooper pair) wave function. Large component: s-wave; small component: d-wave. The Paris force was used in the gap equation and the single particle energies are from Brueckner Hartree-Fock. (From [12]).

rest one has

$$(2\mu - 2\varepsilon_p)\Psi_p = (\bar{n}_p - n_p)\sum_k v_{pk}\Psi_k, \tag{14}$$

where v_{pk} is the two-body interaction in momentum space and the occupation numbers n_p ($\bar{n}_p = 1 - n_p$) are given by

$$n_p = \frac{1}{2}\left[1 - \frac{\varepsilon_p - \mu}{\sqrt{(\varepsilon_p - \mu)^2 + \Delta_p^2}}(1 - 2f_p)\right] \tag{15}$$

with $\Delta_p = \sum_k v_{pk}\Psi_k$ the usual gap function. The chemical potential μ and the density are related by $N/V = 2\sum_p n_p$. We see that for a superfluid state Eq. (10) has turned into Eq. (14). Besides its nonlinearity corresponding to the gap the latter has a similar structure as the linear Eq. (10). One also realizes that for the zero density limit $N/V \to 0$ Eq. (14) turns into the Schrödinger equation for the deuteron. This is demonstrated in Fig. 6 where we show the Fourier transform of Ψ_p, i.e. the Cooper pair wave function for decreasing densities (Fig. 6 is taken from [12]). One clearly sees that the np Cooper pair wave function approaches the free deuteron wave function in the zero density limit.

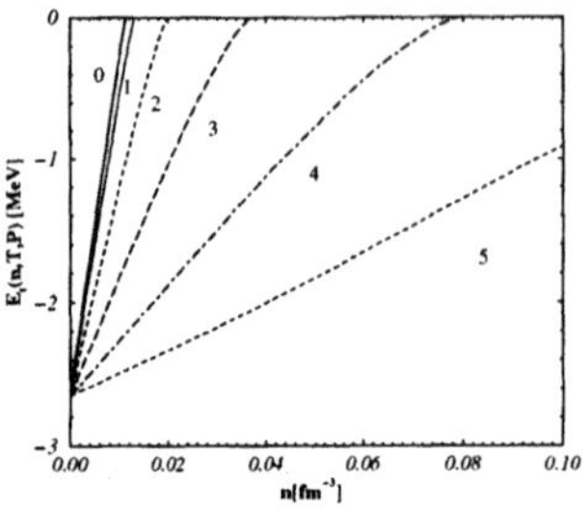

Figure 7. Triton binding energy per nucleon as a function of nuclear density at $T = 10$ MeV, and total momentum relative to the medium. From left to right: $P = 0, 1, 2, 3, 4, 5$ fm^{-1} [13].

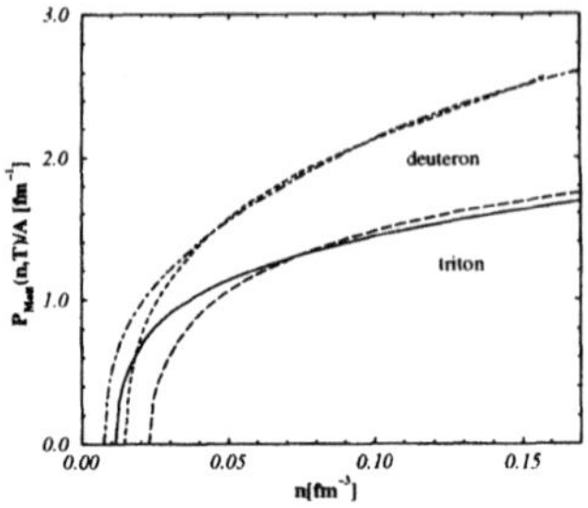

Figure 8. Mott momentum per nucleon P_{Mott}/A for triton - solid ($T = 10$ MeV) and dashed line ($T = 20$ MeV) and deuteron - dashed dotted ($T = 10$ MeV) and short dashed line ($T = 20$ MeV) as a function of nuclear density [13].

4 Three-Body Reactions and Bound States

A first three-body equation for quasiparticles embedded in a medium at zero temperature has been given in [11]. Here we present the equation for the wave function

$$(E - \varepsilon_1 - \varepsilon_2 - \varepsilon_3)\Psi_3(E) = [(\bar{f}_1\bar{f}_2 - f_1f_2)V_2(12) + \text{perm.}]\ \Psi_3(E) \qquad (16)$$

together with the normalization

$$\sum_{123} \tilde{\Psi}^{*\nu}_{123}(f_1f_2f_3 + \bar{f}_1\bar{f}_2\bar{f}_3)\tilde{\Psi}^{\nu'}_{123} = \delta_{\nu\nu'}, \qquad (17)$$

where the dual wave functions are defined by by $\Psi^{\nu}_{123} = (f_1f_2f_3 + \bar{f}_1\bar{f}_2\bar{f}_3)\tilde{\Psi}^{\nu}_{123}$.

An extension to finite temperatures uses the Alt-Grassberger-Sandhas (AGS) formalism to treat scattering phenomena and has been given in [6, 7, 10]. For a numerical solution of the scattering problem we use a separable ansatz for the strong nucleon-nucleon potential [6]. In [6] we have calculated the break up cross section $Nd \to NNN$ and found considerable dependence of the deuteron fluctuation time on the proper treatment of the medium dependence [7], see also Fig. 4. The influence of this refined treatment on the relaxation times is treated in [10]. The solution for the medium modified three-body bound state using the Yamaguchi and the Paris potential for comparison is provided in [13]. Fig. 7 shows the triton binding energy depending on different c.m. momenta as a function of the nuclear density and Fig. 8 shows the Mott momentum. Below the respective lines no deuterons/triton exist as bound states.

A very interesting situation can also happen when a sharp pole at $E = 3\mu$ appears in the in-medium three-body equation (16). Then the original Fermi gas becomes unstable with respect to the new Fermi gas composed out of three-body clusters. For a very low density nuclear matter this could be a Fermi gas

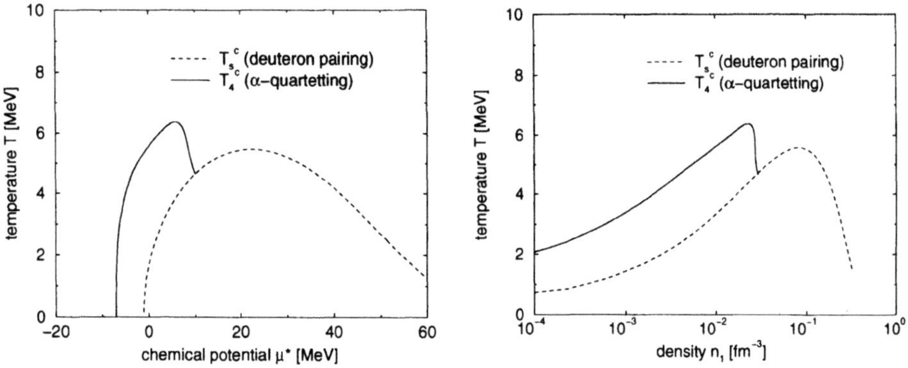

Figure 9. Critical temperatures for the onset of quantum condensation in symmetric nuclear matter, model calculation. The critical temperature of the onset of two-particle pairing T_t^c is compared with T_4^c for the onset of a four-particle condensate, as a function of the chemical potential (left) and as a function of the uncorrelated density (right) [8].

of ^{3}He or tritons. Though this case may be somewhat academic because of the presence of other type of clusters, e.g. α-particles, one may think of another situation where this consideration may become very relevant; suppose one cools down a quark gluon plasma to the point that nucleons (and other hadrons) appear. Then the quark Fermi gas transforms into a nucleon Fermi gas. A relativistic equation describing this transition should have a similar structure as Eq. (16). From such an equation the critical temperature for the transition from quark gas to nucleon gas could be studied.

5 Four-Nucleon Correlations and Quartetting

The four-nucleon correlation is believed to play a significant role for lower densities and temperatures because of the strong binding energy of the α-particle. Exploratory calculations using a simple variational ansatz for the ^{4}He wave function predict an α condensate/quartetting on top of the deuteron condensate/triplet pairing that leads to superfluidity [8].

Bose systems behave quite differently with respect to the occurrence of bound states. The Bose functions enhance the effective residual interaction that might lead to an "opposite Mott effect", i.e. existence of bound states and also pairing (condensate) even if no bound state exist for the isolated case. An example is provided by a pion gas, were a pionic condensate may occur [15] that are discussed, e.g. in the context of neutron stars [3].

In the framework utilized here the four body Schrödinger equation for the bound state problem reads

$$(E - \varepsilon_1 - \varepsilon_2 - \varepsilon_3 - \varepsilon_4)\Psi_4^\nu(E) = [(\bar{f}_1\bar{f}_2 - f_1 f_2)V_2(12) + \text{perm.}]\ \Psi_4^\nu(E) \quad (18)$$

with the normalization [16]

$$\sum_{1234} \tilde{\Psi}^{*\nu}_{1234}(f_1 f_2 f_3 f_4 - \bar{f}_1 \bar{f}_2 \bar{f}_3 \bar{f}_4)\tilde{\Psi}^{\nu'}_{1234} = \delta_{\nu\nu'}, \tag{19}$$

and the dual wave functions $\Psi^{\nu}_{1234} = (f_1 f_2 f_3 f_4 - \bar{f}_1 \bar{f}_2 \bar{f}_3 \bar{f}_4)\tilde{\Psi}^{\nu}_{1234}$.

The solution for $\Psi_4(4\mu)$ provides the critical temperature T_4^c for α condensates as a function of the chemical potential shown in Fig. 9 (left), or of the density as shown in Fig. 9 (right). Using a variational ansatz to model the four-body wave function the transition to quartetting outreaches the transition to isospin singlet pairing if the density is smaller than 0.03 fm^{-3}, see Fig. 9. This may again be relevant in expanding nuclear matter or even in the far tail of the density of a nucleus.

One may suspect that the asymmetry of nuclear matter hinders the formation of np pairing or of quartetting. This is because extra neutrons Pauli-block the formation of Cooper pairs. However, once the density is so low that the corresponding chemical potentials turn negative, then the blocking effect, i.e. the Pauli principle is less effective and formation of deuterons and α particles should not be much unfavored in asymmetric matter with respect to the symmetric case.

6 Conclusion

We have studied the formation of clusters and their condensation in Fermi systems using the Dyson equation approach to many-body Green's functions. They only depend on the c.m. energy of the cluster and therefore the resulting equations are of the Schrödinger type. To lowest order the in-medium n-body equations are – compared to the isolated case – only modified by RPA phase space factors on the two-body matrix elements (leading to self-energy corrections and modified interactions). Such equations have early been written down by the Japanese school [17]. We here investigate them with respect to the Mott effect and Bose-Einstein condensation in homogeneous matter. It is found that deuteron Cooper pairs survive longer than α-particle quartets at high densities but that, not unexpectedly, α-particles win in the low density limit. We give the Mott density for deuterons and tritons and discuss the Pauli principle aspect of the phase transition of a quark Fermi gas to a nucleon Fermi gas. Life time effects of the clusters are also considered. To second order Born approximation they behave like $\Gamma_n^{\downarrow} = a\,(E - n\varepsilon_F)^2 + bT^2$, where n is the nucleon number in the cluster. Therefore n-body clusters at $E_n = n\varepsilon_F$ and $T = 0$ have infinite life time.

Although we have taken examples from nuclear physics, the exact treatment of few-body systems embedded in the medium is not restricted to nuclear physics. Further potential applications are possible, e.g. in the field of semiconductors to treat the formation of excitons or the trion bound state in the dense low dimensional electron plasmas. For recent experiments see e.g. [18]. For highly ionized dense plasmas the impulse approximation may fail for the three-particle break-up cross section needed to calculate the ionization and recombination rates.

Acknowledgement. It is a pleasure for us to thank M. Baldo, C. Kuhrts, U. Lombardo, H. J. Schulze for collaborations and discussions.

References

1. L.P. Kadanoff, G. Baym: *Quantum Theory of Many-Particle Systems.* New York: McGraw-Hill 1962; A.L. Fetter, J.D. Walecka: *Quantum Theory of Many-Particle Systems.* New York: McGraw-Hill 1971
2. H. Stein, A. Schnell, T.Alm, G. Röpke: Z. Phys. **A351**, 295 (1995)
3. D. Pines, R. Tamagaki, S. Tsuruta (eds): *The structure and evolution of neutron stars.* Reading, MA: Addison-Wesley 1992
4. J. Dukelsky, G. Röpke, P. Schuck: Nucl. Phys. **A628**, 17 (1998); P. Schuck: in proceedings of the VI Hispanenisi International Summer School, Sevilla, June 1997, *Many body Theory of Correlated fermion Systems*, eds. J.M. Arias, M.I. Gallardo, M. Lozano (World Scientific).
5. E.O. Alt, P. Grassberger, W. Sandhas: Nucl. Phys. **B2**, 167 (1967)
6. M. Beyer, G. Röpke, A. Sedrakian: Phys. Lett. **B376**, 7 (1996)
7. M. Beyer and G. Röpke: Phys. Rev. **C56**, 2636 (1997)
8. G. Röpke, A. Schnell, P. Schuck, P. Nozieres: Phys. Rev. Lett. **80**, 3177 (1998)
9. M. Schmidt, G. Röpke, H. Schulz: Ann. Phys. **202**, 57 (1990)
10. C. Kuhrts, M. Beyer, G. Röpke: Nucl. Phys **A** in print.
11. P. Schuck, F. Villars, P. Ring: Nucl. Phys. **A208**, 302 (1973)
12. M. Baldo, U. Lombardo, P. Schuck: Phys. Rev. **C52**, 975 (1995)
13. M. Beyer, W. Schadow, C. Kuhrts, G. Röpke: Phys. Rev. **C60**, 034004 (1999)
14. G. Röpke, M. Schmidt, L.Münchow, H. Schulz: Nucl. Phys. **A339**, 587 (1983)
15. T. Alm, G. Chanfray, P. Schuck, G. Welke: Nucl. Phys. **A612**, 472 (1997)
16. P. Danielewicz, P. Schuck: Nucl. Phys. **A567**, 78 (1994)
17. T. Marumori, K. Suzuki: Nucl. Phys. **A106**, 610 (1968); Suppl. Prog. Theor. Phys. **58** (1975)
18. G. Eytan et al.: Phys. Rev. Lett. **81**, 1666 (1998)

Few-Body Systems Suppl. 12, 60–65 (2000)

Few-Body Systems

Ab initio Study of Clusters and Nanotubes in Electric Fields

K. Tada[1,*], K. Watanabe[1,2,3], C. Ishii[1,2]

[1] Department of Physics, Faculty of Science, Science University of Tokyo, 1-3 Kagurazaka, Shinjuku-ku, Tokyo 162-8601, Japan

[2] Frontier Research Center for Computational Sciences, Science University of Tokyo, 1-3 Kagurazaka, Shinjuku-ku, Tokyo 162-8601, Japan

[3] CREST, Japan Science and Technology Corporation.

Abstract. We have investigated the reactions of silicon-carbon mixed clusters, Si_2C_3 and Si_3C_2 and various carbon nanotubes to the applied electric fields by first-principles supercell calculations using the local density approximations and the pseudopotential plane wave method. We manipulated the mixed clusters theoretically with a high electric field. Linear (Si_2C_3) and plane (Si_3C_2) clusters bend in the direction of the field. The field-induced distortion is interpreted using the electronegativity of Si and C atoms. The local field enhancement at the tips of carbon nanotubes and the screening of the electric field inside the nanotubes are considerable for nanotubes with a small gap between the highest occupied band (HOB) and the lowest unoccupied band (LUB) in the electronic states. The result also suggests an important role of the localized states of conducting nanotubes with small electronic gaps in field emissions.

1 Introduction

We studied mixed silicon-carbon Si_nC_m clusters because the interplay between Si and C atoms for stabilizing a cluster is interesting and the small mixed clusters are important components in manufacturing silicon-carbide thin films[1]. Our objective is to explore the possibility of manipulating clusters by using high electric fields. Recent surface-atom manipulation by the scanning tunneling microscopy(STM)[2] showed that the electric field is vital to controlling the forces acting on atoms. In this paper, we manipulate Si_2C_3 and Si_3C_2 clusters, expecting a definite change in the structure caused by the interaction between the electric dipole and the external field.

*E-mail:tada@nano.ph.kagu.sut.ac.jp

Carbon nanotubes have been intensively studied, in anticipation of their application in novel nanoscale materials and device structures.[3, 4] An interesting subject of these applications is field emitters with potential applications in flat panel displays. Theoretically, the role of an electric field in nanotube growth[5] and the electronic properties of open-ended and capped-end SWNTs in strong electric fields[6] have been investigated using a density-functional method. In the latter study,[6] some carbon nanotubes were found to screen the applied electric field, manifesting a metallic nature. The objectives of this study are to investigate, from first principles, the response of four types of carbon nanotubes with different shapes and atomic geometries to an external electric field, and to reveal the properties of the localized states at the tips of nanotubes in the electric field.

2 Method and Model

We used first-principles molecular dynamics method (FPMD)[7] within the density functional theory in the local density approximation. For the exchange correlation potential, we used the Ceperley-Alder functional[8] in the parameterization form given by Perdew and Zunger[9]. We used the norm-conserving pseudopotentials[10] of NCPS95[11] and expand the electronic wave functions in plane waves up to a kinetic energy of 32 Ry and 40 Ry for mixed clusters and 30 Ry for carbon nanotubes. A Γ point is chosen for k sampling in the Brillouin zone. The size of the unit cell is 9.5Å×9.5Å×9.5Åfor mixed clusters and 14.6Å×12.2Å×12.2Å for carbon nanotubes.

3 Results of Calculations

3.1 Manipulating Mixed Clusters with High Electric Fields

We applied an electric field of 1 V/Å to Si_2C_3 and Si_3C_2 clusters to deform and manipulate the structures. The result of Si_2C_3 is shown in Fig. 1(a). The direction of the field is indicated by an arrow. We found that the linear chain bends, and Si (C) atoms move parallel (antiparallel) to the field. We explain bending as follows [Fig. 1(b)]: In general, electrons are transferred more easily from Si atoms to C atoms than vice versa, because the C atom has larger electronegativity than the Si atom. If the linear chain bent upward [below, Fig. 1(b)], induced electric dipole $\boldsymbol{P}$ composed of $C^{-}-Si^{+}$ would be antiparallel to external electric field $\boldsymbol{F}$, increasing the interaction energy positively; $-\boldsymbol{P}\cdot\boldsymbol{F}/2 > 0$. The Si_2C_3 cluster thus bends downward [above, Fig. 4(b)]. Si_3C_2 deformation is similar to that of Si_2C_3 when a field is applied in the z direction [Fig. 1].

An important finding in this manipulation is that mixed clusters of different types of atoms with different electronegativities are structurally controllable using the interaction energy between the electric dipole and external electric field. As the size of the mixed cluster increases, the number of isomers (metastable structures) with similar energy increases significantly. We may thus *switch* from one isomer to another by changing either the direction or strength of the electric field.

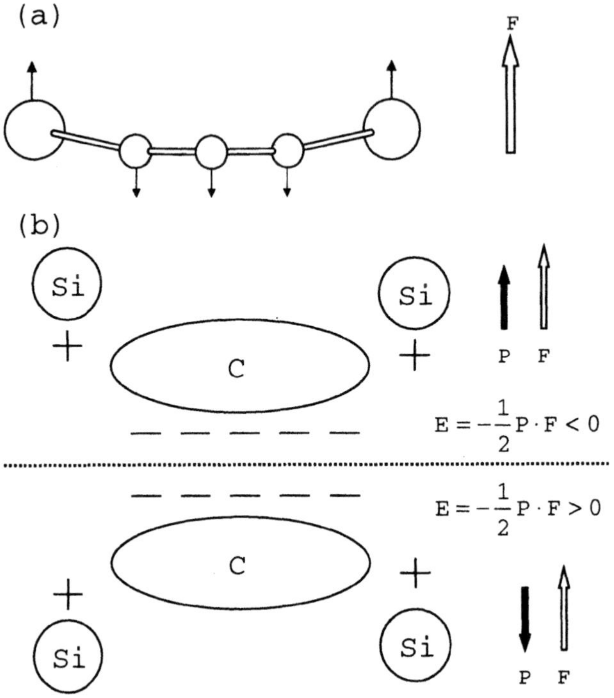

Figure 1. Distortion of Si_2C_3 cluster in an electric field of 1 V/Å. (a) The linear chain bends downward in the direction of the electric field indicated by the arrow. Large circles refer to Si and small to C atoms. (b) The downward bend in (a) is interpreted by the gain (above) in the interaction energy between the electric field and electric dipoles of $C^{-}-Si^{+}$ caused by the large electronegativities of C atoms compared to those of Si atoms.

3.2 Carbon Nanotubes in Electric Fields

In this study, the electronic structures of four types of single wall carbon nanotubes (SWNTs); NT1, NT2, NT3, NT4 in an electric field of 0.2 V/Å are investigated. NT1, NT2, NT3, NT4 are zigzag (6,0), zigzag (4,0), armchair (4,2), armchair (6,3) nanotubes, respectively. The nanotube indices (n, m) used here follow the abbreviations by previous theories.[12, 13] We found that both the local field enhancement at the tips and the screening of the field inside the nanotubes are remarkable for NT1. This is attributed to the electronic states of NT1, because the small electronic gap between the highest occupied band (HOB) and the lowest unoccupied band (LUB) makes a field-induced electron polarization easily occur on the entire tube, resulting in the features of the local electric field obtained.

Let us examine the property of a field-sensitive state in the energy region. The square of the electronic wavefunction of the HOB of NT1 (Fig. 2(a)), which is a DB state of the right-end carbon atoms, is given in Fig. 2(b). The localized property seen in the figure is common to the states in the highest energy region. Thus, the electronic states responding sensitively to the electric field are localized ones represented by the DB states at nanotube tips. This tendency of the localized electronic states in the electric field was also observed in our previous study of the field effect on the surface electronic states of Si(100).[14] More importantly, field emission occurs predominantly from these localized states at the tip because the tunneling probabilities of electrons from the states near the HOB toward the vacuum are substantially high.[14] Therefore, the present result on the field-sensitive localized states supports the expectation of enhanced emission currents due to resonant electron tunneling[15] from a localized state.

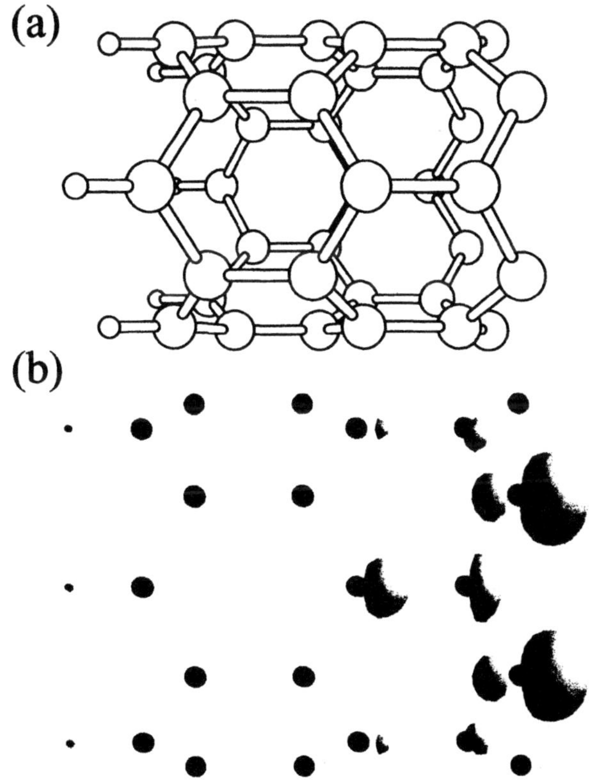

Figure 2. (a)The structure of a (6,0) zigzag nanotube (NT1). Carbon and hydrogen atoms are indicated by large circles and small circles, respectively. (b)The square of the electronic wavefunction of the HOB (DB state) of NT1. The mushroom-shape density is localized at the right edge of the nanotube.

4 Conclusion

We applied a high electric field of 1 V/Å to Si_2C_3 and Si_3C_2 clusters to manipulate the structures, obtaining bending deformation in the direction of the applied field. Deformation is interpreted in terms of the interaction energy between the electric field and electric dipoles $C^-{-}Si^+$. Our simulation thus emphasizes the role of electronegativity of constituent atoms in cluster manipulation.

The electronic responses of various types of carbon nanotubes to the applied electric field have been investigated via *ab initio* electronic structure calculations. Both the local field enhancement around the tip and the screening inside the tube are considerable for nanotubes with a small gap in the electronic structures. We found that the localized DB states at the tip of these nanotubes are strongly influenced by the field and consider that such localized states lower the thresh old field for field emissions. The present results thus enable us to emphasize the role of the localized states at the tip, as well as the shape and the atomic geometry, in the mechanism of field emissions.

Acknowledgement. This work was partially supported by CREST (Core Research for Evolutional Science and Technology) of Japan Science and Technology Corporation (JST). K.T. thanks JST for the use of FACOM VPP300 and the Supercomputer Center, Institute for Solid State Physics, University of Tokyo for the use of FACOM VPP500.

References

1. J.D. Parsons, R.F. Bunshah, O.M. Stafsudd: Solid State Technol. **28**, 133 (1985)
2. H. Kuramochi, H. Uchida, Y. Kuwahara, K. Watanabe, M. Aono: Jpn. J. Appl. Phys. **36**, L1343 (1997)
3. S.J. Tans, M.H. Devoret, H. Dai, A. Thess, R.E. Smalley, L.J. Geerligs, C. Dekker: Nature **386**, 474 (1997)
4. Y. Saito, S. Uemura, K. Hamaguchi: Jpn. J. Appl. Phys. **37**, L346 (1998)
5. A. Maiti, C.J. Brabec, C. Roland, J. Bernholc: Phys. Rev. **B52**, 14850 (1995)
6. L. Lou, P. Nordlander, R.E. Smalley: Phys. Rev. **B52**, 1429 (1995)
7. R. Car and M. Parrinello: Phys. Rev. Lett. **55**, 2471 (1985)
8. D.M. Ceperley and B.J. Alder: Phys. Rev. Lett. **45**, 566 (1980)
9. J.P. Perdew and A. Zunger: Phys. Rev. **B23**, 5048 (1981)
10. N. Troullier and J.L. Martins: Phys. Rev. **B43**, 1193 (1991)

11. K. Kobayashi: To construct a database for norm-conserving pseudopotentials, Proc. CMSMD, 1996, p.96

12. N. Hamada, S. Sawada, A. Oshiyama: Phys. Rev. Lett. **68**, 1579 (1992)

13. R. Saito, M. Fujita, G. Dresselhaus, M.S. Dresselhaus: Appl. Phys. Lett. **60**, 2204 (1992)

14. I. Nakajima, T. Kawai, K. Watanabe: Surf. Sci. **433-435**, 868 (1999)

15. J.-M. Bonard, T. Stöckli, F. Maier, W.A. de Heer, A. Châtelain, J.-P. Salvetat, L. Forró: Phys. Rev. Lett. **81**, 1441 (1998)

Few-Body Systems Suppl. 12, 66–70 (2000)

Few-Body Systems

The Multidimensional Angular Coulomb Function Method in Atomic and Molecular Physics

A.A. Sadovoy*

RFNC–VNIIEF, Sarov 607190, Russia

I. At the last time a certain progress has been achieved in developing unified approaches to studying few-body properties based on non-model solutions of problems of discrete and continuous spectrum. The hyperspherical function method and its versions in the quark model of elementary particles, multinucleon theory of nuclei, atomic physics and quantum chemistry are enough widely spread [1, 2, 3]. However the hyperspherisal functions, used in the MHCF, being solutions of multidimensional Laplace equation, are a subset of wider set of multidimensional angular functions (MAF). In general case, the MAF may be constructed in multidimensional spaces with several collective variables determined so to describe prevailing motions in multipartial systems. New opportunities are opened with using methods of MAF, which may not be the characteristic functions of Laplace operator, what require calculating matrix elements of kinetic energy operator by the MAF. In the report we consider one of indicated total bases of MAF allowing for the most full specifics of multipartial problems with long-range Coulomb interaction, which received the name method of multidimensional angular Coulomb functions (MACF).

II. In the MACF method, used in atomic physics, a collective variable is being introduced $\rho = \sum_{i=1}^{A} |\vec{r}_i|$, the MAF are constructed of uniform multidimensional polynomials in such a way

$$U_{k_j}(\Omega) = \sum_j \Delta_{ij} \sqrt{\frac{\Gamma(2K_j + 3A)}{A!}} \frac{1}{\rho^{k_j}} P_{k_j}(\psi(\vec{r}_i)),$$

here Δ_{ij} — orthogonalization coefficients, $K_j = \sum_\omega (n+l)_\omega$; The multielectron w.f. is being searched in the form of series in the MAF

$$\psi(\vec{r}_{ij} \dots \vec{r}_n) = \sum_{k\nu} \frac{X_{k\nu}(\rho)}{\rho^{\frac{3A-1}{2}}} U_{k\nu}(\Omega_{3A}),$$

* *E-mail address:* Saa@albatross.md08.vniief.ru

that allows from Schrodinger equation

$$\left\{-\frac{h^2}{2m}\sum_{L=1}^{A}\frac{\partial^2}{\partial\vec{r}_i^2}-\sum_{L=4}^{A}\frac{ze^2}{r_i}+\sum_{L>j}^{A}\frac{e^2}{|\vec{r_l}-\vec{r}_j|}-E\right\}\psi\left(\vec{r}_1,\vec{r}_2,\ldots,\vec{r}_A\right)=0$$

to obtain SODE for an expansion amplitude

$$X''(x)+\frac{C}{x}X'(x)+\left(\frac{B}{x^2}+\frac{\rho F}{x}-\frac{1}{4}\right)X(x)=0.$$

In this equation a matrix presentation is used, $x=\alpha\rho$; $\alpha=\sqrt{\frac{8E}{A}}$, 1 - a single matrix. Elements of matrices C and B are expressed through m.e. of kinetic energy and a matrix F through m.e. of potential energy.

With using Shturm functions

$$\psi_{n\beta}(x)=\sqrt{\frac{\Gamma(n+1)}{\Gamma(n+2\beta)}}e^{-x/2}x^{\beta}Ln^{2\beta-1}(x),$$

one can receive an analytical solution of ODE system for the MACF method

$$X_i(x)=\sum_{n=0}^{\infty}J_{in}\psi_{n\beta_i}(x).$$

Coefficients of expansion J and a characteristic value P are being determined by the system $(Pf_{ii}-k-\beta_i)J_{ik}=-\sum_{j\neq i}^{N}\sum_{n=0}^{\infty}M_{ij}^{kn}(p)J_{jn}$, here a matrix M is expressed through m.e. of operators $\frac{1}{y}\frac{d}{dy}$, $\frac{1}{y^2}$, $\frac{1}{y}$ by Shturm functions [4].

The described method was used for calculations of bound energy, electron density, atomic formfactors, atomic radius momentums, electron scattering cross- sections, multielectron w.f. of helium-, lithium-, beryllium-, boron-, carbon-, nitrogen - and oxygenlike ions with $Z<20$ [4].

A high accuracy of MACF method w.f. are illustrated by data on atomic radius moments which for negative $q=-2,-1$ and positive $q=1,2,3,4$ degrees in a large series of heliumlike ions coincide up to fourth significant figure with complex variation calculations of Smith group. The last allow computing the electron density on nuclei with the pinpoint accuracy [5]. By the MACF method it are predicted radial excited levels of collective nature which are absent in calculations with the method of density functional because of inaccuracy account of Paula principle [4]. The electron density of main state O^+ of carbon differs [6] from the one in calculations by Hartree-Fock method owing to inaccuracy of this method asymptotic. Existing computing possibilities allow carrying out a calculation of the main state of oxygen atom with taking into account ten MAF, bound energy - 73,56 at.un. Significant configuration mixing is observed in this case also as it is at smaller number of electrons. For computing properties of heavy nuclei it is possible to proceed from Dirack equation and its solution is being searched in the form

$$\psi(\vec{r}_1, \vec{r}_2, \ldots, \vec{r}_A) = \sum_{k\nu} \rho^{-\frac{3A-1}{2}} \begin{pmatrix} M_{k\nu}(\rho) U_{k\nu}(\Omega) \\ N_{k\nu}(\rho) W_{k\nu}(\Omega) \end{pmatrix}.$$

For w.f. MAF expansion amplitudes it is obtained the system of equations

$$\begin{cases} \dfrac{dN}{d\rho} - \dfrac{B}{\rho} - \alpha \dfrac{A}{\rho} M - \varepsilon_0 M(\rho) = 0; \\ \dfrac{dM}{d\rho} \dfrac{B}{\rho} + \alpha \dfrac{\hat{A}}{\rho} N + \dfrac{\hat{B}}{\rho} M + (\varepsilon_0 - 2z) N(\rho) = 0. \end{cases}$$

were matrices B and $\hat{B}$ are expressed through Dirack operator m.e. and matrices A and $\hat{A}$ - through potential energy m.e. in MAF. At present the numerical methods for solving the given system are developed.

III. In the adiabatic approximation the electron w.f. of molecules are being searched in the form $\psi(\vec{r}_{ij} \ldots \vec{r}_n) = \sum_j \frac{X_j(\rho)}{\rho^{\frac{3A-1}{2}}} U_j(\Omega)$, where the symmetry center of molecules is accepted as zero. MAF $U_j(\Omega)$ must be characterized by quantum numbers of two-atom molecule.

The main term of two-atom homonuclear molecule is ${}^1\Sigma_g^+$ and therefore for estimations in the minimum approximation of MACF method one can obtain for expansion amplitude of electron w. f. the following equation

$$\left\{ \frac{d^2}{d\rho^2} - \frac{15}{4} \frac{1}{\rho^2} \frac{2m}{h^2} \frac{1}{2} \left[-\frac{e^2}{\rho} \frac{25}{16} + U_{en}(\rho) \right] + \frac{2m}{h^2} \frac{E}{2} \right\} X(\rho) = 0,$$

$$U_{en}(\rho) = \begin{cases} \dfrac{4Ze^2}{a}, & \rho < a; \\ \dfrac{Ze^2}{\rho} \left[10 - \dfrac{20a^2}{\rho} + \dfrac{2a^2}{\rho^3} - \dfrac{6a^4}{\rho^4} \right], & \rho > a. \end{cases}$$

If a solution of the equation is being searched in the form

$$X(x) = \sum_{k=2} J_k \psi_k(x), \; x = \alpha\rho, \; \alpha = 2\sqrt{|E|}, \; \psi_k(x) = \sqrt{\frac{\Gamma(n+1)}{\Gamma(n+3)}} e^{-x} x^{5/2} L_n^4(x),$$

expansion coefficients J_k are determined by the system of algebraic equations

$$(\alpha f - k - 5/2) J_k = \sum_{k' \neq k} J_{k'} M_{kk'}.$$

The proposed method has variation properties.

As it follows from numerical data, an equilibrium internuclear distance is equal 0,7 at.un. Analysis of all the received data shows that even in the minimum approximation of MACF method it may be reached a good description of bind energy for the main state of H_2 (-2.31 Ry in comparison with -2.296 Ry according to variation calculations of Weinbaum). The method also predicts new radial excitation levels H_2 of collective nature.

IV. In conclusion we illustrate possibilities of MAF methods for other problems. In the presence of strong magnetic fields, the most effective MAF are the MAF constructed in the multidimensional cylindrical space with two collective variables [7] $Z^2 = \sum_{i=1}^{A} Z_i^2$, $r = \sum_{i=1}^{A} r_i$ (axis is selected along magnetic field). It is interesting to note that the mentioned collective variables are convenient also in research of properties of two-atom molecule in the adiabatic approximation. In this case axis OZ is directed along the straight line connecting the nuclei.

W.f. of two-atom molecule is being looked for in the form

$$\psi(\vec{r}_1, \vec{r}_2, \ldots, \vec{r}_A) = \sum_{RPM} \frac{\varphi_{RPM\alpha}(z,r)}{z^{\frac{A-1}{2}} r^{\frac{2N-1}{2}}} U_{KPM\alpha}(\Omega)$$

here K and P -powers of uniform polynom in Z and r accordingly, M-projection of the total momentum on OZ, α - the rest of two-atom molecule quantum numbers. Unknown amplitudes of this expansion are solutions of the following differential equation

$$\left\{ A\frac{\partial^2}{\partial r^2} + \frac{C}{r}\frac{\partial}{\partial r} + \frac{B}{r^2} + \frac{D}{z}\frac{\partial}{\partial z} + A\frac{\partial^2}{\partial z^2} + \frac{G}{z^2} - 2W(z,r) + 2\varepsilon \right\} \varphi(z,r) = 0.$$

In the equation matrices C, B, D, G are expressed through m.e. of electron kinetic energy operator, W through m.e. of electron-electron interaction. According to preliminary estimations, using two collective variables improves a local convergence of expansion because of the greater flexibility of variation w.f. The MAF method gives also new opportunities for analysis and determination of parameters for the optical model used widely in problems of scattering of composite particles. We have published the results of using the MAF method for finding a real part of optical potential [9]

$$V_{\mathrm{Re}}\left(\vec{R}\right) = \int \psi(\{\vec{r}_i\}) \sum_{i=1}^{A} V_{NN}\left(\vec{r}_i - \vec{R}\right) \psi(\{\vec{r}_i\})\, \Pi_{l=1}^{A} d\vec{r}_l$$

in self-congruent approach, when in compression of $N-N$ potentials and in nuclear density calculated in the minimum approximation of MACF are used the same $N-N$ potentials. We managed to describe satisfactorily differential scattering cross-sections of 14 Mev nucleon by nuclei He, O, Ca. According to estimations it should expected interesting results in using the MACF for computing the optical potential for scattering of composite system with Coulomb interaction.

As another example, it may be mentioned the quantum-mechanical calculations of changing an electron spectrum for the valent electron in spherical cell with the boundary condition which provide going to zero the impulse component normal to the cell surface. On the base of these calculations it was computed elastic energy and elastic pressure in dependence on specific volume, equilibrium density and pressure causing transition of electron in the continuous spectrum (ionization by pressure). Comparison with the calculations for hydrogen carried out in the frames of multipartial theory of metal [10] indicates to satisfactory agreement numerical data up to densities $\sim 100 g/cm^3$.

The MACF method allow studying properties of various exotic systems with Coulomb interaction. To increase accuracy of calculations and to provide correct asymptotic for w.f. it were studied [11] with using Shturm representations of Green functions for two-cluster systems the states with a large orbital moment $L = 30 \div 40$ for exohelium in which one electron is replaced with the antiproton. By these calculations we succeeded in confirmity no theoretical predictions of increasing the exohelium life time up to $1\,s$ at $L = 28$ stipulated with decreasing sharply the m.e. of system dipole moment. It is necessary to take into account more precisely configuration mixing in the calculations.

References

1. C.D. Lin: Phys. Rep. **257**, 1 (1995)
2. P.I. Jibuti, K. V. Shitikova: *Hyperspherical function method in the atomic and nuclei physics*: Moscow 1993
3. D.I. Abramov, L.N. Bogdanova, V.V. Gusev, L.I. Ponomarev: Nucl. Phys. **613**, 520 (1998)
4. A.A. Sadovoy: *Multidimentional Angular Coulomb Function Method in Theoretical and Applied Physics*, p.296. Sarov 1994 (in Russian)
5. B.M. Dzyuba, A.A. Sadovoy: Optics and Spectroscopy, **71**, 709 (1995)
6. T. Koga: J. Chem. Phys. **87**, 4209 (1988)
7. J.C. Slater: *Electronic structure of moleculec*: Moscow 1965
8. A.A. Sadovoy: *New Method for Research of Magnetic Fild Influence on the properties of Multielectron Atoms and Ions.* Proceedings of the 7-th Conf. of the Generation of Megagauss Magnetic Fields, Vol. II, p.857. Sarov 1997
9. I.K. Averyanov. B.M. Dzyaba, A.A. Sadovoy: Vopr. At. Nauki i Tekhniki (ser. Teor. and Prikl. Fiz.), vol. 1, p.25 (1998)
10. A.A. Sadovoy, V.M. Povyshev: Vopr. At. Nauki i Tekhniki (ser. Teor. I Prikl. Fiz.), vol. 1, p.25 (1998)
11. A.A. Sadovoy: *About the structure of antiproton systems.* Proceedings of the 2-nd Int. A.D. Sakharov Phys. Conf., World Scientific, p.73, London 1997

Few-Body Systems Suppl. 12, 71–74 (2000)

Few-Body Systems

Excited States of Positronium Molecule

J. Usukura[1,*], Y. Suzuki[2]

[1] Graduate School of Science and Technology, Niigata University, Niigata 950-2181, Japan

[2] Department of Physics, Niigata University, Niigata 950-2181, Japan

Abstract. We investigated the spectrum of positronium molecule. By calculating all possible states with $L^\pi = 0^\pm, 1^\pm, 2^\pm$ by the stochastic variational method, we confirmed that only four states, $0^+(\mathrm{A}_1)$, $0^+(\mathrm{B}_2)$, $0^+(\mathrm{E})$ and $1^-(\mathrm{B}_2)$, are bound.

Positronium molecule (Ps_2) consisting of two electrons and two positrons has not been found experimentally yet. It is because Ps_2 is electrically neutral and the energy difference of the emitted photon in pair annihilation of Ps_2 and Ps is very small. Several bound states have been predicted by theoretical calculations. We show the energy levels of Ps_2 in Fig. 1. The ground state, ${}^1S^e$, is lower by 0.016 a.u. than the lowest threshold, while the excited states, ${}^1P^o$, ${}^3S^e$ and ${}^1S^{e*}$ [1, 2, 3, 4], are higher than it. The excited states cannot dissociate into two $1S$ states of Ps since they have symmetry different from this threshold. Though the excited states are accessible to Ps($1S$)+Ps($2S, P$) channel, their energies are lower than this channel, which guarantees that they are bound.

The investigation of a spectroscopic study of Ps_2 is not only of academic interest but also helpful to observe Ps_2. For example, we expect that the P state of Ps_2, which was predicted recently by our calculation, is important to confirm Ps_2 experimentally [1, 2]. This state has two main decay modes contrary to the ground state which has only pair annihilation decay mode. One is the pair annihilation, and the other is the electric dipole transition to the ground state. It is difficult to confirm the existence of Ps_2 from the pair annihilation as mentioned above. The photon energy emitted in the dipole transition is, however, unique and could be distinguished from that of the dipole transition of the P state of Ps.

The purpose of this talk is to report on whether or not other bound states exist in Ps_2. We mean by a bound state that the energy of the state is lower than the lowest accessible dissociation channel.

* *E-mail address:* usukura@nt.sc.niigata-u.ac.jp

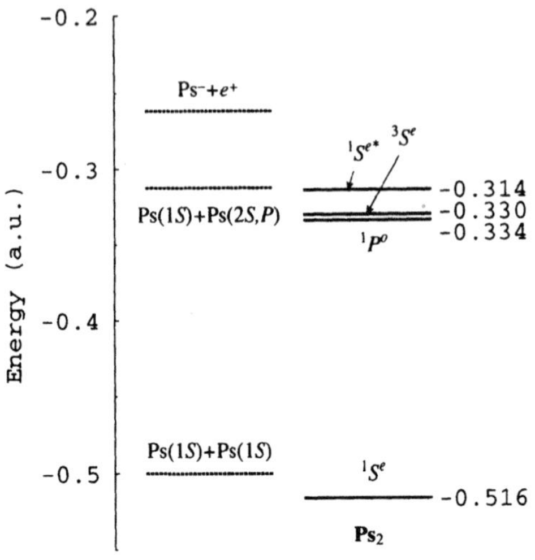

Figure 1. The energy spectrum of Ps_2

The Hamiltonian of Ps_2 is invariant with respect to exchanges of electrons, positrons or charges. Thus the wave function should have the symmetry belonging to D_{2d} point group [4]. The symmetries of the space part of the wave function are built from eight permutation operators as listed in Table 1. Each angular momentum state may have eight symmetry states labeled A, B and E etc, and we calculated the energies of states with $L^\pi = 0^\pm, 1^\pm, 2^\pm$.

To solve a four-body problem, we used the stochastic variational method with a correlated Gaussian basis. The wave function is written as

$$\Psi = \sum_{i=1}^{K} c_i S \psi_i(\mathbf{x}) \chi, \tag{1}$$

where K is the number of basis, c_i is a linear parameter, S is the operator projecting out a state with an appropriate D_{2d} symmetry and χ is a spin function. The space part $\psi_i(\mathbf{x})$ is written in terms of the Jacobe coordinates $\mathbf{x} = \{\mathbf{x}_1, \mathbf{x}_2, \mathbf{x}_3\}$ by

$$\psi_i(\mathbf{x}) = \exp\{-\frac{1}{2} \sum_{j,k=1}^{3} (A_i)_{jk} \mathbf{x}_j \cdot \mathbf{x}_k\} P(\mathbf{x}), \tag{2}$$

with

$$P(\mathbf{x}) = \begin{cases} |\mathbf{v}|^L Y_{LM}(\hat{\mathbf{v}}) & \text{for } \pi = (-1)^L, \\ |\mathbf{v}_1|^L |\mathbf{v}_2| [Y_L(\hat{\mathbf{v}}_1) Y_1(\hat{\mathbf{v}}_2)]_{LM} & \text{for } \pi = (-1)^{L+1} \text{ and } L \neq 0, \\ (\mathbf{x}_1 \times \mathbf{x}_2) \cdot \mathbf{x}_3 & \text{for } L^\pi = 0^-, \end{cases} \tag{3}$$

where A_i is a 3×3 positive-definite symmetric matrix, and the *global vector* $\mathbf{v}$ is defined by $\mathbf{v} = \sum_{i=1}^{N-1} u_i \mathbf{x}_i$ ($\mathbf{v}_1$ and $\mathbf{v}_2$ are difined similarly). $(A_i)_{jk}$ and u_i

Table 1. Classification of Ps_2 according to D_{2d} symmetry. Particle labels 1 and 3 stand for electrons, while 2 and 4 stand for positrons. The parentheses denotes a permutation. S_- (S_+) is a spin of electrons (positrons). C_P is a phase due to exchanging charges.

D_{2d}	symmetry of space part	(S_-, S_+)	C_P
A_1	I+(13)+(24)+(13)(24) +(12)(34)+(1432)+(1234)+(14)(23)	(0, 0)	+
A_2	I−(13)−(24)+(13)(24) −(12)(34)+(1432)+(1234)−(14)(23)	(1, 1)	−
B_1	I+(13)−(24)−(13)(24) +(12)(34)−(1432)−(1234)+(14)(23)	(1, 1)	+
B_2	I+(13)+(24)+(13)(24) −(12)(34)−(1432)−(1234)−(14)(23)	(0, 0)	−
E_{11}	I−(13)(24)+(12)(34)−(14)(23)	(1, 0), (0, 1)	+
E_{12}	(13)−(24)−(1432)+(1234)	(1, 0), (0, 1)	−
E_{21}	(13)−(24)+(1432)−(1234)	(1, 0), (0, 1)	+
E_{22}	I−(13)(24)−(12)(34)+(14)(23)	(1, 0), (0, 1)	−

are nonlinear variational parameters. The basis function (2) remains invariant in its form against the coordinate transformation, which makes the calculation of matrix elements easy. The stochastic nonlinear variational method is a very powerful optimization procedure to look for appropriate nonlinear variational parameters. The details are given in Ref. [5].

The results of calculation are listed in Table 2. If there is no bound state for a given L^π, the variational energy is higher than the lowest accessible threshold and would converge to the threshold energy as the basis dimension increases. On the other hand, if there is a bound state, the energy becomes lower than the threshold. We found that only four states predicted earlier are bound. They are $^1S^e, ^1S^{e*}, ^3S^e$ and $^1P^o$ corresponding to $0^+(A_1)$, $0^+(B_2)$, $0^+(E)$, and $1^-(B_2)$, respectively.

In summary, only four bound states, $0^+(A_1)$, $0^+(B_2)$, $0^+(E)$ and $1^-(B_2)$, have been confirmed theoretically. No other bound states have been obtained. The $1^-(B_2)$ state with the energy of −0.334 a.u. is important for the Ps_2 confirmation because the dipole transition to the ground state is unique. Though no other bound states exist, there may be resonance states. The calculation of the energy and width of a possible resonance remains a future investigation. It is also interesting to study the decay modes distinguishable from Ps and Ps^- decay, for example,

$$Ps_2 \to 2\gamma \quad (E_\gamma = 1\text{MeV}), \qquad Ps_2 \to e^- + e^+ \quad (E_e = 1\text{MeV}).$$

We thank Dr. K. Varga for useful discussions and collaboration at the early stage.

Table 2. The energies of Ps_2. Values in square parentheses are the threshold energies of the lowest accessible channel.

L^π	Symmetry	lowest accessible threshold		energy(a.u.)
0^+	A_1	Ps(1S)+Ps(1S)	[−0.5]	−0.5160037
	A_2	Ps(1S)+Ps(2P)	[−0.3125]	−0.309
	B_1	Ps(1S)+Ps(1S)	[−0.5]	−0.492
	B_2	Ps(1S)+Ps(2P)	[−0.3125]	−0.31449
	E	Ps(1S)+Ps(2S,P)	[−0.3125]	−0.33016
0^-	A,B,E	Ps(2P)+Ps(3P,D)	[−0.0902]	−0.09
1^+	A_1,B_1	Ps(1S)+Ps(3D)	[−0.277]	−0.267
	A_2,B_2	Ps(1S)+Ps(2P)	[−0.3125]	−0.309
	E	Ps(1S)+Ps(2P)	[−0.3125]	−0.312
1^-	A_1,B_1	Ps(1S)+Ps(2S)	[−0.3125]	−0.311
	A_2	Ps(1S)+Ps(2P)	[−0.3125]	−0.3114
	B_2	Ps(1S)+Ps(2P)	[−0.3125]	−0.334408
	E	Ps(1S)+Ps(1S)	[−0.5]	−0.496
2^+	A_1,B_1	Ps(1S)+Ps(1S)	[−0.5]	−0.491
	A_2,B_2	Ps(1S)+Ps(2P)	[−0.3125]	−0.310
	E	Ps(1S)+Ps(2S,P)	[−0.3125]	−0.306
2^-	A_1,B_1	Ps(1S)+Ps(3D)	[−0.277]	−0.26765
	A_2,B_2	Ps(1S)+Ps(2P)	[−0.3125]	−0.3108
	E($C_P = +$)	Ps(1S)+Ps(3D)	[−0.277]	−0.2766
	E($C_P = -$)	Ps(1S)+Ps(2P)	[−0.3125]	−0.3108

References

1. K. Varga, J. Usukura, Y. Suzuki: Phys. Rev. Lett. **80**, 1876 (1998)
2. J. Usukura, K. Varga, Y. Suzuki: Phys. Rev. **A58**, 1918 (1998)
3. Y.K. Ho: Phys. Rev. **A39**, 2709 (1989)
4. D.B. Kinghorn, R.D. Poshusta: Phys. Rev. **A47**, 3671 (1993)
5. Y. Suzuki, J. Usukura, K. Varga: J. Phys. **B31**, 31 (1998)

Few-Body Systems Suppl. 12, 75–79 (2000)

Proton-Hydrogen Charge Exchange and Elastic Scattering in the Faddeev Approach

E.O. Alt[1], A.S. Kadyrov[2], A.M. Mukhamedzhanov[3]

[1] Institut für Physik, Universität Mainz, D-55099 Mainz, Germany

[2] Electronic Structure of Materials Centre, The Flinders University of South Australia, GPO Box 2100, Adelaide 5001, Australia

[3] Cyclotron Institute, Texas A&M University, College Station, TX 77843, USA

Abstract. Results of the application of Faddeev-type integral equations to proton-hydrogen collisions are reported. The approach, realized in the impact parameter representation, incorporates the exact two-particle off-shell Coulomb T-matrices in all 'triangle' contributions to the effective potentials. Calculated total and differential electron-transfer as well as differential elastic scattering cross sections show very good agreement with experiment, over a wide range of incident energies.

1 Introduction

Despite its long history, the theoretical description of even the simplest electron transfer in proton-hydrogen collisions is not yet in a satisfactory state, in particular in the medium to higher energy region. For total cross sections (TCS), two perturbative methods, namely the continuum distorted wave (CDW) [1] and the boundary corrected first Born (B1B) [2] approximations give the best agreement with experimental data. But both fail to describe differential cross sections (DCS) which represent a much more stringent test of the theoretical models. One common feature is that both are one-channel approximations, i.e. neither contribution to the rearrangement channel coming from other reaction channels, nor the interference between different states in a given channel, is taken into account.

In view of these facts a method is still called for which properly takes into account all possible reaction channels on an equal footing and, as a result, is able to explain the TCS as a consequence of the correct description of the elastic and the electron-transfer DCS. We believe that the three-body Faddeev approach, although not having been widely applied to atomic collisions yet, has the desired capacity.

Let us put aside principal questions concerning the validity of integral equations in the presence of (in particular, attractive) Coulomb forces since they are not relevant for the approximate equations used. Independently thereof, also practical applications encounter severe difficulties, mainly due the complicated singularity structure of the two-particle Coulomb T-matrix which is the basic dynamical ingredient. For, as is well known, the off-shell Coulomb T-matrix develops nasty singularities in the on-shell limit and, in case of an attractive interaction, has in addition an infinite number of (bound-state) poles.

These practical difficulties have been overcome in our investigations, as we have developed techniques to *exactly* include the two-particle off-shell Coulomb T-matrices in the first order direct and exchange contributions to the effective potentials [3]. The latter occur as driving terms in the impact parameter Faddeev approach (IPFA) that has been proposed earlier. Let us briefly recapitulate its main features. The starting point is the AGS effective-two-body formulation of the three-body theory, modified to include Coulomb interactions. The resulting set of Lippmann-Schwinger type integral equations, which couple the transition amplitudes for all binary processes, are then written in on-shell approximation, i.e., the exact intermediate-state two-fragment propagators are approximated by the corresponding energy conserving δ-functions. Instead of making a partial wave expansion to further reduce the dimensionality of the equations, transformation to the impact parameter representation is used. This leads to a coupled system of algebraic equations the solutions of which satisfy - at least partially - the constraints from two- and three-particle unitarity.

In initial applications of the IPFA to the calculation of different electron-transfer reactions [4], the (complicated) effective potentials were taken into account only in the lowest-order approximation corresponding to the electron-transfer ('pole') mechanism. Calculations of total and partial electron-transfer cross sections showed on the whole good agreement with available experimental data, over a wide range of reactions and energies. But at high energies this approach overestimated the data, giving only qualitative results, due to the neglect of higher-order terms in effective potentials.

The objective of the present report is to demonstrate that the further development of the IPFA, based upon inclusion of the first-order terms of the effective potentials which contain the two-particle off-shell Coulomb T-matrices, brings about a great improvement, as it leads to a very satisfactory description, not only of TCS but also of DCS for direct and exchange scattering.

2 Results of Calculations

We present some of the results of our calculations of TCS and DCS for electron transfer and of DCS for elastic scattering in proton collisions with hydrogen atoms in their ground state. Here we confine ourselves to the case that only the $1s$ state of hydrogen is retained explicitly in all possible reaction channels. This $1s - 1s$ approximation is believed to give between 85 % in the higher and

100 % in the moderate energy regime, of the complete cross section.

The elastic scattering and electron exchange amplitudes are obtained simultaneously as solutions of the set of coupled linear equations. The (so-called Quasi-Born) expansion of the effective potentials has been truncated after the first order: beside the elementary electron exchange ('pole') term they contain the direct and exchange 'triangle' contributions. Details can be found in [5].

Typical results are presented in Figs. 1 and 2, and compared with calculations by other authors and with experimental data. We restrict ourselves to so-called '1*s*' cross sections where only the '1*s*' final state is taken into account (no summation over final states!). The TCS for electron transfer is displayed in Figs. 1. The fact that inclusion of the exact 'triangle' terms (i.e., calculated with the full Coulomb T-matrices) in the effective potentials is instrumental in reducing the TCS for energies beyond a few keV is seen in Fig. 1a, which shows comparison of the present with the corresponding IPFA results of [4] where only the 'pole' term had been employed. In Fig. 1b, the high-energy part of our TCS is compared with that following from the most sophisticated high-energy models, namely CDW and B1B. Inspection reveals that our approach leads to nearly as good agreement as CDW and B1B. But, in contrast to the CDW and B1B amplitudes, our reaction amplitude is unitary at the two-body (and partially also on the three-body) level. This fact is vital for reducing its value at lower energies, with the result that at the same time we also achieve perfect agreement with the data in the intermediate and low-energy region, as shown

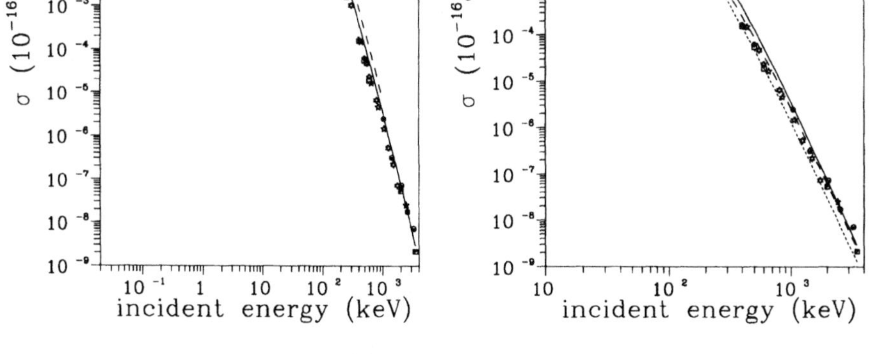

Figure 1. Total cross sections for electron capture by H^+ from $H(1s)$. (a): solid line, present results (1s); dashed line, IPFA [4]. Points are experimental data from different sources, see [5] for details. (b): same as in (a) except: dotted line, CDW [1]; dashed line, B1B [2].

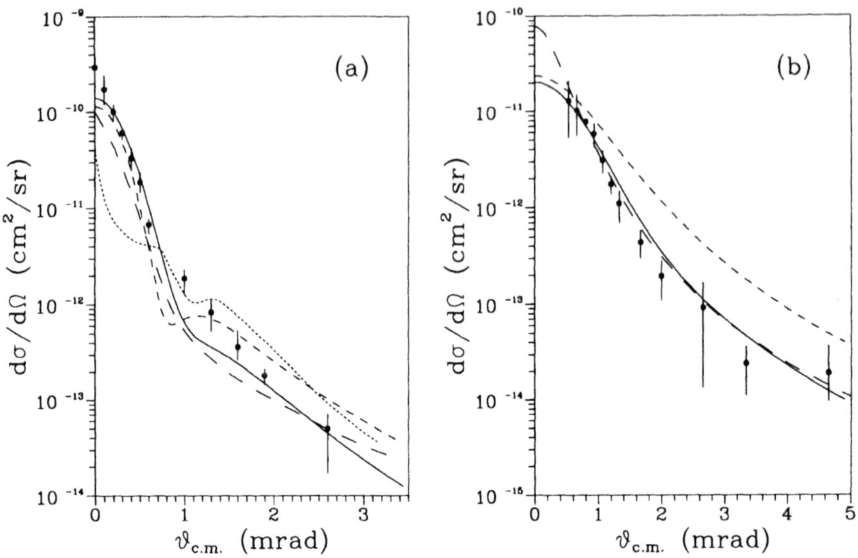

Figure 2. Differential cross section for $H^+ - H(1s)$ collisions at 60 keV. (a) electron capture: solid line, present results (1s); dotted line, CDW (1s) [1]; short-dashed line, B1B (summation over all final states included) [2]; long-dashed line, MS (1s) [6]. (b) elastic scattering: same as in (a) except short-dashed line, FBA. Experimental data are from [7].

in Fig. 1a. Summarizing, both fulfillment of the unitarity constraints and use of the exact 'triangle' amplitudes taken together eventually lead to a very good agreement with experiment, over a wide range of energy.

If the excellent reproduction of the TCS in CDW and B1B were due to a high quality of the transition amplitudes then this should show up also in angular distributions. An example of proton-hydrogen $1s-1s$ DCS for electron-transfer (a) and elastic scattering (b) are presented in Figs. 2, at a projectile energy of 60 keV. As the elastic scattering amplitudes are obtained simultaneously with the transfer amplitudes, both contain information about the other channel through the coupling. Note that the B1B results pertain to DCS summed over all final states. It appears that our approach leads, on the whole, to a physically more realistic reaction amplitude and, consequently, to a significantly better reproduction of the experimental data than either CDW, B1B, or the coupled-multistate (MS) model of Ref. [6]. This success lends additional support to our claim that the reaction amplitudes as calculated in the three-body approach are of high physical significance.

All these results (and see [5] for more) illustrate the great improvement achieved presently over the traditional reaction models. We stress once more that in our approach all interparticle interactions are treated on the same footing, and hence none of them needs to be 'smuggled in' afterwards by means of a phase factor.

3 Conclusions

The main goal of this investigation has been to re-establish the three-body integral equations approach as a valuable and powerful tool for calculating (energetic) atomic collision processes. Its strength derives from one of its most salient and useful features, namely that it allows one, in a natural manner, to systematically include the physically possible and most relevant reaction channels, and thereby to implement two-body and, at least approximately, three-body unitarity. In this way the interesting features of the collision process are exposed in detail. This objective has been exemplified by means of the scattering of protons off hydrogen atoms. Of course, it really does not come as a surprise that proton-hydrogen scattering, being a genuine three-body problem, is best described within the framework of the *ab-initio* three-body formalism.

The promising results obtained encourage further development. For instance, generalization to an arbitrary number of involved bound states does not alter the basic equations although it requires considerable additional numerical effort. Also generalization to multiply-charged projectiles is possible and has, in fact, been done practically in [4] (but with only the 'pole' approximation to the effective potentials taken into account). That is, the general strategy will remain unaltered in such further development.

Acknowledgement. This work was supported by Deutsche Forschungsgemeinschaft, Project 436 USB-113-1-0, by Deutscher Akademischer Austauschdienst, by DOE Grant DE-FG05-93ER0773 and by Australian Research Council.

References

1. Dž. Belkić, R. Gayet, A. Salin: Phys. Rep. **56**, 279 (1979)
2. Dž. Belkić, S. Saini, H.S. Taylor: Z. Phys. **D3**, 59 (1986); Dž. Belkić, S. Saini, H.S. Taylor: Phys. Rev. **A36**, 1601 (1987)
3. E.O. Alt et al.: J. Phys. **B28**, 5137 (1995); E.O. Alt, A.S. Kadyrov, A.M. Mukhamedzhanov: Phys. Rev. **A54**, 4091 (1996); E.O. Alt, A.S. Kadyrov, A.M. Mukhamedzhanov: J. Phys. **B30**, 3659 (1997)
4. G.V. Avakov et al.: J. Phys. **B23**, 2309S (1990); G.V. Avakov et al.: J. Phys. **B23**, 4151 (1990); G.V. Avakov et al.: J. Phys. **B25**, 213 (1992)
5. E.O. Alt, A.S. Kadyrov, A.M.: Mukhamedzhanov: Phys. Rev. **A60**, 314 (1999)
6. R. Shakeshaft: Phys. Rev. **A18**, 307 (1978); J.M. Wadehra and R. Shakeshaft: Phys. Rev. **A26**, 1771 (1982)
7. P.J. Martin et al.: Phys. Rev. **A23**, 3357 (1981); J.T. Park: In *Proceedings of XII ICPEAC*, Gatlinburg, 1981, edited by S. Datz. Amsterdam: North-Holland 1982

Few-Body Systems Suppl. 12, 80–85 (2000)

Few-Body Systems

Impact of Few-Body Reactions on Explosive Nucleosynthesis: The Big-Bang and Supernovae

T. Kajino[1,2,3], K. Otsuki[1], S. Wanajo[1], M. Orito[1], G.J. Mathews[4]

[1] National Astronomical Observatory, Mitaka, Tokyo, 181-8588

[2] Department of Astronomy, University of Tokyo, Tokyo, 113-0033

[3] Department of Astronomical Science, Graduate University for Advanced Studies, Tokyo, 181-8588

[4] Department of Physics, University of Notre Dame, Notre Dame, IN 46556

Abstract. We study the few body reactions in explosive nucleosynthesis of light-to-heavy mass elements in various astrophysical conditions. In neutrino-driven winds of gravitational core collapse of SNeII, several light nuclei as well as heavy unstable nuclei paly significant roles in producing r-process elements. We discuss how the three-body reaction ^{4}He$(\alpha n, \gamma)^9$Be, whose cross section is poorly known, plays very critical role. We make note that several other reactions in the r-process are relevant to a Big-Bang nucleosynthesis. We discuss briefly a Galactic chemical evolution.

1 Introduction

Big-Bang nucleosynthesis in the early Universe is thought to produce the light elements D and ^{3,4}He with small amounts of ^{7}Li, ^{9}Be and ^{11}B. Since these abundances are used as unique observables to determine the cosmological parameter Ω_b, careful studies of the few body reactions of light nuclei are highly desired. Stars with various masses provide a variety of production sites for ^{4}He, ^{12}C and heavier elements. Very massive stars $\geq 10M_\odot$ culminate their evolution by supernova (SN) explosions which are also presumed to be most viable candidate for the still unknown astrophysical site of r-process nucleosynthesis. Even in the nucleosynthesis of heavy elements, initial entropy and density at the surface of proto-neutron stars are so high that nuclear statistical equilibrium (NSE) favors production of abundant light nuclei. In such explosive circumstances few-body reactions play a significant role.

The study of the origin of r-process elements is also critical in cosmology. It is a potentially serious problem that the cosmic age of the expanding Universe derived from cosmological parameters may be shorter than the age of the oldest globular clusters. Since both age estimates are subject to the uncertain cosmological distance scale, an independent method has long been needed. Thorium, which is a typical r-process element and has half-life of 14 Gyr, has recently been detected along with other elements in very metal-deficient stars. If we model the r-process nucleosynthesis in these first-generation stars, thorium can be used as a cosmochronometer completely independent of the uncertain cosmological distance scale.

In this article, we first demonstrate theoretically that supernova explosions of very massive stars could be a viable site for r-process nucleosynthesis. We study the role of few-body reactions and calculate the sensitivity of the r-process yields to the cross section for $^4\mathrm{He}(\alpha n,\gamma)^9\mathrm{Be}$, this rate is poorly known experimentally but is expected to affect the result strongly. Next we note several key reactions in the r-process which are relevant to the Big-Bang nucleosynthesis.

2 Explosive Nucleosynthesis in Supernovae

Recent mesurements using high-dispersion spectrographs with large Telescopes or the Hubble Space Telescope has made it possible to detect minute amounts of heavy elements in faint metal-deficient ([Fe/H] $\leq$ -2) stars. The discovery of r-process elements in these stars has shown that the relative abundance pattern for the mass region 120 $\leq$ A is surprisingly similar to the solar system r-process abundances independent of the metallicity of the star as shown in Fig. 1. Here metallicity is defined by [Fe/H] = log[N(Fe)/N(H)] - log[N(Fe)/N(H)]$_\odot$. It obeys the approximate relation t/10^{10}yr $\sim 10^{[Fe/H]}$. The observed similarity strongly suggests that the r-process occurs in a single environment which is independent fo progenitor metallicity. Massive stars with $10M_\odot \leq M$ have a short life $\sim 10^7$ yr and eventually end up as violent supernova explosions, ejecting material into the intersteller medium early on quickly from the history of the Galaxy. However, the iron shell in SNe is excluded from being the r-process site because of the observed metallicity independence.

Hot neutron stars just born in the gravitational core collapse of SNeII release most of their energy as neutrinos during the Kelvin-Helmholtz cooling phase. An intense flux of neutrinos heat the material near the neutron star surface and drive matter outflow (neutrino-driven winds). The entropy in these winds is so high that the NSE favors a plasma which consists of mainly free nucleons and alpha particles rather than composite nuclei like iron. The equilibrium lepton fraction Y_e is determined by a delicate balance between $\nu_e + n \to p + e^-$ and $\bar{\nu}_e + p \to n + e^+$, which overcomes the difference of chemical potential between n and p, to reach $Y_e \sim 0.45$. R-process nucleosynthesis occurs because there are plenty of free neutrons at high temperature. This is possible only if seed elements are produced in the correct neutron to seed ratio before and during the r-process.

Although Woosley et al. [1] demonstrated a profound possibility that the r-process could occur in these winds, several difficulties were subsequently identified. First, independent non relativistic numerical supernova models [2] have difficulty producing the required entropy in the bubble S/k $\sim$ 400. Relativistic effects may not be enough to increase the entropy dramatically [3, 4, 5]. Second, even should the entropy be high enough, the effects of neutrino absorption $\nu_e + n \rightarrow p + e^-$ and $\nu_e + A(Z, N) \rightarrow A(Z+1, N-1) + e^-$ may decrease the neutron fraction during the nucleosynthesis process. As a result, a deficiency of free neutrons could prohibit the r-process [6]. Third, if neutrinos are massive and have approximate mixing parameters, energetic ν_μ and ν_τ are converted into ν_e due to flavor mixing. This activates the $\nu_e + n \rightarrow p + e^-$ process and results in a deficiency of free neutrons.

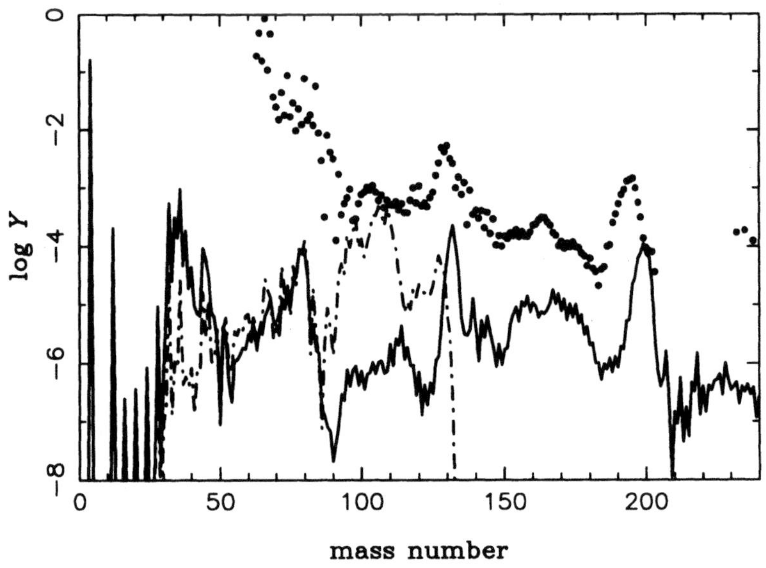

Figure 1. Calculated r-process abundance yields [5] in the neutrino-driven winds as a function of atomic mass number. Filled circles are the solar system r-process abundances [7], shown in arbitrary units.

In order to resolve these difficulties, we have studied neutrino-driven winds in a Schwarzschild geometry under the reasonable assumption of spherical steady-state flow [5]. Then, we carried out r-process nucleosynthesis calculations in our wind model [8]. The nuclear reaction network used for light-mass region is shown in Fig. 2. The full network consists of $\sim$ 3000 elements including radioactive unstable nuclei up to the plutonium isotopes.

We found [5] that the general relativistic effects make expanding dynamic time scale τ_{dyn} much shorter, although the entropy per baryon increases by about 40 % from the Newtonian value of S/k $\sim$ 90. By simulating many supernova explosions, we have found some interesting conditions which lead to successful r-process nucleosynthesis [5, 8], as shown in Fig. 1. The best case is for a relatively large neutron-star mass $M > 1.7M_{\odot}$ and a neutrino luminosity $L_{\nu} > 5 \times 10^{51} erg\ s^{-1}$.

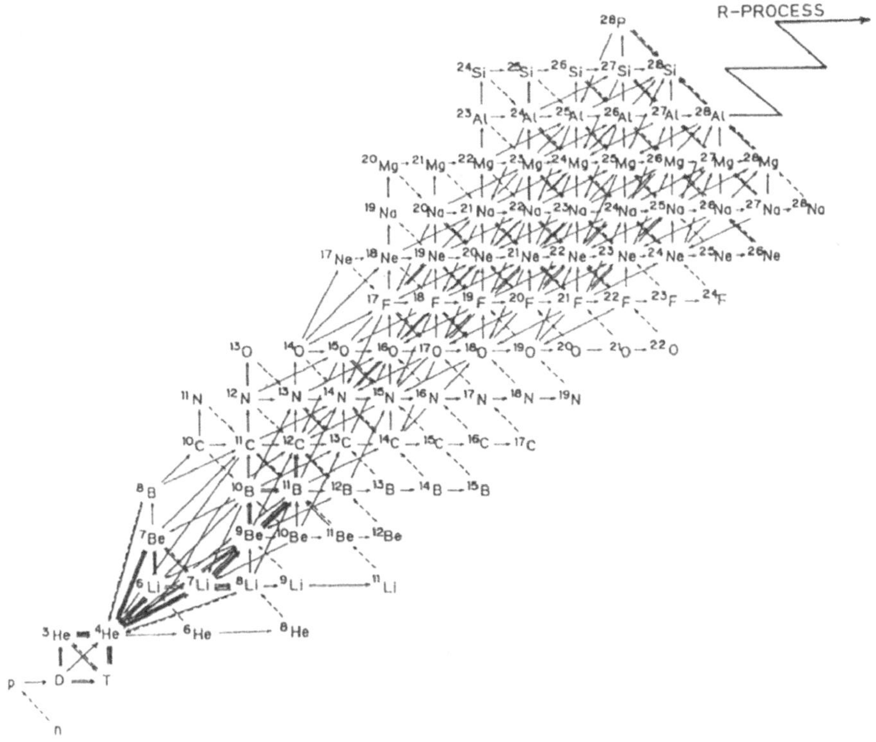

Figure 2. Light mass region of the nuclear reaction network used in the calculations. Solid arrows indicate nuclear reactions in the direction of positive Q–values. Dashed arrows show beta decays.

The key to resolve the first difficulty noted above is the short dynamic time scale $\tau_{dyn} \sim 10$ ms. As the initial nuclear composition of the relativistic plasma consists of neutrons and protons, the α-burning begins when the plasma temperature cools below T $\sim$ 0.5 MeV. The ^{4}He$(\alpha\alpha,\gamma)^{12}$C reaction is too slow at this temperature, and alternative nuclear reaction path ^{4}He$(\alpha n,\gamma)^9$Be$(\alpha,n)^{12}$C triggers explosive α-burning to produce seed elements with A $\sim$ 100. Therefore, the time scale for nuclear reactions is regulated by the ^{4}He$(\alpha n,\gamma)^9$Be. It is given by $\tau_N \equiv \left(\rho_b^2 Y_\alpha^2 Y_n \lambda(\alpha\alpha n \rightarrow^9 \text{Be})\right)^{-1}$. If the neutrino-driven winds fulfill the condition $\tau_{dyn} < \tau_N$, then fewer seed nuclei are produced during the α-process with plenty of free neutrons left over when the r-process begins at T $\sim$ 0.2 MeV. The high neutron-to-seed ratio, $n/s \sim 100$, leads to appreciable production of r-process elements, even for low entropy S $\sim$ 130, producing both the 2nd ($A \sim 130$) and 3rd ($A \sim 195$) abundance peaks and the hill of rare-earth elements ($A \sim 165$) (cf. Fig.1).

The three body nuclear reaction cross section for ^{4}He$(\alpha n,\gamma)^9$Be is poorly determined experimentally. Because two different channels, ^{8}Be + n and ^{5}He + α, contribute to this process it is also a theoretical challenge to understand the reaction mechanism and the resonance structure. We show two calculated results in Fig. 1: The solid line displays the result obtained by using the cross section estimated by Woosley and Hoffman [9], assuming a ^{8}Be + n structure for ^{9}Be. The cross section estimate may only be good to one order of magnitude. Therefore, we also calculated the r-process by multiplying this cross section by a factor of 10 (long-dashed line). This makes a drastic change in the r-process abundance yields. Both theoretical and experimental studies of the ^{4}He$(\alpha n,\gamma)^9$Be reaction are highly desired.

The specific collision time for neutrino-nucleus interactions is given by

$$\tau_\nu \approx 201 \times L_{\nu,51}^{-1} \times \left(\frac{\epsilon_\nu}{\text{MeV}}\right)\left(\frac{r}{100\text{km}}\right)^2\left(\frac{\langle\sigma_\nu\rangle}{10^{-41}\text{cm}^2}\right)^{-1} ms. \qquad (1)$$

$\tau_{dyn} \approx 10$ ms $\ll \tau_\nu \approx 200$ ms holds at the α-burning site, r $\approx$ 100 km. This resolves the second difficulty: Because there is not enough time for ν's to interact with n's in such rapidly expanding winds, the neutron fraction is very insensitive to the neutrino absorption.

We have recently shown that the neutrino flavor oscillation could destroy the r-process condition if the mixing parameters satisfy 0.3 eV$^2 \leq \Delta m^2$ in our wind model. Recent experiments of the atmospheric neutrinos and the missing solar neutrinos have indicated much smaller Δm^2, while the LSND experiment suggests larger Δm^2. We should wait for more experiments.

3 Big-Bang Nucleosynthesis and Galactic Chemical Evolution

We now found that new possible nuclear reaction paths, which were neglected in the previous calculations. In addition to ^{4}He$(\alpha n,\gamma)^9$Be at relatively late epoch in the r-process, there is also ^{4}He$(t,\gamma)^7$Li$(\alpha,\gamma)^{11}$B or ^{7}Li$(t,n)^9$Be$(t,n)^{11}$B, followed by (n,γ) and beta-decays. It is also likely that three-body reactions like ^{9}Li$(2n,\gamma)$, ^{15}B$(2n,\gamma)$, ^{24}O$(2n,\gamma)$, ^{27}F$(2n,\gamma)$, etc., may be important. This has already been pointed out in the inhomogeneous Big-Bang nucleosynthesis model (IBBN) [10, 11] although the physical conditions are different.

These new paths produce ^{9}Be, ^{11}B and even intermediate-mass nuclei $12 \leq A$ in the IBBN, the abundances of which are many orders of magnitude higher than those of SBBN, as displayed in Fig. 3. This figure shows also the Galactic chemical evolution. The rise of the ^{9}Be abundance with increasing [Fe/H] arises from spallation of Galactic cosmic rays after Galaxy formation. Each solid curve reaches a different abundance plateau for Big-Bang nucleosynthesis (SBBN or IBBN) in the limit of [Fe/H] = $-\infty$, i.e. t = 0.

The nuclear physics of the few-body reactions of light neutron-rich nuclei is thus very important not only in studies of r-process nucleosynthesis but in Big-Bang nucleosynthesis and Galactic chemical evolution as well.

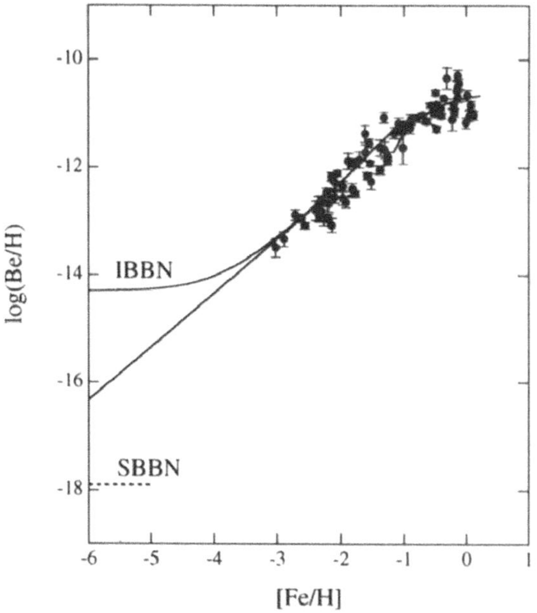

Figure 3. Predicted evolution of N(^{9}Be)/N(H) vs. [Fe/H] with the standard Big-Bang nucleosynthesis model (SBBN; lower dashed plateau) and the inhomogeneous Big-Bang nucleosynthesis model (IBBN; upper solid plateau).

References

1. S.E. Woosley et al.: Astrophys. J. **433**, 229 (1994)
2. J. Witti, H.-Th. Janka, K. Takahashi: Astron. and Astrophys. **286**, 842 (1994)
3. Y.Z. Qian and S.E. Woosley: Astrophys. J. **471**, 331 (1996)
4. C.Y. Cardall and G.M. Fuller: Astrophys. J. **486**, L111 (1997)
5. K. Otsuki, H. Tagoshi, T. Kajino, S. Wanajo: Astrophys. J. (in press)
6. B.S. Meyer: Astrophys. J. **449**, L55 (1995)
7. F. Käppeler, H. Beer, K. Wisshak: Rep. Prog. Phys. **52**, 945 (1989)
8. S. Wanajo, K. Otsuki, T. Kajino, G.M. Mathews: Astrophys. J. (submitted)
9. S.E. Woosley, R.D. Hoffman: Astrophys. J. **395**, 202 (1992)
10. T. Kajino and R.N. Boyd: Astrophys. J. **359**, 267 (1990); T. Kajino et al.: Astrophys. J. **364**, 7 (1990)
11. M. Orito, T. Kajino, G.M. Mathews: Astrophys. J. **488**, 515 (1997)

Few-Body Systems Suppl. 12, 86–91 (2000)

Few-Body Systems

Coulomb Dissociation of Unstable Nuclei – An Application of Breakup Process to Astrophysics –

T. Motobayashi

Department of Physics, Rikkyo University, Nishi-Ikebukuro, Toshima, Tokyo 171-8501, Japan

Abstract. The Coulomb dissociation method has been developed recently to investigate radiative capture processes of astrophysical interest. Its high experimental efficiency enables one to study the systems involving unstable nuclei, which are now available as beams but with relatively weak intensities. We made a series of experiments at RIKEN including the breakup of ^{14}O and ^{8}B.

1 Coulomb Dissociation

In thermonuclear burning processes in stars or in early universe, nuclear reactions at low energies play important roles. Recent development on radioactive (or RI) beam production allows for new approaches to reactions on unstable nuclei, which sometimes dominates in so called explosive burning situation. In such cases both temperature and matter density are high, and nuclear reactions can be faster than their competing β decays.

For studying astrophysical (p,γ) processes involving unstable nuclei, we adopt the Coulomb dissociation method. It uses intermediate-energy RI beams, which are now available at many laboratories with variety of ions. Low-energy cross sections can be extracted indirectly from the data of Coulomb dissociation, for example, at several tens MeV per nucleon. An advantage is availability of thick targets. Targets with several tens or several hundred mg/cm^2 thickness are commonly used in actual experiments. This is larger by an order of 3 or 4 than the typical thickness of targets in low-energy direct measurements, where the thickness is limited by the energy loss of the incident beam. This makes experiments with weak radioactive nuclear beams realistic.

In the Coulomb dissociation method, the residual nucleus B of the capture reaction A(x,γ)B bombards a high-Z target and is Coulomb-excited to an unbound state that decays to the A+x channel. Since the process is regarded as absorption of a virtual photon, *i.e.* B(γ,x)A, the radiative capture (the inverse of the photoabsorption) cross section can be extracted from the dissociation yield. This idea was first proposed by Baur, Bertulani and Rebel [1] based on the virtual photon theory. Topical reviews were given by Baur and Rebel [2]. In addition to the advantage discussed before, the Coulomb dissociation method enhances the original capture cross section by a large factor. This is due to the large virtual-photon number and the phase space factor. The two factors can be in the order of 100 or 1000 in some actual cases like the ^{14}O and ^{8}B dissociation, which will be discussed later in this report.

2 ^{14}O Breakup and Hot CNO Cycle

After the pioneering studies on the stable Li isotopes, $^{6}\mathrm{Li}\rightarrow \alpha$+d and $^{7}\mathrm{Li}\rightarrow \alpha$+t, at around E_{in}=10 MeV/u [3, 4, 5, 6], the first Coulomb dissociation experiments with RI beams were made for the ^{208}Pb(^{14}O,^{13}N p)^{208}Pb [7]. In the core of super massive stars, novae or at the surface of neutron stars, the ^{13}N(p,γ)^{14}O reaction becomes faster than the β^{+} decay of ^{13}N, and the hot CNO cycle starts instead of the regular CNO cycle. This key reaction ^{13}N(p,γ)^{14}O was difficult to be studied because the radioactive nucleus ^{13}N (10 min. half life) is involved. The experimental goal is to determine the E1 electromagnetic-width Γ_{γ} of the first 1^{-} state in ^{14}O, because this resonant state dominates in the hot CNO burning. The experimental accuracy obtained in a few indirect measurements, where a small branching ratio ($\approx 10^{-4}$) of the γ decay was measured, was rather poor. We performed an experiment of Coulomb dissociation [7] hoping that the experimental accuracy for Γ_{γ} might be improved.

An ^{14}O beam was obtained by the projectile-fragmentation scheme with the RIPS separator at RIKEN [8]. It bombarded a thick (350 mg/cm^{2}) ^{208}Pb target. Outgoing particles were detected in coincidence. The detection system consisted of a mosaic of 24 ΔE(Si)-E(CsI(Tl)) counter telescopes "EMRIC" [9] and two (x and y) planes of plastic scintillator hodoscopes.

By comparing the result with the predicted cross section, the radiative width Γ_{γ} was deduced to be 3.1±0.6 eV, which agrees with the width obtained by the direct measurement of the ^{13}N(p,γ)^{14}O reaction performed at Louvain la Neuve [10] almost at the same time as our Coulomb dissociation experiment. Another Coulomb dissociation experiment at 70 MeV/u [11] at GANIL also yields a value consistent with above value within errors.

3 ^{8}B Breakup and Solar Neutrino Problem

The ^{7}Be(p,γ)^{8}B reaction at low energies is relevant to the high energy neutrino production in the sun through the β decay of ^{8}B. A series of experiments has been and is being performed for the Coulomb dissociation of ^{8}B at RIKEN to study this important reaction. It should be noted that the astrophysical process is dominated by the E1 transition to the unbound continuum state of ^{8}B. Therefore the Coulomb dissociation to continuum states should be studied.

The first measurement is described in two articles [12, 13], and the results of the second experiment are reported in Refs. [14] and [15]. The data obtained in the third experiment is now being analyzed. A schematic view of the experimental setup is shown in Fig. 1. Beams of ^{8}B were produced by the ^{12}C+^{9}Be interaction at 92 MeV/u for the first two experiments and 135 MeV for the third one. The ^{8}B energies in the center of the target, 50 mg/cm^2 ^{208}Pb, were approximately 50 MeV/u. We employed the time-of-flight (TOF) technique to determine the energies of fragments. A plastic scintillator hodoscope was set 3-5 m from the target. The outgoing particles of the Coulomb dissociation, ^{7}Be and p, are identified by the ΔE-E method. Their scattering angles are determined by the positions in the 10×16 segments formed by the ΔE and E scintillators. A p-^{7}Be relative energy spectrum could be constructed from the measurement and it was converted to the ^{7}Be(p,γ)^{8}B cross section with the help of a Monte-Carlo simulation calculation on detection efficiency and theoretical calculation for the Coulomb dissociation mechanism.

In the second and third experiments, the DALI setup [16] was also used to measure the deexcitation γ rays from the first excited state of ^{7}Be at 429 keV populated in the dissociation process. The contribution from this process was measured to be about 5% of the Coulomb dissociation yield.

In Fig. 2 the astrophysical S-factors obtained in the first and second experiments are shown together with the ones determined in direct (p,γ) measurements. Our Coulomb dissociation data are consistent within errors with the results by Filippone *et al.* [17] and Vaughn *et al.* [18].

Figure 3 shows the angular distributions for the scattering angle θ_8, the angle of the center-of-mass for the p-^{7}Be system. They are well fitted by the DWBA-type calculation assuming pure E1 transition. This suggests that possible mixture of the E2 amplitude is small enough to be neglected. This may contradict to the asymmetry observed in the parallel momentum spectrum for ^{7}Be from the ^{8}Be breakup [19], calling for further studies.

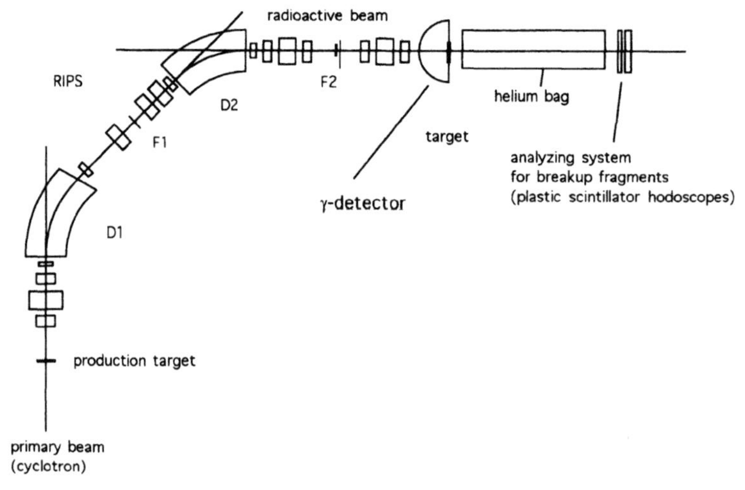

Figure 1. Experimental setup for the ^{8}B Coulomb dissociation experiment.

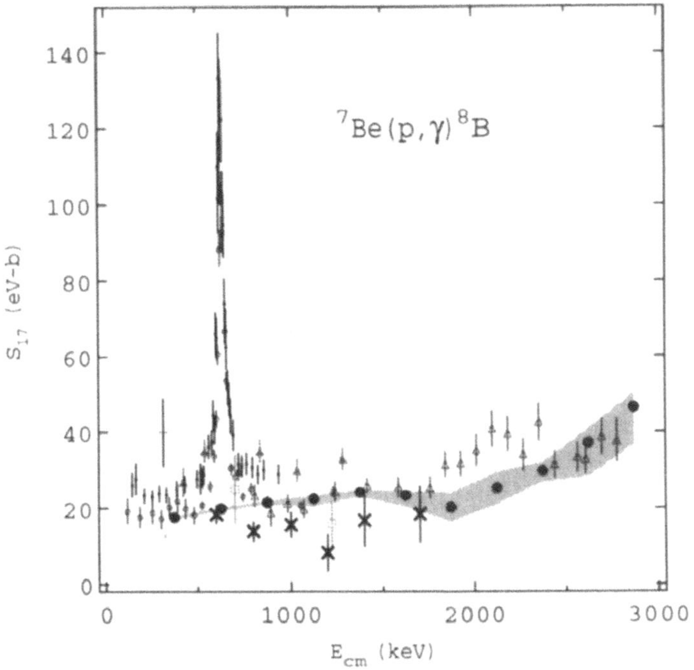

Figure 2. Astrophysical S_{17}-factors for the ^{7}Be(p,γ)^{8}B reaction extracted from the first (thick crosses) and second (large solid dots) experiments. Existing direct (p,γ) data are also shown.

The post Coulomb-acceleration effect is expected to be small owing to the special situation for ^{8}B where the proton binding energy is only 137 keV. Trajectories very far from the target are enough to excite the continuum state of ^{8}B, because a low energy virtual photon is responsible for the excitation. Recent theoretical calculations supports this picture [20, 21]

In general, higher order processes including the post acceleration is less important at higher incident energy. We also performed a Coulomb dissociation experiment at GSI using ^{8}B beams of 254 MeV/u and a ^{208}Pb target [22]. The resultant S_{17} factors are quite similar to the Coulomb dissociation results obtained in the second RIKEN experiment, and therefore to the direct (p,γ) results by Filippone *et al.* [17] and Vaughn *et al.* [18]. The agreement between the two results obtained at the different energies (50 MeV/u at RIKEN and 250 MeV/u at GSI) might indicate smallness of the higher-order contribution.

4 Summary

Since the idea of laboratory experiments for astrophysical thermo-nuclear reactions was proposed in 1950s, much progress has been achieved. Recent development of producing radioactive nuclear beams allows now for access to unstable

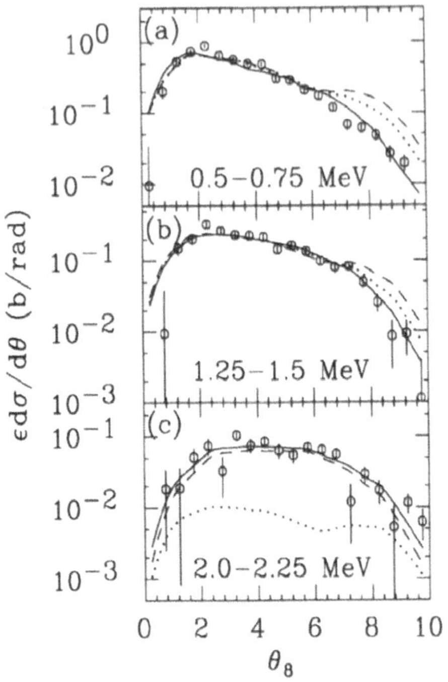

Figure 3. Angular distribution of the ^{8}B Coulomb dissociation reaction.

nuclei important in explosive nuclear burning. So called indirect method using projectiles of unstable nuclei at intermediate-energies can also be applied to "hard-to-access" cases. Especially the Coulomb dissociation method has been studied for several important reactions. The ^{14}O and ^{8}B dissociation provide typical examples of resonant and non-resonant breakups. Their experimental results encourage new measurements of astrophysical reactions. We plan to extend the Coulomb dissociation method to several cases, such as $^{8}B(p,\gamma)^{9}C$ and $^{12}N(p,\gamma)^{13}O$, where the capture cross sections are not known. Further experimental investigation to control the possible ambiguities as well as theoretical works are also desirable.

The works present in this report were performed in collaboration with T. Kikuchi, T. Takei, N. Iwasa, Y. Ando, K. Ieki, M. Kurokawa, S. Moriya, H. Murakami, T. Nishio, J. Ruan (Gen), S. Shirato, S. Shimoura, T. Uchibori, Y. Yanagisawa (*Rikkyo University*), A. Goto, T. Ichihara, N. Inabe, T. Kubo, H. Sakurai, T. Teranishi, Y. Watanabe, M. Ishihara (*RIKEN*), M. Hirai, T. Nakamura (*University of Tokyo*), S. Kubono (*INS*), Y. Furutaka(*TIT*), Y. Futami(*Tsukuba University*), S. Kox, C. Perrin, F. Merchez, D. Rebreyend (*ISN Grenoble*), M. Gai, R. France III, K.I. Hahn, Z. Zhao (*Yale University*), Th. Delbar, P. Lipnik and C. Michotte (*Louvain la Neuve*).

References

1. G. Baur, C.A. Bertulani, H. Rebel: Nucl. Phys. **A458**, 188 (1986)
2. G. Baur and H. Rebel: J. Phys. **G20**, 1 (1994); Ann. Rev. Nucl. and Part. Sci. **46**, 321 (1996)
3. A.C. Shotter, V. Rapp, T. Davinson, D. Brandord, N.E. Sanderson, N.A. Nagarajan: Phys. Rev. Lett. **53**, 1539 (1984); A.C. Shotter, V. Rapp, T. Davinson, D. Brandord: J. Phys. **G14**, L169 (1988)
4. H. Utsunomiya et al.: Phys. Lett. **B211**, 24 (1988); Nucl. Phys. **A511**, 379 (1990); Phys. Rev. Lett **65**, 847 (1990)
5. J. Hesselbarth, S. Khan, Th. Kim, K.T. Knöpfle: Z. Phys. **A331**, 365 (1988); J. Hesselbarth and K.T. Knöpfle: Phys. Rev. Lett. **67**, 2773 (1991)
6. S.B. Gazes, J.E. Mason, R.B. Roberts, S.G. Teichmann: Phys. Rev. Lett. **68**, 150 (1992); J.E. Mason, S.B. Gazes, R.B. Roberts, S.G. Teichmann: Phys. Rev. **C45**, 2870 (1992)
7. T. Motobayashi et al.: Phys. Lett. **B264**, 259 (1991)
8. T. Kubo et al.: Nucl. Instr. Meth. **B70**, 309 (1992)
9. F. Merchez et al.: Nucl. Instr. Meth. **A275**, 133 (1989)
10. P. Decrock et al.: Phys. Rev. Lett. **67**, 808 (1991); Th. Delbar et al.: Phys. Rev. **C48**, 3088 (1993)
11. J. Kiener et al.: Nucl. Phys. **A552**, 66 (1993)
12. T. Motobayashi et al.: Phys. Rev. Lett. **73**, 2680 (1994)
13. N. Iwasa et al.: J. Phys. Soc. Jpn. **65**, 1256 (1996)
14. T. Kikuchi et al.: Phys. Lett. **B391**, 261 (1997)
15. T. Kikuchi et al.: Eur. Phys. J. **A3**, 209 (1998)
16. T. Nishio et al.: RIKEN Accel. Prog. Rep. **29**, 184 (1996)
17. B. Filippone, S.J. Elwyn, C.N. Davids, D.D. Koetke: Phys. Rev. Lett. **50**, 412 (1983); Phys. Rev. **C28**, 2222 (1983)
18. F.J. Vaughn, R.A. Chalmers, D. Kohler, L.F. Chase, Jr.: Phys. Rev. **C2**, 1657 (1970)
19. B. Davids et al.: Phys. Rev. Lett. **81**, 2209 (1988)
20. C.A. Bertulani: Phys. Rev. **C49**, 2688 (1994)
21. S. Typel and G. Baur: Phys. Rev. **C50**, 2104 (1994)
22. N. Iwasa et al.: Phys. Rev. Lett. **83**, 2910 (1999)

Few-Body Systems Suppl. 12, 92–97 (2000)

Few-Body Systems

Crucial Nuclear Reactions of Light Nuclei in Astrophysics

M. Hashimoto[1,*], M. Kamimura[2,†], K. Arai[3,**]

[1] Department of Physics, Kyushu University, Fukuoka, 810-8560, Japan
[2] Department of Physics, Kyushu University, Fukuoka, 812-8581, Japan
[3] Department of Physics, Kumamoto University, Kumamoto, 816-8555, Japan

Abstract. We present several important nuclear reactions of light nuclei which affect crucially the origin of elements, nuclear γ-rays from astronomical objects, stellar evolution scenarios, and the chemical evolution of galaxies. Particular focus on nuclei of $A \lesssim 30$ is given since in this mass region, experiments and/or theories of cluster models concerning nuclear few-body problems have been developed considerably.

1 Primeval Nucleosynthesis

In primordial nucleosynthesis, a standard big-bang model seems to have succeeded in explaining the origin of ^{4}He, D, ^{7}Li. However, recent observations for these elements have been contradictory one another. This problem would be ascribed to some nuclear reaction rates which have caused serious inconsistency with the ^{7}Li observation in low metallicity stars (metallicity is measured usually by the abundance ratio reltive to the solar value, e.g., $[\mathrm{Fe/H}]=\log(\mathrm{Fe/H})_{\mathrm{obs}}$ - $\log(\mathrm{Fe/H})_{\odot}$); this would change the ratio of ^{6}Li/^{7}Li which depends uncertain D$(\alpha,\gamma)^6$Li reaction rate. Fig. 1 shows abundance ratios of ^{6,7}Li, ^{11}B relative to H with the inclusion of neutrino degeneracy (chemical potential μ_i relative to the temperature kT); uncertain ranges due to cross sections of NACRE [1] are illustrated by hatched/band regions; also shown is an observational constraint (region between vertical lines) [2],[3]. Two upper limits deduced from low metallicity stars of [Fe/H] $\simeq -3$ are ^{6}Li/^{7}Li $\leq 0.052 \pm 0.0019$ and B/H

* *E-mail address:* hashi@gemini.rc.kyushu-u.ac.jp
† *E-mail address:* kami2scp@mbox.nc.kyushu-u.ac.jp
** *E-mail address:* arai@aster.sci.kumamoto-u.ac.jp

$\leq (3-6) \times 10^{-12}$; Primordial abundances of ^{6}Li and ^{11}B could be highly affected by neutrino degeneracy through the nucleon to photon ratio η or present baryon density ρ_{m0} in the universe, since it shifts significantly to the higher baryon density side due to the more rapid expansion; Precise determination of cross sections of D$(\alpha, \gamma)^6$Li and ^{7}Li$(\alpha, \gamma)^{11}$B should give an insight into not only primeval abundances of Li-Be-B but the dark baryon density.

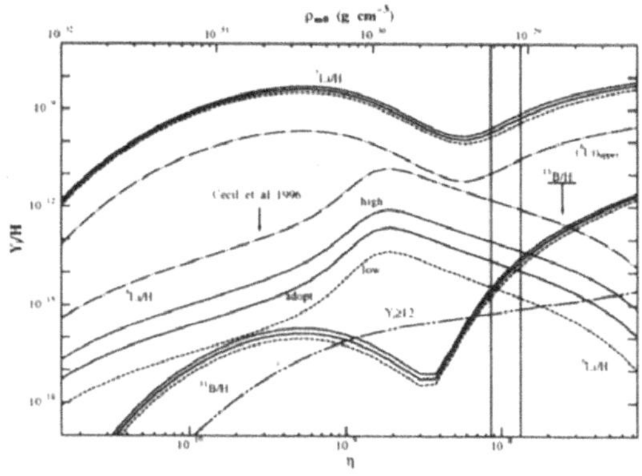

Fig. 1. Primeval abundances of ^{6,7}Li, ^{11}B relative to H with electron and muon ν-degeneracies: $\xi_e = 1.29$ and $\xi_\mu = 20$. Note the critical density of the universe $\rho_c \simeq 10^{-29}$gcm^{-3} for the latest Hubble constant 71 km / (s Mpc).

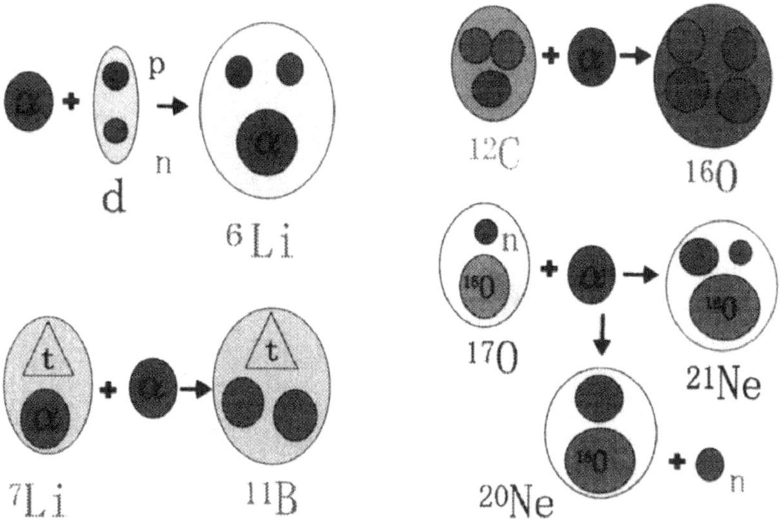

Fig. 2. Cluster model illustration of reactions of D$(\alpha,\gamma)^6$Li and ^{7}Li$(\alpha,\gamma)^{11}$B (left). Those of ^{12}C$(\alpha,\gamma)^{16}$O and ^{17}O$(\alpha,\gamma)^{21}$Ne ^{17}O$(\alpha,$n$)^{20}$Ne (right).

2 Stellar Evolution

Stellar nucleosynthesis is responsible for the production of heavier elements of $A \geq 12$. In the first generation stars, ^{12}C should have been synthesized through triple-alpha reaction. Then, the cross section of $^{12}C(\alpha,\gamma)^{16}O$ reaction which is uncertain by a factor of 2 affects crucially the late stages of stellar evolution. For massive stars of $M \geq 10M_{\odot}$, it controls the final size of Fe-core; explosion mechanism is significantly influenced by the rate. During the stellar evolution, production of ^{12}C and ^{16}O is important; supernovae explosion ejects main elements of $A \leq 40$; it has been recognized that $^{20,21}Ne$, ^{23}Na, ^{27}Al are overproduced appreciably compared to the solar ratio if a lower value of the cross section is used; explosive nucleosynthesis calculations are consistent with NACRE (see e.g., [4]).

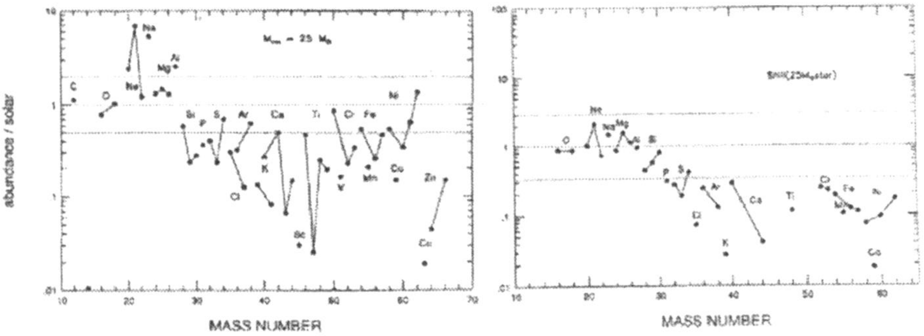

Fig. 3. Overproduction with respect to the solar values; S(300keV) of $^{12}C(\alpha,\gamma)^{16}O$ are 100 (left) and 230 keV·b (right), respectively.

2.1 S-Process

It has been recognized that the s-process nucleosynthesis of $A \lesssim 90$ occurs during the helium burning stage in massive stars where the neutrons are supplied through the reaction $^{22}Ne(\alpha,n)^{25}Mg$ whose rate is still uncertain; its branching ratio to $^{22}Ne(\alpha,\gamma)^{26}Mg$ ranges 1 to 8. To understand the origin of the present amount of the solar s-elements, we need to know the s-process efficiency in low metallicity stars which should suggest abundances in the early stage of star formation. The efficiency in low metallicity massive stars (or 1-st generation stars) has been reduced dramatically due to the newly measured cross section of $^{16}O(n,\gamma)^{17}O$ [5], because during the stage only α-nucleus of $A < 40$ must have been produced appreciable. The neutron recycling is crucial for the s-process efficiency: $^{12}C(n,\gamma)^{13}C(\alpha,n)^{16}O$ is very effective. However it depends on the very uncertain ratio of the reaction rates $^{17}O(\alpha,\gamma)^{21}Ne$ to $^{17}O(\alpha,n)^{20}Ne$ which ranges from 10^{-4} (stars) to 0.1 (squares) as seen in Fig. 3.

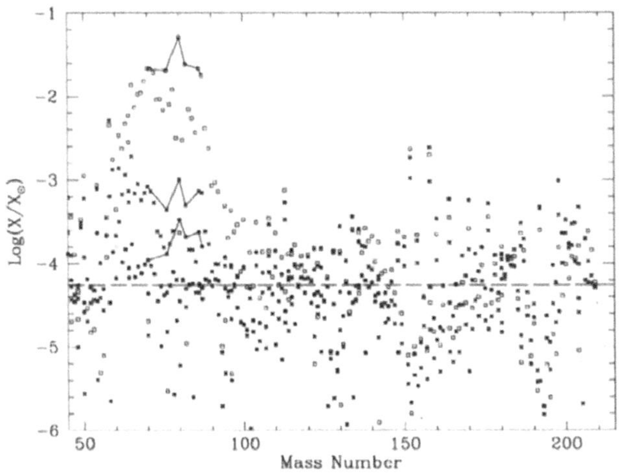

Fig. 4. Overproduction with respect to initial seed of metallicity $Z/Z_{\odot} = 0.001$. Dashed line means seed nuclei. Neutron capture cross sections of ^{16}O : 0.2μb (open squares), 34 μb (filled squares); reduced $^{16}\mathrm{O}(\alpha,\gamma)^{21}\mathrm{Ne}$ rate (stars).

3 X-Ray Bursts: rp-Process

The rapid-proton capture process (rp-process) is known to result from the break out from hot CNO cycle (HCNO) due to high temperature ($T > 3 \times 10^8$ K) environments. Nucleosynthesis up to Ar has been suggested from observations of ONeMg novae. On the other hand, rp-process proceeds beyond ^{56}Ni on the surface of accreting neutron stars; this has been identified as Type I X-ray bursts of which theoretical profile is shown in Fig. 4; the accretion rate: $dM/dt = 3 \times 10^{-9} M_{\odot}/\mathrm{yr}$; the mass and radius of a neutron star: $M = 1.6 M_{\odot}$, and $R = 8$ km; metallicity $Z = 0.02$. Break out begins with either $^{14}\mathrm{O}(\alpha,\mathrm{p})^{17}\mathrm{F}$ and/or $^{15}\mathrm{O}(\alpha,\gamma)^{19}\mathrm{Ne}$ whose rates are very uncertain [6]. Tis causes significant difference in the nuclear energy genertion rate as shown in Fig. 5; it is expected that profiles of X-ray bursts would also be changed. Then the mass/radius relation of neutron stars is modified due to the observation with radius expansion since the black body radius of a neutron star can be in principle related to the neutron star radius.

4 γ-ray Astronomy and ^{26}Al Problem

It has been demonstrated that after the leakage from HCNO, Mg-Al cycle occurs by way of the break out from Ne-Na cycle. In numerical calculations of stellar evolution, Mg-Al cycle should operate during hydrogen burning in asymptotic giant branch stars, massive stars and nova explosion. The production ratio of ^{26}Al could be 1:0.9:0.2, respectively. Produced ^{26}Al must be responsible for the measured 1.8 MeV γ-ray line from the galactic center direction; distribution of this intensity is crucial in constructing the galactic chemical structure. How-

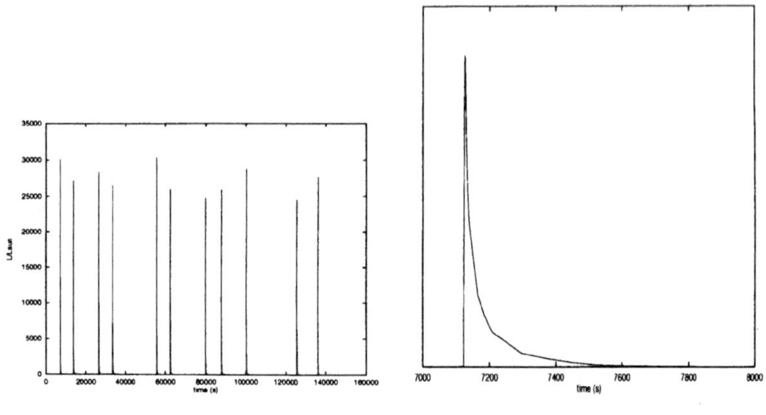

Fig. 5. Successive light curves of X-ray bursts during accretion onto a neutron star (left). The minute profile of the burst in the X-ray bursts (right).

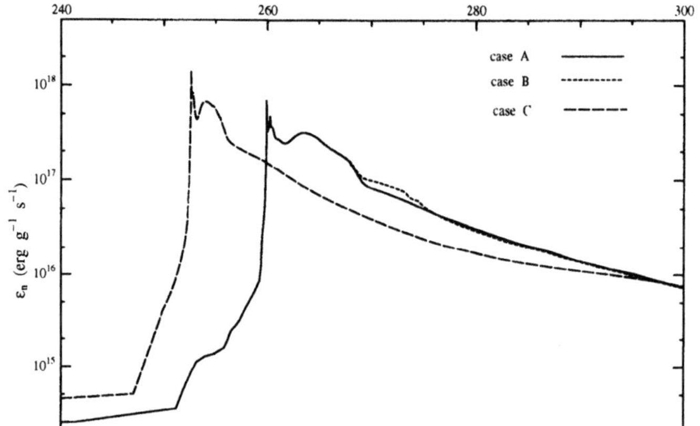

Fig. 6. Comparison of the energy generation rate during the flash. Case A: based on the data of Caltech. Case B: same as case A but different Q-values for waiting nuclei. Case C: based on Thielemann Library [6].

ever, many uncertain reaction rates for the production/destruction of ^{26}Al prohibit reliable prediction in its amount; while in the course of Mg-Al cycle, both the rates of ^{25}Mg(p,γ)^{26}Al and ^{27}Al(p,α)^{24}Mg have been estimated, leakage from the cycle cannot be clear due to unknown rates of ^{25}Al(p,γ)^{26}Si(p,γ)^{27}P and/or ^{26}Al(p,γ)^{27}Si [7].

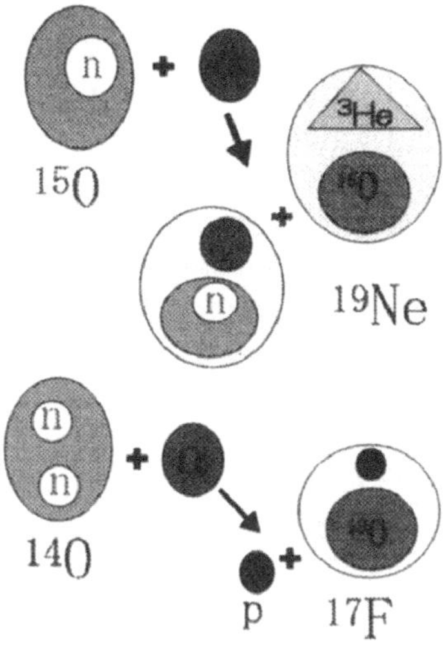

Fig. 7. Cluster model representation of $^{15}O(\alpha,\gamma)^{19}Ne$ (upper) and $^{14}O(\alpha,p)^{17}F$ (low). Neutron holes are illustrated by white disks. Note that ^{19}Ne is considered to be superposition of two nucleus states.

5 Few-body problems in nuclear astrophysics

There are several important nuclear reactions related to nuclear few-body problems. Measurements of cross section σ in astrophysical energy have been often very difficult; a reliable theory could give values of σ with enough accuracy. From cluster models, nuclei with $A < 30$ and $|N - Z| \leq 2$ are required to get reasonable results. We have demonstrated that within this limitation the most crucial reaction rates in astrophysics have not yet been determined. For example, in Fig. 2 and Fig. 6, shown are the typical cluster representaion of a few body reaction crucial in astrophysics. Though investigation of these reactions from the 1-st principle is challenging, it may be difficult to get the matrix elements with plausible nuclear potentials. Nonetheless, during the compilation of NACRE, this has been achived partly as complementary of experimental data [1] though the accuracy of theoretical calculations has not been always guaranteed. Let us hope that both theoretical and experimental approaches would be combined to give bonus to nuclear astrophysics.

References

1. C. Angulo, et al. (the NACRE Collaboration): Nucl. Phys. A, to appear 1999, http://www-astro.ulb.ac.be/
2. S. Burles, D. Kirkman, D. Tytler: Astrophys. J. **519**, 18 (1999)
3. K. Arai, M. Hashimoto, T. Wakita, R. Matsuba: 1999 Annual Meeting by Astronomical Society of Japan, Fukuoka, Oct. 7–9, http://www.tenmon.or.jp/
4. M. Hashimoto: Prog. Theor. Phys. **94**, 663 (1995)
5. M. Rayet, M. Hashimoto: Astron. and Astrophys. submitted 1999
6. O. Koike, M. Hashimoto, K. Arai: Astron. and Astrophys. **342**, 464 (1999)
7. S. Wanajo, M. Hashimoto, K. Nomoto: Astrophys. J. **523**, 409 (1999)

Few-Body Systems Suppl. 12, 98–101 (2000)

Few-Body Systems

Measurements of Radiative Capture Reaction by Recoil Detection in Inverse Kinematics

N. Ikeda*, K. Tsuruta, H. Akiyoshi†, K. Sagara, S. Morinobu

Department of Physics, Kyushu University, 6-10-1 Hakozaki, Fukuoka 812-8581, Japan

Abstract. The analyzing powers of the $^1H(\vec{d},{}^3He)\gamma$ reaction at 17.5 MeV have been measured with enough accuracy to discuss the details of nuclear force. The ^{12}C-α radiative capture reaction of astrophysical interest is planned to be investigated at center-of-mass energies down to 0.7 MeV. Both experiments take advantage of the recoiled particle detection in inverse kinematics to realize the high-efficiency measurement.

1 Introduction

It has been pointed out that some of the radiative capture reactions are of crucial importance in astrophysics. The ^{12}C-α radiative capture is one of the most important astrophysical nuclear reactions. This reaction rate has been suggested to have substantial effects on massive star evolution and the nucleosynthesis [1, 2].

The radiative capture has also been a subject of studies in investigating nuclear interaction. The analyzing powers of the $p + d \to {}^3He + \gamma$ reaction have been expected to provide valuable informations on nuclear force.

On the experimental side, it is most important for the detailed investigation of the radiative capture reactions to realize the high-efficiency measurement, due to their small cross sections. Hitherto, the experimental investigations detecting γ-rays have only been available. Since, however, the γ-ray detection sets a limit to both the intrinsic detection efficiency and the solid angle, the alternative method is desired to be applied to improve the experimental data.

The recoiled particle detection in inverse kinematics is one of the candidates to meet this demand. By detecting the charged particle instead of the γ-ray, the detection efficiency would largely be improved. Further, in the condition of

* *E-mail address:* ikeda@nucl.phys.kyushu-u.ac.jp

† *Present address:* The Institute of Physical and Chemical Research (RIKEN), 2-1 Hirosawa, Wako 351-0198, Japan

the inverse kinematics with heavier projectile and lighter target, the reaction products are emitted into a small cone centering at 0 degree. Both the charged particle detection and concentration of the products in small solid angle ensure the high-efficiency measurement of the reaction events.

To apply this method, a new recoil mass separator which is the hybrid system of the magnetic and electrostatic devices has been developed at Kyushu University tandem accelerator laboratory (KUTL). This separator is designed to achieve a highest possible capability of eliminating a primary beam and background particles associated with the primary beam. In the case that the charge of the recoil ions to be detected is larger than the atomic number of the incident particles, separation of the reaction products from the primary beam can also be realized by only using the magnetic devices.

In this paper, the measurement of the $^1H(\vec{d},^3He)\gamma$ reaction recently carried out at KUTL and a plan of the measurement of the ^{12}C-α radiative capture cross sections using the Kyushu University mass separator are briefly reported.

2 *p*-*d* Radiative Capture

A polarized deuteron beam of 17.5 MeV was focused on the hydrogen gas target which has carbon foil windows. The recoiled ^3He particles were analyzed by a 90°-bending magnet and detected by a silicon detector having 12 vertical strips. The background particles other than ^3He were largely reduced by analyzing the primary beam using the 4°-bending magnet placed at upstream of the target. The angular distributions of the analyzing powers A_{xx}, A_{yy}, A_{zz} and A_y of the $^1H(\vec{d},^3He)\gamma$ reaction in wide center-of-mass angular range were obtained with much higher accuracy than the γ-ray measurements [3, 4].

The experimental angular distributions of the analyzing powers were compared with the calculations using various nucleon-nucleon potentials. It was found that none of the calculations could reproduce all kinds of the experimental analyzing powers simultaneously. The details of the experiment and the theoretical analysis are given in [5].

Recently, Kamada has pointed out [6] that his calculation introducing three-nucleon force (3NF) largely improves the reproducibility of the present analyzing powers. The effect of the 3NF has been expected to become more significant as increasing the incident energy. To investigate the 3NF in more detail, the analyzing powers of the $^1H(\vec{d},^3He)\gamma$ reaction at $E_d = 200$ MeV have been planned to be measured at Research Center for Nuclear Physics, Osaka University.

3 ^{12}C-α Radiative Capture

The astrophysical S-factor of the ^{12}C-α radiative capture is required to be determined at most effective stellar energy of 0.3 MeV. The direct determination of the S-factor at 0.3 MeV ($S(0.3)$) would impracticably be difficult due to the fact that the cross section is only on an order of 10^{-8} nb. At present, the experimental S-factors with sufficient statistics have only been available at center-of-mass energies ($E_{c.m.}$) higher than 1.4 MeV [7, 8].

The theoretical extrapolations using the cross sections of $^{12}C(\alpha,\gamma)^{16}O$, the differential cross sections of the $\alpha + {}^{12}C$ elastic scattering and the energy distribution of the β-delayed α-particles from ^{16}N have been attempted to estimate $S(0.3)$. The recent measurements of the β-delayed α spectrum extended to low energies have been pointed out to largely improve the reliability of the extrapolations [9, 10]. However, it has been suggested [11, 12] that the deduced $S(0.3)$ value may not be definitive unless the precise radiative capture cross sections below 1.4 MeV are included in the analysis.

In order to facilitate observations of low cross section events from the reactions involving H or He in the incident channel, the recoil mass separator together with a new type windowless gas target [13] has been developed at KUTL. A new beam production method of accel-decel strong focusing [14] and a windowless gas stripper have also been developed to derive low energy heavy-ion beams with an intensity of an order of 10 pμA from the tandem accelerator.

The layout of the mass separator is shown in fig. 1. The deflection angle and the electrode gap of the electrostatic deflector (ED) were so determined as to prevent the incident particles from causing additional background by hitting the electrode surface. Beams are analyzed by their velocity at the intermediate focus point (F1) and by their charge-to-mass ratio at the system exit (F2). This feature having two focus points with the aim of effectively eliminating the beam-like particles is unique to the present mass separator. A performance of the separator has been tested to be in reasonable agreement with the prediction of the ion-optical calculation.

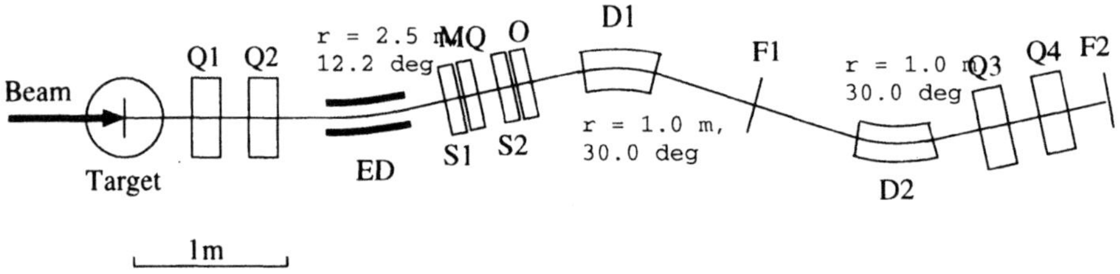

Figure 1. Layout of the new recoil mass separator at Kyushu University tandem accelerator laboratory. Notations: Q1 – Q4, MQ; quadrupole magnet, S1, S2; sextupole magnet. O; octupole magnet, ED; electrostatic deflector, D1, D2; dipole magnet, F1, F2; focus point.

The blow-in method has successfully been applied to the windowless gas target to achieve the high performance in the target thickness, gas confinement and prevention of gas consumption [13]. At present, the maximal target thickness has been achieved to be 10 Torr $\times$ 3 cm. It would be improved to be more than 25 Torr $\times$ 3 cm by cooling ^{4}He gas down to about $-100°$C.

The instruments and methods mentioned above are expected to allow detection of more than 100 ions of ^{16}O with the equilibrium charge state from the $^4He(^{12}C,^{16}O)\gamma$ reaction with the cross section of 1 pb in a month of beam time. The astrophysical S-factors of the ^{12}C-α radiative capture down to $E_{c.m.}$ = 0.7 MeV are therefore estimated to be determined within an accuracy of 10% unless the background particles interfere.

4 Summary

The experimental investigations of the radiative capture reactions taking advantage of the recoil detection in the inverse kinematics have been in progress at KUTL. The $^{1}H(\vec{d},^{3}He)\gamma$ reaction at 17.5 MeV has been studied by detecting full-stripped ^{3}He ions separated from deuterons by the magnetic separator. The angular distributions of the analyzing powers were obtained in wide angular range with enough accuracy to discuss the details of the nuclear force.

The $^{4}He(^{12}C,^{16}O)\gamma$ reaction of astrophysical interest is planned to be investigated at center-of-mass energies down to 0.7 MeV. New equipments have been developed at KUTL aiming to realize the unambiguous experiment to determine the cross sections on an order of 1 pb. The experiment will soon be started.

References

1. W.A. Fowler: Rev. Mod. Phys. **56**, 149 (1984)
2. M. Hashimoto: Prog. Theor. Phys. **94**, 663 (1993)
3. J. Jourdan et al.: Nucl. Phys. **A453**, 220 (1986)
4. G.J. Schmid et al.: Phys. Rev. **C53**, 35 (1996)
5. H. Akiyoshi: Ph.D. Thesis, Kyushu University 1997 (unpublished)
6. H. Kamada: Private Communication
7. A. Redder et al.: Nucl. Phys. **A462**, 385 (1987)
8. J.M.L. Ouellet et al.: Phys. Rev. **C54**, 1982 (1996)
9. R.E. Azuma et al.: Phys. Rev. **C50**, 1194 (1994)
10. Z. Zhao et al.: Phys. Rev. Lett. **70**, 2066 (1993)
11. F.C. Barker: Phys. Rev. **C50**, 2244 (1994)
12. G.M. Hale: Nucl. Phys. **A621**, 177c (1997)
13. K. Sagara et al.: Nucl. Instr. and Meth. **A378**, 392 (1996)
14. K. Sagara et al.: Proc. of the 12th Symp. on Accelerator Science and Technology, Wako, 1999, p. 63

Few-Body Systems Suppl. 12, 102–107 (2000)

Few-Body Systems

Asymptotic Normalization Coefficients and Astrophysical Factors

A.M. Mukhamedzhanov[1], A. Azhari[1], V. Burjan[2], F. Carstoiu[3], H.L. Clark[1], C.A. Gagliardi[1], V. Kroha[2], Y.-W. Lui[1], A. Sattarov[1], L. Trache[1], R.E. Tribble[1]

[1] Cyclotron Institute, Texas A&M University, College Station, Texas 77843, USA

[2] Institute for Nuclear Physics, Prague-Řež, Czech Republic

[3] Institute of Atomic Physics, Bucharest, Romania

Abstract. The S factor for the direct capture reaction ^{7}Be$(p,\gamma)^8$B can be found at astrophysical energies from the asymptotic normalization coefficients (ANC's) which provide the normalization of the tails of the overlap functions for ^{8}B $\rightarrow$ ^{7}Be + p. Peripheral transfer reactions offer a technique to determine these ANC's. Using this technique, the ^{10}B(^{7}Be,^{8}B)^{9}Be and ^{14}N(^{7}Be,^{8}B)^{13}C reactions have been used to measure the asymptotic normalization coefficient for ^{7}Be$(p,\gamma)^8$B. These results provide an indirect determination of $S_{17}(0)$. Analysis of the existing ^{9}Be$(p,\gamma)^{10}$B experimental data withtin the framework of the R-matrix method demonstrates that experimentally measured ANC's can provide a reasonable determination of direct radiative capture rates.

1 Introduction

Nuclear capture reactions such as (p,γ) and (α,γ), play a major role in defining our universe. Until recently, the only reliable method to determine a reaction rate that is dominated by direct capture has been to measure it at laboratory energies with a low energy particle beam and extrapolate the result to astrophysical energies. Often the reaction of interest involves a radioactive target which makes measurements quite difficult. Hence it is important to develop alternative techniques to determine rates. Direct capture reactions of astrophysical interest usually involve systems where the binding energy of the captured proton is low. Hence at stellar energies, the capture proceeds through the tail of the overlap function of the bound states wave functions of the colliding and final nuclei. The shape of the overlap integral in this tail region is completely determined by the Coulomb interaction, so its amplitude, the so-called asymptotic nrmalization coefficient (ANC), alone dictates the rate of the capture reaction. The advantage of the ANC approach is that it provides a method to determine direct capture S factor at zero energy from measurements of nuclear reactions, such as peripheral nucleon transfer reactions, which have cross sections orders of magnitude larger than the direct capture reactions themselves.

The ^{7}Be$(p,\gamma)^8$B reaction is an excellent example of such a direct capture process where new measurements are still needed, as was underscored in a recent review of stellar reaction rates [1] which includes a detailed discussion of the uncertainties in our present knowledge of $S_{17}(0)$ and its importance to the solar neutrino problem. Below we describe a new indirect technique for these measurements.

The asymptotic normalization coefficient (ANC) C^B_{Ap} for A + $p \leftrightarrow$B specifies the amplitude of the radial overlap function $I_{lj}(r)$ of the bound state wave functions of nuclei B, A and p when the core A and the proton are separated by a distance large compared to the nuclear interaction radius R_N:

$$I^B_{Aplj} \stackrel{r \geq R_N}{\approx} C^B_{Ap} \frac{W_{-\eta,l+1/2}(2\kappa r)}{r}. \tag{1}$$

Here, W is the Whittaker function, l and j are the relative angular orbital momentum of A and p in the bound state A + p=B and the total angular momentum of the proton, κ is the bound state A + p=B wave number, η is the Coulomb parameter of the bound state A + p=B. Simultaneously the same ANC defines the residue of the T matrix for the A +p elastic scattering in the pole in the energy plane corresponding to the bound state A + p=B.

In previous communication [2], we have pointed out that astrophysical S factors for peripheral direct radiative capture reactions can be determined through measurements of ANC's using traditional nuclear reactions such as peripheral nucleon transfer. Of course it is extremely important to test the reliability of the technique in order to know the precision with which it can be applied. Determining the S factors for ^{16}O$(p,\gamma)^{17}$F from its ANC's has been recognized as a suitable test for this method [1] because the results can be compared to existing direct measurements of the cross sections [3, 4]. Furthermore, the ^{16}O$(p,\gamma)^{17}$F reaction has substantial similarities to the ^{7}Be$(p,\gamma)^8$B reaction. In addition, we analyse within the framework of the R matrix the S-factor for ^{9}Be$(p,\gamma)^{10}$B using the ANC's determined from ^{9}Be(^{10}B,^{9}Be)^{10}B reaction to calculate the direct capture.

A Test Case

For a peripheral transfer reaction, ANC's are extracted from the angular distributions by comparison to a DWBA calculation. Consider the proton transfer reaction a + A $\rightarrow$ b + B, where $a = b + p$ and B= A + p. The experimental cross section is related to the DWBA according to

$$\frac{d\sigma}{d\Omega} = \sum (C^B_{Ap})^2 (C^a_{bp})^2 \frac{\tilde{\sigma}^{DW}}{b^2_{Ap} b^2_{bp}}. \tag{2}$$

Here, $\tilde{\sigma}$ is the calculated DWBA cross section and the b's are the asymptotic normalization constants for the single particle orbitals used in the DWBA. The sum in Eq. (2) is taken over the allowed angular momentum couplings, and the C's are the ANC's for B $\rightarrow$ A + p and $a \rightarrow c + p$. The normalization of the DWBA cross section by the ANC's for the single particle orbitals makes the extraction of the ANC for B $\rightarrow$ A + p essentially independent of the parameters used in the single particle potential wells. See [5] for additional details.

The relation of the ANC's to the direct capture rate at low energies is straightforward. Allowing for (1) allows one to present the cross section for the direct capture reaction A + $p \rightarrow$ B + γ as

$$\sigma \approx \lambda \left| \langle C \frac{W_{-\eta,l+1/2}(2\kappa r)}{r} \mid \hat{O}(\mathbf{r}) \mid \psi_i^{(+)}(\mathbf{r}) \rangle \right|^2 , \tag{3}$$

where λ contains kinematical factors, $\hat{O}$ is the electromagnetic transition operator, $\psi_i^{(+)}$ is the incident scattering wave. This equation is valid if the dominant contribution to the matrix element comes from outside the nuclear radius. In this case, the ANC's defined in Eq. (2) determine the capture cross section and thus the S factor.

ANC's for ^{17}F $\rightarrow$ ^{16}O + p, corresponding to ground and first excited state transitions, were obtained from the $^{16}O(^3He,d)^{17}F$ reaction which we measured at 29.75 MeV. Using the ANC's, the S factors describing ^{16}O$(p,\gamma)^{17}$F to both states were calculated, with no additional normalization constants. The agreement between the measured S-factors and those calculated from our ^{17}F $\rightarrow$ ^{16}O + p ANC's is quite good indicating that the technique works to better than 9%. Details of the experiment and S-factor determination can be found in [6].

S factor for ^{7}Be$(p,\gamma)^8$B from reactions ^{10}B(^{7}Be,^{8}B)^{9}Be and ^{14}N(^{7}Be,^{8}B)^{13}C

We have measured the (^{7}Be,^{8}B) reaction on ^{10}B and ^{14}N targets in order to extract the ANC for ^{8}B $\rightarrow$ ^{7}Be + p. The radioactive ^{7}Be beam was produced at 12 MeV/A by filtering reaction products from the ^{1}H(^{7}Li,^{7}Be)n reaction in the recoil spectrometer MARS. The reaction was initiated by a primary ^{7}Li beam at 18.6 MeV/A from the TAMU K500 cyclotron incident on an H_2 cryogenic gas target cooled by LN_2. Reaction products were measured by 5 cm $\times$ 5 cm Si detector telescopes consisting of a 100 μm ΔE strip detector, with 16 position sensitive strips, followed by a 1000 μm E counter. A single 1000 μm Si strip detector was used for initial beam tuning. This detector, which was inserted at the target location, allowed us to optimize the beam shape and to normalize the ^{7}Be flux relative to a Faraday cup that measured the intensity of the primary ^{7}Li beam. The angular distributions for the transfer reactions are shown in Fig. 1 along with DWBA predictions from the full finite range code PTOLEMY. Only the dominant components in the transfer reactions are shown, but all are included in the fits. Details can be found in [7, 8]. The ANC for $^8B \to {}^7Be + p$ was extracted based on the fits to the angular distributions and the ANC's for $^{10}B \to {}^9Be + p$ and $^{14}N \to {}^{13}C + p$. These ANC's represent one of the major uncertainties (9% for ^{10}B and 6.4% for ^{14}N) in the extraction of the ANC for $^8B \to {}^7Be + p$. The other major uncertainty in the extracted ANC is due to the optical model parameters (10%) for both targets. The S factor for $^7Be\ (p,\gamma)^8B$ has been determined from the ANC's. The values that we find for $S_{17}(0)$ are 18.4 ± 2.5 eV·b and 16.6 ± 1.9 eV·b for the ^{10}B and ^{14}N targets, respectively. Both values are in good agreement with the recommended value of 19^{+4}_{-2} eV· b [1].

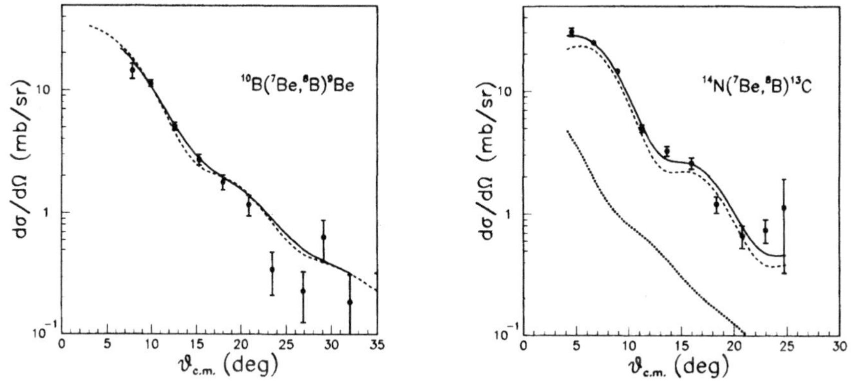

Figure 1. Angular distributions for the transfer reactions. Only the dominant components in the reaction on the ^{14}N target have been shown. DWBA calculations have been smoothed to match the angular resolution in the data.

Astrophysical factor for $^9Be(p,\gamma)^{10}B$

The reaction $^9Be(p,\gamma)^{10}B$ plays an impotant role in primordial nucleosynthesis of light elements in the p shell. There is one previous measurement of the $S(E)$ for $^9Be(p,\gamma)^{10}B$ at low energies [9] and measurement of the analyzing power [10]. The energy dependence of this capture reaction over the range important to nuclear astrophysics is quite complex because it includes contributions from direct capture and several resonances [9] and they interfere with each other. Analysis of the behavior of $S(E)$ has been performed in [9, 10]. However, all these analyses have been done with some inconsistencies. We reanalyse the $^9Be(p,\gamma)^{10}B$ experimental data within the framework of the R-matrix method. The direct capture part of the S factor is calculated using the experimentally measured ANC's for $^{10}B \to {}^9Be + p$. The simple relationship between ANC's and direct capture rates that exist for $^7Be(p,\gamma)^8B$ and $^{16}O(p,\gamma)^{17}F$ is not applicable to direct capture in $^9Be(p,\gamma)^{10}B$ because of the relatively tight binding energy of the last proton in $^{10}B(\varepsilon = 6.586\,MeV)$. Nonetheless, it is interesting to investigate how well one can predict the $^9Be(p,\gamma)^{10}B$ direct capture rate from the measured ANC's, in order to test the relationship between them under adverse circumstancies. The R-matrix approach relates a direct capture rate to a radial integral which is taken from the channel radius to ∞, so only the peripheral part of the channel overlap function is needed. The absolute normalization of the peripheral part of this overlap function is specified by the corresponding ANC. When calculating the S factor we took into account all for low lying resonances of ^{10}B with the parameters taken or close to those given in [11]. The result of the fit is shown in Fig. 2. Details and parameters of the fit are presented in [?].

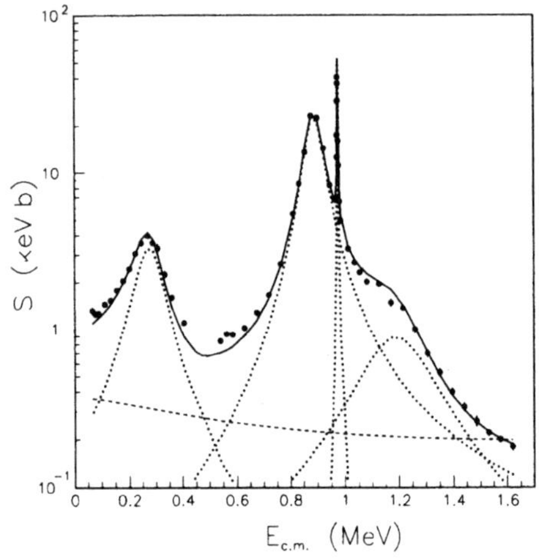

Figure 2. The S(E) factor for the reaction $^9Be(p,\gamma)^{10}B$. The points are experimental data from [9]. The solid line is our fit. The dashed line is our calculated contribution for the direct radiative capture and dotted lines show our calculated contribution for the resonances.

Acknowledgement. This work was supported in part by the U.S. Department of Energy under Grant number DE-FG03-93ER40773 and by the Robert A. Welch Foundation.

References

1. E.G. Adelberger et al.: Rev. Mod. Phys. **70(4)**, 1265 (1998)
2. H.M. Xu et al.: Phys. Rev. Lett. **73**, 2027 (1994)
3. R. Morlock et al.: Phys. Rev. Lett. **79**, 3837 (1997)
4. H.C. Chow, G.M. Griffith, T.H. Hall: Can. J. Phys. **53**, 1672 (1975)

5. A.M. Mukhamedzhanov et al.: Phys. Rev. **C56**, 1302 (1997)

6. C.A. Gagliardi et al.: Phys. Rev. Lett. **C59**, 1149 (1999)

7. A. Azhari et al.: Phys. Rev. Lett. **82**, 3960 (1999)

8. A. Azhari et al.: Phys. Rev. C (in press)

9. D. Zahnow et al.: Nucl. Phys. **A589**, 95 (1995)

10. E.A. Wulf et al.: Phys. Rev. **C58**, 517 (1998)

11. F. Ajzenberg-Selove: Nucl. Phys. **A490**, 1 (1988)

Few-Body Systems Suppl. 12, 108–113 (2000)

Few-Body Systems

Superfluidity of Λ-, Σ^--, Ξ^--Hyperons in Neutron Star Cores

T. Takatsuka[1,*], S. Nishizaki[1,†], Y. Yamamoto[2,**], R. Tamagaki[3,††]

[1] Faculty of Humanities and Social Sciences, Iwate University, Morioka 020-8550, Japan

[2] Physics Section, Tsuru University, Tsuru 402-0054, Japan

[3] Kamitakano Maeda-Cho 26-5, Sakyo-ku, Kyoto 606-0097, Japan

Abstract. Superfluidity of Λ-hyperons admixed in neutron star cores is discussed on the basis of a realistic approach to use the $\Lambda\Lambda{}^1S_0$-pairing interaction tested by the data of double Λ hypernuclei and to take account of the density-dependence of Λ effective-mass. It is found that Λ-superfluid can be realized, though in a restricted density region. Also the study extended to Σ^- and Ξ^- hyperons strongly shows the possibility of their superfluidities. These results suggest that the so-called "hyperon cooling scenario" would be one of the candidates for a rapid cooling of neutron stars compatible with observations.

1 Introduction

It is very likely that various hyperons (Y) join in the constituents of neutron star matter as the density (ρ) increases toward the center of neutron stars. The studies [1, 2, 3, 4, 5, 6, 7] so far done on this aspect show that this Y-mixing begins to occur at not so high-density although the population $y_Y(\rho)$ depends strongly on the many-body approach and the interactions assumed; for example, the threshold densities are estimated as $\rho_t(\Lambda) \simeq \rho_t(\Sigma^-) \sim 2\rho$ and $\rho_t(\Xi^-) \sim (3-4)\rho_0(\rho_0 \equiv 0.17$ nucleons/fm$^3 \simeq 2.8 \times 10^{14}$ g/cc being the nuclear density) which are well below the central density $\rho_c \sim (5-10)\rho_0$ of typical neutron stars. This means that the inclusion of Y-effects is essentially important in discussing the properties of dense neutron star cores.

In this paper, we address whether hyperons admixed in neutron star cores could be superfluid or not [8]. This problem is of particular interest in relation

*E-mail: takatuka@iwate-u.ac.jp
†E-mail: nisizaki@iwate-u.ac.jp
**E-mail: yamamoto@tsuru.ac.jp
††E-mail: tamagaki@yukawa.kyoto-u.ac.jp

to the cooling of neutron stars. It is because the surface temparature observations of some neutron stars suggest the necessity of a rapid cooling mechanism such as the one due to the hyperon-mixing and at the same time the hyperon superfluidity controlling its "too rapid" cooling [9, 10, 11].

First we focus attention on the Λ case since the uncertainties in the $\Lambda\Lambda$ interaction are relatively small compared to those in the $\Sigma^-\Sigma^-$ and $\Xi^-\Xi^-$ ones, due to the check using the data from double Λ hypernuclei $({}^{10}_{\Lambda\Lambda}Be, {}^{13}_{\Lambda\Lambda}B)$. Then we extend our study to Σ^- and Ξ^- cases by applying the Baryon-Baryon interaction constrained by the SU(3) symmetry and discuss their possible superfluidities in reference to the Λ superfluidity.

2 Outline of Approach

Although hyperons appear at high-densities ($\rho \gtrsim 2\rho_0$), the Fermi energies of respective hyperons are rather low because their contaminations are not so large, at most $y_Y \sim (10-25)\%$. This means that the pairing interaction responsible for superfluidity should be the one in the ${}^1S_0$ pair state which is most attractive at low scattering energies. Then our main concern is to treat the energy gap equation of a well-known ${}^1S_0$-type [12].

First we discuss the Λ-case. To be realistic, we pay attention to the points: [A] We use a realistic "bare" $\Lambda\Lambda$ potential $V_{\Lambda\Lambda}({}^1S_0)$ for the pairing interaction, since to use an "effective" one $\tilde{V}_{\Lambda\Lambda}({}^1S_0)$ based on the G-matrix, as in Balberg-Balnea's work [13], is not justified, leading to too large energy gaps. In practice, we choose two OBEP-type $V_{\Lambda\Lambda}({}^1S_0)$. One (abbreviated here to as ND-Soft) is a soft-core version of the Nijmegen-D hard core potential [14] (ND) and expressed by the gaussian superposition,

$$V_{\Lambda\Lambda}({}^1S_0) = \Sigma_{i=1}^{3}\{V_i^c + \boldsymbol{\sigma}_1 \cdot \boldsymbol{\sigma}_2 V_i^{ss}\}e^{-(r/\beta_i)^2} \tag{1}$$

with $V_i^c = \{-21.337, -187.01, 10853.3\}$ MeV and $V_i^{ss} = \{0.19321, 32.166, 2035.1\}$ MeV for $\beta_i = \{1.342, 0.777, 0.35\}$ fm, which is constructed so as to reproduce the t-matrix from the original ND potential. The other (Ehime) is the $\Lambda\Lambda$ potential from Ehime group [15], which is based on the framework of nonet mesons plus SU(3) symmetry and has a soft repulsive core and the velocity-dependent short-range interaction. The ND-Soft potential has a stronger short-range repulsion and a deeper medium-range attraction as compared to the Ehime potential. Both of two potentials have been tested by the double Λ hypernuclei data and are expected to cover the present uncertainties in the $\Lambda\Lambda$ ${}^1S_0$ interaction.

[B] We take account of the ρ- and y_Y- dependences of the effective mass parameter m^*_Λ of Λ. Our m^*_Λ is derived from the G-matrix calculations performed for $\{n+\Lambda\}$- matter with the Λ fraction $y_\Lambda (\equiv \rho_\Lambda/\rho)$ and by using the ND potential. The resulting m^*_Λ are large (~ 0.8) in contrast with those of protons admixed in usual neutron star matter, e.g., $m^*_p \simeq (0.7 \to 0.5)$ for $\rho = (1 \to 3)\rho_0$. This suggests that as far as m^*_Λ is concerned, the Λ-superfluidity is more likely than the p-superfluidity which has been shown to disappear at high-densities $\rho > 3\rho_0$.

For the Σ^- and Ξ^- cases, we treat the $^1S_0$ gap equation similarly to the Λ case. However, unfortunately our present knowledge of $\Sigma^-\Sigma^-$ and $\Xi^-\Xi^-$ interactions is less certain compared to $\Lambda\Lambda$ one since we lack the double hypernuclei data to check these interaction directly. Here we are content to using, in the light of SU(3) symmetry, the $^1S_0$ $\Sigma^-\Sigma^-$ and $\Xi^-\Xi^-$ pairing interactions from the ND-Soft and the Ehime potentials for which the $\Lambda\Lambda$ case has been checked by hypernuclear data. Here the $^1S_0$ $\Sigma^-\Sigma^-$ and $\Xi^-\Xi^-$ potentials corresponding to the ND-Soft are constructed similarly to the $\Lambda\Lambda$ case, but in a simple manner to multiply $v_i^{(\gamma)}(\gamma \equiv c, ss)$ in Eq.(1) by the factors C_a and C_r; $v_i^{(\gamma)} \to C_a v_i^{(\gamma)}$ for $i = 1, 2$ and $v_i^{(\gamma)} \to C_r V_i^{(\gamma)}$ for $i = 3$ with $C_a = 1.37(0.88)$ and $C_r = 1.85(1.08)$ for $\Sigma^-\Sigma^-(\Xi^-\Xi^-)$. In the next section, the results for Σ^-- and Ξ^-- superfluidities are shown by focusing attention on a comparison with Λ case for the same m_Y^* and y_Y parameters.

3 Results and Discussion

First we discuss the Λ-superfluidity.
(i) Fig.1 shows the critical temperature given by $T_c \simeq 0.66\Delta \times 10^{10}$ K with the energy gap Δ in MeV, calculated for the ND-Soft potential. We see a sensitivity of T_c to m_Λ^* (indicated by an arrow), i.e., a remarkable reduction of T_c from the simple case setting $m_\Lambda^* = 1$ to the case taking account of the ρ- dependence of m_Λ^* (numbers attatched to $y_\Lambda = 5\%$ curve, as an example). Also we observe a notable ρ- and y_Λ- dependences of T_c. This feature indicates that the Λ- superfluidity disappears ($T_c < T_i$) for large y_Λ and/or ρ since the internal temperature T_i is around 10^8 K.
(ii) We obtain similar results for the Ehime potential, except that the T_c is considerably larger for the same parameters, ρ, y_Λ and m_Λ^*, indicating a wider density-region of superfluidity. This comes mainly from the fact that the effect of short-range repulsion growing with $\rho_\Lambda (= y_\Lambda \rho)$ is relatively small for the Ehime potential due to the weaker repulsive core in $V_{\Lambda\Lambda}(^1S_0)$.
(iii) A $y_\Lambda - \rho$ relation (namely, a hyperon core model), together with the ρ- and y_Λ- dependent T_c, determines the density-region for the existence of superfluidity. For example, if we take the hyperon core model from Pandaripande [1], where y_Λ increases with increasing ρ and has a maximum value of about 10%, the region is obtained as $\rho \simeq (2\rho_0 - \rho_d)$ with $\rho_d \simeq 3.7(4.6)\rho_0$ for the ND-Soft (the Ehime) potential.
(iv) How much portion of neutron star interiors is occupied by the superfluid depends on the mass M since ρ_c depends on M. It is of great interest to note that for neutron stars with larger M a non-superfluid region takes part in their core, removing a mechanism necessary to suppress "too rapid" hyperon cooling. This suggests that neutron stars compatible with hyperon cooling scenario would not so massive, for example, $M \lesssim 1.5 M_\odot$ in the case of the Λ- superfluid from the ND-Soft potential and the neutron star models from the BJ-1H equation of state [16].

The T_c for Σ^- and Ξ^- hyperons are shown in Fig.2 in comparison with

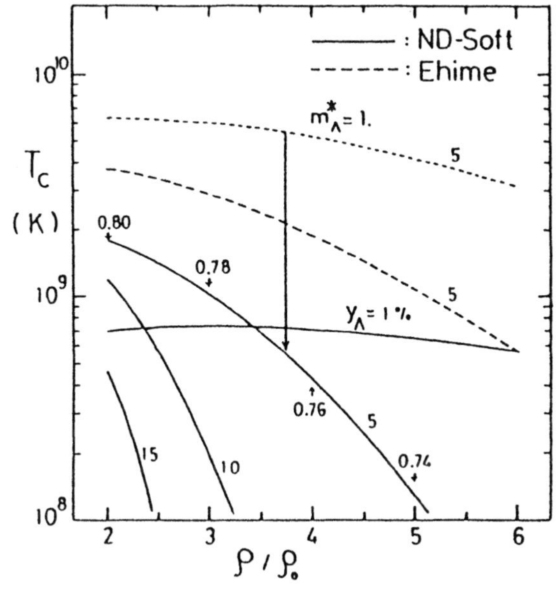

Figure 1. Density ρ-dependence of the critical temperature T_c for the Λ-superfluidity for several mixing ratio y_Λ. Attatched numbers are the effective mass parameter m^*_Λ shown for reference. ρ_0 denotes the nuclear density.

those of Λ, by fixing parameters as $m^*_Y = 1.0$ and $y_Y = 5\%$.
(v) In the case of the Ehime potential, the T_c both for Σ^- and Ξ^- are much larger than that of Λ. For the ND-Soft potential, the T_c for Σ^- is in a similar situation as the Ehime, but the T_c for Ξ^- is somewhat smaller than that of Λ. However, our preliminary results of m^*_Y in medium, calculated by the use of the ND potential, strongly suggest that $m^*_{\Sigma^-}$ and $m^*_{\Xi^-}$ are by far larger than m^*_Λ; $m^*_{\Sigma^-} \gtrsim 1$ and $m^*_{\Xi^-} \gtrsim 1$ whereas $m^*_\Lambda \sim 0.8$, and therefore there is a strong possibility of $T_c(\Xi^-) > T_c(\Lambda)$ also for the ND-Soft $V_{\Xi^-\Xi^-}(^1S_0)$. From these results, we can say that both of the Σ^-- and Ξ^-- superfluids would be realized with the T_c comporable to or larger than that of Λ-superfluid, although a conclusion in more quantitative level has to await the results taking account of the ρ- and y_Y- dependences of m^*_Y.

4 Conclusion

The Λ-superfluidity has been investigated by a realistic approach and found to occur in neutron star cores, though in a restricted density-region. This restriction interestingly limits the mass of a neutron star to which the hyperon cooling scenario is applicable. Also it has been shown by an extended study that Σ^- and Ξ^- hyperons would be more likely in a superfluid state than Λ ones, due to the stronger attraction in the pairing interactions and /or the larger m^*_Y. Thus

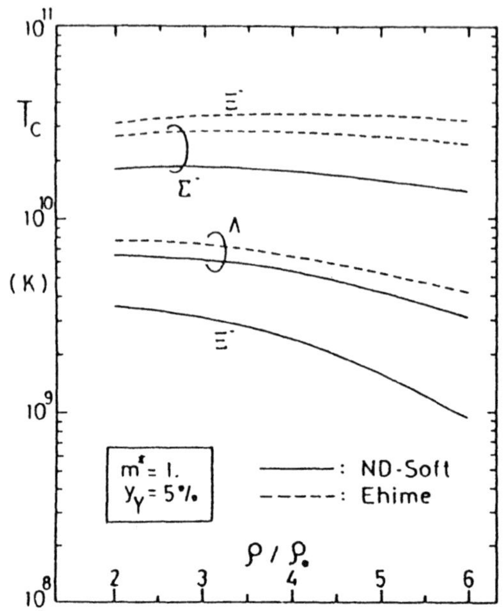

Figure 2. T_c of Σ^-- and Ξ^-- superfluidities in comparison with that of Λ- one.

we can say that the hyperon cooling scenario would be one of the candidates for a rapid cooling scenario compatible with observations. Finally we wish to remark that the Λ- superfluidity is realized by the $\Lambda\Lambda$ interaction attractive enough to account for the binding energy of double Λ hypernuclei.

Acknowledgments

This work is indebted to a Grant-in-Aid for Scientific Research from the Ministry of Education, Science, Sports and Culture (No.11640245).

References

1. V.R. Pandharipande: Nucl. Phys. **A178**, 123 (1971)

2. N.K. Glendenning: Phys. Lett. **B114**, 392 (1982); Nucl. Phys. **A493**, 521 (1989)

3. F. Weber and M.K. Weigel: Nucl. Phys. **A505**, 779 (1989)

4. Y. Sugahara and H. Toki: Prog. Theor. Phys. **92**, 803 (1994)

5. F. Schaffner and I.N. Mishustin: Phys. Rev. **C53**, 1416 (1996)

6. S. Balberg and A. Gal: Nucl. Phys. **A265**, 435 (1997)

7. M. Baldo, G.F. Burgio, H.J. Schultz: Phys. Rev. **C58**, 3688 (1998)

8. T. Takatsuka and R. Tamagaki: Proc. Int. Sym. on *Physics of Hadrons and Nuclei*, Tokyo, Dec.14-17, 1998, ed. Y. Akaishi, O. Morimatsu and M. Oka (to appear in Nucl. Phys. A); Prog. Theor. Phys. **102**, No.5 (1999)

9. S. Tsuruta: Phys. Reports **292**, 1 (1998)

10. D. Page: Astrophys. J. **428**, 250 (1994)

11. Madappa Prakash, Manju Prakash, J.M. Lattimer, C.J. Pethick: Astrophys. J. **390**, L77 (1992)

12. T. Takatsuka and R. Tamagaki: Prog. Theor. Phys. Suppl. **No.112**, 27 (1993)

13. S. Balberg and N. Barnea: Phys. Rev. **C57**, 409 (1998)

14. M.M. Nagels, T.A. Rijken and J.J. de Swart: Phys. Rev. **D15**, 2547 (1977)

15. T. Ueda, K. Tominaga, M. Yamaguchi, N. Kijima, D. Okamoto, K. Miyagawa, T. Yamada: Prog. Theor. Phys. **99**, 891 (1998)

16. R.C. Malone, M.B. Johnson, H.A. Bethe: Astrophys. J. **199**, 741 (1975)

Few-Body Systems Suppl. 12, 114–117 (2000)

Few-Body Systems

Measurement of the $^{8}Li(\alpha, n)^{11}B$ Reaction with the Multiple-Sampling and Tracking Proportional Chamber (MSTPC) for Astrophysical Interest

Y. Mizoi[1,*,†], T. Fukuda[1], Y. Matsuyama[1], T. Miyachi[1,**], H. Miyatake[1], N. Aoi[2], N. Fukuda[2], M. Notani[2,*], Y.X. Watanabe[2,*], K. Yoneda[2], M. Ishihara[2,3], H. Sakurai[3], Y. Watanabe[3], A. Yoshida[3]

[1] Institute for Nuclear Study (INS)[††], University of Tokyo, 3-2-1 Midori-chou, Tanashi Tokyo 188-8501, Japan

[2] Department of Physics, University of Tokyo, 7-3-1 Hongo, Bunkyou-ku Tokyo 113-8654, Japan

[3] The Institute of Physical and Chemical Research (RIKEN), 2-1 Hirosawa, Wako Saitama 351-0198, Japan

Abstract. We have developed a Multiple-Sampling and Tracking Proportional Chamber (MSTPC) for the measurement of reactions involving unstable nuclei. We have measured the $^{8}Li(\alpha, n)^{11}B$ reaction directly and exclusively by using the MSTPC, and obtained a new data of the total cross section.

1 Introduction

Recently, inhomogeneous Big-Bang models (IM) have been proposed and intensively studied [1]. According to the IMs, the baryon distribution in the early universe was inhomogeneous, and the distribution of proton and neutron was very unbalanced, while the standard Big-Bang models (SM) assume the homogeneous distribution. Primordial nucleosynthesis in the IM universe is considered to proceed with involving much more unstable nuclei than the SM universe.

* *Present address:* The Institute of Physical and Chemical Research (RIKEN), 2-1 Hirosawa, Wako Saitama 351-0198, Japan

† *E-mail address:* mizoi@rikaxp.riken.go.jp

** *Present address:* The Japan Research Institute, Limited, 2-5 Ushijima-chou, Nishi-ku Nagoya 451-0046, Japan

†† *Present address:* High Energy Accelerator Research Organization (KEK), 1-1 Oho, Tsukuba Ibaraki 305-0801, Japan

One feature of the IM is an enhancement of abundances of the heavy elements in the early universe. While the predicted abundances of the elements up to $A = 7$ in the IM universe are similar to predicted ones in the SM universe, a big difference is found for elements above $A = 12$. The abundances predicted by the IMs are much larger than those predicted by the SMs by a few orders of magnitude, because the unstable nuclei play the important roles to synthesize them. In particular, the $^{8}\mathrm{Li}(\alpha, n)^{11}\mathrm{B}$ reaction is the key reaction to produce elements which is heavier than $A = 12$.

In order to estimate the abundances of the elements synthesized in the early universe, the reaction cross sections involving unstable nuclei should be known. However it is difficult to measure their cross sections, because beam intensity of unstable nuclei is generally limited due to the short life time of them. In spite of experimental difficulties due to the short life time of ^{8}Li ($T_{1/2} = 840$msec), two experiments were performed. One experiment measured the inverse reaction, $^{11}\mathrm{B}(n, \alpha)^{8}\mathrm{Li}$, and obtained the $^{8}\mathrm{Li}(\alpha, n)^{11}\mathrm{B}$(ground state)-reaction cross section by using the principle of the detailed balance [2]. The other experiment performed a direct-inclusive measurement and obtained the total cross section of the $^{8}\mathrm{Li}(\alpha, n)^{11}\mathrm{B}$ reaction [3, 4]. The latter cross section is five-times larger than the former one at energy regions where their data overlap. Discrepancy between them is attributed to a difference in the final states of ^{11}B; the former includes only the ground-state transition, whereas the latter includes the excited states as well. However, the latter measurements have not directly observed the individual final states of the $^{8}\mathrm{Li}(\alpha, n)^{11}\mathrm{B}$ reaction. Therefore, we performed a direct-exclusive measurement with identifying the individual final states of the $^{8}\mathrm{Li}(\alpha, n)^{11}\mathrm{B}$ reaction using the Multiple-Sampling and Tracking Proportional Chamber (MSTPC) [5], which was designed for the measurement of reactions involving low-energy unstable-nuclear beams, and neutron counters.

The ^{8}Li ions were produced from 70MeV/A ^{12}C beam by a projectile-fragmentation reaction. The ^{8}Li beams were selected by the RIKEN projectile-fragment separator, RIPS, and their energies were degraded by an aluminum plate down to 3 $\sim$ 5 MeV/A from about 45 MeV/A. The beam energy was measured by the time-of-flight method.

The low-energy ^{8}Li beams were injected into the MSTPC with an intensity of 1 $\sim$ 2 kcps. The MSTPC is a kind of gas counter which can measure the stopping power and the 3-dimensional track for the multiple charged-particle events. By utilizing a counter gas as a target, we can detect events in the detector with high efficiency, and use a sufficiently-thick target without losing the energy information, which compensates limited intensities of unstable-nuclear beams. The ^{4}He gas with an admixture of iso-butane gas was filled in the MSTPC at a pressure of 400 torr. In order to estimate the effect of the admixture gases, the 5%- and 10%-mixing rate of iso-butane gas were used.

The neutron counters surrounded the MSTPC from 0 to 90 degrees at laboratory frame. The intrinsic detection efficiency of the neutron counter is about 10% for the neutron energies of between 1 MeV to 15 MeV. The solid angle covered by twelve neutron counters is about 20% of 4π.

Table 2.1. Branching ratio to the individual state of ^{11}B.

State (MeV)	gs (0.00)	1st (2.12)	2nd (4.44)	3rd (5.02)
Ratio(%)	24.32 ± 12.55	15.17 ± 10.44	4.17 ± 7.17	6.13 ± 5.74
	4th (6.74)	5th (6.79)	6th (7.29)	7th (7.98)
	4th+5th=13.01 ± 8.03		12.45 ± 10.38	10.56 ± 7.10
	8th (8.56)	9th (8.92)		
	0.0 ± 2.12	14.22 ± 6.69		

The production of ^{11}B was identified by observing the change of the energy loss of ^{8}Li beams in the MSTPC. By measuring the reaction energy with the MSTPC, and measuring the neutron energy and momentum with the neutron counters, the final state of ^{11}B can be determined according to a kinematic calculation. The $^{9}\mathrm{Be}(\alpha,n)^{12}\mathrm{C}$ reaction was also measured for a calibration of the present system.

2 Result and Discussion

Table 2.1 gives the branching ratio to the individual states of ^{11}B. They are the averaged ratio for the energy regions of between 1.5 MeV to 7.0 MeV at the center-of-mass system.

Figure 1 shows the total cross sections of the $^{8}\mathrm{Li}(\alpha,n)^{11}\mathrm{B}$ reaction. The black circles show the present results. For comparison, the results of previous experiments are also shown. The black dots show the ground-state cross sections obtained by the inverse reaction [2]. The while stars show the total cross sections measured by the direct-inclusive measurement at RIKEN [3], and the black stars show those at Notre Dame [4]. The present total cross sections are smaller than the previous ones by a factor of about 1/2 over the entire energy region. On the other hand, comparing the ground-state cross section averaged between the energy region of 1.5 MeV and 5.0 MeV, the present ground-state cross section is 55.52 ± 30.85 mb, and that measured by the inverse reaction is 74.07 ± 0.33 mb. They agree well within the statistical error. The reason why two results of the total cross section disagree may be due to systematic errors relevant to previous measurements. We made any systematic errors in our experiment as small as possible by utilizing a better performance of the MSTPC in cooperation with the neutron counters, which gave us a sufficiently high achievement of event selection.

The present experimental method is applicable for measurement of other $\mathrm{I}(\alpha,n)\mathrm{J}$-type reactions involving the unstable nuclei. In addition, by utilizing ^{1}H and ^{2}H gases, we will also be able to measure $\mathrm{I}(p,n)\mathrm{J}$- and $\mathrm{I}(d,n)\mathrm{J}$-type reactions involving the unstable nuclei. These reactions are also important for estimating the abundances of elements synthesized in the IM Big Bang as well

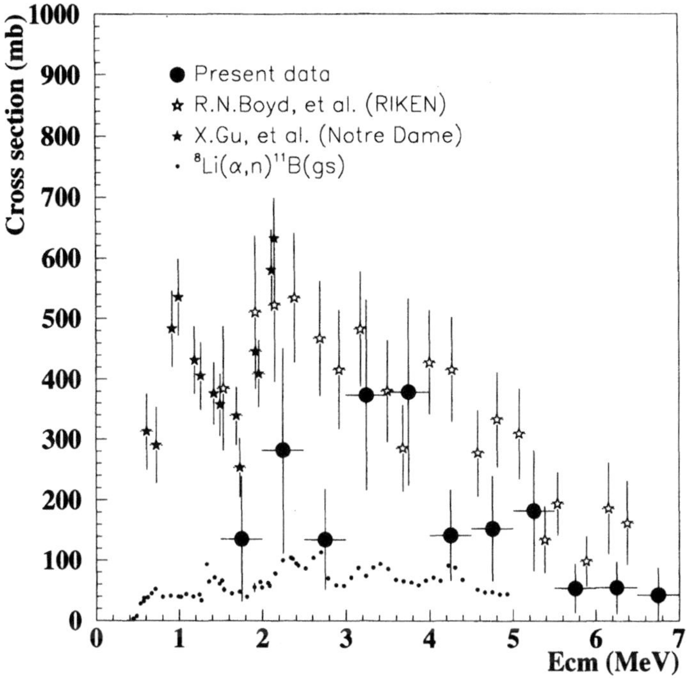

Figure 1. Cross sections of the $^{8}Li(\alpha, n)^{11}B$ reaction.

as Supernova explosion and Stellar burning.

References

1. T. Kajino: Nucl. Phys. **A570**, 161c (1994)
2. T. Paradellis et al.: Z. Phys. **A337**, 211 (1990)
3. R. N. Boyd et al.: Phys. Rev. Lett. **68**, 1283 (1992)
4. X. Gu et al.: Phys. Lett. **B343**, 31 (1995)
5. Y. Mizoi et al.: Nucl. Instrum. Methods **A431**, 112 (1999)

Few-Body Systems Suppl. 12, 118–124 (2000)

Few-Body Systems

Complex Scaling Methods for the Study of Light Unstable Nuclei

K. Katō[1], S. Aoyama[2], T. Myo[1], H. Masui[3], T. Yamada[4], K. Ikeda[5]

[1] Division of Physics, Graduate School of Science, Hokkaido University, Sapporo 060-0810, Japan

[2] Information Processing Center, Kitami Institute of Technology, Kitami 090-8507, Japan

[3] Meme Media Laboratory, Hokkaido University, Sapporo 060-8628, Japan

[4] Laboratory of Physics, Kanto Gakuin University, Yokohama 236-8501, Japan

[5] The Institute of Physical and Chemical Research (RIKEN), Wako 351–198, Japan

Abstract. We point out that the complex scaling method is one of the most successful framework to study many-body resonances. As recent developments of the complex scaling method, we present several ways to analyze the properties of resonant states; the matrix elements associated with resonant states, the extended completeness relation and partial widths of resonances. We also discuss the binding mechanism and excited resonant structure of the Borromean systems: ^{4}He+n+n and ^{9}Li+n+n. We show that the pairing correlation between valence neutrons and between core neutrons plays an important role in neutron-rich nuclei.

1 Introduction

Recent developments of the experimental studies using radioactive nuclear beams have revealed a new facet of the drip-line nuclei far from the stability valley. The most basic properties, for instance, the saturation of nuclear density, the magic numbers and so on, observed in stable nuclei have been found to be different and changed in the drip-line nuclei. For example, the most popular rule of the nuclear r.m.s. radii, which is proportional to $A^{-1/3}$, does not hold for the drip-line nuclei [1].

Typical examples have been found in the Borromean systems of neutron-rich nuclei [1]. These weakly bound systems have an exotic structure with a neutron halo, and have been studied in terms of a core+n+n three-body model which has no bound states between any two-body sub-systems. To investigate the ground state and the excited states of such an exotic system, we must solve the unbound state and the resonant state problmes of two- and three-body systems as well as the problem of the extremely-weak binding. For these studies, the complex scaling method [2] presented in detail by the previous speaker [3] is very powerful.

The resonance energies and widths are obtained by a diagonalization procedure of the complex-scaled Hamiltonian in the same way as it is done in solving the bound state problems. In this paper, we present the recent developments of the complex scaling method for the analysis of the properties of resonant states. Using the calculational method of matrix elements associated with resonant states [4], we discuss the strength distribution of the electromagnetic transition [5]. The strength can be expressed in the form separated into the resonance and continuum contributions based on the extended completeness relation. The partial decay widths, which are also very useful quantities in order to see properties of the resonant states, are shown to be calculated by using the Jost function method for the coupled channel systems [6].

Applying the complex scaling method to ^{4}He+n+n, we study the binding energy of ^{6}He and the excited three-body resonant states [7]. It is shown that the pairing correlation between valence neutrons plays a very important role in explaining the binding energy of the 0^{+} ground state. The 2^{+} resonant states are also well reproduced comparing with recent experiments [8].

The ^{9}Li+n+n model is known to have a serious problem, inasmuch as the binding energy of ^{11}Li cannot be explained even by applying the more attractive ^{9}Li-n interaction which is derived on the basis of recent experiment of ^{10}Li [9]. We propose a new idea to explain the low-lying neutron s-orbit in the Li-isotopes. The essential point is that the Pauli-blocking play an important role in the neutron pairing correlation of the ^{9}Li core due to the addition of a valence neutron in the case of ^{10}Li. We discuss that the neutron-pairing correlation in ^{9}Li is also important in the case of ^{11}Li.

2 Analyses of Resonant States within the Complex Scaling Method

The complex scaling method is very useful for calculating resonance widths in addition to energies. However, it is more convenient if we can calculate further physical quantities for resonant states. In the framework of the complex scaling method, the wave functions are also obtained as eigenstates for the complex resonant eigenvalues. Since the wave function is scaled by the transformation $U(\theta)r = r \cdot e^{i\theta}$, the θ-independent values are obtained by taking the matrix elements of an operator $\hat{O}(\theta) = U(\theta)\hat{O}U(\theta)^{-1}$ [4, 10].

By using the matrix elements, we can evaluate the distribution of the transition strength associated with resonant states. We are interested in the strength contributions from resonances and from the continuum separately. For this purpose, the pole expansion method of Green's function can be applied, as it has been discussed by many authors [11]. In the complex scaling method, the separation of bound state poles, resonant poles and the continuum can be performed without any ambiguity by the scaling angle θ. As Myo explained [5], we can construct the extended completeness relation for the bound, resonant and continuum states:

$$\mathbf{1} = \sum_{B}^{n_B} |\Phi_B^\theta(r,k_B)\rangle\langle\tilde{\Phi}_B^\theta(r',k_B)| + \sum_{R}^{n_R(L_\theta)} |\Phi_R^\theta(r,k_R)\rangle\langle\tilde{\Phi}_R^\theta(r',k_R)| + \int_{L_\theta} dk_\theta |\Phi^\theta(r,k_\theta)\rangle\langle\tilde{\Phi}^\theta(r',k_\theta)|. \quad (1)$$

In order to see the structure of resonances, the partial decay widths are also very useful. It is desired to calculate the partial widths in addition to the total decay width in the framework of the complex scaling method. We apply the Jost function method developed by Sofianos and Rakityansky [12] to calculate the partial widths in the coupled-channel system. We solve the coupled channel equation of the Jost function method given as

$$\frac{\partial F_{nm}^{(\pm)}(E,r)}{\partial r} = \pm \frac{\mu_n}{i\hbar^2 k_n} H_n^{(\mp)} \sum_{n'=1}^{N} V_{nn'} \left(H_{n'}^{(+)} F_{n'm}^{(+)}(E,r) + H_{n'}^{(-)} F_{n'm}^{(-)}(E,r) \right), \quad (2)$$

where $F_{nm}^{(\pm)}$ are unknown functions and $H_n^{(\mp)}(k_n, r)$ are the homogeneous solutions for the n-channel momentum k_n. The resonance energy is obtained from the following equation:

$$det||F^{(-)}(E, r = \infty)|| = 0. \quad (3)$$

Since the ratios of partial widths are expressed by those of the residues of the S-matrices at the resonance energy $E_{\rm res}$, we can calculate them as

$$\frac{\Gamma_n}{\Gamma_{n'}} = \left\| \frac{\sum_m F_{nm}^{(+)}(E_{\rm res},\infty) G_{mn}(E_{\rm res})}{\sum_m F_{n'm}^{(+)}(E_{\rm res},\infty) G_{mn'}(E_{\rm res})} \right\|, \quad (4)$$

where $G_{mn}(E_{\rm res})$ is the adjoint matrix of $F_{nm}^{(-)}(E_{\rm res},\infty)$. ¿From these ratios and the total width $\Gamma = \sum_n \Gamma_n$, we obtain every partial width.

The above ratio of the partial widths is also shown to be equal to the atio of the current densities [13]:

$$\frac{\Gamma_n}{\Gamma_{n'}} = \left| \frac{A_n}{A_{n'}} \right|^2, \quad (5)$$

where A_n is the asymptotic channel amplitude of the resonance wave function $\psi_{\rm res} = \sum_{n=1}^{N} \phi_n(r)$, and the channel wave function $\phi_n(r)$ is expressed as

$$\phi_n(r) \to A_n \jmath_n(r); \quad r \to \infty, \qquad \jmath_n(r) = \sqrt{\frac{\mu_n}{k_n}} H_n^{(+)}(k_n, r). \quad (6)$$

3 Borromean Systems: ^{6}He and ^{11}Li

3.1 $^6He{=}^4He{+}n{+}n$

We apply the ^{4}He+n+n model [7] to ^{6}He, and solve the three-body problem in a variational framework using the hybrid-TV model [14] of the V-type cluster-orbital-shell-model (COSM) [15] and the T-type extended cluster model (ECM) [14] configurations. We employ the KKNN-potential [16] for ^{4}He-n and the Minnesota potential [17] for n-n, which describe well the observed properties of ^{4}He-n and n-n systems, respectively. The result of the ^{4}He+n+n binding energy calculations is shown in Fig.1, where the convergences of COSM and hybrid-TV model calculations are compared. In COSM, the convergence is very slow for the channel number $[n]{=}[(l_n j_n)^2]_{J^\pi=0^+}$. On the other hand, when we take $[l = 0,\ L = 0]_{J^\pi=0^+}$ in ECM of the hybrid-TV model, we obtain a very rapid convergence. This result indicates that the pairing correlation between valence neutrons is very important in the binding of the ^{4}He+n+n system.

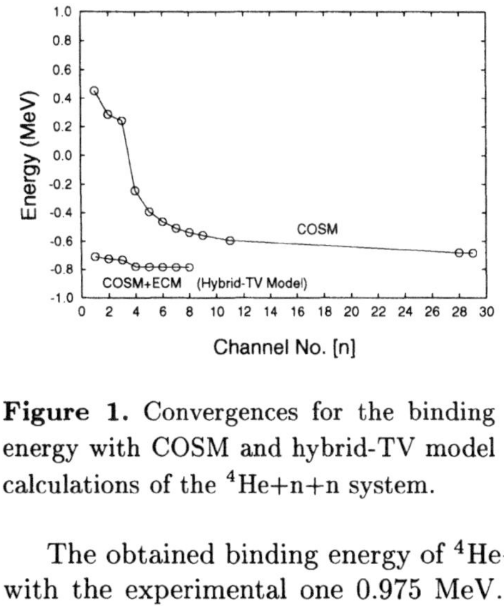

Figure 1. Convergences for the binding energy with COSM and hybrid-TV model calculations of the ^{4}He+n+n system.

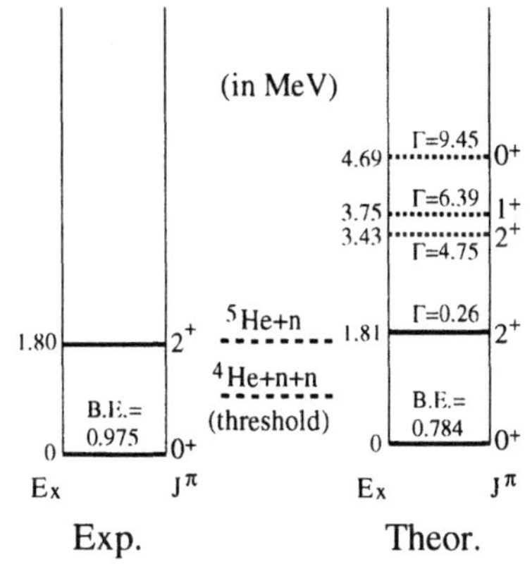

Figure 2. Energy levels of ^{6}He.

The obtained binding energy of ^{4}He+n+n is 0.784 MeV, which agrees well with the experimental one 0.975 MeV. Nevertheless, the theoretical value is smaller than the experiments by about 0.2 MeV. This insufficiency may be explained by considering the core excitation effect of ^{4}He. However, this effect is very small in comparison with the case of ^{9}Li+n+n, as will be shown later.

Applying the complex scaling method to this model, we obtain the excited states of ^{6}He, which are all resonant states, as shown in Fig. 2. The important result is that all the possible configurations of the $(p)^2$-shell model are obtained as resonant states, though their widths are rather large. Recently, the observation of the second 2^+ state was reported [8].

3.2 $^{11}Li{=}^9Li{+}n{+}n$

To ^{11}Li, we apply the ^{9}Li+n+n model [18] in the same framework as it was presented for the ^{4}He+n+n system. Using the ^{9}Li-n interaction reconstructed based on the new spectroscopic data of the ^{10}Li [19], we calculated the binding energy of ^{9}Li+n+n. However, the result [21] indicates that the observed ^{11}Li binding energy is still not explained. The calculated energy value is smaller by about 1 MeV.

In association with the halo structure of ^{11}Li, it has been discussed that the s-orbit configuration of the valence neutrons plays an important role [20]. Neverthless, there is no reasonable explanation why the s-orbital configuration gets a large binding and causes a degeneracy with the p-orbit in Li-isotopes. Here we propose a new explanation of the origin of the low-lying s-orbit. Our idea is based on the Pauli-blocking of the pairing correlation in the ground state of ^{9}Li. To see this effect, we performed the coupled-channel calculation of $[^{9}\mathrm{Li}((p_{3/2})^4_\nu)+\mathrm{n}]+[^{9}\mathrm{Li}((p_{3/2})^2_\nu(p_{1/2})^2_\nu)+\mathrm{n}]+[^{9}\mathrm{Li}((p_{3/2})^2_\nu(s_{1/2})^2_\nu)+\mathrm{n}]$ [22].

The Pauli-blocking provides a strong repulsive potential for the p-wave neutron, but gives a small effect on other partial waves, as shown in Fig. 3. The s-wave neutron also has a very small effect due to the small overlap with the core-neutron wave functions. Therefore, the $p_{1/2}$-orbit neutron is described by a shallow potential with the ^{9}Li core, while deep potentials act on the s- and d-orbit neutrons. As a result, the $s_{1/2}$-neutron shows a virtual resonance behaviour. We can see a considerable improvement of the resonance spectrum in ^{10}Li in comparison with the experiments and with a single-channel calculation as shown in Fig. 4.

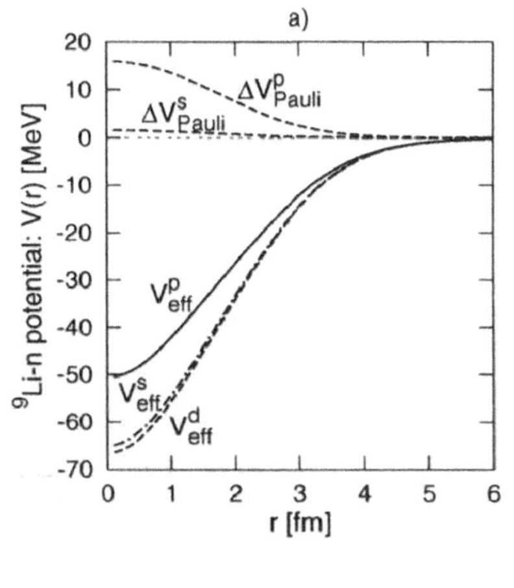

Figure 3. Effective ^{9}Li-n potentials and the Pauli-blocking potentials for s-, p- and d-wave neutrons.

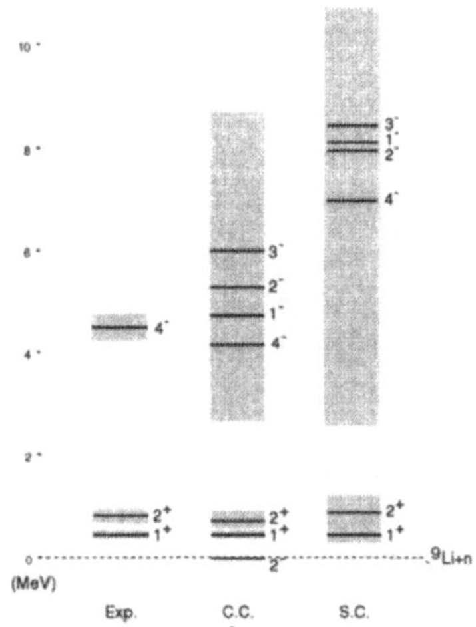

Figure 4. Energy levels of ^{10}Li.

4 Conclusion

In the problems of unstable nuclei, we must treat the three-body resonant states of the Borromean systems which have no bound states in any two-body subsystems. The complex scaling method proved to be a very useful method in describing unbound states together with bound states in a unified way.

We showed that the matrix elements associated with resonant states are calculated in the framework of the complex scaling method and are useful for investigating the character of the resonant states. Using the matrix elements for the resonant states and introducing the extended completeness relation, we presented the strength distribution of transitions to resonant and continuum states separately. The partial decay widths, which are also important in the analysis of resonances, can be calculated easily by applying the Jost function method to the coupled channel problems.

We applied the complex scaling method to the neutron rich nuclei of He- and Li-isotopes. The pairing correlation was shown to play a very important role in the binding mechanism of the ^{4}He+n+n and ^{9}Li+n+n systems. In the ^{4}He+n+n system, the three-body resonances are well described. Furthermore, we have shown that we must solve the coupling between the pairing correlation of the ^{9}Li core and the motion of valence neutrons above the core.

References

1. I. Tanihata: Nuc. Phys. **A488**, 113c (1988), and this conference

2. J. Aguilar and J.M. Combes: Commun. Math. Phys. **22**, 269 (1971); E. Balslev and J.M.Combes: Commun. Math. Phys. **22**, 280 (1971)

3. Y.K. Ho: this conference

4. M. Homma, T. Myo, K. Katō: Prog. Theor. Phys. **97**, 561 (1997)

5. T. Myo, A. Ohnishi, K. Katō: Prog. Theor. Phys. **99**, 801 (1998), and this conference

6. H. Masui, S. Aoyama, T. Myo, K. Katō: Submitted in Prog. Theor. Phys., and this conference

7. S. Aoyama, S. Mukai, K. Katō, K. Ikeda: Prog. Theor. Phys. **93**, 99 (1995); **94**, 343 (1995)

8. J. Jänecke et al.: Phys. Rev. **C54**, 1070 (1996)

9. H. G. Bohlen et al.: Z. Phys. **344**, 381 (1993)

10. B. Gyarmati and T. Vertse: Nucl. Phys. **A160**, 523 (1971)

11. J. Humblet and L. Rosenfeld: Nucl. Phys. **26**, 529 (1961)

12. S.A. Sofianos and S.A. Rakityansky: J. Phys. A: Math. Gen. **30**, 3725 (1997); *ibid.*, **31**, 5149 (1998)

13. M. Moiseyev and U. Peskin: Phys. Rev. **A42**, 255 (1990)

14. K. Ikeda: Nucl. Phys. **A538**, 355c (1992)

15. Y. Suzuki and K. Ikeda: Phys. Rev. **C38**, 410 (1988)

16. H. Kanada, T. Kaneko, S. Nagata, M. Nomoto: Prog. Theor. Phys. **61**, 1327 (1979)

17. D.R. Thompson, M. LeMere, Y. C. Tang: Nucl. Phys. **A286**, 53 (1977)

18. Y. Tosaka and Y. Suzuki: Nucl. Phys. **A512**, 46 (1990); Y. Tosaka, Y. Suzuki, K. Ikeda: Prog. Theor. Phys. **83**, 1140 (1990)

19. K. Katō and K. Ikeda: Prog. Theor. Phys. **89**, 623 (1993)

20. I. J. Thompson and M. V. Zhukov: Phys. Rev. **C49**, 1904 (1994)

21. S. Mukai, S. Aoyama, K. Katō, K. Ikeda: Prog. Theor. Phys. **99**, 381 (1998)

22. K. Katō T. Yamada, K. Ikeda: Prog. Theor. Phys. **101**, 119 (1999)

Few-Body Systems Suppl. 12, 125–129 (2000)

Few-Body Systems

Novel Features of Unstable Nuclei Studied with AMD

H. Horiuchi[1], Y. Kanada-En'yo[2], A. Doté[1], M. Kimura[1]

[1] Department of Physics, Kyoto University, Kyoto 606-8502, Japan
[2] Institute of Particle and Nuclear Studies, High Energy Accelerator Research Organization, 3-2-1 Midori-cho Tanashi Tokyo 188, Japan

Abstract. Formulation of AMD is briefly explained. It is explained that AMD can describe both mean field structure and cluster structure and also their coexistence. As an example to show this fact the results of the study of ^{12}C are reported. Then it is shown that AMD is useful for the study of unstable nuclei which show many unexpected features different from stable nuclei. Nobel features of unstable nuclei we discuss include molecular orbits in Be isotopes, novel type of clustering in B isotopes, and different deformation of proton and neutron distributions in C isotopes.

1 Introduction

Nucleons and nuclei easily assemble and disassemble. It is due to the saturation property of nuclear binding energy. The formation of clusters is a fundamental aspect of nuclear many body dynamics together with the formation of mean field [1, 2]. Clustering aspects appear abundantly in many problems in both nuclear structure and nuclear collisions. The coexistence of the clustering aspect and the single particle aspect is a unique feature of nuclear many body system [3].

In the region of very light stable nuclei with A less than about 10, the clustering shows up in ground states. Recently an ab initio calculation by the group of Pandharipande could verify by the use of realistic nuclear force the well-known pronounced α-α dumbbell structure of ^{8}Be. In heavier mass region of stable nuclei, however, the mean field is formed in ground states and the clustering appears in excxited states.

How is the clustering in unstable nuclei where the density saturation is broken, and neutron and proton density distributions are no more proportional ? In order to make theoretical studies of unfamiliar properties of neutron-rich nuclei, it is desirable to use a theoretical framework that can give good description of ordinary stable nuclei. However, for most theoretical frameworks for light nuclei, it is not easy to make systematic research on the isotopes ranging from ordinary nuclei to neutron-rich nuclei. First of all, the applicability of mean field approches is not necessarily assured in light nuclei because of the existence of the cluster structure. On the other hand, needless to say, the cluster model can not be applied if the nucleus does not have a postulated cluster structure or if the nucleus does not have any clustering nature at all. This argument shows that in order to make systematic research on the isotopes ranging from ordinary nuclei to neutron-rich nuclei, we have to use a theoretical framework which can describe both mean-field-like states and clustering states. The method of the Antisymmetrized Molecular Dynamics (AMD) [1, 2, 4, 5, 6, 7] is the very theoretical framework that satisfies this requirement and enables us to study isotopes of light nuclei systematically without any model assumptions.

In this talk, we first explained the formulation of AMD and then discussed the results of the application of the AMD model to various isotopes. In the discussion, many problems were treated which included novel type of clustering near neutron dripline, opposite deformation of neutron and proton density distributions, breaking of the neutron magic numbers N=8, and so on. All these problems are typical examples of novel features of unstable nuclei. The discussion was not necessarily only for the ground states or ground rotational bands but also for the excited states or excited rotational bands in the case of Be isotopes.

2 Formulation

In AMD the total wave function is a superposition of Slater determinants of single particle wave functions described by Gaussian wave packets. We project out parity before energy variation while the projection of angular momentum is made before or after energy variation depending on nuclear systems. From the obtained AMD wave function we can extract single particle orbits and energies by solving the Hartree-Fock equation within the functional space spanned by the adopted Gaussian wave packets. This means that we can discuss both mean field nature and clustering nature of the system with the AMD framework. An important point is that we need no model assumption in AMD; namely whether the system has mean field structure or cluster structure is determined by solving AMD variational calculation.

3 Coexistence of Mean Field Structure and Cluster Structure

In order to demonstrate the powerful ability of AMD in describing the coexistence of mean field structure and cluster structure and also the structure change between these two kinds of structure, we reported the results of the study of ^{12}C [8]. We discussed the the coexistence of the mean field structure of the ground band states and the 3-α cluster structure of excited rotational band states in ^{12}C. We showed that the beta transition from the ^{12}N ground state to both mean field structure states and cluster structure states of ^{12}C are well reproduced by AMD. We also showed the E2 transitions between these two kinds of states are also well reproduced.

4 Be Isotopes

We then reported the studies of Be isotopes which present us a very good example of the importance of clustering in neutron-rich nuclei. AMD calculations show that in all Be isotopes from ^{8}Be up to the neutron dripline nucleus ^{14}Be there exists the α-α core. We explained that the motion of neutrons is well understood by the concept of the molecular orbits around α-α core. Long time ago Seya, Kohno, and Nagata extended this model of molecular orbit up to the neutron dripline in Be and B isotopes [9]. The *ab initio* model AMD now confirmed theoretically the formation of the $\alpha - \alpha$ core in Be isotopes up to the neutron dripline [5].

We showed that an important valence orbit for neutrons is the so-called σ orbit coming down from the sd-shell due to the clustering deformation. Especially in ^{10}Be we could classify the observed states into four rotational bands; $K^\pi = 0_2^+$ and $K^\pi = 0_1^-$ bands were interpreted to have the neutron configulations with two neutrons and one neutron in σ orbit, respectively while $K^\pi = 0_1^+$ and $K^\pi = 2_1^+$ bands were interpreted to have neutron configulations with no neutrons in σ orbit. We showed that the Gamow-Teller transition strengths from the ^{10}B ground state to $K^\pi = 0_1^+$ and $K^\pi = 2_1^+$ band states are well reproduced by AMD [10].

5 B Isotopes

We reported the AMD study of B isotopes [6]. According to AMD, B isotopes near neutron dripline nucleus ^{19}B were shown to have prominent di-cluster density distribution. Here five protons are divided spatially into two groups with two and three protons, respectively, which are surrounded by neutrons. Reliability of such AMD results was assured by good reproduction of data for ^{13}B $\sim$ ^{19}B which include binding energies, radii, electric quadrupole moments, and magnetic moments. We notice an important fact that the clustering in ^{17}B and ^{19}B is entirely different from the clustering feature in stable nuclei. For example, if we regard ^{19}B as having a di-cluster structure of ^{8}He + ^{11}Li or ^{10}He + ^{9}Li, we notice that any of clusters ^{8}He, ^{10}He, ^{9}Li, and ^{11}Li does not satisfy the " cluster condition " that the cluster should be a stable nucleus. Furthermore we notice that the threshold rule of the Ikeda diagram [11, 12] is heavily violated, because ^{19}B is deeply bound by 12.73 MeV and 14.17 MeV from the ^{8}He + ^{11}Li and ^{10}He + ^{9}Li thresholds, respectively. (Also the neutron dripline nucleus ^{14}Be is deeply bound by 9.16 MeV and 11.13 MeV from the ^{8}He + ^{6}He and ^{10}He + α thresholds, respectively.) Thus we are forced to regard that the clustering in ^{17}B and ^{19}B is of a novel type different from the clustering in stable nuclei. We presented a conjecture that in general a novel type of clustering is a candidate of nuclear strucure near neutron dripline. By definition the neutron dripline nucleus has the structure which accommodates maximum number of neutrons for a given number of protons. Such structure is realized by making the volume of the neutron skin region maximum by avoiding the energy loss due to the Pauli principle and shell effects. When we notice that the thickness of the neutron skin cannot be larger than the nuclear force range because neutrons without protons cannot be bound, we see that deformed shape has bigger volume of neutron skin than spherical shape and that clustering shape with spatially divided protons has even bigger volume of neutron skin.

6 Different Deformation of Proton and Neutron Distributions

We briefly discussed about the possibility of the opposite deformation of neutron and proton density distributions in ^{11}C and ^{10}C [13]. Experimantal data of the quadrupole moments of ^{11}C and ^{11}B show that $Q(^{11}\mathrm{C})/Q(^{11}\mathrm{B}) = 0.84$ in spite of the fact that the charge of ^{11}C is larger than that of ^{11}B. Also experimantal data of the E2 transition rates in ^{10}C and ^{10}Be show that $B(E2, 2^+ \to 0^+; ^{10}\mathrm{C})/B(E2, 2^+ \to 0^+; ^{10}\mathrm{Be}) = 1.17$ in spite of the fact that the squared charge ratio of ^{10}C and ^{10}Be is 2.25. We explained that these problems can be solved by AMD since AMD gives different deformation of neutron and proton density distributions in ^{11}C and ^{10}C.

7 Summary

We explained that AMD can describe both mean field structure and cluster structure and also their coexistence. As an example to show this fact we reported the results of the study of ^{12}C. We then showed that AMD is useful for the study of unstable nuclei which show many unexpected features different from stable nuclei. Nobel features of unstable nuclei we explained are molecular orbits in Be isotopes, novel type of clustering in B isotopes, and different deformation of proton and neutron distributions in C isotopes.

References

1. H. Horiuchi: *Proc. NATO Advanced Study Institute on Correlations and Clustering Phenomena in Subatomic Physics*, Dronten (1996), edited by M.N. Harakeh, J.H. Koch, O. Scholten, (NATO ASI Series B: Physics Vol.**359**), p.29. Plenum Press 1997

2. H. Horiuchi: *Proc. Int. Symp. in Honor of Akito Arima; Nuclear Physics in 1990's*, Santa Fe (1990), edited by D.H. Feng, J.N. Ginocchio, T. Otsuka D. Strottman, Nucl. Phys. **A522**, 257c (1991)

3. A. Arima: *Proc. IV Int. Conf. on Clustering Aspects of Nuclear Structure and Nuclear Reactions*, Chester (1984), edited by J.S. Lilley and M.A. Nagarajan, p.399. Dordrecht, Boston, Lancaster: Reidel 1985

4. H. Horiuchi, T. Maruyama, A. Ohnishi, S. Yamaguchi: *Proc. Int. Conf. on Nuclear and Atomic Clusters*, Turku (1991), edited by M. Brenner, T. Lönnroth, F.B. Malik, p.512. New York: Springer 1992); *Proc. Int. Symp. on Structure and Reactions of Unstable Nuclei*, Niigata (1991), edited by K. Ikeda and Y. Suzuki, p.108. Singapore: World Scientific 1992

5. Y. Kanada-En'yo, H. Horiuchi, A. Ono: Phys. Rev. C **52**, 628 (1995)

6. Y. Kanada-En'yo and H. Horiuchi: Phys. Rev. C **52**, 647 (1995)

7. A. Doté, H. Horiuchi, Y. Kanada-En'yo: Phys. Rev. C **56**, 1844 (1997)

8. Y. Kanada-En'yo: Phys. Rev. Lett. **81**, 5291 (1998)

9. M. Seya, M. Kohno, S. Nagata: Prog. Theor. Phys. **65**, 204 (1981)

10. Y. Kanada-En'yo, H. Horiuchi, A. Doté: Phys. Rev. C, to be published

11. K. Ikeda, N. Takigawa, H. Horiuchi: Prog. Theor. Phys. Supplement Extra Number, 464 (1968)

12. H. Horiuchi and K.Ikeda: *Cluster Model of the Nucleus*, International Review of Nuclear Physics, Vol.**4**, ed. T.T.S. Kuo and E. Osnes, pp.1 $\sim$ 258. Singapore: World Scientific 1986

13. Y. Kanada-En'yo and H. Horiuchi: Phys. Rev. C **55**, 2860 (1997)

Few-Body Systems Suppl. 12, 130–136 (2000)

Few-Body Systems

Gambling in Physics—The Stochastic Variational Method: Recent Applications—

Y. Suzuki[1], K. Varga[2,*]

[1] Department of Physics, Niigata University, Niigata 950-2181, Japan

[2] Physics Division, Argonne National Laboratory, 9700 South Cass Avenue, Argonne, 60438 Illinois, USA

Abstract. This paper is an overview of the recent developments and the future promises of the stochastic variational method.

1 Introduction

The variational method is the simplest approach to solve few-particle problems in physics. One starts with a suitably chosen trial function (or a linear combination of those) and calculates the energy and other properties of the system. The problem is to find the most appropriate trial function (which gives the best upper bound of the ground state energy). The stochastic variational method (SVM) [1, 2] sets up the wave function with a random trial and error procedure.

The basic idea is best illustrated by the following example: Imagine that you are chosen to be the coach of the national soccer team of a small country. You have never been in that country before, and you do not know anything about the soccer players there. To set up the best team you invite players by random selection and let them play a game against the national team of a neighbouring country. Depending on their performances you keep some of the players and replace some others by randomly selected new players. You let the team play new games and try to improve team by replacing the players with better ones. This gambling-like selection might obviously not be the smartest way, but if there is no other information available one may have to try something like this.

The stochastic variational method uses a very similar procedure by selecting those randomly chosen basis functions which improve the energy. One limitation is clear in the above example: If the "country" is too big, one needs too many

* *On leave from:* Institute for Nuclear Research of the Hungarian Academy of Sciences, (ATOMKI),Debrecen, Hungary

trials and the chance to find the best basis functions gets too small. We have applied the above procedure to many different problems in few-body physics with great success. Some of the examples of application are overviewed in this paper.

2 Basics

Our aim is to solve the N-particle Schrödinger equation

$$H\Psi = E\Psi, \qquad H = \sum_{i=1}^{N} T_i + \sum_{i=1}^{N} U_i + \sum_{i<j} V_{ij} + \dots. \tag{1}$$

The Hamiltonian is a sum of the kinetic energy operator T_i and the two-body interaction V_{ij} but it may also contain single-particle potential (like an external magnetic field) or three-body interaction. The variational solution is taken as

$$\Psi(\mathbf{r}_1, \dots, \mathbf{r}_N) = \sum_{i=1}^{K} c_i \Phi(\mathbf{a}_i; \mathbf{r}_1, \dots, \mathbf{r}_N), \tag{2}$$

where $\mathbf{a}_i$ is a set of variational parameters. The method gives an upper bound to the ground state energy and this upper bound crucially depends on these parameters. The variational upper bound to the exact energy is given by solving the generalized eigenvalue problem

$$\sum_{i=1}^{K} (H_{ij} - E_K A_{ij}) c_i = 0, \qquad (j = 1, ..., K), \tag{3}$$

where the matrix elements are defined by

$$O_{ij} = \langle \Phi(\mathbf{a}_i; \mathbf{r}_1, \dots, \mathbf{r}_N) | O | \Phi(\mathbf{a}_j; \mathbf{r}_1, \dots, \mathbf{r}_N) \rangle. \tag{4}$$

To facilitate the analytical calculation of the matrix elements we have chosen the Correlated Gaussians (CG) as basis functions:

$$\Phi(A; \mathbf{r}_1, \dots, \mathbf{r}_N) = \exp\Big(-\frac{1}{2} \tilde{\mathbf{x}} A \mathbf{x} \Big) \equiv \exp\Big(-\frac{1}{2} \sum_{i=1}^{N-1} \sum_{j=1}^{N-1} A_{ij} \mathbf{x}_i \cdot \mathbf{x}_j \Big)$$

$$= \exp\Big\{ -\frac{1}{2} \sum_{j>i=1}^{N} \alpha_{ij} (\mathbf{r}_i - \mathbf{r}_j)^2 \Big\}. \tag{5}$$

The matrix elements A_{ij} or, equivalently, α_{ij} are nonlinear parameters of the basis. The vectors $\mathbf{x} = (\mathbf{x}_1, ..., \mathbf{x}_{N-1})$ are a set of relative (Jacobi) coordinates. The advantage of the CG basis is that it can be easily generalized for an N-body system, the matrix elements are analytically available, the permutational (or other) symmetry can be easily taken care of and they are flexible enough

Table 3.1. Comparison of the energies of different systems calculated by SVM and other methods. The energies of the Coulombic systems (first three rows) are given in atomic units. The energies of the triton and the alpha particle, given in MeV, are calculated with the Malfliet-Tjon [6] interaction.

system	SVM	other method	
(e^+,e^-,e^-)	−0.262005070226	−0.2620050702328	[3]
(e^+,e^+,e^-,e^-)	−0.5160037890	−0.51601±0.00001	[4]
(p,e^+,e^-,e^-)	−0.7891965	−0.7891967	[5]
triton	−8.2527	−8.2527	[7]
alpha	−31.36	−31.36	[8]

to approximate even rapidly changing functions. The spin part of the wave function is represented by successively coupled spin functions. The orbital part can either be described by (1) a partial wave expansion

$$\vartheta_L(x) = \left[\left[[Y_{l_1}(\hat{\mathbf{x}}_1) Y_{l_2}(\hat{\mathbf{x}}_2)]_{l_{12}} Y_{l_3}(\hat{\mathbf{x}}_3) \right]_{l_{123}} \cdots \right]_{LM},$$

or (2) a global vector representation

$$\vartheta_L(x) = [[Y_{l_1}(\hat{\mathbf{v}}_1) Y_{l_2}(\hat{\mathbf{v}}_2)]_{LM}, \qquad \mathbf{v}_1 = \sum_{i=1}^{N} u_{1i}\mathbf{r}_i, \qquad \mathbf{v}_2 = \sum_{i=1}^{N} u_{2i}\mathbf{r}_i.$$

The partial wave expansion is generally used in few-body physics but for larger systems it becomes very tedious. The global vector representation [2] contains a set of variational parameters $(u_{11}, ..., u_{1N}, u_{21}, ..., u_{2N})$. The best parameters are selected by the SVM.

3 Calibration

To test the method we have calculated many well-known quantum-mechanical systems, especially Coulombic systems (small atoms and molecules) and compared our results to those of other methods. Table 3.1 shows a few examples.

The results calculated by the SVM are in excellent agreement with those of other methods. The accuracy of the SVM is often better than that of other approaches.

4 Exotic Molecules

We have investigated the stability and structure of systems comprising a negative ion and positrons [9] by the SVM. It is shown that the H^- and the Li^- ions can bind not only one but two positrons. The binding energies of these double positronic atoms, $E(H^-,e^+,e^+)$=0.57 eV and $E(Li^-,e^+,e^+)$=0.15 eV, are somewhat smaller than those of their single positronic counterparts (E(HPs)=1.06

Table 4.1. Binding energies of exotic molecules (in a. u.)

system	basis size	E_{total}	$E_{binding}$
(p,Ps_2)	400	−0.810152	0.02096
(Li^+,Ps_2)	800	−7.80510	0.00918
(p^+,$2e^+$,$4e^-$)	800	−1.05542	0.00423

Figure 1. The (p^+,e^+,e^+,e^-,e^-,e^-,e^-) system is bound but the (p^+,e^+,e^+,e^-,e^-,e^-) system is not. The dotted line is the HPs+Ps, the dashed line is the H^-+Ps_2, the long dashed line is the HPs+Ps^- threshold.

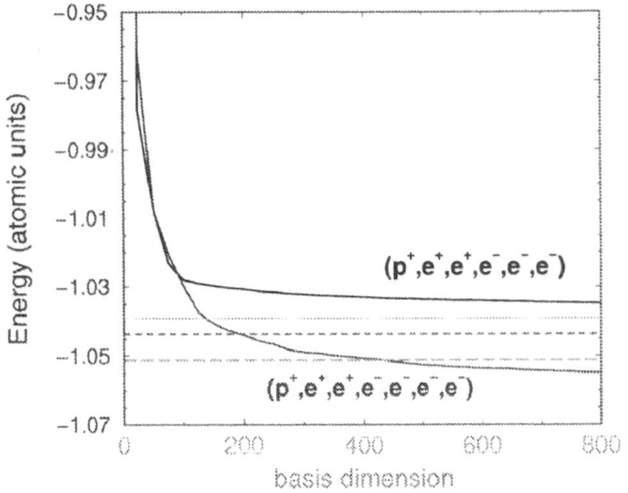

eV and E(LiPs)=0.32 eV). We have also found that a system comprising Ps^-, Ps^- and a proton is bound. (See Table 4.1 and Fig. 1.)

5 Light Nuclei

Our primary goal is to use the SVM to calculate the energies and other properties of light nuclei by using realistic interactions. This is an obviously challenging application due to the noncentral forces and strong repulsive core of the nuclear interaction. We have calculated the energies of the triton and the alpha particle by different realistic interactions [14]. The energies obtained were in good agreement with other methods. In particular the SVM result agreed with the results of the Green Function Monte Carlo (GFMC) calculations [10] within its statistical error (± 0.2MeV). The GFMC calculations have been greatly improved in the last few years and the statistical error for the alpha particle is about 0.01MeV. We have repeated our calculation for the alpha particle im-

Table 5.1. Energy (in MeV) of the alpha particle with the AV6 part of the AV14 [13] interaction.

	SVM	VMC [11]	GFMC [12]
$<T>$	101.89	101.19(0.28)	100.61(0.21)
$-<V>$	127.42	125.32(0.27)	126.14(0.20)
$-<H>$	25.52	24.14(0.02)	25.53(0.01)

Figure 2. Convergence of the energy of the alpha particle calculated by using the AV6 part of the AV14 interaction. The lower solid line is the energy obtained by the GFMC method. The upper solid line indicates the energy given by our previous calculation.

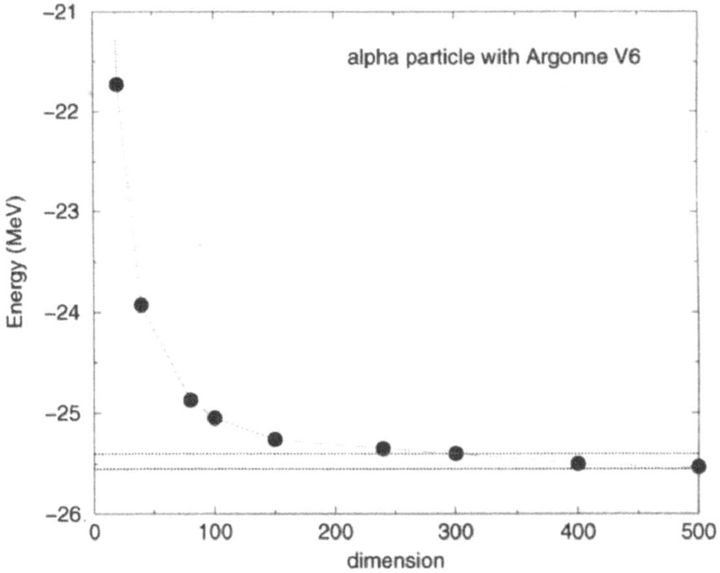

proving our approach in two respects. We have used the full "A" matrix (see. eq. (5)), combined with a global vector representation for the orbital part. In the previous calculation the matrix of the nonlinear parameters was chosen to be diagonal, and a (truncated) partial wave expansion was used. In these calculations not only the nonlinear parameters but the spin-isospin components and the orbital quantum numbers are also randomly chosen. Our results are improved by about 150 keV (see Tables 5.1 and 5.2 and Fig. 2). There is a nice agreement between SVM and the Faddeev-Yakubovski calculation for the expectation values of the different components of the interactions (the GFMC is only exact for the energies, the expectation values are based on an approximation). Other methods [16] give similar results.

Calculations for A=6 nuclei are under way. The convergence of the energy requires considerebly larger basis for these systems. At present the energy of the

Table 5.2. Energy (in MeV) of the alpha particle with the AV8 part of the AV18 interaction.

	VMC [11]	GFMC [12]	FY [15]	SVM
$\langle K \rangle$	92.9	97.12	99.23	99.21
$\langle V_4 \rangle$	−41.41	−42.55	−42.55	−42.55
$\langle T \rangle$	−73.78	−78.61	−79.45	−79.40
$\langle LS \rangle$	−2.92	−2.76	−2.96	−2.95
E	−25.06	−25.70	−25.71	−25.67

^{6}Li is below the alpha particle threshold but it is not bound with respect to the alpha-deuteron threshold. The ^{6}He is likely to be unbound without three-body interaction.

6 Summary

The SVM is a very simple and powerful approach to solve few-body problems in physics. Flurry of examples in atomic, nuclear and solid state systems shows its effectiveness and usefulness. One of the major limitations of the method is that by increasing the number of particles the evaluation of the matrix elements gets longer and longer on the computer. That prevents us from trying the necessary number of random candidates and lowers the chance of finding the most appropriate parameters.

Acknowledgement. This work was supported in part by a Grant-in-Aid for Scientific Research (No. 10640255) of the Ministry of Education, Science, Sports and Culture (Japan), the Department of Energy, Nuclear Physics Division, under contracts W-31-109-ENG-38, and OTKA grant T029003 (Hungary).

References

1. V.I. Kukulin and V.M. Krasnopol'sky: J. Phys. **G3**, 795 (1977)
2. Y. Suzuki and K. Varga: *Stochastic Variational Approach to Quantum Mechanical Few-body Problems.* Springer 1988; K. Varga and Y. Suzuki: Phys. Rev. **C52**, 2885 (1995)
3. Y. K. Ho: Phys. Rev. **A48**, 4780 (1993)
4. D. Bressanini, M. Mella, G. Morosi: Phys. Rev. **A55**, 200 (1997)
5. Zong-Chao Yan and Y. K. Ho: Phys. Rev. **A59**, 2697 (1999)
6. R. A. Malfliet and J. A. Tjon: Nucl. Phys. **A127**, 161 (1969)
7. J. L. Friar, B. F. Gibson, G. L. Payne: Phys. Rev. **C24**, 2279 (1981)

8. M. Kamimura and H. Kameyama: Nucl. Phys. **A508**, 17c (1990)

9. K. Varga: submitted for publication

10. J. Carlson: Phys. Rev. **C38**, 1879 (1988)

11. R. B. Wiringa: Phys. Rev. **C43**, 1585 (1991)

12. S. Pieper: private communication

13. R. B. Wiringa, V. G. J. Stoks, R. Schiavilla: Phys. Rev. **C51**, 38 (1995)

14. K. Varga, Y. Ohbayashi, Y. Suzuki: Phys. Lett. **B396**, 1 (1997)

15. A. Nogga and W. Glöckle: privata communication

16. M. Viviani, A. Kievsky, S. Rosati: Few-Body Systems **18**, 25 (1995)

Few-Body Systems Suppl. 12, 137–140 (2000)

Few-Body Systems

Microscopic Coupled-Channel Study of $^{12}C+^{12}C$ Resonance Reactions

M. Ito[1], Y. Sakuragi[1], Y. Hirabayashi[2]

[1] Department of Physics, Osaka City University, Osaka 558-8585, Japan

[2] Center for Information Processing Education, Hokkaido University, Sapporo 060, Japan

Abstract. Resonance structures recently observed in the $^{12}C+^{12}C$ inelastic scattering leading to multi-cluster exit channels, having $3\alpha+^{12}C$ and $3\alpha+3\alpha$ configuration, are studied theoretically by the microscopic coupled-channel calculation. We use the microscopic 3α cluster model wave functions of ^{12}C and the density dependent nucleon–nucleon interaction. The results of CC calculation consistently reproduce the resonance structures of the excitation functions not only in the multi-cluster exit channels but also in those of the well-known exit channels having di-nucleus configuration. The reaction mechanism leading to the resonance formation in the multi-cluster channels is also discussed in some details in connection to the various molecular bands of $^{12}C+^{12}C$ system existing at more than 40 MeV excitation energies of ^{24}Mg.

1 Introduction

Recently, pronounced resonance structures have been observed in the $^{12}C+^{12}C$ inelastic scattering at $E_{c.m.} = 32.5$ MeV, in which both of the colliding ^{12}C nuclei are excited to the 0_2^+ state [1]. The observed resonance energy corresponds the excitation energy of about 46 MeV with respect to the ground state of compound system, ^{24}Mg. Since the 0_2^+ state is known to have a well-developed 3α structure [2], the observed resonance suggests the existence of a quasi-stable molecular state consisting of six α particles. The resonance state was initially discussed in connection with the so-called *6α linear-chain structure* [3].

However, Hirabayashi et al. have performed the microscopic coupled-channels (CC) calculation [4] using the double-folding model (DFM) interactions based on the microscopic 3α-cluster-model wave functions and the realistic nucleon-nucleon interaction. The calculation succeeded in reproducing the angular distributions and the excitation functions of this $0_2^+ + 0_2^+$ channel. The calculation also reproduced the famous resonance in the $0_1^+ + 0_2^+$ channel around $E_{c.m.} = 29.5$ MeV [5] and the resonances in other reaction channels such as the $3_1^-+0_2^+$ and $0_3^+ + 0_2^+$ ones around the same energy region [6]. The strong coupling model such as the *6α linear-chain* one cannot explain those resonances and the results of Hirabayashi et al. suggest that the resonance in the $0_2^+ + 0_2^+$ channel has a different character from the *6α linear-chain* state.

Meanwhile, new experimental data of the various $^{12}C+^{12}C$ inelastic channels such as the $3_1^-+3_1^-$ and $3_1^-+2_1^+$ channels have recently been reported around $E_{c.m.}$= 32.5 MeV [7]. Moreover, it is well known that the pronounced resonance states exist in the $0_1^++2_1^+$, $2_1^++2_1^+$ and $0_1^++3_1^-$ channels over a wide energy range from $E_{c.m.}$=10 MeV to 50 MeV [5]. These resonances are known to have a di-nuclear configuration [8].

In this paper, we present the latest results of our extensive study of the $^{12}C+^{12}C$ resonances observed in various inelastic channels based on the microscopic CC calculation and discuss the reaction mechanism leading to the most exotic channels, such as the $0_1^++0_2^+$ and $0_2^++0_2^+$ ones, having the $3\alpha+3\alpha$ and $^{12}C+3\alpha$ structures, respectively.

2 Framework of Microscopic Coupled-Channels Method

The real part of the nucleus-nucleus interactions are calculated by DFM and they are written symbolically as

$$V_{\alpha,\beta}(\boldsymbol{R}) = \int \rho_{ik}^{(^{12}C)}(\boldsymbol{r}_1)\, \rho_{j\ell}^{(^{12}C)}(\boldsymbol{r}_2)\, v_{\rm NN}(\boldsymbol{r}_1, \boldsymbol{r}_2, \boldsymbol{R})\, d\boldsymbol{r}_1\, d\boldsymbol{r}_2 \quad , \tag{1}$$

where, $\rho_{ik}^{(^{12}C)}(\boldsymbol{r})$ is the transition density of ^{12}C between the states i and k, while $v_{\rm NN}$ is the density-dependent nucleon-nucleon interaction called DDM3Y. We adopt the transition densities calculated by Kamimura in the 3α RGM and these densities well describe the electro-magnetic transitions in ^{12}C [2].

We perform a CC calculation using these folding-model interactions. In the calculation, we consider the coupling among the ground state (0_1^+) and the 2_1^+, 3_1^-, 0_2^+, 2_2^+ and 0_3^+ excited states. The former three states are known to have a spatially compact structure such as the "shell-model-like" structure, namely the 3α particles are tightly bound to each other, while the latter three states are known to have a "well-developed 3α-cluster" structure.

In addition to the real potentials obtained by DFM, we introduce an imaginary part in the diagonal part, $V_{\alpha\alpha}(\boldsymbol{R})$. We adopt a J-dependent imaginary potential [8] having an Woods-Saxon form factor and their parameters are taken to be common to all the channels. We choose the parameter values so as to optimize the fit to the experimental data.

3 Results of CC Calculation

Figure 1 shows the comparison of the calculated cross sections for the $0_1^++2_1^+$, $2_1^++2_1^+$ and $0_1^++3_1^-$ channels with the corresponding experimental data [5]. The calculation well reproduces the well-known resonance structure of the excitation functions of these channels.

The same calculation also reproduces the new experimental data [7] on the characteristic resonance structures in other channels including the $0_2^+ + 0_2^+$ and $3_1^- + 3_1^-$ ones, as shown in Fig. 2. It is seen that the essential feature of the well-correlated, newly-found resonances in those channels is reproduced well. It should be emphasized that the calculation correctly reproduces not only the resonance structure but also the absolute magnitudes of the cross sections within a factor of five in various channels shown in Figs.1 and 2, which range from a few μb to a hundred mb. It is also interesting to note that the calculation predicts additional structures outside the energy range of the present experimental data. A distinct structure in the energy range of $E_{\rm c.m.} \simeq 28 - 30$ MeV in the $0_2^+ - 0_1^+$ channel is consistent with the famous resonance observed around $E_{\rm c.m.}$=29 MeV in this channel [5]. A further experiment in various channels over a wider energy range is strongly expected.

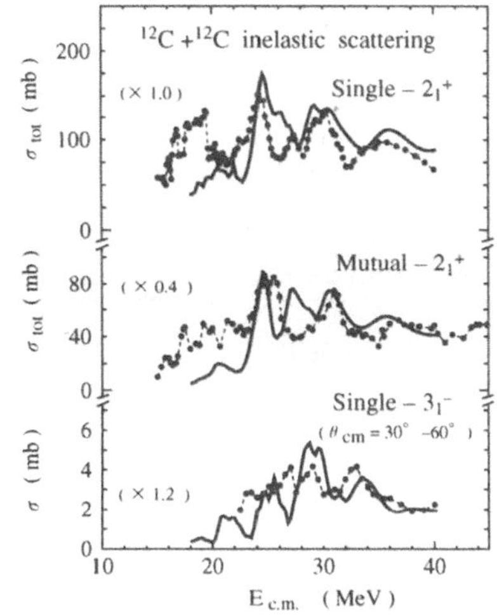

Figure 1

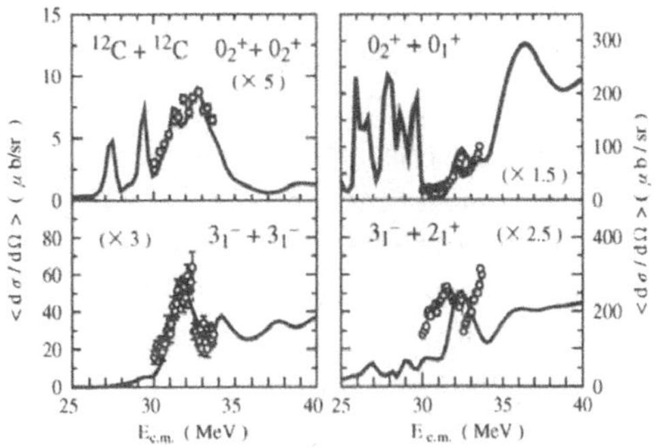

Figure 2

Figure 1: Comparison of the calculated cross sections for the $0_1^+ + 2_1^+$, $2_1^+ + 2_1^+$ $0_1^+ + 3_1^-$ channels with the experimental data [5]. The cross section in the $0_1^+ + 3_1^-$ channel is integrated over the limited angular range indicated in Fig. 1.

Figure 2: Comparison of the calculated and newly measured excitation functions. The ordinate denotes the differential cross section averaged over $\theta_{\rm c.m.} = 70° \sim 105°$. The calculation should be multiplied by the factor indicated. The experimental data are taken from Ref. [7]. All the calculated curves have been shifted toward the high-energy side by 2.2 MeV to be compared with the experimental data.

4 Reaction Mechanism leading to Multi-Cluster Exit Channels

A key for understanding the resonance formation in the $0_2^+ + 0_2^+$ and $0_1^+ + 0_2^+$ channels exists in the characteristic nuclear structure of ^{12}C. As we mentioned, two groups of states having a very different nuclear structure coexist in ^{12}C; namely, the shell-model-like states (the 0_1^+, 2_1^+ and 3_1^- states) and the cluster states (the 0_2^+, 2_2^+ and 0_3^+ ones). Because of the difference of nuclear structure, the transition between the states belonging to the different groups is very weak. On the other hand, the transitions among the state belonging to the same groups, especially those between the 0_1^+ and 2_1^+ states and between the 0_2^+ and 2_2^+ ones, are very strong and important.

Taking account of the characteristic nuclear structure of ^{12}C, we can classify the ^{12}C+^{12}C reaction channels into three groups. The first we call the "shell-group" (SG) consists of the 0_1^++0_1^+, 0_1^++2_1^+ and 2_1^++2_1^+ channels, the second we call the "cluster-group" (CG) consists of the 0_2^++0_2^+, 0_2^++2_2^+ and 2_2^++2_2^+ channels, and the third we call the "hybrid-group" (HG) consists of the 0_1^++0_2^+, 2_1^++0_2^+, 0_1^++2_2^+ and 2_1^++2_2^+ channels. Because of the strong coupling within the same group, "eigen-states" are obtained as the solution of the CC equations in each group and they form distinct rotational bands. Here, we call the rotational bands "eigen bands" which are shown in Fig. 3. These three types of eigen bands have very different moment of inertia, which reflects the large difference of the interaction range between two ^{12}C nuclei in these three kinds of bands.

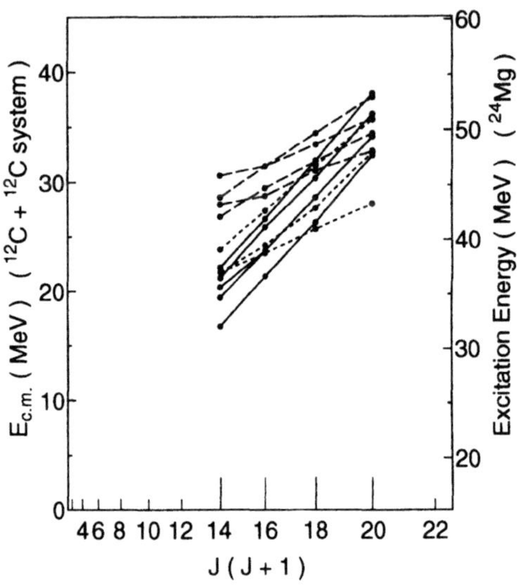

Figure 3: Diagram of Eigen bands. The solid lines represent the eigen bands in SG, while the dotted- and long-dashed lines represent those in the HG and CG, respectively. See text for details.

The components of the 0_1^++0_1^+, 0_1^++0_2^+ and 0_2^++0_2^+ channels of interest here are rather small in those bands. Since the channel coupling among the different kind of eigen bands is weak, it can be interpreted that the inelastic scattering to the 0_1^++0_2^+ (0_2^++0_2^+) channel occurs through the weak transitions from the elastic component of the eigen bands in SC to the 0_1^++0_2^+ (0_2^++0_2^+) component of those in HC (CC).

References

1. A.H. Wuosmaa et al.: Phys. Rev. Lett. **68**, 1295 (1992)
2. M. Kamimura: Nucl. Phys. **A351**, 456 (1981)
3. W.D.M. Rae et al.: Phys. Rev. Lett. **68**, 3709 (1992)
4. Y. Hirabayashi et al.: Phys. Rev. Lett. **74**, 4141 (1995)
5. B. R. Fulton et al.: Phys. Rev. **C21**, 198 (1980)
6. S. Chappell et al.: Phys. Rev. **C51**, 695 (1995)
7. R.A. Le Marechal et al.: Phys. Rev. **C55**, 1881 (1997)
8. Y. Abe et al.: Suppl. Prog. Theor. Phys. **68**, 303 (1980)

Few-Body Systems Suppl. 12, 141–144 (2000)

Few-Body Systems

Continuum Level Density in the Microscopic Cluster Model

K. Arai[1,*], A.T. Kruppa[2]

[1] Department of Physics, Niigata University, Niigata 950-2181, Japan
[2] Institute of Nuclear Research of the Hungarian Academy of Sciences, Debrecen, P. O. Box 51, H–4001, Hungary

Abstract. Positions and widths of nuclear resonance states of the nucleus ^{8}Be have been calculated in the microscopic cluster model using real square integrable basis. The imposition of Gamow or scattering asymptotic boundary condition onto the wave function is avoided. The continuum level density smoothed by the Strutinsky averaging procedure is calculated by making use of the eigenvalues of the full and the free Hamiltonian matrices. The continuum level density is connected to the S-matrix and has a Breit-Wigner peak around the resonance energy. This approach is compared with the complex scaling method and with the scattering phase shift calculation.

1 Introduction

The study of resonance properties is one of the most important subjects in few-body physics. Some methods to describe resonances by the use of the bound state type technique were developed, e.g. the L^2-stabilization method[1, 2] and the complex scaling method[3, 4]. The merit of these approach is that the sophisticated models and computer codes which deal with the bound state problem are applied to the resonance state problem.

Recently we have generalized [5] the stabilization method of Mandelshtam et al. [2] employing the notion of the continuum level density (CLD). We showed that an approximate CLD, smoothed by the Strutinsky averaging procedure, can be determined without the box quantization or without the imposition of the scattering asymptotic boundary condition onto the wave function. The smoothed CLD can be calculated by making use of an arbitrary real square integrable basis. The CLD is connected to the scattering S-matrix and gives a Breit-Wigner type peak around the resonance energy and so the resonance parameters can be determined.

* *E-mail address:* arai@nt.sc.niigata-u.ac.jp

This method is applied to determine nuclear resonance states of ^{8}Be which is described by the microscopic two-cluster model α+α[6]. We have found that the approximate smoothed CLD is consistent with the exact density derived from the scattering phase shift. The parameters of the low lying resonance states predicted by the CLD method are compared with other two methods, the complex scaling method and scattering phase shift calculation.

2 Formulation

The models used for this nucleus is the microscopic cluster model of the type of the resonanting group method (RGM). The relative wave function between the clusters is expanded over the shifted Gaussian basis(Tohsaki-Suzuki's new GCM). The effective N-N interaction used is Volkov No.1(m=0.6). The size parameter of the α-particle is ν=0.235fm^{-2}. The size parameter of new GCM basis is γ=0.4fm^{-2} and the mesh of generator coordinate $\boldsymbol{s}$ is Δs=1.0fm.

The continuum level density $\Delta(E)$ is defined with the help of the Green-operators or the scattering phase shift $\delta(E)$,

$$\Delta(E) = -\frac{1}{\pi} Im\{Tr[\hat{G}(E+i0) - \hat{G}^0(E+i0)]\} = \frac{1}{\pi}\frac{d\delta(E)}{dE}, \tag{1}$$

where $\hat{G}(z) = (z-\hat{H})^{-1}$, $\hat{G}^0(z) = (z-\hat{H}_0)^{-1}$ are the Green-operator of the full and the free Hamiltonian, respectively. Since the CLD is proportional to the energy derivative of the scattering phase shift, this CLD has a Breit-Wigner peak around the resonance energy. If N square integrable basis function are used to solve the eigenvalue problem of the Hamiltonian, then the approximate CLD $\Delta^N(E)$ can be turned into the following form[5],

$$\Delta^N(E) = \sum_{i=1}^{N} \delta(E-e_i) - \sum_{i=1}^{N} \delta(E-e_i^0), \tag{2}$$

where e_i and e_i^0 are eigenvalues of the full and the free Hamiltonian matrices, respectively. In the free Hamiltonain a cluster relative Hamiltonian is given by the kinetic energy and the Coulomb term. Only direct term of basis function should be used for the calculation of matrix elements of the free Hamiltonain. To avoid the use of Dirac δ functions, the approximate CLD is smoothed by the Strutinsky averaging procedure[7]. After the Strutinsky smoothing procedure the smoothed approximate CLD is expressed as the difference of the two densities,

$$\bar{\Delta}^N(E) = \sum_{i=N-N_B}^{N} \frac{1}{\Gamma} w\left(\frac{E-e_i}{\Gamma}\right) - \sum_{i=1}^{N} \frac{1}{\Gamma} w\left(\frac{E-e_i^0}{\Gamma}\right), \tag{3}$$

where $N_B (N_B \leq N)$ denotes the number of the bound state eigenvalues of the full Hamiltonian matrix and $w(x)$ is the weight function of smoothing procedure and Γ is the range parameter of the smoothing. Here we take the Lorentzian

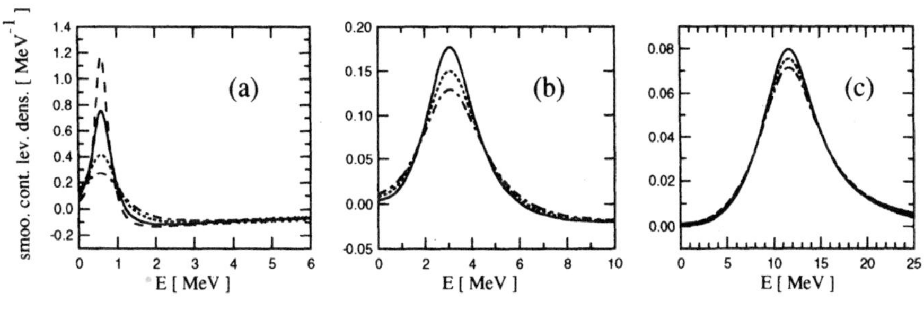

Figure 1. The smoothed approximate continuum level densities of the system α+α for the states J^π=0^+(a), 2^+(b) and 4^+(c). The dashed, solid, dotted and solid-dotted lines denote the result with Γ=0.2, 0.4, 0.8 and 1.2 MeV, respectively.

shape for the weight function. The smoothed CLD has also a peak around the resonance energy and the resonance parameters are determined from the fitting procedure with a smoothed Breit-Wigner form and the background term(second order polynomial in E).

3 Results and Discussion

Fig.1 shows the approximate smoothed CLD of 0^+, 2^+, 4^+ α+α states with a few smoothing parameters. The 0^+, 2^+, 4^+ densities have a resonance peak around 0.6 MeV, 3.1 MeV and 12 MeV, respectively. The dependence of the peak position on the smoothing parameter Γ is weak, but the peak becomes more pronounced (sharper) as the Γ parameter decreases. The basis dimensions used for the states J^π=0^+, 2^+ and 4^+ are N=500, 800 and 800, respectively. These dimensions are obviously much larger ones than what we used in the complex scaling method or in the variational method for the scattering calculation. Our approximate way of the calculation of the CLD has slow convergence.

The smoothed exact CLD can be derived from the scattering phase shift which are calculated with the exact asymptotic boundary condition. The smoothed exact CLD are compared with the smoothed approximate CLD and these two densities are agree with each other very well. It should be stressed that the good agreement what we have found holds not only around the resonance peak but also at non-resonant energies.

Table I shows the resonance parameters determined by the present method together with complex scaling method (CSM) and the scattering phase shift method. In scattering phase shift method, energy derivative of the phase shift are assumed to have a Breit-Wigner term plus the background term around a resonance energy and the resonance parameters are determined by fitting procedure. The good agreements found between the methods CLD, CSM and the phase shift method show that our approximate way of the calculation of the CLD is a good tool to determine the nuclear resonance states. It should be noted that the good agreements are found not only for the sharp resonance

Table 1. The resonance energy E_r [MeV] and width Γ_r [MeV] of ^{8}Be derived from the smoothed continuum level densities (CLD), the complex scaling method (CSM) and the scattering phase shift methods.

J^π	0^+		2^+		4^+	
	E_r	Γ_r	E_r	Γ_r	E_r	Γ_r
CLD	0.60	0.25	3.07	2.38	11.60	6.99
CSM	0.59	0.24	3.06	2.38	11.60	6.99
phase shift	0.60	0.25	3.07	2.38	11.60	7.00

but also for the broad resonance.

4 Summary

We have applied the continuum level density method to determine the well known low lying two-cluster resonance states of nucleus ^{8}Be. The CLD, smoothed by the Strutinsky averaging procedure, can be calculated approximately by making use of the eigenvalue of the full and the free Hamiltonian matrices. These eigenvalues are determined using finite number of real square integrable basis functions. The smoothed approximate CLD is found to be in good agreement with the smoothed exact density which is derived from the scattering phase shift with the exact scattering boundary condition. The CLD is connected to the scattering S-matrix, and has a Breit-Wigner peak around a resonance energy. The resonance parameters determined by this approximate CLD give good agreement with CSM and the scattering phase shifts. Our results shows that the present CLD method works well to calculate the nuclear resonance parameters.

References

1. A.U. Hazi, H.S. Taylor: Phys. Rev. **A1**, 1109 (1970)
2. V.A. Mandelshtam, T.R. Ravuri, H.S. Taylor: Phys. Rev. Lett. **70**, 1932 (1993)
3. Y.K. Ho: Phys. Rep. **99**, 1 (1993); N. Moiseyev: Phys. Rep. **302**, 211 (1998)
4. A.T. Kruppa, R.G. Lovas, B. Gyarmati: Phys. Rev. **C37**, 383 (1988)
5. A.T. Kruppa and K. Arai: Phys. Rev. **A59**, 3556 (1999)
6. K. Arai and A.T. Kruppa: Phys. Rev. **C** (in print)
7. V. M. Strutinsky: Nucl. Phys. **A95**, 420 (1967)

Few-Body Systems Suppl. 12, 145–148 (2000)

Few-Body Systems

Structures of Be Nuclei by a New Molecular Orbital Approach

N. Itagaki[1], S. Okabe[2]

[1] RI beam science laboratory, RIKEN, Wako, Saitama 351-0198, Japan

[2] Center for Information and Multimedia Studies, Hokkaido University, Sapporo 060-0810, Japan

Abstract. We discuss the importance of clustering in light neutron-rich nuclei. As examples, we analyse ^{10}Be and ^{12}Be based on a microscopic α+α+valence n model and show that the $\alpha - \alpha$ clustering is enlarged due to valence neutrons.

In ^{10}Be, all the observed positive parity bands and negative parity bands are described within the model (α+α+n+n), and the second 0^+ state is shown to have a large $\alpha - \alpha$ distance. Here, the two valence neutrons occupy along the $\alpha - \alpha$ axis and they enlarge the $\alpha - \alpha$ distance. The configuration of the valence neutrons corresponds to $(1/2^+)^2$ (σ-orbit in terms of MO), and these neutrons reduce their kinetic energy caused by the Pauli principle as the increase of the $\alpha - \alpha$ distance.

This mechanism is related to the breaking of $N = 8$ magic number in ^{12}Be. In ^{12}Be, this $(1/2^+)^2$ component plays an important role in the ground state. The calculated energy of this configuration is lower than the closed p-shell configuration for neutrons. Due to this effect, the observed slow β-decay probability from ^{12}Be to ^{12}B can be explained.

1 Introduction

Recently, contributions of higher shells were analyzed in $N = 8$ nuclei. A calculation based on the shell model has shown that the slow β-decay of ^{12}Be to ^{12}B can be explained due to admixture of the sd-shell in ^{12}Be ($N = 8$), where the single particle energy is artificially modified[1]. This shows that the concept of magic number is vague in ^{12}Be.

There have been pioneering studies for Be isotopes based on the molecular orbit (MO) model[2, 3, 4], however, the ground state of ^{12}Be have been analyzed to be dominantly the closed p-shell configuration for the neutrons. Here, we propose a new α+α+n+n+$\cdots$ model based on MO, where the model space for the valence neutrons is vastly extended, and we show the breaking of p-shell closed configuration in ^{12}Be.

In MO, the orbits for valence neutrons are expressed as linear combination of orbits around each α-cluster. When a linear combination of two p-orbits around the left-α-cluster and the right-α-cluster is summed up by the same sign, the resultant MO also has negative parity and one node. It cannot spread along the z-axis where two α-clusters are already located. If a linear combination of p-orbit is summed up by the different sign, the resultant MO has two nodes and positive parity. The optimal direction of this MO becomes $\alpha - \alpha$ direction. In this way, three basic orbits $K^\pi = 3/2^-, 1/2^-$ (one node) and $1/2^+$ (two nodes) for the valence neutrons are introduced.

The model space employed in the present model is extended from the traditional MO models. In the previous MO models, the valence neutrons have been treated as linear combinations of pure p-orbits around each α-cluster. We, however, solve the neutron motion around the α-clusters and their wave functions are expressed by superpositions of local Gaussians. In this framework, the total wave function is fully antisymmetrized, and the angular momentum and parity are projected to good quantum numbers. The Coulomb interaction is included and the center-of-mass motion is correctly eliminated.

As effective two-body $N-N$ interactions, we introduce Volkov No.2 potential with $M = 0.6, B = H = 0.125$ for the central part, and G3RS potential with the strength parameter V=2000 MeV for the spin-orbit part. These parameters are determined from the $\alpha - \alpha$, $\alpha - N$ and $N - N$ scattering phase-shifts.

First, we apply the model to ^{10}Be (α+α+n+n) [5], and the calculated energy levels are displayed in Fig. 1. The ground 0^+ state (-61.1 MeV) has

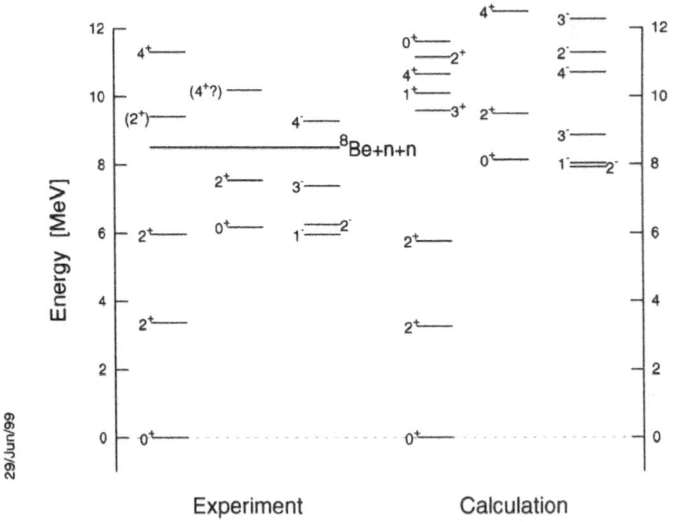

Figure 1. Calculated and experimental energy levels of ^{10}Be. The calculated levels sorted into five columns. From left to right: states whose dominant component is $K^\pi = 0^+$, $K^\pi = 2^+$, $K^\pi = 1^+$. The fourth column is the second 0^+ band characterized by the σ-orbits. The fifth column is the negative parity band.

dominantly the $(3/2^-)^2$ component for the two valence neutrons, and the 0_2^+ state at 8.1 MeV (experimentally 6.263 MeV) has dominantly not the $(1/2^-)^2$ but the $(1/2^+)^2$ component. Thus, the level inversion of *p*-shell and *sd*-shell holds also in ^{10}Be. The second 0^+ state has much larger charge radius (2.93 fm, proton radius 0.813 fm is used) than the ground state (2.51 fm), since the two valence neutrons occupy higher nodal orbit along the $\alpha - \alpha$ axis and increase the distance between two α-clusters to reduce the kinetic energy of the system. The state is a band head of cluster state.

This $(1/2^+)^2$ configuration strongly mixes also in the ground state of ^{12}Be. To analyse this effect, we introduce α+α+n+n+n+n model for ^{12}Be. As seen in Fig. 2, when the $\alpha - \alpha$ distance is small for example 2fm, the dominant configuration of the last two valence neutrons is $(1/2^-)^2$ for the ground state (dotted line), which corresponds to the *p*-shell closed configuration at the limit of $\alpha - \alpha$ distance zero. However, the $(1/2^+)^2$ configuration for the last two valence neutrons becomes lower as increase of the $\alpha - \alpha$ distance (solid line). At the $\alpha - \alpha$ distance of 3 fm, which is the optimal distance of ^{12}Be, the ground 0^+ state has squared overlap with $(1/2^-)^2$ by 0.41 and with $(1/2^+)^2$ by 0.64. Therefore, in the ground state of ^{12}Be, two different configurations contribute, and the neutron magic number $N = 8$ disappears.

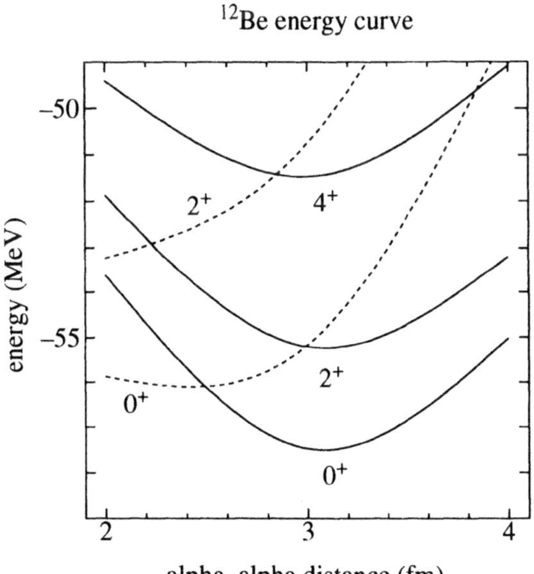

Figure 2. The energy curve with respect to the $\alpha - \alpha$ distance in ^{12}Be. The solid line has dominantly the $(1/2^+)^2$ configuration for the valence two neutrons.

One evidence of the mixing of $(1/2^+)^2$ component is the β-decay probability from ^{12}Be to ^{12}B. Using the present model, the log(ft) value of $^{12}\mathrm{Be}(0^+)\rightarrow{}^{12}\mathrm{B}(1^+)$ is calculated to be 3.81, and this value agrees with the experimental value of 3.8. However, if we restrict the configuration of the last two valence neutrons to $(1/2^-)^2$, the log(ft) value becomes 3.51. Therefore, the mixing of $(1/2^+)^2$ component suppresses this transition and increases the log(ft) value. The coincidence of the experimental and the theoretical log(ft) values suggests that the breaking of $N = 8$ magic number is realized in ^{12}Be.

References

1. T.Suzuki and T.Otsuka: Phys. Rev. **C56**, 847 (1997)
2. Y. Abe, J. Hiura, H. Tanaka: Prog. Theor. Phys. **49**, 800 (1973)
3. H. Furutani, H. Kanada, T. Kaneko, S. Nagata, H. Nishioka, S. Okabe, S. Saito, T. Sakuda, M. Seya: Prog. Theor. Phys. Suppl. 68, 193 (1980)
4. M. Seya, M. Kohno, S. Nagata: Prog. Theor. Phys. **65**, 204 (1981)
5. N. Itagaki and S. Okabe: Phys. Rev. **C** (to be published)

Few-Body Systems Suppl. 12, 149–152 (2000)

Analyses of Resonances by Using the Complex Scaling Method

T. Myo*, K. Katō

Division of Physics, Graduate school of Science, Hokkaido University, Sapporo 060-0810, Japan

Abstract. We discuss the role of unbound resonant and continuum states on the transition strength by applying the complex scaling method (CSM). We decompose the transition strength into the contributions from bound, resonance and continuum terms, and as a specific example we analyze the mechanism of the Coulomb breakup reaction of ^{11}Be. It is confirmed that the breakup is direct but does not take place through resonances.

1 Introduction

Many states appearing in unstable nuclei are observed as resonances due to their small binding energy with respect to the lowest threshold. Therefore, we can say that the studies of unstable nuclei are closely related to the resonance phenomena. Theoretically, it is indispensable to investigate the properties of resonances for getting knowledge on the structure of unstable nuclei.

We here discuss the analyses of resonances using the complex scaling method (CSM)[1], which is a very powerful method in solving resonance energies and decay widths of not only two-body but also few-body systems, with a familiar technique for bound state problems.

In addition to the decaying widths of the resonances, other physical quantities such as transition probabilities are expected to play an important role in understanding the properties of resonances in detail, and for this purpose the extension of the applicability of CSM is necessary.

Establishing a framework for the investigating the effects of resonant and continuum states will provide us with a deeper understanding of the excitation mechanism of the system. For example, in an actual system, the observed transition strengths in the Coulomb breakup reactions of some neutron-rich nuclei such as ^{11}Be and ^{11}Li, show an enhancement at the low energy region and it is needed to clarify which states determine the observed behavior.

* *E-mail address:* myo@nucl.sci.hokudai.ac.jp

2 Method

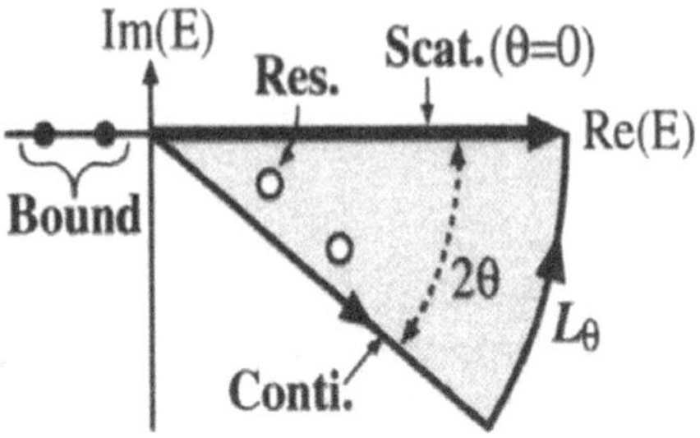

Figure 1. Schematic energy eigenvalue distribution with CSM.

In CSM, we transform spatial coordinates and linear momenta into complex numbers as;

$$\begin{aligned} \boldsymbol{r} &\rightarrow \boldsymbol{r}_\theta = \boldsymbol{r}\ e^{i\theta} \\ \boldsymbol{k} &\rightarrow \boldsymbol{k}_\theta = \boldsymbol{k}\ e^{-i\theta} \end{aligned} \tag{1}$$

where θ is a real parameter. Eigenvalue distribution in a complex energy plane is schematically displayed in Fig. 1. Furthermore, we proposed an extended completeness relation with CSM for bound, resonant and continuum states[2, 3]:

$$\mathbf{1} = \sum_B^{N_B} |\Phi_B^\theta\rangle\langle\widetilde{\Phi}_B^\theta| + \sum_R^{N_R(L_\theta)} |\Phi_R^\theta\rangle\langle\widetilde{\Phi}_R^\theta| + \int_{L_\theta} dk_\theta |\Phi_{k_\theta}^\theta\rangle\langle\widetilde{\Phi}_{k_\theta}^\theta|, \tag{2}$$

where the bra-states with tilde indicate biorthogonal ones of the ket-states. Eq. (2) provides us with a useful expression of the Green function;

$$\mathcal{G}^\theta(E,\boldsymbol{r},\boldsymbol{r}') = \mathcal{G}_B^\theta(E,\boldsymbol{r},\boldsymbol{r}') + \mathcal{G}_R^\theta(E,\boldsymbol{r},\boldsymbol{r}') + \mathcal{G}_{k_\theta}^\theta(E,\boldsymbol{r},\boldsymbol{r}'), \tag{3}$$

which enables us to study various kinds of strength distributions including resonances and back ground continuum states[3].

The reliability of the matrix elements for resonances with CSM is also discussed. Gyarmati and Vertse[4] proved that matrix elements calculated with CSM is equivalent to those with the convergence factor method which is another way to calculate the matrix elements.

$$\langle\widetilde{\Phi}_1|\widehat{O}|\Phi_2\rangle = \lim_{\alpha\to 0}\int dr e^{-\alpha r^2}\widetilde{\Phi}_1^*\widehat{O}\,\Phi_2 = \int dr\{\widetilde{\Phi}_1^*\}^\theta\widehat{O}^\theta\,\Phi_2^\theta = \langle\widetilde{\Phi}_1^\theta|\widehat{O}^\theta|\Phi_2^\theta\rangle \tag{4}$$

We numerically checked the equivalence of the matrix elements for resonances between two methods[5], and found it to be valid.

3 Application

We apply our method to a typical one-neutron halo nucleus ^{11}Be. One of the interesting question in relation with this nucleus is the mechanism of Coulomb breakup reaction into ^{10}Be and a neutron in association with the possibility of the soft dipole resonance. The dipole strength has been observed by Nakamura *et. al.*[6], which shows a strong enhancement at the low energy.

We concentrate our attention to the problem whether the dipole strength has any contributions from resonances or not. We employ a simple ^{10}Be+n model where ^{10}Be has a $(0s_{\frac{1}{2}})^4(0p_{\frac{3}{2}})^6$ configuration. In Fig. 2, we show the pole distribution of the final states $(\frac{1}{2}^-, \frac{3}{2}^-)$ in ^{11}Be at $\theta = 37.4°$. In the present simple model, low energy resonance poles are not found. In Table 1, we show the contribution of bound, resonance and continuum states to the energy weighted sum rule values. Small deviations of total values from the unity are considered to be a numerical error. The result shows that every state contributes comparably to the sum rule value, which requires the investigation of the contributions from both resonant and continuum states.

The decomposed dipole strength is shown in Fig. 3. We can easily recognize that the effect of resonances is negligible, and it is very hard to find any difference between the continuum term and the original distribution. This indicates that the Coulomb breakup of ^{11}Be is caused directly but not through specific resonances.

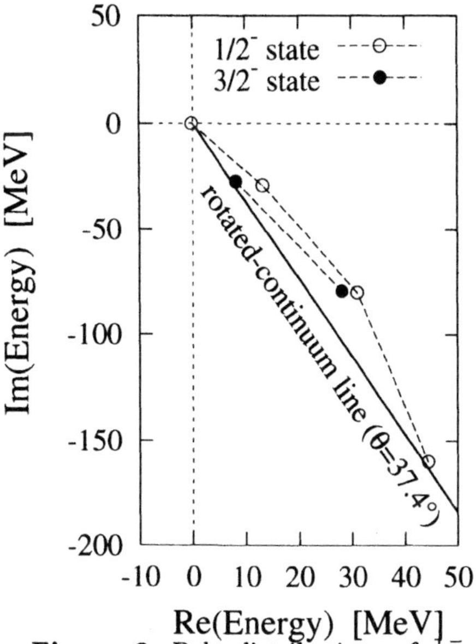

Figure 2. Pole distributions of $\frac{1}{2}^-$ and $\frac{3}{2}^-$ states with CSM ($\theta = 37.4°$).

As we have shown in Table 1 and Fig. 3. the contributions of resonances to the energy weighted sum rule value and the strength function are quite different, however, this result is easily understood as follows. The sum rule values is defined by the energy-weighted transition matrices which are enhanced for resonances with large energies and decay widths. On the other hand, the strength function (Green function) is defined as the matrix elements divided by the energy.

At least the enhanced peak of the strength function at the low energy is caused through a direct breakup and indicates the spatially extended wave function of the ground state of ^{11}Be.

Table 1. Sum rule values of the $E1$ transition ($\frac{1}{2}^+ \to \frac{1}{2}^-, \frac{3}{2}^-$) at the scaling angle $\theta = 37.4°$. All values are represented in the unit of $S_1(E1, \frac{1}{2}^-) = 2.62\ e^2\text{fm}^2\text{MeV}$ and $S_1(E1, \frac{3}{2}^-) = 3.17\ e^2\text{fm}^2\text{MeV}$. Last row is a sum of the values of each term.

	$J_f^\pi = \frac{1}{2}^-$		$J_f^\pi = \frac{3}{2}^-$	
	real	imaginary	real	imaginary
bound	0.1001	0.0000	—	—
1st resonance	0.5189	0.2869	0.2642	0.4051
2nd resonance	−0.2289	0.2263	−0.1716	0.0211
3rd resonance	0.3493	−0.2595	—	—
continuum	0.2616	−0.2540	0.9079	−0.4254
total	1.0009	−0.0003	1.0005	0.0008

4 Summary

Since unbound states are expected to play an essential role in unstable nuclear physics, we have to know as much information associated with these states as possible. It is necessary to develop a framework which enables us to examine the effects of resonances in the real observed physical quantities in direct way. In this paper, we presented a method to solve this problem by applying the complex scaling method and an extended completeness relation. We performed the separation of the transition strength into bound, resonance and continuum terms.

For ^{11}Be, the mechanism of the Coulomb breakup reaction is analyzed. The present calculation explicitly separates the contributions both from resonances and continuum states to the physical quantities. And the interpretation of the direct breakup is justified.

In the future, more theoretical developments of our framework including multi-channel and many-body cases are expected to be done.

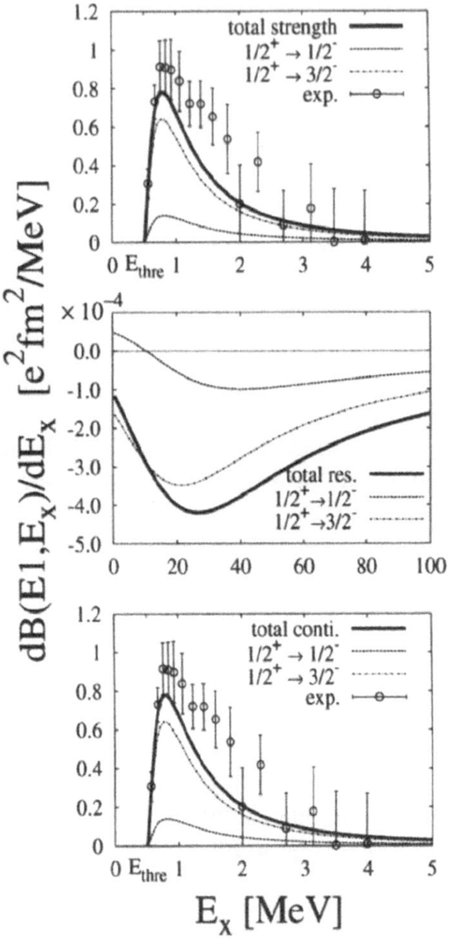

Figure 3. Separation of the original dipole strength function (upper) into resonance (middle) and continuum (lower) terms at $\theta = 37.4°$. Dashed, dash-dotted and solid lines correspond to the transition for $\frac{1}{2}^+ \to \frac{1}{2}^-, \frac{3}{2}^-$ and total, respectively. Open circles with errorbars are the experimental data.

References

1. J. Aguilar and J.M. Combes: Commun. Math. Phys. **22**, 269 (1971); E. Balslev and J.M. Combes: Commun. Math. Phys. **22**, 280 (1971)
2. T. Myo and K. Katō: Prog. Theor. Phys. **98**, 1275 (1997)
3. T. Myo, A. Ohnishi, K. Katō: Prog. Theor. Phys. **99**, 801 (1998)
4. B. Gyarmati and T. Vertse: Nucl. Phys. **A160**, 523 (1972)
5. M. Homma, T.Myo, K.Katō: Prog. Theor. Phys. **97**, 561 (1997)
6. T. Nakamura et. al: Phys. Lett. **B331**, 296 (1994)

Few-Body Systems Suppl. 12, 153–156 (2000)

Few-Body Systems

Measurement of Deuteron Elastic Scattering at E_d = 270 MeV and a Comparison with Sudden Approximation

T. Ohnishi[1], H. Sakai[1], H. Okamura[2], S. Ishida[3], H. Otsu[4], N. Sakamoto[5], T. Uesaka[5], T. Wakasa[6], Y. Satou[5]

[1] Department of Physics, University of Tokyo, Tokyo, Japan

[2] Department of Physics, Saitama University, Urawa, Saitama, Japan

[3] Soei International Patent Firm, Chuo-ku, Tokyo, Japan

[4] Department of Physics, Tohoku University, Sendai, Miyagi, Japan

[5] The Institute of Physical and Chemical Research, Wako, Saitama, Japan

[6] Research Center for Nuclear Physics, Osaka University, Ibaraki,Osaka, Japan

Abstract. Differential cross section and analyzing powers, A_y and A_{yy}, for the elastic scattering of polarized deuteron from ^{12}C, ^{40}Ca, ^{90}Zr, and ^{208}Pb targets have been measured at $E_d^{\mathrm{lab}} = 270$ MeV.

A sudden approximation has been applied to explain the experimental results. The calculation reproduces the phase of the momentum transfer dependence, but not the amplitude.

1 Introduction

The experimental results of the deuteron elastic scattering at intermediate energy have been compared to those of the proton elastic scattering [1]. From the comparison, the simple relation between the deuteron and the proton elastic scattering was obtained. The differential cross section of the deuteron elastic scattering resembles to the proton elastic scattering in both of the magnitude and the momentum transfer dependence. The same relation was obtained for the vector analyzing power. It means that the deuteron elastic scattering can be described by using the elastic scattering of the constituent nucleons, proton and neutron.

To examine such a complex nucleus scattering, the sudden approximation is one of the useful approach. In this approximation, the projectile is considered as a weakly bound system. An impulse approximation is employed. Thus the complex nucleus scattering is replaced by the composite of the constituent nucleons scattering. Y. Sakuragi and M. Tanifuji applied this approximation to the deuteron elastic scattering for ^{16}O at 400 MeV, ^{40}Ca and ^{58}Ni at 400 and 700 MeV [2]. The experimental data were reproduced with this model. Since their comparison was performed for the particular incident energy and the target combination, more systematic examination is desirable to check the usefulness of the sudden approximation. In this work, the deuteron elastic scattering has been measured at E_d = 270 MeV for various targets. The sudden approximation calculations have been performed.

2 Experiment and Results

This measurement was performed at the RIKEN Accelerator Research Facility (RARF). The polarized deuteron was accelerated to 270 MeV by the AVF and the Ring cyclotrons. The vector and tensor polarizations of the beam were measured by two polarimeters using d+p elastic scattering [3]. The beam polarizations were about 60 – 70% of the theoretical maximum values throughout the present experiment.

The differential cross section and the analyzing powers, A_y and A_{yy}, have been measured for ^{12}C, ^{40}Ca, ^{90}Zr, and ^{208}Pb targets over the angular range $\theta_{\rm lab}$=4°–40°. The scattered deuterons were momentum analyzed by the magnetic spectrograph SMART [4] with an energy resolution of 300 keV(FWHM).

The momentum transfer q dependence of the experimental results is displayed in figure 1. The uncertainties indicated in figure 1 are statistical only and they are typically about 1% and 0.03 for the differential cross section and the analyzing powers, respectively.

3 Comparison with the Sudden Approximation

In the sudden approximation, the scattering matrix of the deuteron elastic scattering, $M^{(d)}$ is described by using the single and double collision terms. In the single collision term, one nucleon is scattered by the target and the other nucleon plays as a spectator. In the double collision term, both nucleons are scattered. When the single collision term is only considered, the relations between the elements of $M^{(d)}$ and those of the proton and the neutron scattering matrices are obtained. Finally, the observables of the deuteron elastic scattering can be written as

$$\frac{d\sigma}{d\Omega}^{(d)} = K(q)^2\left(|a^{(p)}+a^{(n)}|^2+\frac{2}{3}|b^{(p)}+b^{(n)}|^2\right), \tag{1}$$

$$A_y^{(d)} = \frac{2\mathrm{Im}((a^{(p)}+a^{(n)})(b^{(p)}+b^{(n)})^*)}{|a^{(p)}+a^{(n)}|^2+\frac{2}{3}|b^{(p)}+b^{(n)}|^2}, \tag{2}$$

$$A_{yy}^{(d)} = \frac{2}{3}\frac{|b^{(p)}+b^{(n)}|^2}{|a^{(p)}+a^{(n)}|^2+\frac{2}{3}|b^{(p)}+b^{(n)}|^2}, \tag{3}$$

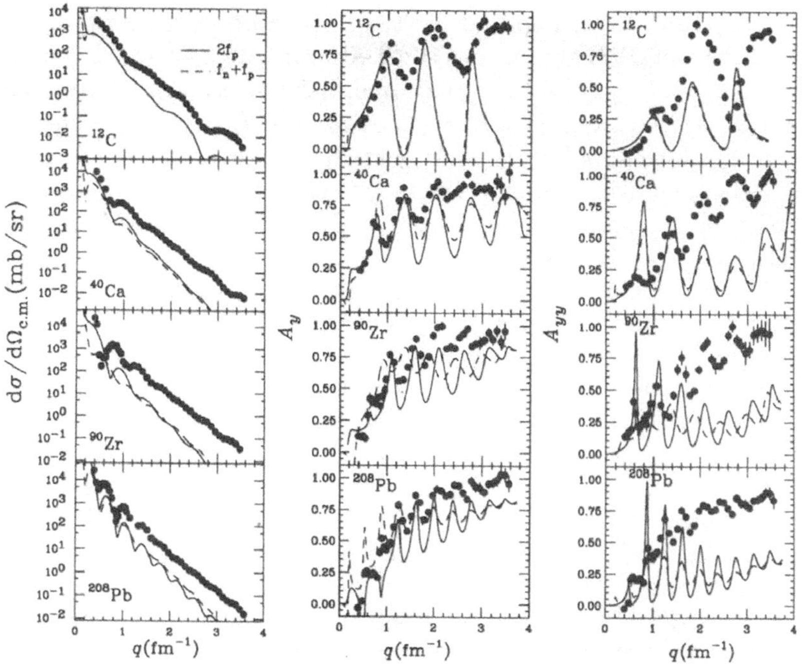

Figure 1. The momentum transfer dependence of the differential cross section and the vector and tensor analyzing powers, A_y and A_{yy}, on ^{12}C, ^{40}Ca, ^{90}Zr, and ^{208}Pb. The solid and dashed lines are for the $2f_p$ type and $f_p + f_n$ type calculations, respectively.

$$M^{(j)} = \begin{array}{c} \\ \nu'_j = \frac{1}{2} \\ -\frac{1}{2} \end{array} \begin{array}{c} \begin{array}{cc} \nu_j = \frac{1}{2} & -\frac{1}{2} \end{array} \\ \left(\begin{array}{cc} a^{(j)} & b^{(j)} \\ -b^{(j)} & a^{(j)} \end{array} \right) \end{array} . \tag{4}$$

$M^{(j)}$ is the proton or the neutron scattering matrices. ν_j is the z-component of the spin of the particle j. $K(q)$ is the deuteron form factor.

To calculate the proton scattering amplitude, the global optical potential [5] is used. As for the neutron scattering amplitude, the proton optical potential without the Coulomb term is used. The deuteron form factor is calculated by using the Hulthen wave functions. The effect of the double collision term is included by modifying the momentum transfer as $q = (1 - \alpha)q_p$ where q_p is the momentum transfer for the proton elastic scattering. The factor α is a adjustable parameter and in this calculation the value 0.07 is used, because it was predicted in the previous calculation [2].

In this calculation, two type calculations are performed. One is referred as $f_p + f_n$ type, which uses the proton and neutron optical potentials. f_p (f_n) represents the proton (neutron)-nucleus optical potential. The other is $2f_p$ type, which uses only the proton optical potential. The difference between two type calculations appears in the amplitude of the Rutherford cross sections. The Rutherford cross section is proportional to the factor $(\mu Z_1 Z_2)^2$ where μ is the reduced mass, Z_1 and Z_2 are the projectile and the target charge, respectively. Factors in case of the deuteron optical potential, the $f_p + f_n$ type, and the $2f_p$ type, are equals to $(2\mu_p Z_2)^2$, $(\mu_p Z_2)^2$, and $(2\mu_p Z_2)^2$, respectively. μ_p is the reduced mass between the proton and the target. Thus, the $2f_p$ type calculation is more preferable to reproduce the deuteron's Rutherford cross section than the $f_p + f_n$ type.

The solid and dashed lines in figure 1 are the results of the $2f_p$ type and the $f_p + f_n$ type calculations, respectively. The amplitudes of the cross section calculated in both types are underestimated the experimental values. At the small momentum transfer, the $2f_p$ type calculation reproduces the momentum transfer dependence of the cross section better than the $f_p + f_n$ type does. The momentum transfer dependence of the analyzing power A_y is also reproduced better in the $2f_p$ type than in the $f_p + f_n$ type. However the amplitude of the analyzing power A_y is not reproduced in both calculation. Furthermore, the tensor analyzing power A_{yy} at the large momentum transfer region is not reproduced. That is probably because the effect of the deuteron D-state is not included in the sudden approximation. Consequently, the sudden approximation reproduces the momentum transfer dependence, but not the amplitude. The $2f_p$ type calculation is better than the $f_p + f_n$ type calculation.

4 Summary

In summary, the differential cross section and the analyzing powers, A_y and A_{yy}, have been measured for the deuteron elastic scattering for ^{12}C, ^{40}Ca, ^{90}Zr, and ^{208}Pb targets at $E_d = 270$ MeV. The calculations using the sudden approximation are performed and compared with the experimental data. Though the amplitudes are underestimated, the momentum transfer dependences are almost reproduced. To improve the calculation, the inclusion of the deuteron D-state and more exact treatment of the double collision are needed. Finally, we thank Prof. M. Tanifuji for his advice of the calculation, and also thank S. Fujita and T. Nonaka for their help doing the experiment.

References

1. Nguyes van sen et al.: Nucl. Phys. **A464**, 717 (1987)
2. Y. Sakuragi and M. Tanifuji: Nucl. Phys. **A560**, 945 (1993)
3. N. Sakamoto et al.: Phys. Lett. **B367**, 60 (1996)
4. T. Ichihara et al.: Nucl. Phys. **A569**, 287c (1994)
5. P. Schwandt et al.: Phys. Rev. **C26**, 55 (1982)

Few-Body Systems Suppl. 12, 157–162 (2000)

Few-Body Systems

Chiral SU(3) Quark Model and Dibaryon $(\Omega\Omega)_{J^\pi=0^+}$

Z.Y. Zhang[1], Y.W. Yu[1], C.R. Ching[2], T.H. Ho[2]

[1] Institute of High Energy Physics, 100039, Beijing, P. R. China
[2] Institute of Theoretical Physics, 100080, Beijing, P. R. China

Abstract. The structure of a new dibaryon $(\Omega\Omega)_{0^+}$ is studied and predicted in the framework of the chiral SU(3) quark model by solving a resonating group method (RGM) equation. The binding energy of this dibaryon is around 100 MeV, the mean square root of the distance between two Ω is 0.84 fm and its mean life time is about 10^{-10} sec. All these interesting properties, and also the two negative charge units it carries could make it to be easily identified in the experiments.

1 Introduction

As is well known, in the ordinary strong interation world there are only baryons consisting of three valence quarks and mesons of quark-antiquark pair. However, for more than twenty years people have discusseed and explored the possible existence of some exotic multi-quark states and gluonic states. Not only because it provides a good place to examine the Quantum Chromodynamics (QCD) theory and to display the quark-gluon behavior in short distance, but also the very existence of such systems would open a new area for studying many new physical phenomena we have not known before. The theoretical investigations have concentrated mainly on H dihyperon [1, 2, 3] and d' particle [4, 5], and up to now there is no convincing experimental evidence for the existence of these particles. This indicates, in seeking for multi-quark system, or so-called dibaryon, one probably should go beyond these few candidates to the multi-strangeness systems.

Recently we have developed a chiral SU(3) quark model[6, 7] in which the coupling between the chiral fields and quarks is considered to describe the medium range non-perturbative QCD effect. Since this model reproduced the energies of the baryon ground states, the nucleon-nucleon $(N-N)$ scattering phase shifts and the hyperon-nucleon $(Y-N)$ cross sections correctly, making an extrapolation in the same framework without introducing any new parameters to predict dibaryons' structure is feasible and reliable to some extent. Using this model, [8], we found that some six quark states with high strange number have more attraction from the chiral quark coupling. Of particular interest is the $(\Omega\Omega)_{0^+}$. It is a deeply bound state and its mean life time is quite long because it can only undergo through weak decay. Since this dibaryon has charge -2 and enough long mean life time, and therefore it could be a new interesting candidate of dibaryon.

2 A Brief Introduction of Chiral $SU(3)$ Quark Model

In the usual quark potential model, the quark-quark interation has only two parts: one-gluon-exchange(OGE) and confinement potential. This simple model is quite successful in explaining the heavy quarkonia. However, if applied to the light quark system, one finds, that the medium range non-perturbative effect has to be added, and the constituent quark mass must be put in phenomenologically. An alternative approach is to employ the $SU(2)$ σ model. In order to study the systems with strangeness, we have generalized the idea of the $SU(2)$ σ model, to the flavor $SU(3)$ case, in which, a unified coupling between quarks and all scalar and pseudo-scalar chiral fielfs is introduced. With this generalization, the interacting Hamiltonian between quarks and chiral fields is now written as:

$$H_I^{ch} = g_{ch}F(q^2)\bar{\psi}(\sum_{a=0}^{8}\sigma_a\lambda_a + i\sum_{a=0}^{8}\pi_a\lambda_a\gamma_5)\psi, \tag{1}$$

where λ_0 is a unitary matrix, $\lambda_1...\lambda_8$ the Gell-Mann matrix of flavor $SU(3)$ group, $\sigma_0...\sigma_8$ the scalar nonet fields and $\pi_0...\pi_8$ the pseudo-scalar nonet fields. The chiral field coupling constant g_{ch} can be fixed by the relation with $g_{NN\pi}$ and $\frac{g_{NN\pi}^2}{4\pi}$ is taken to be the experimental value.

It is easy to show that H_I^{ch} is invariant under the $SU(3)$ infinitesimal chiral transformation, and from H_I^{ch}, the chiral-field-induced quark-quark potential V_{ij}^{ch} between two quarks is derived, which describes the medium range non-perturbative effect.

The total Hamiltonian of the chiral $SU(3)$ quark model is obtained by adding V_{ij}^{ch} as:

$$H = \sum_i T_i - T_G + \sum_{i<j} V_{ij}, \tag{2}$$

and

$$V_{ij} = V_{ij}^{conf} + V_{ij}^{OGE} + V_{ij}^{ch}. \tag{3}$$

Where $\sum\limits_i T_i - T_G$ is the kinetic energy of the system, and V_{ij} is the interaction between two quarks. V_{ij}^{conf} is the confinement potential, taken as quadratic form, V_{ij}^{OGE} the OGE interaction and V_{ij}^{ch} includes scalar meson exchange and pseudo-scalar meson exchange potentials. Their expressions can be found in Ref.[6].

We fixed the parameters by fitting the baryons' masses and their stability conditions. Once the parameters are determined, all the Octet and Decuplet baryons' masses can be reproduced in our model [8]. Equipped with this model with all the parameters thus determined, we studied the two baryon systems on quark level dynamically by solving the resonating group method (RGM) equation of the Hamiltonian Eq.(2). In the RGM calculation, the trial wave function is taken to be

$$\Psi_{ST} = \sum_i c_i \Psi_{ST}^{(i)}(\vec{s_i}), \tag{4}$$

with

$$\Psi_{ST}^{(i)}(\vec{s_i}) = \mathcal{A}(\phi_A(\vec{\xi_1}, \vec{\xi_2})_{S_A T_A} \phi_B(\vec{\xi_4}, \vec{\xi_5})_{S_B T_B} \chi(\vec{R}_{AB} - \vec{s_i}) \mathcal{R}_{CM}(\vec{R}_{CM}))_{ST}, \tag{5}$$

where A and B describe two clusters, and ϕ, χ and $\mathcal{R}$ represent internal, relative and center of mass motion wave functions respectively. $\vec{s_i}$ is the generator coordinate and $\mathcal{A}$ is the anti-symmetrization operator,

$$\mathcal{A} = 1 - \sum_{i \in A, j \in B} P_{ij}, \tag{6}$$

where P_{ij} is the permutation operator of ith and jth quarks.

The results show that all the calculated $N - N$ scattering phase shifts, the $Y - N$ scattering and reaction cross-sections are in agreement with the experimental data. Therefore, an unified description of the baryon structure and the baryon-baryon interaction is achieved in the frame-work of our model.

In this paper, we extend our study to the dibaryon systems by solving a bound state equation. We found that among all possible candidates, of particular interest is the two Ω dibaryon $(\Omega\Omega)_{0^+}$.

3 Main Properties of $(\Omega\Omega)_{0^+}$ Dibaryon

We calculated the eigen energy of the two Ω system by taking the same set of parameters which we used in the $N - N$ and $Y - N$ scattering calculations [6]. The results showed, that $(\Omega\Omega)_{0^+}$ is a deeply bound state with small quark distribution distance.

$$B_{(\Omega\Omega)_{0^+}} = -(E_{(\Omega\Omega)_{0^+}} - 2M_\Omega) = 116 MeV, \quad rms = \sqrt{< r_{ij}^2 >} = 0.66 fm, \tag{7}$$

where $B_{(\Omega\Omega)_{0^+}}$ denotes the binding energy and rms the root of mean square distance between two quarks. Moreover, the relative motion wave function between two Ωs, $\chi(R)$ is also separated and drawn in Fig.1. And the root of mean square distance between $\Omega - \Omega$, $RMS = \sqrt{< R_{\Omega\Omega}^2 >} = 0.84 fm$ is obtained.

These results have shown that $(\Omega\Omega)_{0^+}$ is a very interesting six quark state, not only because of its huge binding energy , but also of the quite short distance between $\Omega - \Omega$. This property, combined with the long life-time, makes it most favorable candidate for looking for dibaryon.

To understand the very physics underlying the curious features, and to examine to what extent the results are dependent on the model with all its parameters, we have carried out an analysis to see the reason.

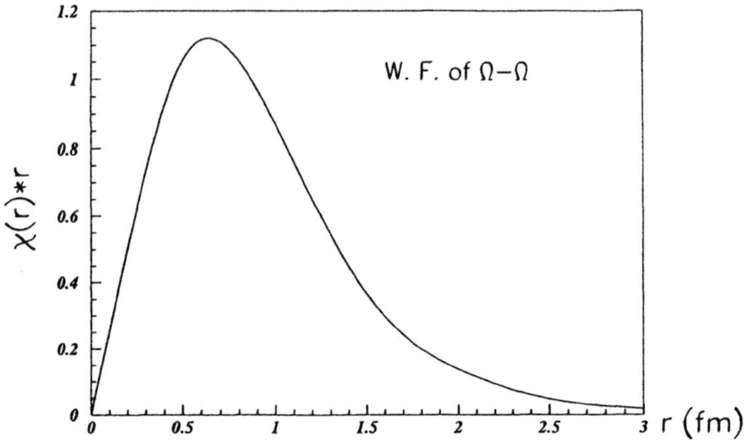

Figure 1. The relative motion wave function of $\Omega - \Omega$

First let us discuss the symmetry property of this system. As is well known, the compositeness of the baryon and the color magnetic interaction in the *OGE* potential govern the short range region. In the $(\Omega\Omega)_{0^+}$ case, although the OGE interaction offers repulsion, but its symmetry property is very special. According to the transformation relations between the physical states of two baryons and the symmetry bases (group chain classification bases) of six quark system [9, 10], one sees that among all 280 physical bases only six of them, namely, $(\Delta\Delta)_{ST=30}$, $(\Delta\Delta)_{ST=03}$, $(\Delta\Sigma^*)_{ST=3\frac{1}{2}}$, $(\Delta\Sigma^*)_{ST=0\frac{5}{2}}$, $(\Xi^*\Omega)_{ST=0\frac{1}{2}}$, and $(\Omega\Omega)_{ST=00}$, have the largest component of $[6]_r$ symmetry in the orbital space of the symmetry bases. And what is remarkable, that $(\Omega\Omega)_{0^+}$ is the only state among them, which is stable under the strong interaction.

On the other hand, the eigenvalue of the normalization kernel in the spin-flavor-color space $< \mathcal{A}^{\sigma fc} >=< 1 - \sum_{i\in A, j\in B} P_{ij}^{\sigma fc} >$ can serve as a direct measure on how the Pauli principle works in the relevant state [11]. When $< \mathcal{A}^{\sigma fc} >= 0$, it is a forbidden state for forming $[6]_r$ symmetry in the orbital space as a consequence of the Pauli exclusion principle. $< \mathcal{A}^{\sigma fc} >= 1$ means the quark exchange effect between two baryons is unimportant, the two baryons can be regarded as two clusters without any quark exchange. When $< \mathcal{A}^{\sigma fc} >$ larger than 1, the quark exchange effect is also cardinal, and tends to drag two clusters closer. Our calculation showed that for $(\Omega\Omega)_{0^+}$, $< \mathcal{A}^{\sigma fc} >= 2$, so that $(\Omega\Omega)_{0^+}$ has relatively high antisymmetry in the spin-flavor-color space and symmetry in the orbital space, therefore is favored by forming the $[6]_r$ symmetry basis.

This effect can be seen more clearly in the generator coordinate method (GCM) matrix elements of the kinetic energy $V_{kin}(s)$. In Fig.2, the solid curve represents the case of $(\Omega\Omega)_{0^+}$, and the dashed one is for the case of Deuteron (only central part). For comparison, the corresponding potential without quark exchange is also shown as dotted curve. From Fig.2, one sees that, in the contrary of the Deuteron case, the $(\Omega\Omega)_{0^+}$ state has very strong attraction in the kinetic energy. This is because $< \mathcal{A}^{\sigma fc} >_{\Omega\Omega} = 2$, but $< \mathcal{A}^{\sigma fc} >_{deu} = 10/9$, the attraction arising from the quark exchange in the kinetic energy part for $(\Omega\Omega)_{0^+}$ is quite large, but for Deuteron it is very small. Since the kinetic energy itself is spin-flavor-color independent, the big difference between $V_{kin}^{deuteron}(s)$ and $V_{kin}^{\Omega\Omega}(s)$ is totally arising from the symmetry property.

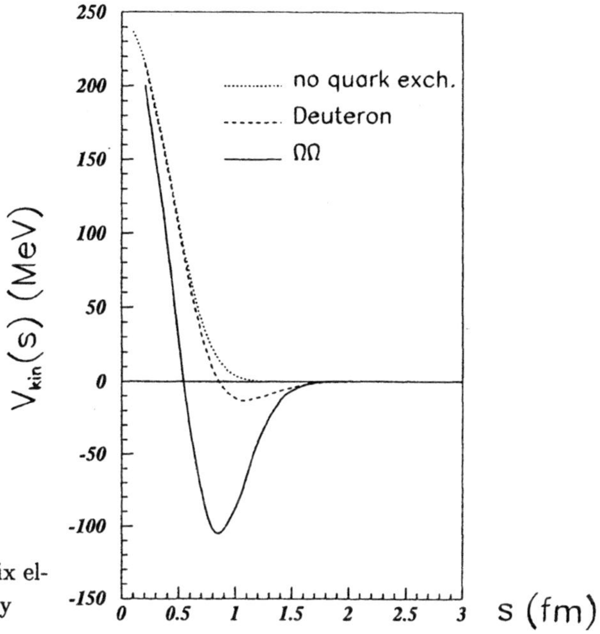

Figure 2. The GCM matrix elements of the kinetic energy

For the chiral field coupling effect, first we have made a chiral $SU(2)$ quark model calculation to see the contribution from the σ_0 exchange quantitatively. The binding energy of $(\Omega\Omega)_{0^+}$ in this case is larger than $50 MeV$ (see case (III) in Table 1), while for deuteron case the binding energy is only of 2 MeV as the experimental data have dictated. That means, even in the framework of SU(2), the theoretical calculation predicts that the binding energy of $(\Omega\Omega)_{0^+}$ is almost twenty times bigger than that of Deuteron's. Then we include the exchange of all nonet pseudo-scalar mesons plus σ_0. The result shows, that when π, K, η, η' and σ_0 chiral fields are added, $B_{(\Omega\Omega)_{0^+}} = 74 MeV$ (see Table 1 case (II)). Finally, with the inclusion of the exchange of all nonet pseudo-scalar and scalar mesons in our calculation, i.e. the chiral SU(3) quark model, the binding becomes even more stronger. One sees that besides σ_0, the effect from the other scalar mesons (σ', κ and ϵ) also enhance the binding in the $(\Omega\Omega)_{0^+}$ system. The total binding energy reaches 116 MeV when these scalar mesons are included (see Table 1, case (I)). All these analyses point to the fact, that both the specific symmetry property of the system $(\Omega\Omega)_{0^+}$ in the short range, and the nonperturbative QCD effect described by chiral-quark coupling in the medium range, are responsible for forming a deeply bound state of $(\Omega\Omega)_{0^+}$.

Table 3.1. Binding Energies and RMS of $(\Omega\Omega)_{0^+}$ for different cases

	Chiral SU(3) Quark Model	$\pi, K, \eta, \eta' + \sigma_0$ Model	Chiral SU(2) Quark Model
	(I)	(II)	(III)
$B_{(\Omega\Omega)_{0^+}}(MeV)$	116	74	54
$RMS(fm)$	0.84	0.92	0.98

4 Conclusions

In the preceding sections we have shown, that the $(\Omega\Omega)_{0^+}$ is a deeply bound dibaryon with the binding energy, which is about forty times bigger than that of the Deuteron's. Its mean square root distance between the quarks is only about half of that of the Deuteron. Such remarkable features predominantly are stemmed from its special symmetry property. The symmetry property further coupled with meson exchange potentials, and finally resulted in about 100 MeV binding energy for $(\Omega\Omega)_{0^+}$. In this sense $(\Omega\Omega)_{0^+}$ is not a normal two-baryon's bound state, it is a six-quark state. Given the binding energy, another important and intersting property is the life time of $(\Omega\Omega)_{0^+}$. Because of the large strangeness quantum number of this system, only the weak decay is allowed. Its life-time is around 10^{-10} sec. As a conclusion, we believe, $(\Omega\Omega)_{0^+}$ is a best candidate for looking for the long-sought dibaryon.

References

1. R.J. Jaffe: Phys. Rev. Lett. **38**, 195 (1977)
2. K. Shimizu and M. Koyama: Nucl. Phys. **A646**, 211 (1999)
3. U. Straub, Zong-ye Zhang, K. Bräuer, A. Faessler, S.B. Khadkikar: Phys. Lett. **B200**, 241 (1988); P.N. Shen, Z.Y. Zhang, Y.W. Yu, X.Q. Yuan, S. Yang: *H Dihyperon in Quark Cluster Model*, J. Phys. **G-Nu 25**, (1999)
4. A.J. Buchmann et al.: Prog. Part. Nucl. Phys. **36**, 383 (1996)
5. H. Clement et al.: Prog. Part. Nucl. Phys. **36**, 369 (1996)
6. Z.Y. Zhang, Y.W. Yu, P.N. Shen, L.R. Dai: Nucl. Phys. **A625**, 59 (1997)
7. P.N. Shen, Y.B. Dong, Y.W. Yu, Z.Y. Zhang, T.S.H. Lee: Phys. Rev. **C55**, 2024 (1997)
8. Y.W.Yu, Z.Y.Zhang, X.Q.Yuan: Commeun. Theor. Phys. **31**, 1 (1999); Z.Y. Zhang, Y.W. Yu, X.Q.Yuan: *A Study of Six Quark Cluster States in Chiral SU(3) Quark Model*, Nucl. Phys. **A** (to be published)
9. M. Harvey: Nucl. Phys. **A352**, 301 (1981)
10. F. Wang, J.L. Ping, T. Goldman: Phys. Rev. **C51**, 1648 (1995)
11. K. Shimizu: Prog. Phys. **52**, 1 (1989)

Few-Body Systems Suppl. 12, 163–170 (2000)

Few-Body Systems

Baryon Structure and the Chiral Symmetry of QCD

L.Ya. Glozman*

High Energy Accelerator Research Organization (KEK), Tanashi Branch, Tanashi, Tokyo 188-8501, Tokyo, Japan

Abstract. I begin with the key problem of light and strange baryon spectroscopy which suggests a clue for our understanding of underlying dynamics. Then I discuss spontaneous breaking of chiral symmetry in QCD, which implies that at low momenta there must be quasiparticles - constituent quarks with dynamical mass, which should be coupled to other quasiparticles - Goldstone bosons. Hence in the low-energy regime the underlying dynamics in baryons is due to Goldstone boson exchange (GBE) between constituent quarks. Using as a prototype of the microscopical quark-gluon degrees of freedom the instanton-induced 't Hooft interaction I show why the GBE is so important. When the 't Hooft interaction is iterated in the qq t-channel it inevitably leads to a pole which corresponds to GBE. I show how the explicitly flavour-dependent short-range part of the GBE interaction between quarks, perhaps in combination with the vector-meson exchange interaction, solves the key problem of baryon spectroscopy and present spectra obtained in a simple analytical calculation as well as in exact semirelativistic three-body approach.

1 Where is the Key Problem?

If one considers a model with an effective confining interaction between quarks in light and strange baryons, which is flavour- and spin - independent and assuming that there are no residual interactions, then the spectrum of lowest baryons should be arranged into successive bands of positive and negative parity, see Fig. 1. In Nature, however, the lowest levels in the spectra of nucleon, Δ-resonance and Λ-hyperon, which are shown on Fig. 2, look pretty different.

Typically models pay an attention to the octet-decuplet splittings. Within a quark picture one needs a spin-spin force between valence quarks with a proper sign. Then, adjusting a strength of this spin-spin force one can explain why Δ is heavier than nucleon, or why Σ is heavier than Λ. When QCD appeared, it has been immediately suggested that such a spin-spin force is supplied by the

* *Present address:* Institute for Theoretical Physics, University of Graz, Universitätsplatz 5, A-8010 Graz, Austria

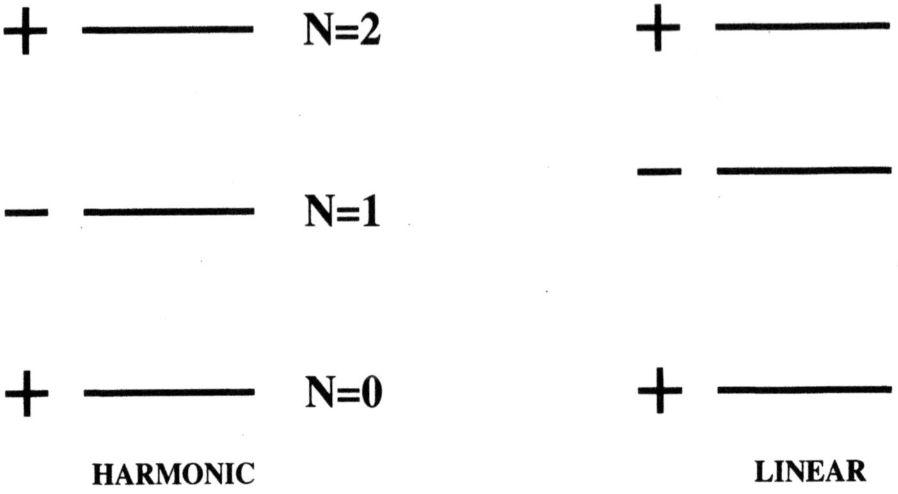

Figure 1. A sequence of positive and negative parity levels with harmonic and linear confining interactions.

colour-magnetic component of the one gluon exchange (OGE) [1, 2, 3]. At the price of a very large strong coupling constant, $\alpha_s \sim 1$, one can then fit $\Delta - N$ mass difference. Clearly that such a picture is self-contradictory, because a big value of α_s is not compatible with the perturbative treatment of QCD.

The crucial point, however, is that the perturbative gluon exchange is sensitive only to spin (and colour) degrees of freedom of quarks and there is no sensitivity at the operator level to the flavour of quarks (in the u,d,s quark sector there is only a very weak sensitivity via different masses of quarks which, however, completely vanishes in the chiral limit). The spin structure of all baryons in N and Λ spectra, depicted in Fig. 2, is the same, it is described by the mixed permutational symmetry. This means that the contribution of the colour-magnetic interaction to leading order is the same in all these baryons (up to some small difference in baryon orbital wave functions), which is in apparent conflict with the *opposite* orderings of the lowest levels in N and Λ spectra. The only difference between N and Λ system is that one light quark is substituted by a strange one. It immediately hints that the physics, responsible for Fig. 2, should be explicitly flavour dependent. In addition, a colour magnetic interaction cannot shift the $N = 2$ states $N(1440)$ and $\Lambda(1600)$ below the $N = 1$ states $N(1535) - N(1520)$ and $\Lambda(1670) - \Lambda(1690)$, respectively, because to leading order its contribution is the same in all these states. In the Δ spectrum the situation is even more dramatic as the colour - magnetic interaction shifts the $N = 2$ state $\Delta(1600)$ up, but not down, with respect to the $N = 1$ pair $\Delta(1620) - \Delta(1700)$. These facts rule out perturbative gluon exchange picture as a source of the hyperfine interactions in the light and strange baryons.

The other possible source of the hyperfine interactions, the 't Hooft instanton induced interaction [4] between valence quarks also could generate the

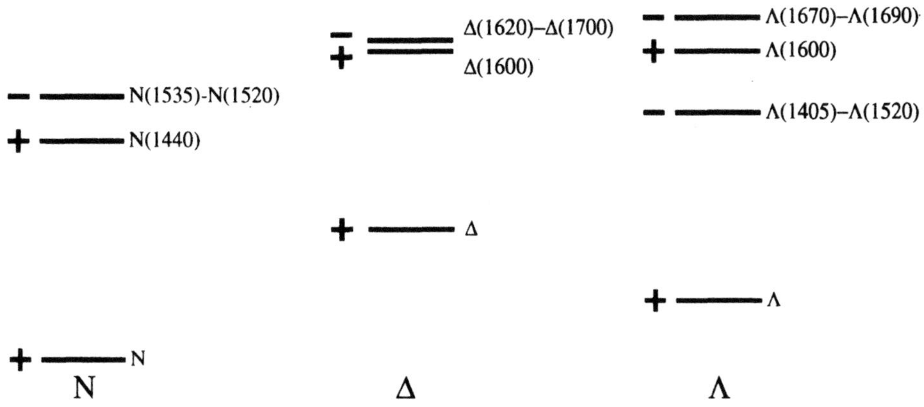

Figure 2. Low-lying spectra of nucleon, Δ-resonance and Λ-hyperon.

octet-decuplet splittings [5, 6, 7] when its strength is adjusted. However, it is easy to see from its operator structure that it also fails to explain Fig. 2 as far as N and Λ parts are concerned. But the most convincing evidence comes from the Δ spectrum, where the 't Hooft interaction between valence quarks is identically zero (it is absent in flavour - symmetric states). So according to this scenario the Δ spectrum should be exclusively due to confining interaction, which is ruled out by comparison of Figs. 1 and 2.

Thus the key problem is to explain *at the same time* both the octet-decuplet splittings and the pattern of Fig. 2.

2 Spontaneous Breaking of Chiral Symmetry and its Implications

The $SU(3)_{\mathrm{L}} \times SU(3)_{\mathrm{R}}$ chiral symmetry of QCD Lagrangian is spontaneously broken down to $SU(3)_{\mathrm{V}}$ by the QCD vacuum. There are two important generic consequences of the spontaneous breaking of chiral symmetry (SBCS). The first one is an appearance of the octet of pseudoscalar mesons of low mass, π, K, η, which represent the associated approximate Goldstone bosons. The second one is that valence (practically massless) quarks acquire a dynamical mass, which has been called historically constituent mass. Indeed, the nonzero value of the quark condensate, $< \bar{q}q > \sim -(250MeV)^3$, itself implies at the formal level that there must be at low momenta rather big dynamical mass, which should be a momentum-dependent quantity. Such a dynamical mass is now directly observed on the lattice [8]. Thus the constituent quarks should be considered as quasiparticles whose dynamical mass at low momenta comes from the nonperturbative gluon and quark-antiquark dressing. The flavour-octet axial current conservation in the chiral limit tells that the constituent quarks and Goldstone bosons should be coupled with the strength $g = g_A M/f_\pi$, which is a quark analog of the famous Goldberger-Treiman relation.

We have recently suggested that in the low-energy regime, below the chiral symmetry breaking scale, $\Lambda_\chi \sim 1$ GeV, the low-lying light and strange baryons

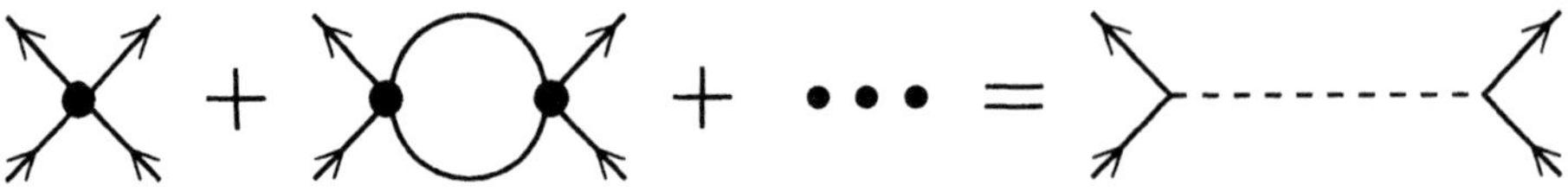

Figure 3. Iteration of the instanton-induced 't Hooft interaction in the t-channel. Black filled circle means a bare gluonic vertex.

should be predominantly viewed as systems of 3 constituent quarks with an effective confining interaction and a chiral interaction mediated by a Goldstone boson exchange (GBE) between the constituent quarks [9].

3 Why the Goldstone Boson Exchange is so Important?

Consider as example of a microscopical QCD nonperturbative interaction the instanton-induced 't Hooft interaction for two light flavours

$$H = -G[(\bar{\psi}\psi)^2 + (\bar{\psi}i\gamma_5\boldsymbol{\tau}\psi)^2 - (\bar{\psi}\boldsymbol{\tau}\psi)^2 - (\bar{\psi}i\gamma_5\psi)^2]. \tag{1}$$

This interaction is known to lead to chiral symmetry breaking, i.e. to creation of the quark condensate and dynamical (constituent) mass m of quarks. It happens because of the first term in (1), which represents a scalar part of the interaction. The interquark interaction in the pseudoscalar-isovector $\bar{q}q$ systems is driven by the second term, which is attractive and so strong that when it is iterated it exactly compensates the $2m$ energy supplied by the first term, and thus there appear $T = 1, J^P = 0^-$ mesons with zero mass - Nambu-Goldstone bosons. The first two terms in the Hamiltonian above form a classical Nambu and Jona-Lasinio model [10]. The fourth term in (1), which is repulsive, contributes only in the flavour-singlet $\bar{q}q$ pair (η'), making this meson heavy - contrary to π - and solving thus the $U(1)_A$ problem (note that the perturbative gluon exchange force cannot solve it). There is no interaction term which can contribute in vector mesons. This means that the masses of vector mesons, ρ and ω, should be approximately $2m$, which is well satisfied empirically. The interaction (1), extended to the u, d, s sector, also naturally explains completely different mixing between the octet and singlet components in the pseudoscalar and vector mesons [11].

As I said, a direct application of this instanton-induced interaction between valence quarks in baryons does not solve problems. But what happens when this interaction is iterated in qq t-channel, see Fig. 3 [12]? Specifically, the second term in (1) will imply the following amplitude

$$T_P(q^2) = 2G + 2GJ_P(q^2)2G + ... = \frac{2G}{1 - 2GJ_P(q^2)}, \tag{2}$$

where $J_P(q^2)$ is a bubble with a pseudoscalar vertex (vacuum polarization in the pseudoscalar channel). The expression (2) has a pole in the chiral limit at $q^2 = 0$, which can be identified as a pion-exchange. The eq. (2) defines a "running amplitude" and a negative sign in the denominator implies its antiscreening behavior. In essence this antiscreening is some kind of asymptotic freedom: at sufficiently large space-like momenta the interaction is represented by a bare 't Hooft vertex, but at $q^2 \to 0$ it becomes *infinitely* enhanced in the channel with GBE quantum numbers. So, if a typical momentum transfer is not large, which is the case in baryons in the low-energy regime, the pole contribution dominates. It explains why the GBE is so crucially important both in baryons and baryon-baryon systems. Thus the GBE interaction between constituent quarks is an effective representation of the pole contribution in (2), which is provided by the original quark-gluon degrees of freedom.

In fact any pairwise gluonic interaction between quarks in the local approximation will necessarily contain the first and second terms of (1) with fixed relative strength. This is because of chiral invariance. Thus all our conclusions on $\pi - \rho$ mass splitting and Goldstone boson exchange interaction in baryons are rather general and do not rely necessarily on 't Hooft interaction.

4 The Goldstone Boson Exchange Interaction and the Structure of the Baryon Spectrum

The coupling of the constituent quarks and the pseudoscalar Goldstone bosons will – to lowest order – give rise the $\sim \boldsymbol{\sigma} \cdot \boldsymbol{q} \boldsymbol{\lambda}^{\mathrm{F}}$ structure of the meson-quark vertex, where $\boldsymbol{q}$ is meson momentum. This type of vertex implies spin-spin and tensor interactions between constituent quarks, mediated by Goldstone bosons. The spin-spin force has a traditional long-range Yukawa part, which is important for nuclear force. But at short range the spin-spin force is much stronger and its sign is *opposite*. This short-range interaction has a form [9]

$$H_\chi \sim -\sum_{i<j} \frac{V(\boldsymbol{r}_{ij})}{m_i m_j} \boldsymbol{\lambda}_i^{\mathrm{F}} \cdot \boldsymbol{\lambda}_j^{\mathrm{F}} \boldsymbol{\sigma}_i \cdot \boldsymbol{\sigma}_j , \tag{3}$$

where a radial behavior of this short-range interaction is unknown. *It is this short-range part of the GBE interaction between the constituent quarks that is of crucial importance for baryons: it has a sign appropriate to reproduce the level splittings and strongly dominates over the Yukawa tail towards short distances.* Note that this spin-spin force is explicitly flavour-dependent, which reflects the fact that the GBE interaction is a flavour-exchange one. It is also significant that this short-range part of the interaction appears at the leading order within the chiral perturbation theory (i.e. in the chiral limit) [13], while the Yukawa part of the interaction vanishes in this limit. This simple observation has by far-going consequences: while the physics of baryons does not change much in the chiral limit (e.g. the $\Delta - N$ mass splitting persists), the long-range nuclear spin-spin force vanishes. This means that in some sense the short-range part of the pion exchange interaction is "more fundamental" than its Yukawa part. Note that the short-range quark-quark interaction of the same form can also be generated by the ρ - like exchange between the constituent quarks [14], which can be considered as a representation of the correlated two GBE interaction

[15].

Consider first, for the purposes of illustration, a schematic model which neglects the radial dependence of the potential function $V(r)$ in (3), and assume a harmonic confinement among quarks as well as $m_{\mathrm{u}} = m_{\mathrm{d}} = m_{\mathrm{s}}$. In this model

$$H_\chi = -\sum_{i<j} C_\chi \, \boldsymbol{\lambda}_i^{\mathrm{F}} \cdot \boldsymbol{\lambda}_j^{\mathrm{F}} \, \boldsymbol{\sigma}_i \cdot \boldsymbol{\sigma}_j. \tag{4}$$

The Hamiltonian (4) reduces the $SU(6)_{\mathrm{FS}}$ symmetry down to $SU(3)_{\mathrm{F}} \times SU(2)_{\mathrm{S}}$. Let us now see how the pure confinement spectrum of Fig. 1 becomes modified when the Hamiltonian (4) is switched on. The leading $SU(6)$ wave functions are known for all low-lying baryons and we thus can evaluate analytically the expectation values of the operator (4) [9].

The constant C_χ may be extracted from the N$-\Delta$ splitting to be 29.3 MeV. The oscillator parameter $\hbar\omega$, which characterizes the effective confining interaction with this schematic model, may be determined as one half of the mass differences between the first excited $\frac{1}{2}^+$ states and the ground states of the baryons, which have the same flavour-spin, flavour and spin symmetries (e.g. N(1440) - N, $\Lambda(1600)$ - Λ, $\Sigma(1660)$ - Σ), to be $\hbar\omega \simeq 250$ MeV. The two free parameters of this simple model are fixed and one can make predictions.

The mass difference between the lowest excited $\frac{1}{2}^+$ states (N(1440), $\Lambda(1600)$, and $\Sigma(1660)$) and the $\frac{1}{2}^- - \frac{3}{2}^-$ negative parity pairs (N(1535) - N(1520), $\Lambda(1670)$ - $\Lambda(1690)$, and $\Sigma(1750)$ - $\Sigma(?)$, respectively) will then be

$$\mathrm{N}, \Lambda, \Sigma: \quad m(\frac{1}{2}^+) - m(\frac{1}{2}^- - \frac{3}{2}^-) = 250\,\mathrm{MeV} - C_\chi(14-2) = -102\,\mathrm{MeV}, \tag{5}$$

whereas for the lowest state $\frac{1}{2}^+$ in the Λ system, $\Lambda(1600)$, and the lowest negative parity pair $\frac{1}{2}^- - \frac{3}{2}^-$ ($\Lambda(1405)$ - $\Lambda(1520)$) it should be

$$\Lambda: \quad m(\frac{1}{2}^+) - m(\frac{1}{2}^- - \frac{3}{2}^-) = 250\,\mathrm{MeV} - C_\chi(14-8) = 74\,\mathrm{MeV}. \tag{6}$$

At last, the lowest positive parity state $\frac{3}{2}^+$ in the Δ spectrum $\Delta(1600)$ [1] should be approximately degenerate with the lowest negative parity $\frac{1}{2}^- - \frac{3}{2}^-$ excitations $\Delta(1620) - \Delta(1700)$

$$\Delta: \quad m(\frac{3}{2}^+) - m(\frac{1}{2}^- - \frac{3}{2}^-) = 250\,\mathrm{MeV} - C_\chi(4+4) = 15\,\mathrm{MeV}. \tag{7}$$

One recovers precisely the spectrum shown in Fig. 2. It is astonishing that such a crude model predicts not only a general structure of the low-lying spectrum, but also the absolute values for splittings.

[1]Note that the experimental uncertainties for the mass of this baryon are 1550 - 1700 MeV.

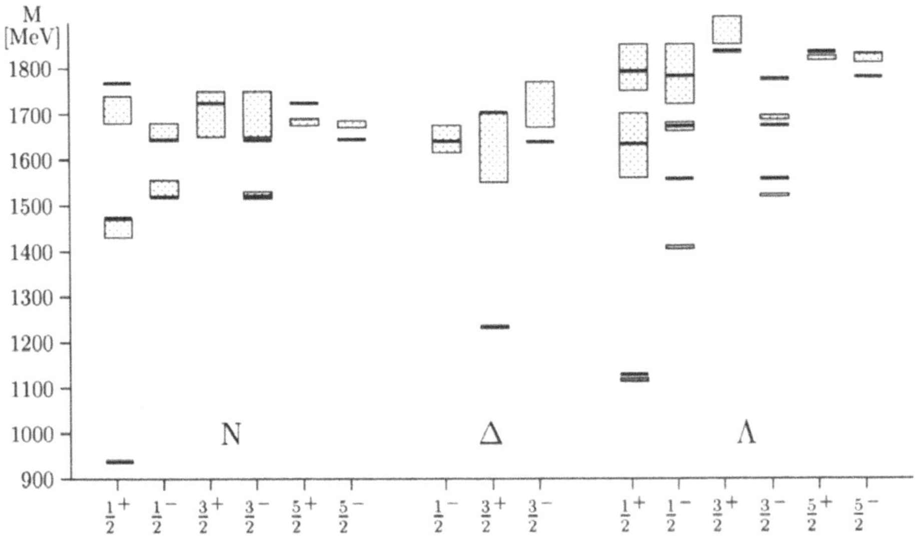

Figure 4. Energy levels of the low-lying light- and strange-baryon states with total angular momentum and parity J^P (solid lines). The shadowed boxes represent the experimental values with their uncertainties.

If one goes beyond the schematic - but analytical - calculation above, one should parameterize the short range parts of the interaction (the long range parts are fixed by meson masses), extract approximate meson-quark coupling constants from the known meson-baryon ones and solve with computer 3 - body equations. Such a program, with a semirelativistic Hamiltonian (i.e. kinetic energy is taken in a relativistic form) and with the linear confinement, has been realized in ref. [16] . In ref. [16] only the spin-spin force of GBE interaction is included. The extended versions includes ps-, vector- and scalar-exchanges with spin-spin, tensor and central force components. The spectra in both cases look pretty much the same, which is achieved by a slight readjustment of the cut-off parameters in the latter case, see Fig. 4.

References

1. A. DeRujula, H. Georgi, S.L. Glashow: Phys. Rev. **D12**, 147 (1975)

2. A. Chodos et al.: Phys. Rev. **D9**, 3471 (1974)

3. N. Isgur and G. Karl: Phys. Rev. **D18**, 4187 (1978); ibid. **D19**, 2653 (1979)

4. G. 't Hooft: Phys. Rev. **D14**, 3432(1976)

5. N.I. Kochelev: Sov. J. Nucl. Phys. **41**, 291 (1985)

6. E.V. Shuryak and J. L. Rosner: Phys. Lett. **B218**, 72 (1989)

7. W. H. Blask et al.: Z. Phys. **A337**, 327 (1990)

8. S. Aoki et al.: Phys. Rev. Lett. **82**, 4392 (1999)

9. L.Ya. Glozman and D.O. Riska: Phys. Reports **268**, 263 (1996)

10. Y. Nambu and G. Jona-Lasinio: Phys. Rev. **122**, 345 (1961)

11. G. 't Hooft: hep-th/9903189

12. L.Ya. Glozman and K. Varga: hep-ph/9901439, Phys. Rev. D - in print

13. L.Ya. Glozman: Phys. Lett. **B459**, 589 (1999)

14. L.Ya. Glozman: Surveys in High Energy Physics **14**, 109 (1999)

15. D.O. Riska and G.E. Brown: Nucl. Phys. **A653**, 251 (1999)

16. L.Ya. Glozman, W. Plessas, K. Varga, R. Wagenbrunn: Phys. Rev. **D58**, 094030 (1998)

Few-Body Systems Suppl. 12, 171–176 (2000)

Few-Body Systems

Effective Theory Approach to Two-Nucleon Systems

K. Kubodera*†

Department of Physics and Astronomy, University of South Carolina, Columbia, SC 29208, U.S.A.

Abstract. There is growing interest in nuclear physics applications of effective field theory (in the modern sense), because this approach is believed to provide a useful framework for connecting nuclear physics with the symmetries and dynamics of the underlying theory of strong interactions, QCD. I sketch here the basic ideas of this approach and present a few examples of its latest applications.

1 Introduction

Establishing a connection between nuclear dynamics and the fundamental QCD is one of the major challenges in nuclear physics today. Effective field theory (EFT) [1, 2] is considered to offer a natural and informative framework for this purpose. The physical picture of EFT is simple. When we want to describe phenomena characterized by a typical energy-momentum scale Q, we may expect that we need not explicitly include in our Lagrangian those degrees of freedom with energy scales much larger than Q. Then, introducing a cut-off scale Λ that is significantly larger than Q, we may separate our fields (generically denoted by Φ) into a high-energy part Φ_H and a low-energy part Φ_L. *Integrating out* Φ_H leads to an effective Lagrangian that only involves Φ_L. Thus the original Lagrangian $\mathcal{L}$ and the effective Lagrangian $\mathcal{L}_{\rm eff}$ are related to each other as

$$\int [d\Phi] e^{i\int d^4x \mathcal{L}(\Phi)} = \int [d\Phi_L] e^{i\int d^4x \mathcal{L}_{\rm eff}(\Phi_L)} \qquad (1)$$

This construction results in $\mathcal{L}_{\rm eff}$ that inherits all the symmetries (and the patterns of symmetry breaking, if any) of the original $\mathcal{L}$, which means that $\mathcal{L}_{\rm eff}$ consists of all possible monomials of Φ_L and their derivatives that are consistent with the symmetry requirements. Because a term involving n derivatives scales like $(Q/\Lambda)^n$, we have perturbative expansion in terms of a small parameter

*Supported in part by the National Science Foundation (USA), Grant No. PHYS-9602000
†*E-mail address:* kubodera@sc.edu

Q/Λ. In a nuclear-physics application of EFT, the original Lagrangian $\mathcal{L}$ is the QCD Lagrangian, whereas, in the energy-momentum regime $Q \ll \Lambda_{\rm QCD} \sim 1$ GeV, $\mathcal{L}_{\rm eff}$ is expected to involve hadrons rather than quarks and gluons. Furthermore, for $Q \lesssim m_\pi$, it seems sufficient to retain only the pions and nucleons as explicit degrees of freedom. Meanwhile, chiral symmetry of QCD (and its small violation due to the quark mass terms) must be inherited by our $\mathcal{L}_{\rm eff}$. An effective theory that results from these considerations is called chiral perturbation theory (χPT) [2, 3]. A problem with including the nucleon in χPT is that the nucleon mass $m_{\rm N}$ is comparable to $\Lambda_{\rm QCD}$. The heavy-baryon chiral perturbation formalism (HBχPT)[4, 5] allows us to avoid this difficulty. In HBχPT, effective degrees of freedom are the pions and the large components of nucleons, and $\mathcal{L}_{\rm ch}^{\rm HB}$ involves expansion in $\partial_\mu/\Lambda_{\rm QCD}$, $m_\pi/\Lambda_{\rm QCD}$ and $\partial_\mu/m_{\rm N}$. As $m_{\rm N} \approx \Lambda_{\rm QCD}$, it is convenient to lump together chiral and heavy-baryon expansions. Then $\mathcal{L}_{\rm ch}^{\rm HB}$ can be organized in terms of the chiral order index $\bar\nu$ defined by $\bar\nu = d + (n/2) - 2$, where n is the number of fermion lines participating in a vertex, and d is the number of derivatives (with m_π counted as one derivative). For example, the leading-order $\nu = 0$ term is given as [5]

$$\begin{aligned} \mathcal{L}^{(0)} &= \frac{f_\pi^2}{4}{\rm Tr}[\partial_\mu U^\dagger \partial^\mu U + m_\pi^2(U^\dagger + U - 2)] \\ &\quad + \bar{B}(iv\cdot D + g_A S\cdot u)B - \frac{1}{2}\sum_A C_A(\bar{B}\Gamma_A B)^2 \end{aligned} \tag{2}$$

Here $U(x)$ is an SU(2) matrix field related non-linearly to the pion field, $\xi \equiv \sqrt{U}$, $u_\mu \equiv i(\xi^\dagger\partial_\mu\xi - \xi\partial_\mu\xi^\dagger)$, $S_\mu = i\gamma_5\sigma_{\mu\nu}v^\nu/2$, and D_μ is the covariant derivative for the nucleon. Weinberg's counting rule [3] states that a Feynman diagram consisting of A nucleons, N_E external fields, L loops and C-disjoint pieces is order of $\mathcal{O}(Q^\nu)$ with $\nu = 2L + 2(C-1) + 2 - (A + N_E) + \sum_i \bar\nu_i$. This counting scheme works well for cases with zero or one nucleon but, for nuclear systems ($A \geq 2$), its straightforward application fails, because purely nucleonic intermediate states that occur in a nucleus can have very low excitation energies and thereby spoil the ordinary chiral counting rule [3]. To avoid this difficulty, one classifies Feynman diagrams into two types. Those diagrams in which every intermediate state contains at least one meson in flight are called irreducible diagrams, and all others are called reducible diagrams. Weinberg proposes to apply the chiral counting only to irreducible diagrams. The contribution of all the irreducible diagrams (up to a specified chiral order) is then used as an effective operator acting on the nucleonic Hilbert space. Summing up infinite series of irreducible diagrams (either solving the Schrödinger equation or the Lippman-Schwinger equation), one can incorporate the contributions of reducible diagrams [3, 6]. We refer to this two-step procedure as *nuclear* χPT . The past several years have witnessed a flurry of research work on nuclear χPT. (for a recent review, see. *e.g.* Refs.[7, 8].) I describe here only a few examples of these works, primarily concentrating on the work of Park, Min, Rho and myself.

2 A Hybrid Approach to Nuclear χPT

When we apply nuclear χPT to cases that involve external probes, we equate a nuclear transition operator $\mathcal{T}$ to the set of all the irreducible diagrams (up to a given chiral order ν) with the external current inserted [10]. In a formally consistent χPT approach, $\mathcal{T}$ should be sandwiched between the initial and final nuclear states that are *distorted* by the nuclear interactions representing irreducible A-nucleon diagrams of up to ν-th order. In practice, however, one often uses initial and final nuclear wavefunctions obtained from the Schrödinger equation that involves phenomenological nucleon-nucleon interactions. This eclectic method may be called a *hybrid approach* to nuclear χPT. The hybrid approach has proved highly successful in many cases. As a notable example let me discuss testing of the validity of the "chiral filter mechanism" [9, 10]. This "mechanism" is the statement that, unless kinematically suppressed, soft-pion exchange diagrams should give a dominant exchange-current contribution. According to [9], the space component of the vector current ($\mathbf{V}$) and the time component of the axial current (A_0), are expected to exhibit this dominance. As for $\mathbf{V}$, a detailed calculation based on hybrid HBχPT was carried out for the isovector M1 transition amplitude in the $n(\text{thermal}) + p \to d + \gamma$ reaction [11]. The result indicates that the soft-pion exchange current indeed gives a dominant contribution and that, with the next order corrections added, agreement with the data is perfect. As regards A_0, the enhancement factors for first-forbidden ($\Delta J = 0$, $\Delta\pi =$ yes) β-decay transitions were calculated in hybrid HBχPT [12]. The results indeed demonstrate that the bulk of the empirical enhancement factors [13] can be attributed to the soft-pion exchange current, and that the next-order corrections are a small fraction of it. Park, Min, Rho and myself (PKMR) [14] recently evaluated in HBχPT the cross sections for the solar proton fusion reaction $p + p \to d + e^+ + \nu_e$. The *pp* fusion rate obtained in HBχPT essentially agrees with the rate used in the standard solar model [15].

3 Attempts at *ab initio* Calculations

Despite the impressive success of the hybrid approach, it is desirable to formally justify or go beyond this approximation. We (PKMR) therefore have carried out a calculation wherein the transition operators and the nuclear interactions are treated on the same footing [16, 17]. This formally consistent approach may be called an *ab initio* calculation. I summarize here some salient features of our calculation.

We have carried out next-to-leading-order (NLO) *ab initio* calculations for the two-nucleon systems both with and without the pion field. The purpose of considering these two cases is to study whether the "pionless" and "pionful" effective theories, with different cut-off scales, exhibit behaviors that are generally expected for EFT. Because no loops appear to NLO, we can simply work with a potential. The *bare* potential $\mathcal{V}$ has the form

$$\mathcal{V}(\mathbf{q}) = -\tau_1 \cdot \tau_2 \frac{g_A^2}{4f_\pi^2} \frac{\sigma_1 \cdot \mathbf{q} \sigma_2 \cdot \mathbf{q}}{\mathbf{q}^2 + m_\pi^2} + \frac{4\pi}{m_N} \left[C_0 + (C_2 \delta^{ij} + D_2 \sigma^{ij}) q^i q^j\right] \quad (3)$$

where $\sigma^{ij} = 3/\sqrt{8}[(\sigma_1^i \sigma_2^j + \sigma_1^j \sigma_2^i)/2 - (\delta^{ij}/3)\sigma_1 \cdot \sigma_2]$, with $\mathbf{q}$ the momentum transfer. The first term coming from Goldstone-boson (pion) exchange is uniquely dictated by χPT. The (local) terms in the square brackets encode the effects of those degrees of freedom that have been integrated out. Use of the potential Eq.(3) in the Lippman-Schwinger equation engenders an infinite series of divergent terms. A great deal of controversy exists as to how to regularize this divergence [21, 18, 7]. PKMR use the momentum cut-off scheme [19] and introduce a cutoff: $V(\mathbf{r}) \equiv \int [d^3\mathbf{q}/(2\pi)^3]\, e^{i\mathbf{q}\cdot\mathbf{r}}\, S_\Lambda(\mathbf{q}^2)\, \mathcal{V}(\mathbf{q})$, with $S_\Lambda(\mathbf{q}^2) = \exp(-\mathbf{q}^2/2\Lambda^2)$. For a given value of Λ we can determine the parameters C's and D's in (3) by relating them (after renormalization) to the low-energy two-nucleon observables; the scattering length and the effective range for the scattering channels or to a selected set of the deuteron observables. Then we can make prediction for the N-N scattering phase shifts and the low-energy properties of the deuteron (other than those used as input). Moreover, it is possible to make *parameter-free* estimation of electroweak transition amplitudes, although in our NLO calculation this is limited to the 1-body (impulse-approximation) contributions. PKMR [17] evaluated one-body contributions to the charge radius r_d, the quadrupole moment Q_d, and the magnetic moment μ_d of the deuteron. In addition, we have computed the one-body M1 matrix element $\mathcal{M}_{\rm M1}$ for the *np* capture process and the Gamow-Teller matrix element $\mathcal{M}_{\rm GT}$ for the *pp* fusion process. The main results of PKMR's *ab initio* calculations are as follows. (1) All the calculated quantities are in good agreement with the empirical information in both *pionless* and *pionful* cases. (2) The results are stable against the variation of the cut-off parameter Λ, so long as it lies within a reasonable range; this *reasonable* range is found to be Λ =100 - 300 MeV (Λ =200 - 500 MeV) in the absence (presence) of the pion, values that are consistent with the general EFT consideration. (3) Inclusion of the pion brings a noticeable improvement in the accuracy of the prediction, markedly reducing the cutoff dependency. (4) Good agreement between the numerical results of our *ab initio* calculation and those of the hybrid approach gives justification to the hybrid approach.

You may ask, however, what is the difference between the familiar effective-range formula and our prediction. Well, in the pionless case, the roles of the two low-energy constants, C_0 and C_2 in Eq.(3) are indeed practically equivalent to those of the effective-range expansion parameters, a and r_e. But, with introduction of the pion, there is no such trivial correspondence, and χPT expansion contains more physics in it. Moreover, in χPT one can apply a unified expansion scheme not only to N-N scattering but also to nuclear transition processes.

PKMR [17] have used a momentum cut-off to regularize the divergence that appears in the Lippman-Schwinger equation with a potential of the type Eq.(3). Kaplan, Savage and Wise [20] proposed another regularization method, called the polynomial divergence subtraction (PDS). PDS has been used extensively to calculate various observables in the two-nucleon systems [21, 22]. PDS has the advantage that it preserves chiral invariance, whereas the cut-off regularization loses manifest chiral invariance. Although this feature is not consequential in the NLO calculation discussed above, it does become relevant in higher-order calculations. On the other hand, the PDS scheme generates a counter term that is of the same order as the leading-order pion-exchange current [22]. As the strength of this counter term must be determined from the data, there is significant loss of predictive power. By contrast, in the cut-off scheme, the leading-order pion-exchange contribution is a prediction of the theory, and the problem of unknown counter terms occurs only in the loop corrections to the one-pion exchange term [11, 12].

The above examples are concerned with observables that are governed by the lowest order contributions. To illustrate that the usefulness of χPT is not limited to these cases, let me talk about the spin observables in $\vec{n}$ (thermal)+ $\vec{p} \rightarrow d+\gamma$. It turns out that the photon circular polarization, P_γ, and the photon anisotropy, η_γ, are highly sensitive to the small isoscalar M1 and E2 transition matrix elements, and that these matrix elements are controlled by terms much higher in chiral counting than considered so far. There have recently been two χPT calculations for these spin observables, one based on the cut-off approach [23] and the other based on PDS [24]. The important point is that, although higher-order calculations involve many unknown low-energy constants, only a certain combination of them features in the relevant matrix elements so that χPT still has predictive power. The results of Refs.[23],[24] essentially agree with each other and with the existing experimental value of P_γ. It is expected that the on-going measurement of η_γ [25] will soon test the theoretical prediction for η_γ.

I have briefly surveyed the recent progress in χPT treatments of low energy-momentum transfer observables in the two-nucleon systems. It seems that we are entering the first stage of *ab initio* calculations based on HBχPT.

References

1. S. Weinberg: Physica **6A**, 327 (1979)
2. J. Gasser and H. Leutwyler: Ann. Phys. **158**, 142 (1984); J. Gasser, M. Sainio, S. Švarc: Nucl. Phys. **B307**, 779 (1988)
3. S. Weinberg: Phys. Lett. **B251**, 288 (1990); Nucl. Phys. **B363**, 3 (1991); Phys. Lett. **B295**, 114 (1992)
4. E. Jenkins and A.V. Manohar: Phys. Lett. **B255**, 558 (1991)
5. For a review, see e.g., V. Bernard, N. Kaiser, U.-G. Meissner: Int. J. Mod. Phys. **E4**, 193 (1995)
6. U. van Kolck: Phys. Rev. Lett. **72**, 1982 (1994)

7. R. Seki, U. van Kolck, M. Savage (eds.): *Nuclear Physics with Effective Field Theory*, Proceedings of the Joint Caltech/INT Workshop (February 1998). World Scientific 1998

8. U. van Kolck: Prog. Part. Nucl. Phys. **43**, 409 (1999)

9. K. Kubodera, J. Delorme, M. Rho: Phys. Rev. Lett. **40**, 755 (1978)

10. M. Rho: Phys. Rev. Lett. **66**, 1275 (1991)

11. T.-S. Park, D.-P. Min, M. Rho: Phys. Rev. Lett. **74**, 4153 (1995); Nucl. Phys. **A596**, 515 (1996)

12. T.-S. Park, D.-P. Min, M. Rho: Phys. Reports, **233**, 341 (1993); T.-S. Park, I.S. Towner, K. Kubodera: Nucl. Phys. **A579**, 381 (1994)

13. E.K. Warburton: Phys. Rev. Lett. **66**, 1823 (1991)

14. T.-S. Park, K. Kubodera, D.-P. Min, M. Rho: Astrophys. J. **507**, 443 (1998)

15. J.N. Bahcall and R. May: Astrophys. J. **155**, 501 (1969); M. Kamionkowski and J.N. Bahcall: Astrophys. J. **420**, 420 (1994)

16. T.-S. Park, K. Kubodera, D.-P. Min, M. Rho: Phys. Rev. **C58**, R637 (1998)

17. T.-S. Park, K. Kubodera, D.-P. Min, M. Rho: Nucl. Phys. **A646**, 83 (1999)

18. S.R. Beane, T.D. Cohen, D.R. Phillips: Nucl. Phys. **A632**, 445 (1998); J.V. Steele and R.J. Furnstahl: Nucl. Phys. **A637**, 46 (1998); J. Gegelia: Phys. Lett. **B429**, 227 (1998); D.R. Phillips and T.D. Cohen: nucl-th/9906091

19. G.P. Lepage: nucl-th/9706029

20. D.B. Kaplan, M.J. Savage, M.B. Wise: Phys. Lett. **B424**, 390 (1998); Nucl. Phys. **B535**, 329 (1998)

21. D.B. Kaplan, M.J. Savage, M.B. Wise: Phys. Rev. **C59**, 617 (1999)

22. M. Savage, K.A. Scaldeferri, M.B. Wise: Nucl. Phys. **A652**, 273 (1999)

23. T.-S. Park, K. Kubodera, D.-P. Min, M. Rho: nucl-th/9904053; nucl-th/9906005

24. J.-W. Chen, G. Rupak, M.J. Savage: Phys. Lett. **B464**, 1 (1999)

25. T.M. Mueller et al.: *Proc. International Workshop on Particle Physics with Slow Neutrons*, ILL, Grenoble, October 1998

Few-Body Systems Suppl. 12, 177–182 (2000)

Few-Body Systems

Charge Symmetry Breaking in Inelastic Nucleon-Nucleon Scattering

A.K. Opper*

Ohio University, Department of Physics and Astronomy, Athens OH 45701 USA

Abstract. Investigations of charge symmetry breaking (CSB) in the strong interaction allow the measurement of effects that are ultimately due to the mass difference between u and d quarks, although meson-exchange models describe the existing data most consistently. A high precision measurement of CSB in the strong interaction has been undertaken at TRIUMF to complement those data and lead to a richer understanding of the dynamical quark effects in the NN system. The observable of interest is the forward–backward asymmetry (A_{fb}) in np $\rightarrow$ dπ^0, which must be zero in the center-of-mass if charge symmetry is conserved and is predicted to be approximately -35×10^{-4} with the dominant contributions being an order of magnitude larger than those of the elastic scattering CSB measurements carried out at TRIUMF and IUCF.

The measurement used a 279.5 MeV neutron beam, a liquid hydrogen target, and the SASP spectrometer positioned at 0°. With these kinematics and the large acceptance of SASP the full deuteron distribution was detected in one setting of the spectrometer thereby eliminating many systematic uncertainties. A measurement of the pp $\rightarrow$ dπ^+ distribution accompanied the primary measurement as a robust test of the analysis and simulation codes. The experimental challenges and status are presented.

1 Introduction

Discovering conservation principles and the associated symmetries is a powerful tool in understanding the forces of nature. However, a more subtle and intriguing situation arises when a fundamental symmetry is broken. Within QCD, charge symmetry breaking (CSB) and charge independence breaking (CIB) arise from differences between the current masses of the u and d quarks, and from electromagnetic interactions between the quarks. CSB has been clearly observed in the nucleon-nucleon system, most recently in the difference between the neutron and proton analyzing powers in np elastic scattering [3, 4, 5].

* *E-mail address:* opper@ohiou.edu

Another indication of CSB in the strong interaction would be an asymmetry about 90° in the center-of-mass angular distribution of deuterons (or pions) in the reaction np $\to$ dπ^0.

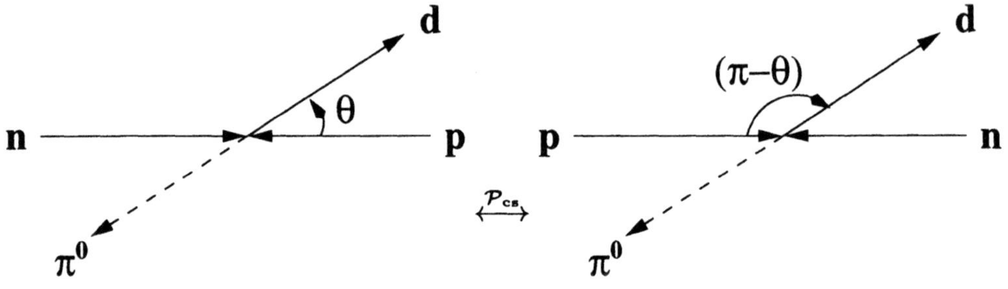

Figure 1. Charge symmetric sketches of np $\to$ dπ^0 in the center-of-mass frame.

Fig. 1 shows in an intuitive manner that charge symmetry requires the deuteron distribution in np $\to$ dπ^0 be symmetric about 90° in the center of mass. Both sketches show the np $\to$ dπ^0 reaction in the center of mass and are related by the charge symmetry operator P_{CS}. In going from one sketch to the other, P_{CS} interchanges neutrons and protons. Given that these two figures are invariant under charge symmetry, the deuteron yield at an angle θ in the drawing on the left must be equal to the yield at an angle $(\pi - \theta)$ in the charge symmetric drawing on the right. Said another way, a non-zero forward–backward asymmetry can only occur if charge symmetry is violated. This forward–backward asymmetry in the deuteron distribution is defined as

$$A_{fb}(\theta) \equiv \frac{\sigma(\theta) - \sigma(\pi - \theta)}{\frac{1}{2}[\sigma(\theta) + \sigma(\pi - \theta)]}. \tag{1}$$

Since the deuteron distribution from pp $\to$ dπ^+ must be symmetric in the center of mass due to the indistinguishability of the two protons, this experiment has undertaken a measurement of the pp $\to$ dπ^+ distribution [6] as a test case to verify the analysis and simulation codes.

Niskanen's model of pion production includes CSB effects and he has extended his calculations to energies near threshold [7]. These calculations include contributions from the *np* mass difference (δ), $\eta - \pi^0$ mixing, $\eta' - \pi^0$ mixing, and $\rho^0 - \omega$ mixing. This calculation also includes the mixing contributions from both the production vertex and the transition potential and 16 time-ordered diagrams. Fig. 2 shows values for A_{fb} extracted from these calculations. Although $\rho^0 - \omega$ mixing is the dominant contribution to the analyzing power difference in elastic *np* scattering measured at 183 MeV, the contribution from this to A_{fb} near 280 MeV is on the same order as the δ contributions and is not plotted in this figure. Electromagnetic effects are comparably small and are also not plotted. The symbol plotted on this figure indicates the *predicted* value of A_{fb} at the energy of this experiment with *anticipated* error bars. As the figure shows, when only δ contributions are included, A_{fb} is positive and ranges from (1 to 5)$\times 10^{-4}$ for the different neutron energies. However, adding the contributions from the $\eta - \pi^0$ and $\eta' - \pi^0$ mixing, neither of which enters in the elastic *np* scattering analyzing power results, changes the sign, shape, and magnitude of A_{fb} for all neutron energies. It is interesting to note in fig. 2 that the predicted value for A_{fb} increases by only a factor of two as the neutron energy increases from 279.5 MeV to 350 MeV. This is due to the decrease in the vertex contributions above 300 MeV. In fact, both the η and δ contributions change sign around 400 MeV.

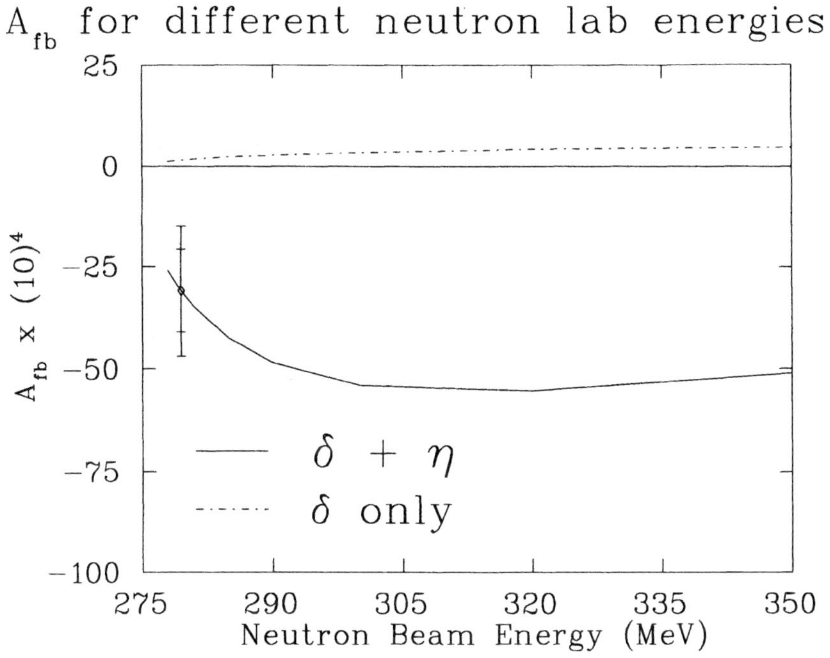

Figure 2. Contributions near threshold to A_{fb} from δ, $\eta - \pi^0$ mixing, and $\eta' - \pi^0$ mixing. The open diamond indicates the ***predicted*** value of A_{fb} at the energy of this experiment with ***anticipated*** error bars [7].

The potential for the exchange of an $\eta - \pi$ ($\eta' - \pi$), isospin mixed meson is linearly dependent on the ηNN coupling constant and the $\eta - \pi$ ($\eta' - \pi$) mixing amplitude. The mixing matrices are well known from analysis of η and η' decay [8] but the value of the η-nucleon coupling constant ranges from $(g^2_{\eta NN}/4\pi) =$ 0.2 to 6.2 [7, 9]. With this CSB measurement dominated by η effects, it will be an effective tool to constrain the ηNN coupling.

2 Experimental Apparatus

The TRIUMF CHARGEX facility [10] was used to produce the neutron beam. A 283-MeV primary proton beam striking a ^{7}Li target produces neutrons of 279.5 MeV by (p,n) reaction. Unscattered beam is deflected 20° towards an external dump by a sweeping magnet. Neutrons are shaped to a 4 cm by 6 cm beam in a tungsten-alloy collimator, in which are set two veto scintillators to eliminate protons scattered from the Li target and deuterons generated by (n,d) reactions on the collimator surfaces. The neutron beam impinges on a liquid hydrogen target of nominal thickness 2 cm. Deuterons from reactions in the target pass through a thin trigger scintillator and three pairs of multiwire proportional counter planes before entering the Second Arm Spectrometer (SASP) [11]. Vertical drift chambers and three layers of plastic scintillators are located at and beyond the SASP focal plane. Energy loss and time-of-flight allow a very clean separation of deuterons from protons.

The pp $\rightarrow$ dπ^+ reaction must have A_{fb}=0 due to the identity of beam and target particles, and may be used to check for many of the potential systematic errors in the np $\rightarrow$ dπ^0 experiment. Interleaved measurements of the pp $\rightarrow$ dπ^+ reaction are carried out with the following changes from neutron beam mode: the ^{7}Li target is retracted and the sweeping magnet turned off; a 1-mm thick Ta foil is inserted in the beamline approximately 13 m upstream of the liquid hydrogen target; the primary beam intensity is lowered below 1 nA and its energy changed to account for the difference in reaction thresholds for the neutral and charged pion; SASP fields are increased to put the deuteron locus at the same position in the focal plane; veto counters are used in coincidence mode.

3 Extracting A_{fb} from Lab Variables

Near threshold the deuterons from np $\rightarrow$ dπ^0 form a small forward cone centered about the beam direction and a kinematic locus with a narrow range of momentumand scattering angle in the lab. This locus is easily seen in fig. 3. The background surrounding the locus is from (n, d) reactions on nuclei such as carbon in materials near the target. The deuterons with higher (lower) momentum are produced in the forward (backward) direction in the center-of-mass frame so that comparing the top half with the bottom half of the locus gives an indication of A_{fb}.

In the center of mass frame the np $\rightarrow$ dπ^0 cross section can be described by three parameters, A_0, A_1, and A_2:

$$d\sigma/d\Omega = A_0 + A_1 P_1(\theta) + A_2 P_2(\theta), \tag{2}$$

where θ is the deuteron centre-of-mass angle and P_i are Legendre polynomials [12]. In this expression, the second term describes any charge symmetry breaking with the coefficient A_1 giving the asymmetry.

All the experimental features (target, detectors, magnetic fields, ...) are described in a GEANT simulation with the np $\rightarrow$ dπ^0 center-of-mass cross section given in equation 1. After the deuteron and π are produced, their positions and momenta are converted to the lab frame, they are tracked through the detector system and a kinematic locus is produced. This simulated locus is compared to data and the coefficients of the Legendre polynomials extracted through

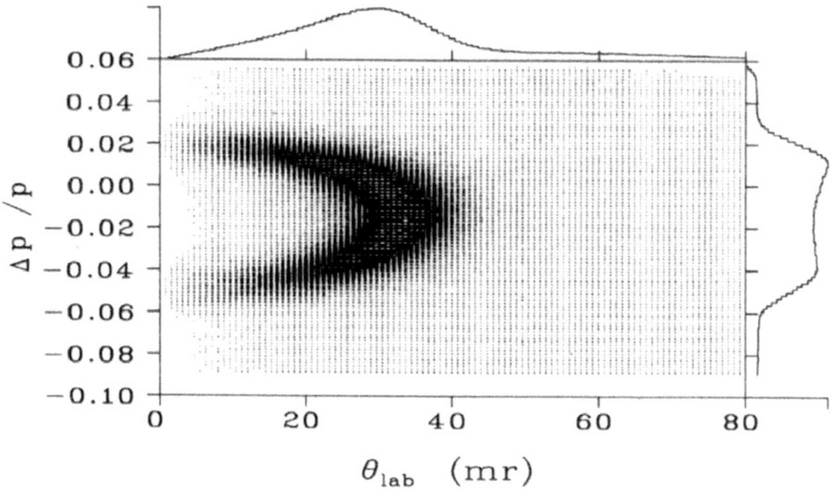

Figure 3. Data for np → dπ^0 through Summer 1998 showing the kinematic locus of momentum *vs* scattering angle in the lab.

a χ^2 minimization procedure. This gives the CSB observable since the angle integrated A_{fb} is essentially equal to A_1/A_0.

4 Status

As of August 1999, almost seven million np → dπ^0 events and 10 million pp → dπ^+ events have been put to tape. Additionally, calibration mesurements of detector efficiencies and various sources of background have been carried out. Analysis of these data should be finished by the end of 1999. Fig. 4 shows a comparison of the GEANT simulation as of summer 1999 and all np → dπ^0 and pp → dπ^+ data (target full and empty) taken through 1998. The simulation shows descrepancies with the data at the 0.1% level with the suspected sources of error being in the description of the spectrometer acceptance and the the dipole magnetic field. By the spring of 2000 the GEANT code should be fine tuned to allow generation of the final sets of simulation data and a result available by the end of 2000. Given the large number of events on tape, the statistical uncertainty in A_{fb} is 9.4×10^{-4}; as many of the systematic effects are themselves determined by statistics we anticipate a systematic uncertainty of 12×10^{-4}.

References

1. G.A. Miller and W.T.H. Van Oers: In: *Fundamental Symmetries,* edited by W.C. Haxton and E.M. Henley, p.127. Singapore: World Scientific 1995

2. G.A. Miller, B.M.K. Nefkens, I. Slaus: Phys. Rep. **194**, 1 (1990)

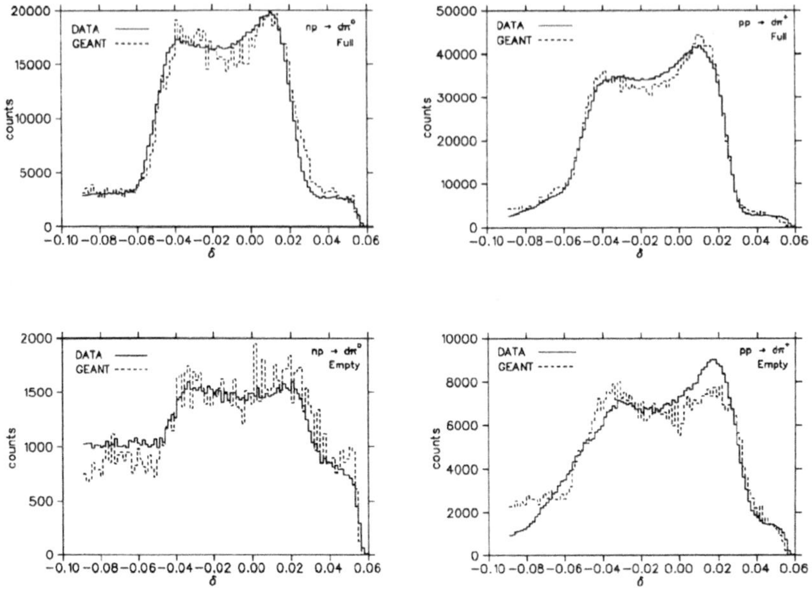

Figure 4. The plots on the left (right) show the best fit to the np $\rightarrow d\pi^0$ (pp $\rightarrow d\pi^+$) data. The top (bottom) row details the comparisons for target full (empty) comparison.

3. R. Abegg et al.: Phys. Rev. **D39**, 2464 (1989)
4. S.E. Vigdor et al.: Phys. Rev. **C46**, 410 (1992)
5. J. Zhao et al.: Phys. Rev. **C57**, 2126 (1998)
6. E. Korkmaz et al.: Nucl. Phys. **A535**, 637 (1991)
7. J.A. Niskanen: *nucl-th/9809009*
8. S.A Coon, BHJ McKellar, MD Scadron: Phys. Rev. **D34**, 2784 (1986)
9. Benmerrouche, Mukhopadhyay, Zhang: Phys. Rev. **D51**, 3237 (1995)
10. R. Helmer et al.: Can J. Phys. **65**, 588 (1987)
11. P. Walden et al.: to be published in Nucl. Instr. and Meth.
12. D.A. Hutcheon et al.: Nucl. Phys. **A535**, 618 (1991)

Few-Body Systems Suppl. 12, 183–190 (2000)

Few-Body Systems

Exchange Terms in the Quark-Meson Coupling Model

G. Krein*

Instituto de Física Teórica, Universidade Estadual Paulista,
Rua Pamplona, 145 - 01405-900 São Paulo, SP, Brazil

Abstract. In this talk we report on recent progress in implementing exchange terms in the quark-meson coupling model. Exchange effects are related to the Pauli exclusion principle. We discuss exchange effects at the nucleon level and at the quark level. We also address the incorporation of chiral symmetry and Δ degrees of freedom in the model.

1 Introduction

The quark-meson coupling (QMC) [1] model provides the simplest extension of quantum-hadrodynamics (QHD) [2] to incorporate explicitly the substructure of the hadrons. In the QMC model the nucleons still exchange mesons, but the mesons couple directly to the confined quarks, not to point-like nucleons as in QHD. Since the original version of the model, there have been several ameliorations and extensions. These include the treatment of the nucleon center-of-mass motion [3, 4], treatment of finite nuclei [5, 6, 7], change of the bag constant in medium [8] and inclusion of finite temperature effects [9]. Here, we will report on recent extensions of the model to include the effects of exchange terms [10, 11].

Exchange effects in the present context are related to the Pauli exclusion principle. It is important to consider such effects not only as a matter of principle, but also because it is only through exchange terms that the explicit role of pions can be assessed; at the mean field level, the pion field simply averages to zero because of parity. One would like to learn about pions in matter because they are the (pseudo) Goldstone bosons of the dynamical chiral symmetry breaking in quantum chromodynamics. The enforcement of the exclusion principle at the quark level in a many-hadron system gives rise to quark exchange effects. The inclusion of such effects in a many-body system makes the problem very complicated. One of the major difficulties is the simultaneous presence in the system of composite-particle (hadrons) degrees of freedom and

* *E-mail address:* gkrein@ift.unesp.br

of constituent (quarks) degrees of freedom. In order to deal with such a situation, we employ a many-body formalism for composite particles developed in atomic physics [12] and recently adapted to quark models [13]. These effects are interesting because it has been shown in a variety of models that they are an important source of short-range repulsion in the nucleon-nucleon interaction. However, quark exchange between different hadrons are very difficult to treat within the MIT bag model, because of the sharp surface of the bag, and a potential model with constituent quarks seems more appropriate [10].

In the next section we present the model and discuss the origin of quark exchange. In section 3 we discuss nucleon exchange effects, i.e. the Pauli principle is enforced at the nucleon level; following the approach of Ref. [11]. These give rise to the familiar Fock terms. In section 4, we discuss quark exchange effects on the NN interaction and on nuclear matter saturation properties, within the perspective of the formalism of Ref. [13].

2 The Model

We start formulating the model in the language of an effective meson-nucleon Hamiltonian following the ideas of the Cloudy Bag Model (CBM) [14]. The Lagrangian density of the model is

$$\begin{aligned}\mathcal{L}_{\rm QMC} &= \mathcal{L}_{\rm MIT} + \frac{1}{2}\left(\partial_\mu\sigma\partial^\mu\sigma - m_\sigma^2\sigma^2\right) - \frac{1}{4}F_{\mu\nu}F^{\mu\nu} + \frac{1}{2}m_\omega^2\,\omega_\mu\omega^\mu \\ &+ \frac{1}{2}\left(\partial_\mu\pi\cdot\partial^\mu\pi - m_\pi^2\pi^2\right) - \frac{1}{4}G_{\mu\nu}\cdot G^{\mu\nu} + \frac{1}{2}m_\rho^2\,b_\mu\cdot b^\mu \\ &+ \bar{\psi}\left(g_\sigma^q\sigma - g_\omega^q\gamma_\mu\omega^\mu - \frac{1}{2f_\pi}\gamma_\mu\gamma_5\tau\cdot\partial_\mu\pi - g_\rho^q\frac{1}{2}\gamma_\mu\tau\cdot b^\mu\right)\psi\,\theta_V \end{aligned} \quad (1)$$

where $\mathcal{L}_{\rm MIT}$ is the Lagrangian density of the MIT Bag Model,

$$\mathcal{L}_{\rm MIT} = \left[\bar{\psi}\left(i\gamma_\mu\partial^\mu - m_q\right)\psi - B\right]\theta_V - \frac{1}{2}\bar{\psi}\psi\,\delta_S \quad (2)$$

θ_V is one inside the bag and zero outside, δ_S is a delta function on the bag surface, $F_{\mu\nu} = \partial_\mu\omega_\nu - \partial_\nu\omega_\mu$, and $G_{\mu\nu} = \partial_\mu b_\nu - \partial_\nu b_\mu$.

The σ and ω^μ field operators have constant expectation values in symmetric nuclear matter, σ_0 and ω_0. The b^μ field has a nonzero expectation value only in asymmetric nuclear matter, b_0. The pion field π has zero expectation value in both symmetric and asymmetric matter, because of parity considerations. We begin by solving the single-nucleon problem in the presence of the constant mean-fields.

Let $B_\lambda^\dagger(p)$ and $B_\lambda(p)$ denote the creation and annihilation operators for single-nucleon states with spin-isospin λ and momentum p in the presence of the mean fields. That is, the operator $B_\lambda^\dagger(p)$ creates the state

$$|p,\lambda\rangle = B_\lambda^\dagger(p)|0\rangle = B_\alpha^\dagger|0\rangle = \frac{1}{\sqrt{3!}}\Psi_\alpha^{\mu_1\mu_2\mu_3}q_{\mu_1}^\dagger q_{\mu_2}^\dagger q_{\mu_3}^\dagger|0\rangle \quad (3)$$

where $\alpha = \{\lambda, p\}$ denotes the quantum numbers of the nucleon (including space-time coordinates), $\mu_1, \mu_2, \ldots$ the ones of the quarks and $\Psi_\alpha^{\mu_1\mu_2\mu_3}$ is the Fock-space baryon amplitude. The convention of a summation over repeated indices is assumed.

Using the quark anticommutation relations, and assuming the Ψ_α's orthonormalized, it can easily be shown that the baryon operators satisfy the following *noncanonical* anticommutation relations [13]

$$\{B_\alpha, B_\beta^\dagger\} = \delta_{\alpha\beta} - \Delta_{\alpha\beta}, \qquad \{B_\alpha, B_\beta\} = 0 \tag{4}$$

where

$$\Delta_{\alpha\beta} = 3\Psi_\alpha^{*\mu_1\mu_2\mu_3}\Psi_\beta^{\mu_1\mu_2\nu_3} q_{\nu_3}^\dagger q_{\mu_3} - \frac{3}{2}\Psi_\alpha^{*\mu_1\mu_2\mu_3}\Psi_\beta^{\mu_1\nu_2\nu_3} q_{\nu_3}^\dagger q_{\nu_2}^\dagger q_{\mu_2} q_{\mu_3}. \tag{5}$$

In addition,

$$\{q_\mu, B_\alpha^\dagger\} = \sqrt{\frac{3}{2}}\Psi_\alpha^{\mu\mu_2\mu_3} q_{\mu_2}^\dagger q_{\mu_3}^\dagger, \qquad \{q_\mu, B_\alpha\} = 0. \tag{6}$$

The presence of the term $\Delta_{\alpha\beta}$ in Eq. (4) is obviously due to the internal structure of the baryons, and it vanishes as the separation between the centers-of-mass of the two baryons becomes much larger than the sum of the radii of the two baryons. In the same way, the nonzero value of the anticommutator $\{q_\mu, B_\alpha^\dagger\}$ is a manifestation of the lack of kinematic independence of the hadron operators from the quark operators.

Physically, the origin of these effects is the fact that quarks of different hadrons are identical particles, and this identity cannot be neglected when the the quark cores of two hadrons overlap appreciably. Among other consequences, the identity of the quarks gives rise to quark exchange effects, which play an important role in the description of the short-range part of the nucleon-nucleon interaction in some models. When quark exchange can be neglected, one is still left with exchange effects, which are related to the identity of the hadrons and give rise to the usual Fock terms.

In the next section we discuss the inclusion of Fock terms in the context of the QMC model and in Section 4 we discuss quark-exchange effects in a model with massive constituent quarks.

3 Fock Terms in the QMC Model

Once the mean meson fields are separated from the full quantum fields, the fluctuating meson fields are then used to construct an effective quark-meson Hamiltonian. The Hamiltonian is then projected onto the space of the single-nucleon states. The single-nucleon states are not the bare MIT bag states, but the bag states under the influence of the mean-meson fields, as explained previously. The Hamiltonian resulting from this procedure is similar to the usual CBM Hamiltonian, with the difference that the effective meson-nucleon vertices in the present case are evaluated with density dependent wave functions.

The resulting effective meson-nucleon Hamiltonian can be written as

$$H_{\text{eff}} = H_{\text{MMF}} + H_0 + W + W^\dagger \tag{7}$$

where H_{MMF} is the Hamiltonian of the meson mean fields,

$$H_{\text{MMF}} = \frac{1}{2}m_\sigma^2\sigma_0^2 - \frac{1}{2}m_\omega^2\,\omega_0^2 - \frac{1}{2}m_\rho^2\,b_0^2 \tag{8}$$

with

$$\boldsymbol{H_0 = \sum_\lambda \int dp\,\varepsilon(p,\lambda)\,B_\lambda^\dagger(p)\,B_\lambda(p) + \sum_{j=\sigma,\omega,\pi,\rho} \int dk\,\omega_j(k)\,a_j^\dagger(k)\,a_j(k)} \tag{9}$$

where $\varepsilon(p,\lambda) = E^*(p) + 3g_\omega^q\,\omega_0 + \frac{1}{2}g_\rho^q\,\langle\lambda|\tau_3|\lambda\rangle\,b_0$, with $\boldsymbol{E^*(p)} = \sqrt{\boldsymbol{p^2 + M^{*2}}}$, M^* is obtained by solving the MIT bag equations, $\omega_j(k) = (k^2 + m_j^2)^{1/2}$, and $a_j^\dagger(k)$ and $a_j(k)$ are the meson creation and annihilation operators. The meson-nucleon interacting Hamiltonian W is given by

$$W = \int dp\,dp' \sum_{j,\lambda\lambda'} \frac{(2\pi)^{3/2}}{\sqrt{2\omega_j(k)}} W_{\lambda'\lambda}^j(p',p)\,B_{\lambda'}^\dagger(p')\,B_\lambda(p)\,a_j(k) \tag{10}$$

with $k = p' - p$, and $W_{\lambda'\lambda}^j(p',p)$ are the vertex functions

$$\boldsymbol{W_{\lambda'\lambda}^j(p',p) = \langle p'\lambda'|\bar{\psi}(0)\Gamma^j\psi(0)|p\lambda\rangle} \tag{11}$$

with the explicit form of the Γ^j given in Ref. [11]. The effects of the internal structure of the nucleons are contained in these vertices; the approximation scheme to calculate these in the MIT bag model is discussed in Ref. [11].

Given the vertices, the energy density of symmetric nuclear matter up to second order in the W's can be written as

$$\mathcal{E}_{NM} = 4\int_0^{k_F} \frac{dp}{(2\pi)^3}\,E^*(p) + 3\,g_\omega^q\omega_0\,\rho_N + \frac{1}{2}m_\sigma^2\sigma_0^2 - \frac{1}{2}m_\omega^2\omega_0^2 + \sum_j \mathcal{E}_{\text{Fock}}^j \tag{12}$$

where $\rho_N = 2k_F^3/3\pi^2$ is the nucleon density, and $\mathcal{E}_{\text{Fock}}^j$ are the Fock energies, the second-order exchange contributions

$$\mathcal{E}_{\text{Fock}}^j = \frac{1}{2}\int_0^{k_F} \frac{dp}{(2\pi)^3} \int_0^{k_F} \frac{dp'}{(2\pi)^3} \frac{\sum_{\lambda\lambda'} |W_{\lambda\lambda'}^j(p,p')|^2}{(p'-p)^2 + m_j^2} \tag{13}$$

$$= \frac{1}{2}\int_0^{k_F} \frac{dp}{(2\pi)^3} \sum_\lambda \bar{u}(p,\lambda)\,U_{\text{Fock}}^j(p)u(p,\lambda), \tag{14}$$

where the $u(p,\lambda)$ the nucleon Dirac spinors with mass M^* [11], and $U_{\text{Fock}}^j(p)$ are the Fock potentials, or the exchange nucleon self-energies, with Dirac structure as

$$\boldsymbol{U_{\text{Fock}}^j(p) = U_s^j(p) + \gamma^0\,U_0^j(p) + \gamma\cdot\hat{p}\,U_v^j.} \tag{15}$$

The final step is to determine the scalar mean field, σ_0, which satisfies the self-consistency equation

$$\sigma_0 = \frac{g_\sigma}{m_\sigma^2} C(\sigma_0)\, 4 \int_0^{k_F} \frac{dp}{(2\pi)^3} \frac{M^*}{E^*(p)} + \frac{1}{m_\sigma^2} \frac{\partial}{\partial \sigma_0} \sum_j \mathcal{E}^j_{\text{Fock}} \tag{16}$$

where $g_\sigma = 3 g_\sigma^q S(0)$, with $C(\sigma_0) = \frac{S(\sigma_0)}{S(0)}$, and $S(\sigma_0)$ is the in-medium quark scalar density [7]. The mean vector field, ω_0, is given in terms of the baryon density, as usual.

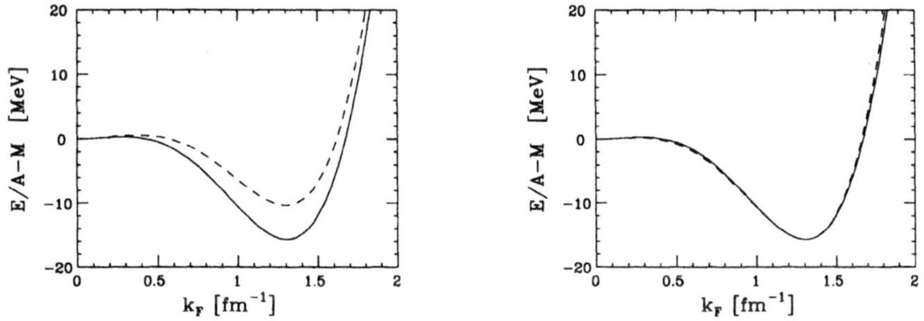

Figure 1. Energy per particle as a function of the Fermi momentum. Left: Effect of the Fock terms, the dashed line is the result of adding the Fock terms due to σ and ω to the mean-field result (solid). Right: Result including all mesons (see text).

In Fig. 1 we present our results for the binding energy of nuclear matter. On the left panel, the dashed line indicates the effect of the Fock terms on the top of the mean-field result (solid line), using only the σ and ω mesons. As seen, the effect is repulsive and of the order of 5 MeV. On the right panel, the solid line is the result including all mesons. As usual the σ and ω coupling constants have been readjusted (from their Hartree values) to fit the nuclear saturation point. We have also subtracted the short-distance contact interactions from the π and ρ couplings, in order to get a better assessment of their effect in nuclear matter; in a more complete treatment that goes beyond the Hartree plus Fock approximation, these short-distance effects will be absent due to the short-range correlations.

The effective nucleon mass and radius remain almost identical to their values in the Hartree approximation. The Fock terms increase the values of the average scalar and vector potentials [11], through the Fock potentials U_s and U_0 of Eq. (15). This feature is important for the spin-orbit splittings is finite nuclei. Another important result of our study is that we have obtained for the symmetry energy of nuclear matter the value $a_4 = 30$ MeV (experimental value is 33 MeV), without adjusting g_ρ. This result is stable under the variations of bag radii and quark masses.

4 Quark-Exchange

When there is a large probability that the quark cores of the hadrons have considerable overlap in the system, the term $\Delta_{\alpha\beta}$ in Eq. (4) cannot in general be neglected. Among other effects, this term gives rise to quark exchange forces. Here we use the Fock-Tani representation [12] to treat such effects. This is a formalism based on the ideas of mapping of composite particles onto elementary particles. The mapping is implemented through a unitary transformation. When the unitary transformation is applied to the quark-model Hamiltonian, one obtains effective Hamiltonians that describe all possible hadron-hadron and hadron-quark interactions [13].

As we remarked above, the MIT bag model is not a convenient model to treat quark exchange effects, because of the sharp surface of the bag. In view of this, we use a semi-relativistic constituent quark model. We use the contact piece of the one-pion exchange between two quarks to derive an effective nucleon-nucleon (NN) interaction [10]. The details of the derivation of the NN interaction can be found in Ref. [13], here we simply present the final results. In Fig. 2, we compare the effective NN potential with the one-boson-exchange Bonn potential [15] for ω and ρ exchanges. We use here a local approximation [13] to the full, nonlocal potentials. Because of the Yukawa form of the meson-exchange potentials, we have multiplied the potentials by r^2. It is seen that the two potentials have roughly the same ranges, but have very different (volume) strengths.

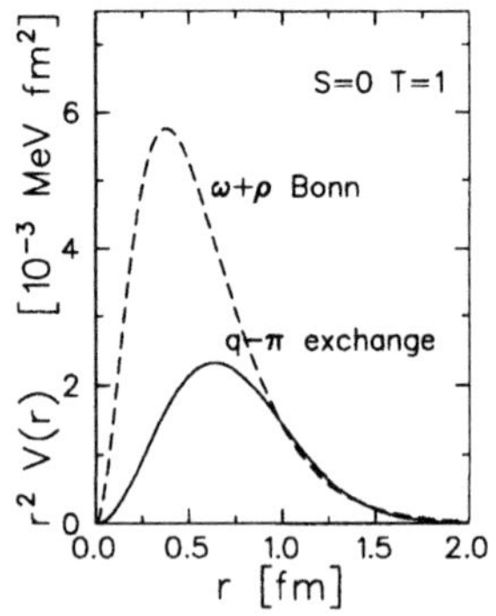

Figure 2. The quark-exchange NN potential (solid) and the $\omega + \rho$ (dashed) Bonn potential.

Next, we calculate the binding energy of nuclear matter in the Hartree-Fock approximation using the following strategy. For the short-range part of the NN interaction, we use the quark-exchange potential just discussed. For the medium- and long-range parts, we use sigma and (noncontact) one-pion-exchange, but do not allow for quark-exchange. The idea is of course, to replace vector meson exchange as the main source of the short range part of the NN interaction. We find [10] that nuclear matter becomes overbound at the empirical saturation density, i.e. there is no value of g_σ^q that gives a binding close to -15 MeV at the saturation density. This is because the short-range repulsion is not strong enough, as already seen in Fig. 2; the quark-exchange repulsion is weaker that the traditional $\omega + \rho$.

We then couple the ω, but do not allow quark-exchange to happen as the ω is exchanged between two nucleons, i.e. the ω couples to the nucleon through the form factor generated within the quark model. We then refit g_σ^q and g_ω^q to obtain the saturation of nuclear matter at the right place. The interesting result of this calculation is that the coupling constant of the ω to the nucleon is very close to the quark-model SU(6) symmetry prediction $g_\omega^2/4\pi \approx 9g_\rho^2/4\pi = 9 \times 0.55$, which is much smaller as the coupling constants used in QHD and in one-boson-exchange models.

5 Future Perspectives

We presented results of the implementation of exchange terms in the QMC model. This is the first step towards a more complete calculation to include short-range correlations and the implementation of chiral symmetry in the model. Another important direction is the study of quark deconfinement with this model.

Regarding chiral symmetry, the model we are using here is the lowest order truncation to the Cloudy Bag Model [14]. There is therefore a natural way to proceed, with great opportunities for interesting studies [16]. The modifications of the properties of the $\Delta(1232)$ in medium is of particular interest.

Acknowledgement. Different parts of the work presented here have been done in collaboration with M.E. Bracco, D. Hadjimichef, M. Nielsen, S. Szpigel, A.W. Thomas, K. Tsushima and J.S. da Veiga. Work partially supported by CNPq and FAPESP.

References

1. P.A.M. Guichon: Phys. Lett. **B200**, 235 (1988)
2. B.D. Serot and J.D. Walecka: Adv. Nucl. Phys. **16**, 1 (1986)
3. S. Fleck, W. Bentz, K. Shimizu, K. Yazaki: Nucl. Phys. **A510**, 731 (1990)
4. K. Saito and A.W. Thomas: Phys. Lett. **B327**, 9 (1994)
5. P.A.M. Guichon, K. Saito, E. Rodionov, A.W. Thomas: Nucl. Phys. **A601**, 349 (1996)
6. P.G. Blunden and G.A. Miller: Phys. Rev. **C54**, 359 (1996)

7. K. Saito, K. Tsushima, A.W. Thomas: Nucl. Phys. **A609**, 339 (1996)

8. X. Jin and B.K. Jennings: Phys. Lett. **B374**, 13 (1996); Phys. Rev. **C54**, 1427 (1996); H. Müller and B.K. Jennings, Nucl. Phys. **A640**, 55 (1988)

9. P. K. Panda, A. Mishra, J. M. Eisenberg, W. Greiner: Phys. Rev. **C56**, 3134 (1997); I. Zakout and H.R. Jaqaman: Phys. Rev. **C59**, 968 (1999)

10. M.E. Bracco, G. Krein, M. Nielsen: Phys. Lett. **B432**, 258 (1998)

11. G. Krein, A.W. Thomas, K. Tsushima: Nucl. Phys. **A610**, 313 (1999)

12. M.D. Girardeau: Phys. Rev. Lett. **27**, 1416 (1971); J. Math. Phys. **16**, 1901 (1975)

13. D. Hadjimichef, G. Krein, S. Szpigel, J.S. da Veiga: Phys. Lett. **B367**, 317 (1996); Ann. Phys. (NY) **267**, 105 (1998); M.D. Girardeau, G. Krein, D. Hadjimichef: Mod. Phys. Lett. **A11**, 1121 (1996)

14. S. Théberge et al.: Phys. Rev. **D22**, 2838 (1980); **D23**, 2106 (1981); A.W. Thomas: Adv. Nucl. Phys. **13**, 1 (1984); G.A. Miller: Int. Rev. Nucl. Phys. **2**, 190 (1984)

15. R. Machleidt: Adv. Nucl. Phys. **19**, 189 (1989)

16. J. Delorme, M. Ericson, P.A.M. Guichon, A.W. Thomas: LANL archives nucl-th/9904070

Few-Body Systems Suppl. 12, 191–196 (2000)

Few-Body Systems

Structure of the Nucleon based on a Relativistic 3-Body Approach to the NJL Model

W. Bentz[1,*], H. Mineo[1], N. Ishii[2], H. Asami[3], K. Yazaki[4]

[1] Department of Physics, Univ. of Tokyo, Tokyo 113-0033, Japan

[2] Institut für Theoretische Physik III, Universität Erlangen, Erlangen, Germany

[3] High Energy Accelerator Research Organization (KEK), Tanashi Branch, Tokyo 188-0002, Japan

[4] Department of Physics, Faculty of Humanities and Sciences, Tokyo Woman's Christian University, Tokyo 167-8585, Japan

Abstract. After a brief review of the relativistic Faddeev approach to the nucleon in the NJL model, some results for the nucleon structure functions and the nuclear matter equation of state are discussed.

1 The Relativistic Faddeev Approach to the Nucleon

As an effective quark theory of QCD in the low energy region, the Nambu-Jona-Lasinio model [1] provides a powerful tool to investigate the structure of hadrons. In recent years exact numerical solutions to the relativistic 3-quark Faddeev equation have been obtained in the ladder approximation [2, 3]. This approach incorporates the relativistic dynamics and spontaneous breaking of the chiral symmetry.

The relativistic Faddeev equation in the NJL model can be reduced to a Bethe-Salpeter equation for a quark a 'diquark' interacting via quark exchange. In the actual calculation of baryon masses we include the scalar (0^+) and axial vector (1^+) diquark channels. The effective coupling constants in these channels (g_s and g_a) depend on the form of the chirally symmetric 4-fermion interaction which characterizes the NJL model. Here we treat the ratios $r_s = g_s/g_\pi$ and $r_a = g_a/g_\pi$ as free parameters, reflecting different possible forms of the 4-fermi interaction. g_π is the effective coupling constant in the pionic $q\bar{q}$ channel, which is fixed by the pion mass. For the constituent quark mass we use $M = 400MeV$, and the Euclidean sharp cut-off scheme is employed.

* *E-mail address:* bentz@phys.s.u-tokyo.ac.jp

As an example we show in fig.1 the nucleon and delta masses as functions of r_s for different values of r_a. (The delta mass depends only on r_a.)

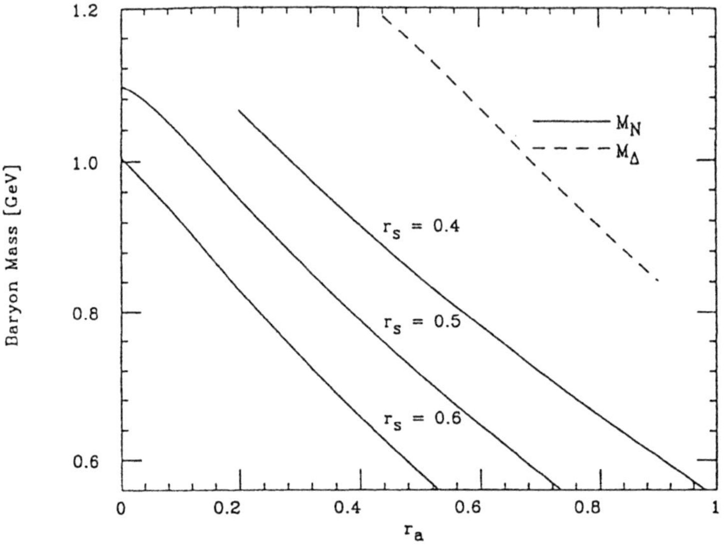

Figure 1. The solid lines show the nucleon mass as a function of r_a for three values of r_s, and the dashed line shows the delta mass as a function of r_a.

Using these results one can construct a lagrangian which can describe simultaneously the masses of the pion, the nucleon and the delta: $\mathcal{L}_I = -g\left[\sum_{c=1}^{8}\left(\bar{\psi}\gamma^\mu\frac{\lambda_c}{2}\psi\right)^2 + \kappa\,(\bar{\psi}\gamma^\mu\gamma_5\psi)^2\right]$. For $\kappa = 0$ ('color current lagrangian': $r_s = 0.5, r_a = 0.25$) we obtain $M_N \simeq 900 MeV$, but no bound state for the delta. For $\kappa = 0.27$ ($r_s = 0.37$, $r_a = 0.44$), we obtain reasonable masses $M_\Delta = 1190 MeV$, and $M_N = 920 MeV$. The main mechanism for the delta-nucleon mass difference is the interaction in the scalar diquark channel: Increasing r_s lowers the nucleon mass but leaves the delta mass unchanged. ($M_\Delta - M_N \simeq (1.1 r_s - 0.06 r_a - 0.18)\, GeV$.)

Using the relativistic Faddeev wave functions, the static properties of the nucleon have been calculated [4]. If for simplicity one neglects the axial vector diquark channel, one obtains $g_A = 1.33$, $< r^2 >_p = 0.81 fm^2$, $< r^2 >_n = -0.08 fm^2$, $\Sigma_{\pi N} = 35.1 MeV$, $\mu_p = 1.78$, and $\mu_n = -1.34$, if r_s is adjusted to reproduce the nucleon mass. Comparing these values with the experimental ones ($g_A = 1.25$, $< r^2 >_p = 0.74 fm^2$, $< r^2 >_n = -0.12 fm^2$, $\Sigma_{\pi N} = 45 \pm 7 MeV$, $\mu_p = 2.79$, and $\mu_n = -1.91$), one sees that the axial vector coupling constant, the charge radii and the pion nucleon sigma-term are reasonable, but the magnetic moments are too small in magnitude, leaving room for further contributions due to the axial vector diquark channel.

2 Quark Light Cone Momentum Distributions

In the Bjorken limit, the spin independent structure functions F_1 and F_2 of the nucleon can be expressed in terms of the twist 2 quark light cone momentum distributions [5] $q(x) = \frac{1}{4\pi}\int dz^- e^{i(p_- x)z^-} \langle p|T\left(\overline{\psi}(0)\gamma^+\psi(z^-)\right)|p\rangle$, where x is the Bjorken variable. This is nothing but the quark propagator (traced with γ^+) inside the nucleon with a fixed quark momentum component $k_- = p_- x$, while the other components are integrated out.

As a first step, we employ here the 'static approximation' to the Faddeev equation [6], in which the momentum dependence of the quark exchange kernel is neglected. We will also restrict ourselves to the scalar diquark channel in the nucleon wave function. In order to estimate also the sea quark distributions $\overline{q}(x)$, we take into account the pionic cloud around the constituent quarks by using the familiar convolution formalism. We compare the results obtained by using the 'invariant mass regularization' [7] (LB scheme) with the 'transverse momentum regularization' [8] (TR scheme). The calculated distributions are evolved [9] from the NJL low energy scale Q_0^2, which is treated as a parameter, up to $Q^2 = 4GeV^2$, where empirical parametrizations exist [10].

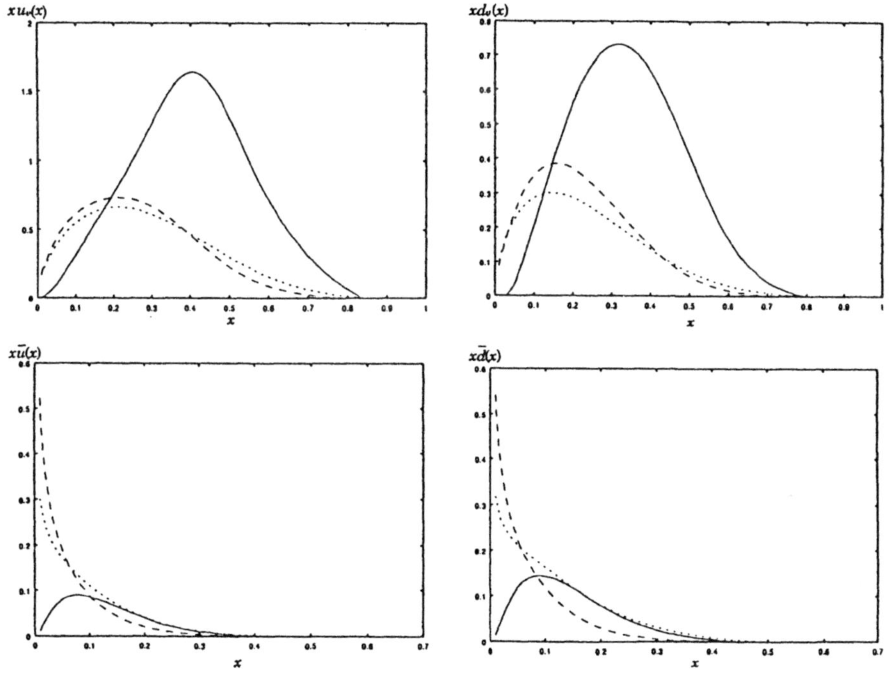

Figure 2. The valence quark (upper figures) and sea quark (lower figures) distributions in the nucleon obtained with the LB regularization scheme. The solid lines show the NJL input distributions ($Q_0^2 \equiv 0.16GeV^2$), the dashed lines the distributions evolved to $Q^2 = 4GeV^2$, and the dotted lines the empirical distributions at $Q^2 = 4GeV^2$.

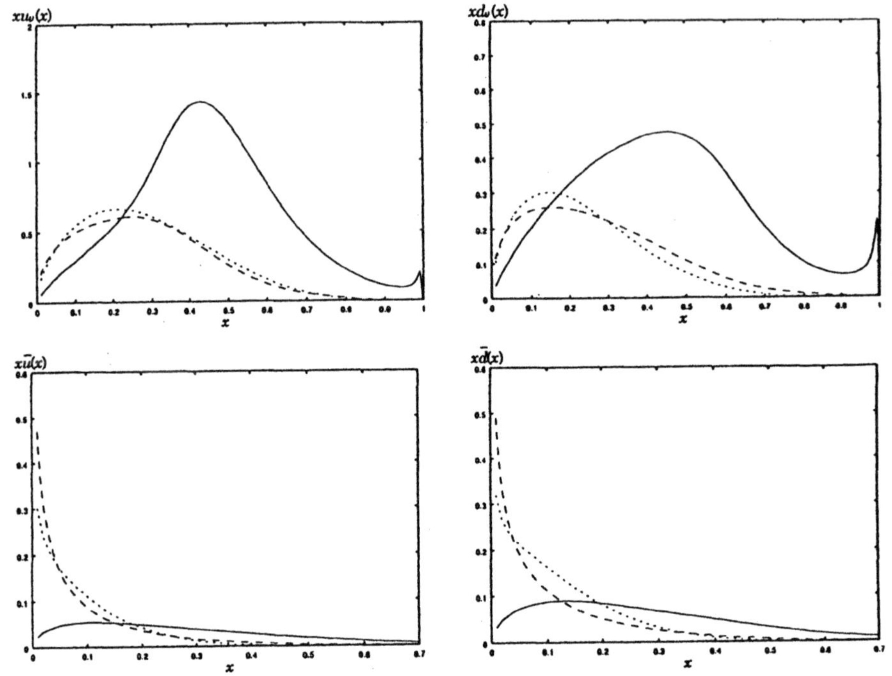

Figure 3. Same as fig.2 for the TR regularization scheme.

These results [11] show that this simple model leads to reasonable valence quark distributions, in particular in the TR regularization scheme. We expect that the inclusion of higher mass qq states (axial vector diquark channel) and the use of exact Faddeev wave functions (softer input distributions) will allow for a higher value of Q_0^2. We find that at the low energy scale the valence quarks carry 92% (87%) of the nucleon momentum in the LB (TR) scheme. The pionic effects lead to the well known asymmetry $\bar{d} > \bar{u}$, although it is too small compared to the observed one. In our calculation, the probability of the quark state without the pion cloud is 84% (83%) for the LB (TR) scheme, which corresponds to a reduction of the Gottfried sum from the naive (valence quark) value $\frac{1}{3}$ to 0.262 (0.257), compared to the experimental value of 0.235 ± 0.026 [12].

3 Nuclear Matter Equation of State

It would be interesting to extend the calculations of nucleon properties described in the previous sections to finite density. For this purpose, however, one first has to construct an equation of state. Here we consider a kind of 'hybrid model' for nuclear matter, in which the Dirac sea is made of quarks and the Fermi sea consists of nucleons. The single nucleon is constructed as a solution of the Faddeev equation, where again we restrict ourselves for simplicity to the 'static approximation' and the scalar diquark channel. The quarks in the Dirac sea and within the nucleon couple to self consistent mean fields (quark bilinears) of σ and ω meson type. The energy density of the system then consists of the vacuum part ('Mexican hat' potential) $\epsilon_V = -12 \int^{\Lambda} \frac{d^3k}{(2\pi)^3} \left(\sqrt{k^2 + M^{*2}} - \sqrt{k^2 + M^2}\right) + \frac{M^{*2} - M^2}{4g_\pi}$, where $M^* = m - 2g_\pi \langle \overline{\Psi}\Psi \rangle$ is the effective quark mass and g_π the 4-fermi coupling constant in the pionic channel, a repulsive ω meson like term (ϵ_ω), and the energy due to the Fermi motion of the composite nucleons in the scalar field (ϵ_F). The resulting expression for the energy density $\epsilon = \epsilon_V + \epsilon_\omega + \epsilon_F$ is similar to the familiar linear sigma model expression (with $M_\sigma = 2M$) for elementary nucleons, but (i) the vacuum polarization term is smaller since the relevant scale is now the quark mass instead of the nucleon mass, and (ii) the effective nucleon mass does not behave linearly as a function of the scalar potential due to the composite nature of the nucleon. Due to (i), the vacuum polarization term cannot prevent the well known instability due to the transition to the Lee-Wick abnormal state [13]. Our numerical results show that the normal state does not saturate before the quark condensate drops to small values.

It is known that instabilities of this type in sigma or NJL type theories can be prevented by including higher order interaction terms [14]. For example, the effect of a 8-fermi interaction term of the form $\mathcal{L}_8 = g_8 \left[(\overline{\psi}\psi)^2 - (\overline{\psi}\gamma_5 \tau \psi)^2 \right] (\overline{\psi}\gamma^\mu \psi)^2$ can be incorporated into an effective coupling constant in the vector channel which increases with increasing scalar field. This cures the instability and leads to a saturating equation of state. As an example of such a calculation, we show in fig. 4 the binding energy per nucleon and the effective nucleon and quark masses. The details of this calculation will be discussed elsewhere [15].

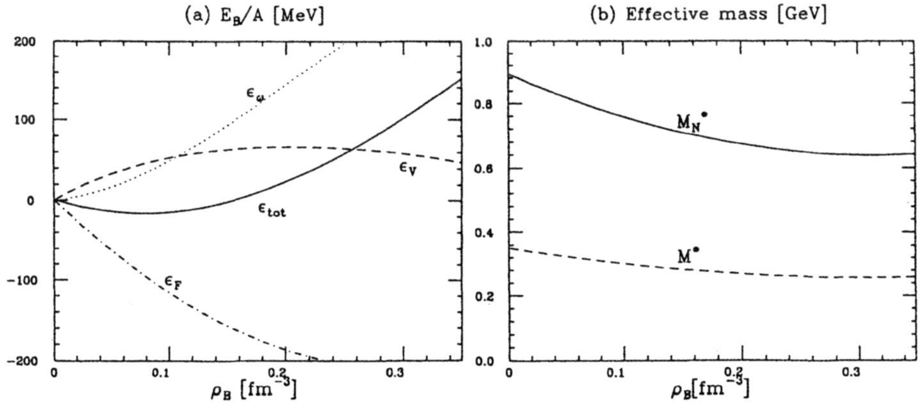

Figure 4. An example of a saturating nuclear matter equation of state based on the quark-diquark picture of the single nucleon. (a) shows the various contributions to the binding energy per nucleon, where the ω meson like repulsion ϵ_ω includes the effect of the 8-fermi term, and (b) shows the effective quark and nucleon masses.

ACKNOWLEDGEMENTS
The results discussed in sect. 3 are based on a collaboration with A.W. Thomas. The authors express their thanks to him for many helpful discussions on structure functions, and to M. Miyama and S. Kumano for the Q^2 evolution program (ref. [9]).

References

1. Y. Nambu and G. Jona-Lasinio: Phys. Rev. **122**, 345 (1960); **124**, 246 (1961)

2. N.Ishii, W. Bentz, K. Yazaki: Nucl. Phys. **A587**, 617 (1995); N. Ishii: Phys. Lett. **B431**, 1 (1998)

3. S. Huang and J. Tjon: Phys. Rev. **C49**, 1702 (1994); U. Zückert, R. Alkofer, H. Weigel, H. Reinhardt: Phys. Rev. **C55**, 2030 (1997)

4. H. Asami, N. Ishii, W. Bentz, K. Yazaki: Phys. Rev. **C51**, 3388 (1995)

5. R.L. Jaffe: 1985 Los Alamos school on Relativistic Dynamics and Quark Nuclear Physics, eds. M.B. Johnson and A. Pickleseimer. New York: Wiley 1985

6. A. Buck, R. Alkofer, H. Reinhardt: Phys. Lett. **B286**, 29 (1992)

7. G.P. Lepage and S.J. Brodsky: Phys. Rev. **D22**, 2157 (1980)

8. F. Lenz, K. Ohta, M. Thies, K. Yazaki: (to be published)

9. M. Miyama and S. Kumano: Comp. Phys. Commun. **94**, 185 (1996)

10. A.D. Martin, W.J. Stirling, R.G. Roberts: Phys. Rev. **C50**, 6734 (1994)

11. H. Mineo, W. Bentz, K. Yazaki: nuclth-9907043, to appear in Phys. Rev. **C** (Nov. 1999)

12. M. Arneodo et al.: Phys. Rev. **D50**, R1 (1994)

13. T.D. Lee and G.C. Wick: In: *Mesons in Nuclei*, eds. D.H. Wilkinson and M. Rho, p.869. Amsterdam: North-Holland 1979

14. V. Koch, T. Biro, J. Kunz, U. Mosel: Phys. Lett. **B185**, 1 (1987); J. Boguta: Phys. Lett. **B 120**, 34 (1983)

15. W. Bentz, A.W. Thomas, K. Yazaki: (in preparation)

Few-Body Systems Suppl. 12, 197–200 (2000)

Few-Body Systems

Effects of Coupling with Nucleon Resonances on NN Interaction

I. Maeda[1], M. Arima[1], K. Masutani[2]

[1] Department of Physics, Osaka City University, Osaka 558-8585, Japan
[2] Faculty of Engineering, Yamanashi University, Kofu 400-8511, Japan

Abstract. Nucleon-nucleon S-wave scattering is studied by using the resonating group method which includes not only the nucleon and Δ but also their excited states. As quark-quark potentials, the one-gluon-exchange and one-pion-exchange potentials are employed in addition to phenomenological confinement potential. It is found that, besides the well-known ${}^3D_1NN$ and ${}^5D_0N\Delta$, several channels play important roles owing to the tensor coupling induced by the one-pion-exchange potential.

1 Introduction

There have been many attempts to explain the nucleon-nucleon (NN) interaction within constituent quark models [1-5]. In these works, the one-gluon-exchange potential (OGEP) and/or the one-meson-exchange potential (OMEP) have been mainly used as quark-quark (qq) potentials, supplemented with phenomenological confinement potential.

Two main features of the NN interaction, the repulsive core and the medium-range attractive part, have been successfully explained by the Pauli Exclusion Principle for constituent quarks , the effect of various types of OMEP (π, σ, η ...), etc. Several calculations with the various combination of qq and baryon-baryon potentials provide reasonably good results in agreement with the experimental NN scattering data [1-3].

On the other hand, there still remains issue to be clarified; i.e. the dynamical effect of N^* and Δ^* in the NN scattering is generally obscured if only the N and Δ are considered and the potential parameters are varied so as to fit experimental data. Due to the technical difficulty, systematic study of these effects has not been appeared yet. Furthermore, adiabatic estimations have shown that the effect of N^* and Δ^* becomes more important if the OMEP is used as qq potential [4], whereas it is not large if the OGEP is employed [5]. Thus, examination of these dynamical effects is not only interesting task itself, but also expected to give us a better understanding of the NN interaction on the quark level.

The purpose of this work is to clarify the dynamical effect of N^* and Δ^* in the NN interaction. To this end, we systematically study the NN S-wave scattering by using the coupled-channel resonating group method (RGM) within the non-relativistic constituent quark model.

2 Model

We used the one-pion-exchange potential (OPEP) and the OGEP as qq interactions. Their explicit forms are written as follows;

$$\begin{aligned}
V_{ij} &= V_{ij}^{OGEP} + V_{ij}^{OPEP}, \\
V_{ij}^{OGEP} &= \frac{\alpha_s}{4}(\boldsymbol{\lambda}_i \cdot \boldsymbol{\lambda}_j)\left[\frac{1}{r_{ij}} - \frac{2\pi}{3m^2}(\boldsymbol{\sigma}_i \cdot \boldsymbol{\sigma}_j)\left(\frac{\kappa}{\pi}\right)^{\frac{3}{2}} e^{-\kappa r_{ij}^2} - \frac{1}{4m^2}\frac{1}{r_{ij}^3}S_{ij}\right], \\
V_{ij}^{OPEP} &= \frac{g_{\pi qq}^2}{4\pi}\frac{1}{12m^2}(\boldsymbol{\tau}_i \cdot \boldsymbol{\tau}_j)\left[S_\pi(r_{ij})(\boldsymbol{\sigma}_i \cdot \boldsymbol{\sigma}_j) + T_\pi(r_{ij})S_{ij}\right], \\
S_\pi(r_{ij}) &= m_\pi^2 \frac{e^{-m_\pi r_{ij}}}{r_{ij}} - 4\pi\left(\frac{\kappa_\pi}{\pi}\right)^{\frac{3}{2}} e^{-\kappa_\pi r_{ij}^2}, \\
T_\pi(r_{ij}) &= \left(\frac{3}{r_{ij}^3} + \frac{3m_\pi}{r_{ij}^2} + \frac{m_\pi^2}{r_{ij}}\right) e^{-m_\pi r_{ij}}, \qquad (1)
\end{aligned}$$

where m is the constituent u- and d-quark mass, $g_{\pi qq}$ and m_π are the quark-pion coupling constant and the pion mass, respectively. The SU(3) color generator for the i-th quark is denoted by $\boldsymbol{\lambda}_i$, and the distance between the i-th and j-th quarks by r_{ij}. The short-range interactions, originally stemming from the δ function parts in the OGEP and OPEP, are smeared out by introducing the range parameters κ and κ_π with the proper normalization.

To make the RGM calculation tractable, we assume that the single baryon wave function is an eigenfunction of a harmonic oscillator potential with an oscillator parameter β and is corresponded to actual resonance which has same spin, isospin and parity quantum numbers [6]. All parameters included in this model are determined by using following condition; 1) Experimental values are used for $g_{\pi qq}$ and m_π, 2) Variational condition of the nucleon mass is satisfied by β, 3) Mass splittings of N-Δ and N-N_1^* (the first excited state of nucleon) are fitted by rest parameters.

We consider following two-baryon sets in this work; NN,$N\Delta$,NN^*,$N^*\Delta$,$N\Delta^*$, $\Delta\Delta^*$, NN^{**}, N^*N^*,$N^{**}\Delta$,$N^*\Delta^*$, $N\Delta^{**}$,$\Delta\Delta^{**}$ and $\Delta^*\Delta^*$ (number of asterisks shows energy shell to which each resonance belongs). There are 42 (34) channels coupled to the ${}^3S_1NN$ (${}^1S_0NN$) channel in this model space. Because of consideration of these resonances, our model space has richer six-quark symmetry patterns than previous RGM calculations. This is one of the advantage of our work.

3 Result

We first select several channels that may play important roles in the NN scattering; we solve the two-channel RGM equation of motion, which contains one of the 42 (34) channels and the ${}^3S_1NN$ (${}^1S_0NN$) channel, and compare the result with the single-channel calculation. Results of this calculations are summarized in Table 1. Large effect is observed in the case of the ${}^3D_1NN$ (${}^5D_0N\Delta$) in the ${}^3S_1NN$ (${}^1S_0NN$) scattering. This fact has been already known on the baryon level and has been also reproduced on the quark level [2, 3]. Table 1 also shows that other important channels, which have never been pointed out explicitly in literatures. Especially, effects of the ${}^7D_1\Delta\Delta$, ${}^3D_1NN_1^{**}$ and ${}^3S_1\Delta\Delta_2^{**}$ (${}^5D_0\Delta\Delta$ and ${}^1S_0\Delta\Delta_2^{**}$) on the ${}^3S_1NN$ (${}^1S_0NN$) are remarkable. These coupling effects are mainly induced by the tensor interaction of the OPEP. Role of the OGEP is small.

Results of full calculations which include all channels listed in Table 1 are represented in Figure 1. It shows that although the effect of the ${}^3D_1NN$ (${}^5D_0N\Delta$) on the ${}^3S_1NN$ (${}^1S_0NN$) is large but new finding channels listed in Table 1 can not be neglected. It is remarkable that these effects appear even at low energy region. In order to show coupling effects at low energy region, the scattering length and the binding energy of the deuteron are calculated and results of them are represented in Table 2. It shows large effects of new finding channels on the low energy scattering observables.

As a result of our analysis, we conclude that the dynamical effect of N^* and Δ^* are important in the NN interaction and they have to be considered to obtain better understanding of the NN interaction even in low energy region.

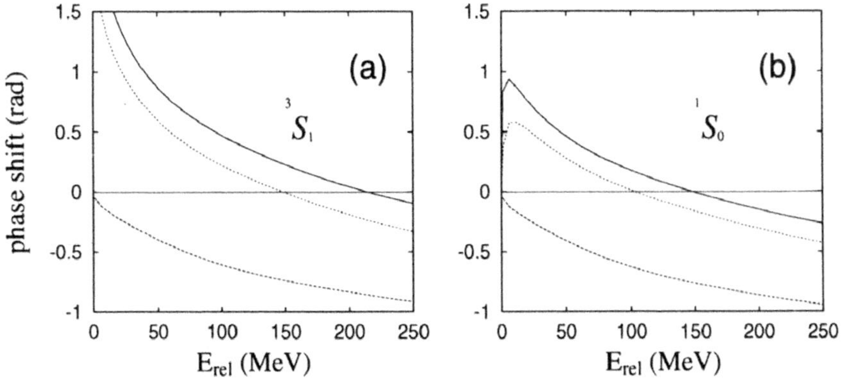

Figure 1. (a): ${}^3S_1NN$ phase shifts as a function of E_{rel}, the energy of the relative motion in the center of mass system. The dashed line corresponds to the one-channel (${}^3S_1NN$) calculation. Short dashed line shows result of ${}^3S_1NN$-${}^3D_1NN$ and solid line represents one of full calculation. (b): Same as (a) but for ${}^1S_0NN$. The dashed line corresponds to the one-channel (${}^1S_0NN$) calculation. Short dashed line shows result of ${}^1S_0NN$-${}^5D_0N\Delta$ and solid line represents one of full calculation.

Table 1. List of selected channels. Resonances which belong to same energy shell are distinguished by their mass and labeled by subscript from the lightest one. Strength of coupling effects are expressed by number of asterisks.

$^3S_1$-channel			$^1S_0$-channel		
Baryons	State	Effect	Baryons	State	Effect
NN	$^3S_1$	—	NN	$^1S_0$	—
NN	$^3D_1$	* * * * **	$\Delta\Delta$	$^1S_0$	*
$\Delta\Delta$	$^3S_1$	*	$\Delta\Delta$	$^5D_0$	* * **
$\Delta\Delta$	$^3D_1$	*	$N_1^*N_2^*$	$^1S_0$	*
$\Delta\Delta$	$^7D_1$	* * * * *	$\Delta\Delta_2^{**}$	$^1S_0$	* * *
NN_1^*	$^1P_1$	*	$N\Delta$	$^5D_0$	* * * * **
NN_2^*	$^1P_1$	*	$N_1^{**}\Delta$	$^5D_0$	*
$N_1^*N_1^*$	$^3S_1$	*	$N\Delta_1^{**}$	$^5D_0$	*
$N_2^*N_2^*$	$^3S_1$	*			
NN_1^{**}	$^3D_1$	**			
$\Delta\Delta_1^{**}$	$^7D_1$	*			
$\Delta\Delta_2^{**}$	$^3S_1$	* * **			

Table 2. Deuteron binding energy (B.E.) and scattering length (a_0)

State	Coupled channel	B.E. (MeV)	a_0 (fm)
$^3S_1$	^3D_1NN	0.70	9.1
	full	5.6	3.8
	Exp.	2.2	5.4
$^1S_0$	$^5D_0N\Delta$	—	−2.8
	full	—	−8.8
	Exp.	—	$-16 \sim -24$

References

1. M. Oka and K. Yazaki: In: *Quarks and nuclei* (International Review Nuclear Physics, vol. 1), p.489. Singapore: World Scientific 1984
2. K. Bräuer et al.: Nucl. Phys. **A507**, 599 (1990)
3. A. Valcarce, A. Faessler, F. Fernández: Phys. Lett. **B345**, 367 (1995)
4. Fl. Stancu, S. Pepin, L.Ya. Glozman: Phys. Rev. **C56**, 2779 (1997)
5. S. Ohta et al.: Phys. Lett. **B119**, 35 (1982); **B123**, 477 (E) (1983)
6. I. Maeda, M. Arima, K. Masutani: (in preparation)

Few-Body Systems Suppl. 12, 201–204 (2000)

Few-Body Systems

The Interaction between H-Dibaryons and the Property of H-Dibaryon Matter

T. Sakai[1,*], J. Mori[2], A.J. Buchmann[3], K. Shimizu[2], K. Yazaki[4], H. Suganuma[1]

[1] Research Center for Nuclear Physics, Osaka University, 10-1 Mihogaoka, Ibaraki, Osaka 567-0047, Japan

[2] Department of Physics, Sophia University, Tokyo 102-0094, Japan

[3] Institut für Theoretische Physik, Universität Tübingen, Auf der Morgenstelle 14, D-7400, Tübingen, Germany

[4] Tokyo Women's Christian University, Tokyo 167-8585, Japan

Abstract. The interaction between H-dibaryons and the property of H-dibaryon matter are investigated using the quark cluster model and the Skyrme model on a hypersphere. It is shown that the H-H interaction has attractive nature and causes the swelling of the H-dibaryon, and that at high densities H-matter transforms to the deconfinement phase and the chiral symmetry restores.

1 Introduction

During two decades, the H-dibaryon has been elaborately investigated as a most plausible candidate of exotic states beyond mesons and baryons. In spite of experimental efforts, no conclusive result on the existence of the H-dibaryon has been obtained. However, the lower bound of its mass is put by the binding energy of $\Lambda\Lambda$ in a double hypernucleus. Reported double hypernuclear events restrict the H-dibaryon mass within at most thirty MeV below the $\Lambda\Lambda$ threshold. It is worth noting that it is not known whether these double hypernuclei are $\Lambda\Lambda$-nuclear states, H-nuclear states or mixing of them. In Ref.[1], we pointed out that some of low-lying states of a double hypernucleus may have appreciable ammounts of H-dibaryon component as well as $\Lambda\Lambda$ component.

The H-dibaryon attracts attention also from astrophysical viewpoints. For instance, it is pointed out that H-dibaryon matter may exist in the core of a

* *E-mail address:* tsakai@rcnp.osaka-u.ac.jp

high-density neutron star [2]. Several authors have attempted to explain some cosmic ray events like Cygnus X-3 event [3] or ultra-high energy cosmic rays beyond GZK cutoff [4] on the basis of the neutrality and the expected stability of the H-dibaryon. Unfortunately, the H-dibaryon does not seem to be so long-lived as their expectation.

We have studied the interaction between H-dibaryons and the behavior of dense H-dibaryon matter. We use a quark cluster model for the former, and a Skyrme model for the latter. We will show that some of the properties of H-matter are closely related to the attractive nature of the H-H interaction.

2 The Interaction between H-Dibaryons

We made a microscopic calculation of the H-H interaction [5] employing the quark cluster model, which has been successful in describing NN and NY interactions. In this model, the internal wave functions of particles are described by harmonic oscillator wave functions, and the relative wave function between particles is obtained by solving RGM equation. Here, we employ the one-gluon-exchange potential (OGEP) and the quadratic two-body confinement potential as interactions between quarks, and an effective meson exchange potential (EMEP) as an interaction between dibaryons. The strength of EMEP is determined so that the H-dibaryon mass is consistent with the double hypernucleus data, while the parameters in OGEP are determined from baryon masses.

The following results are obtained. (i) The interaction between H-dibaryons gives deeply bound state with the binding energy 100$\sim$170 MeV. (ii) The main properties of the interaction between H-dibaryons are the short-range repulsion due to Pauli blocking and the color magnetic interaction in OGEP, and the medium-range attraction due to σ-meson exchange. (iii) Because of the repulsive core, two H-dibaryons separate by 0.8$\sim$0.9 fm.

Here, we comment on an implication to possible H-dibaryon matter in neutron star. In Tamagaki's work[2], the critical density of phase transition from neutron matter to H-matter is estimated to be six to nine times normal nuclear matter density using a simple potential between H-dibaryons, which consists of hard core and square well atractive potential outside the core. However, the depth of the attractive potential is rather weak comparing with our microscopically determined potential. Therefore, the phase transition will occur at a lower critical density if a more attractive potential like ours is used.

3 H-Dibaryon Matter in the Skyrme Model on S^3

The Skyrme model was used to study the H-dibaryon first by Balachandran et al. [6], who described the H-dibaryon as a soliton with baryon number two and SO(3) configuration. The H-dibaryon mass lower than the mass of two baryons was obtained in the flavor symmetric limit. We have studied the property of dense H-dibaryon matter [7] following the method proposed by Manton and Ruback [8], in which they investigated the properties of nuclear matter at high densities by regarding an SU(2) hedgehog Skyrmion placed on an S^3 hypersphere with radius R as an approximation of nuclear matter with density $1/(2\pi^2 R^3)$. They showed that their model can mimic the transition to deconfinement phase and the chiral symmetry restoration at a critical density, which are also shown in a calculation with Skyrmions placed on a lattice[9].

In Fig.1, we show the baryon number density distribution on S^3. Here, a point on the S^3 hypersphere is indicated by polar coordinates (μ, θ, φ). At low densities (large R), the baryon number density localizes around a pole and the Skyrmion behaves as a particle. On the other hand, the baryon number density spreads all over the hypersphere at high densities (small R) and the Skyrmion loses its profile as a particle. The latter situation can be regarded as deconfinement phase. At high densities, (angle averaged) chiral order parameter also becomes small, which may be a signal of the chiral symmetry restoration.

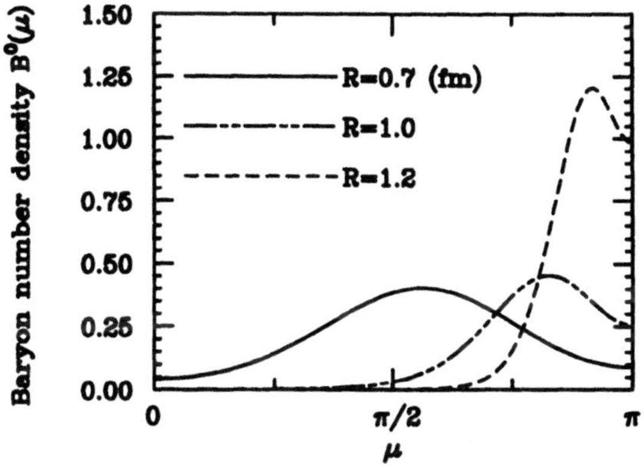

Figure 1. Baryon number density $B^0(\mu)$ vs. the angle μ of the hypersphere for several hypersphere radii R.

The root mean square radius of the H-dibaryon is shown in Fig.2. It is remarkable that an enlargement of rms radius appears just before the transition to deconfinement phase. This swelling phenomenon of the H-dibaryon is a reflection of the attractive nature of the interaction between H-dibaryons. Swelling phenomenon is also known in various studies on the properties of dense nuclear matter. The linear decrease of the size at high densities in deconfinement phase is another reflection of the lost particle-like profile of the Skyrmion. There, the average of r^2 vanishes because the distribution of the baryon number density becomes symmetric about the equator $\mu = \pi/2$.

4 Summary

We have studied the interaction between H-dibaryons and the behavior of the dense H-dibaryon matter.

Quark cluster model calculation shows that the interaction between H-dibaryons consists of the short range repulsion and the medium range attrac-

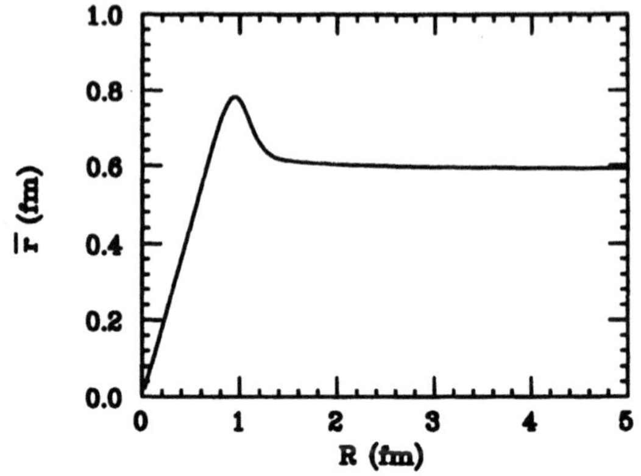

Figure 2. The H-dibaryon rms radius $\bar{r}$ as a function of the hypersphere radius R.

tion. This attractive force binds two H-dibaryons deeply.

A study of the H-dibaryon matter by Skyrme model á là Manton shows that the deconfinement and the chiral symmetry restoration occurs at high densities. Swelling phenomenon implies the attractive nature of the interaction between H-dibaryons.

Acknowledgement. One of the authors (T.S.) is grateful for the support by Special COE grant of the Ministry of Education, Science and Culture of Japan.

References

1. T. Sakai, K. Yazaki, K. Shimizu: Nucl. Phys. **A594**, 247 (1995)
2. R. Tamagaki: Prog. Theor. Phys. **85**, 321 (1991)
3. G. Baym et al.: Phys. Lett. **B160**, 181 (1985)
4. N.I. Kochelev: Preprint hep-ph/9905333
5. T. Sakai, J. Mori, A.J. Buchmann, K. Shimizu, K. Yazaki: Nucl. Phys. **A625**, 192 (1997)
6. A.P. Balachandran et al.: Phys. Rev. Lett. **52**, 887 (1984); A.P. Balachandran et al.: Nucl. Phys. **B256**, 525 (1985)
7. T. Sakai and H. Suganuma: Phys. Lett. **B430**, 168 (1998)
8. N.S. Manton and P.J. Ruback: Phys. Lett. **B181**, 137 (1986); N.S. Manton: Commun. Math. Phys. **111**, 469 (1987)
9. L. Castillejo et al.: Nucl. Phys. **A501**, 801 (1989)

Few-Body Systems Suppl. 12, 205–208 (2000)

Few-Body Systems

Analysis of Axially Symmetric $B = 2$ Solution in the Chiral Quark Soliton Model

N. Sawado

Department of Physics, Faculty of Science and Technology, Science University of Tokyo, Noda, Chiba 278-8510, Japan

Abstract. The baryon-number-two ($B = 2$) solution in the SU(2) chiral quark soliton model (χQSM) is solved, including fully the sea quark degrees of freedom. The axially symmetric pion configurations yield the energy minimum for the $B = 2$ state in the χQSM when taking into account quark dynamics. The calculation is done by using the eigenfunctions of a deformed axially symmetric oscillator. By using these basis, the numerical convergence is a significant improvement over the previous one, using the plane wave basis.

Since QCD is the fundamental theory of the strong interactions, we would like to consider hadronic systems as built up by quark degrees of freedom. However, QCD has difficulty in describing low- and medium-energy nuclear phenomenology and it is desirable to formulate an effective, tractable theory for the strong interaction. The most important feature of QCD in the low energy region is chiral symmetry and its spontaneous breakdown. A simple quark model that incorporates the above features is chiral quark soliton model (χQSM), which is characterized by the following formula [1]:

$$Z = \int \mathcal{D}\pi \mathcal{D}\psi \mathcal{D}\psi^{\dagger} \exp\left[i \int d^4x \bar{\psi}(i\partial\!\!\!/ - MU^{\gamma_5})\psi\right] . \tag{1}$$

In the Skyrme model, the minimal energy pion configuraion for baryon-number-two ($B = 2$) has axial symmetry [2, 3, 4]. If we choose the z axis as the symmetry axis, pion fields U with a winding number m are given in Manton's conjecture [5],

$$U(\boldsymbol{x}) = \cos F(\rho, z) + i\boldsymbol{\tau} \cdot \hat{\boldsymbol{n}} \sin F(\rho, z) , \tag{2}$$

where

$$\hat{\boldsymbol{n}} = (\sin\Theta(\rho, z)\cos m\varphi, \sin\Theta(\rho, z)\sin m\varphi, \cos\Theta(\rho, z)). \tag{3}$$

In this paper, we investigate the $B = 2$ soliton with axially symmetric pion fields in the χQSM. First, we consider the one particle Dirac eigen-problem:

$$H(U^{\gamma_5})\phi_\mu(\boldsymbol{x}) = E_\mu \phi_\mu(\boldsymbol{x}) , \qquad H(U^{\gamma_5}) = -i\boldsymbol{\alpha} \cdot \nabla + \beta M U^{\gamma_5} . \tag{4}$$

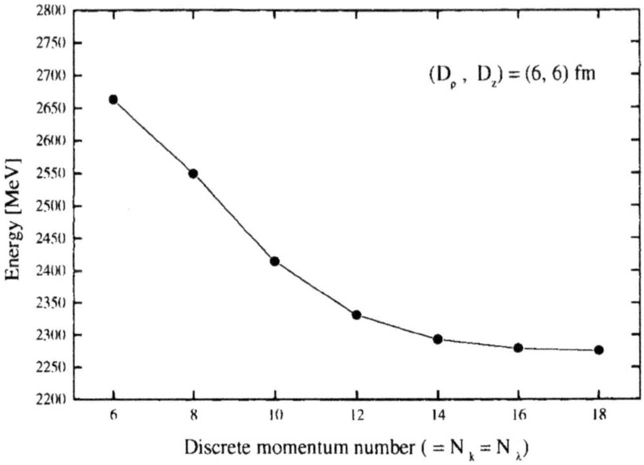

Figure 1. The convergence property of the total energy of the soliton as a function of the number of basis.

In order to solve this eigen-problem we introduce basis as a set of eigenfunctions of Dirac hamiltonian. In previous paper, we use a plane-wave basis which are the eigenstates of the free hamiltonian in a cylindrical box with heihgt $2 \times D_z$ and radius D_ρ [6]. Using these basis, we are enable to diagonalize our hamiltonian. Figure 1 shows a convergence property of the total energy of the soliton as a function of the number of basis. To be energy converge, we need at least $N_k = N_\lambda = 18$. Since the computation times increase rapidly with the number of basis, the calculations have to be restricted to smaller values of them. Therefore, in this paper we introduce different type of basis, that is the eigenfunctions of a deformed axially symmetric oscillator potential [7]. The upper- and the lower-component of the Dirac spinors are expanded separately in terms of these functions,

$$\phi_\mu(\boldsymbol{x}) = \begin{pmatrix} f_\mu(\boldsymbol{x}) \\ ig_\mu(\boldsymbol{x}) \end{pmatrix} = \begin{pmatrix} \sum_a f_{\mu a}\Phi_a(\boldsymbol{x}, s) \\ i\sum_{\tilde{a}} g_{\mu\tilde{a}}\Phi_{\tilde{a}}(\boldsymbol{x}, s) \end{pmatrix} \chi_{m_t}(\tau).$$

Where $\Phi_a(\boldsymbol{x}, s)$ are the eigenfunctions of a deformed harmonic oscillator,

$$\Phi_a(\boldsymbol{x}, s) = \phi_{m,n_r}(\rho)\phi_{n_z}(z)\frac{1}{\sqrt{2\pi}}e^{im\varphi}\chi_{m_s}(s), \tag{5}$$

and

$$\phi_{m,n_r}(\rho) = \left(\frac{2\alpha_\rho n_r!}{(n_r + m)!}\right)^{\frac{1}{2}} (\sqrt{\alpha_\rho}\rho)^{|m|} e^{-\frac{1}{2}\alpha_\rho\rho^2} L_{n_r}^{|m|}(\alpha_\rho\rho^2), \tag{6}$$

Table 1. The total energy and the root mean sqeare radius of the soliton for the volume factor α and the ratio q, with $N_{rmax} = N_{zmax} = 8$.

q	E [MeV]	$\sqrt{\langle r^2 \rangle}$ [fm]	E [MeV]	$\sqrt{\langle r^2 \rangle}$ [fm]
	$\alpha = 2$		$\alpha = 4$	
0.6	2296	0.831	2237	0.819
0.8	2285	0.831	2238	0.822
1.0	2290	0.830	2234	0.825
1.2	2310	0.830	2231	0.826
1.4	2341	0.831	2232	0.826

$$n_r = 0, 1, \cdots, N_{rmax},$$

$$\phi_{n_z}(z) = \left(\frac{1}{2^{n_z} n_z! (\pi/\alpha_z)^{\frac{1}{2}}} \right)^{\frac{1}{2}} e^{-\frac{1}{2}\alpha_z z^2} H_{n_z}(\sqrt{\alpha_z} z), \tag{7}$$

$$n_z = 1, 3, \cdots, 2N_{zmax} + 1 \quad \text{or} \quad 0, 2, \cdots, 2N_{zmax}.$$

The functions $L_{n_r}^{|m|}(\alpha_\rho \rho^2)$ and $H_{n_z}(\sqrt{\alpha_z} z)$ are the associated Laguerre polynomials and the Hermite polynomials. The constants α_ρ and α_z are related with the oscillator frequency of the potential. We impose the volume conservation and introduce the "volume" $\alpha^3 = \alpha_\rho \sqrt{\alpha_z}$ and the "ratio" $q = \sqrt{\alpha_\rho}/\sqrt{\alpha_z}$. We solve the eigen-problem for various values of α and q . The N_{rmax} and N_{zmax} are the number of eigenfunctions in the r and z direction. It is found that for $N_{rmax} = N_{zmax} \geq 2$, the self consistent solutions are emerged, and for $N_{rmax} = N_{zmax} > 8$, the total energy is almost convergent. Then the deformed harmonic oscillator basis has better convergent property than the plane wave basis.

For detailed description of the numerical procedures are given in Ref. 6. As in Ref. 6, we calculate the baryon number density and the root mean square radius of the soliton. In Fig. 2 we present result for the baryon number density. From Fig. 2 it is found that the baryon number density has a toroidal in shape. The result is consistent with the other chiral invariant models using axially symmetric pion fields, such as the Skyrme model [4]. The total energy and the root mean square radius for $\alpha = 2$ and 4 and for $q = 0.6$ to 1.4 are given in Table 1. The whole calculations are done with $M = 400$ MeV. For $\alpha = 4$ and $q = 1.2$ we obtained $\langle r^2 \rangle^{1/2} = 0.826$ fm , which is in qualitative agreement with the Skyrme model value 0.92 fm [4].

The solutions obtained here were classical ones which have no definite spin, isospin quantum numbers corresponding to physical particles. The quantization of our solution using the cranking procedure in SU(2) is done and various mass spectrum of the two-baryon system are obtained [8].

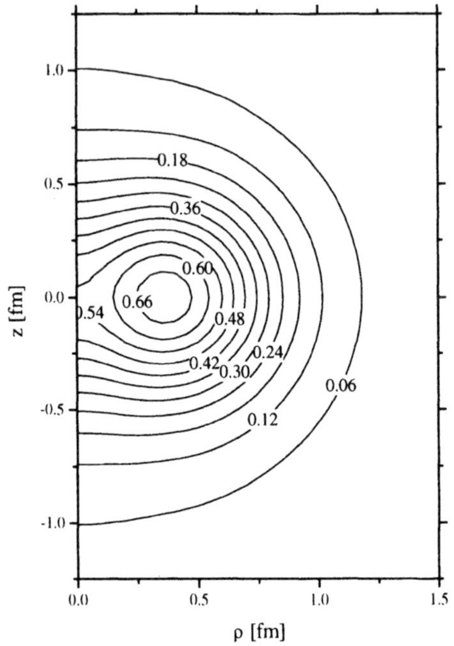

Figure 2. The baryon number density $B(\rho, z)$, with $\alpha = 4$, $q = 1$ and $N_{rmax} = N_{zmax} = 8$.

References

1. D.I. Diakonov and V.Yu. Petrov, P.V. Pobylitsa: Nucl. Phys. **B306**, 809 (1988)
2. V.B. Kopeliovich and B.E. Stern: Pis'ma v ZhETF **45**, 165 (1987); J. Verbaarschot: Phys. Lett. **B195**, 235 (1987)
3. H. Weigel, B. Schwesinger, G. Holzwarth: Phys. Lett. **B168**, 321 (1986)
4. E. Braaten and L. Carson: Phys. Rev. **D38**, 3525 (1988)
5. N.S. Manton: Phys. Lett. **B192**, 177 (1987)
6. N. Sawado and S. Oryu: Phys. Rev. **C58**, R3046 (1998)
7. Y.K. Gambhir, P. Ring, A. Thimet: Ann. Phys. **198**, 132 (1990)
8. N. Sawado: (to be published)

Few-Body Systems Suppl. 12, 209–213 (2000)

Few-Body Systems

Roles of Quark-Pair Correlations in Baryon Structure and Non-Leptonic Weak Transitions of Hyperon

K. Suzuki[1,*], E. Hiyama[2], H. Toki[1], M. Kamimura[3]

[1] RCNP, Osaka University, Osaka 567-0047, Japan
[2] RIKEN, Wako, Saitama 351-0198, Japan
[3] Department of Physics, Kyushu University, Fukuoka 812-8581, Japan

Abstract. Roles of quark-pair correlations in the baryon structure and the hyperon non-leptonic weak decay are studied within the non-relativistic constituent quark model. We construct the SU(3) ground state baryons by solving the three body problem rigorously with the confinement force and the short range spin-dependent attraction. We emphasize the importance of the $s = 0$ quark-quark correlation to reproduce the $\Delta I = 1/2$ enhancement of the hyperon decay, and demonstrate that resulting static properties as well as the decay amplitudes agree with the experiments, if we deal with the $s = 0$ correlation properly. Special attention is also put on the consequences of the SU(6) spin-flavor symmetry breaking due to the $s = 0$ correlation. Calculated magnetic moments are the almost same as the naive SU(6) predictions in spite of the existence of the strong correlations.

1 Introduction

Properties of light baryons have been extensively studied by various models based on the constituent quark picture. Their results are consistent with experiments including applications for the one and two nucleon systems, although this model involves several adjustable parameters. Despite the success of this approach, it is obscure whether or not these models correctly describe the quark distributions in the baryons, and in fact understandings of the non-leptonic weak hyperon decay and its $\Delta I = 1/2$ rule are still incomplete.

The $\Delta I = 1/2$ rule implies dominance of the $\Delta I = 1/2$ transitions on the non-leptonic hyperon decay[1]. It is known that relative magnitudes of parity conserving decay amplitudes are reasonably described within the baryon pole approximation, where the weak decay takes place as the two quark transition

*e-mail: ksuzuki@rcnp.osaka-u.ac.jp

process: a us-pair in the initial hyperon with their total spin s being 0 changes to a ud-pair in the final state baryon. However, if one calculates them using the constituent quark models, the absolute value of the amplitudes is about a half of the experimental data at most[1, 2]. We emphasize here, because of the heavy W-boson mass, the weak matrix elements are quite sensitive to the short range quark-quark correlations[3]. Consequently, the failure of the constituent quark model to describe the non-leptonic weak transition may indicate the lack of the quark correlation in the $s = 0$ channel.

On the other hand, it was pointed out that, from both theoretical and phenomenological points of view[4], there exists a strong correlation between quarks in the $s = 0$ channel.

In this work we try to clarify the roles of the quark correlation in the baryon structure and hyperon non-leptonic weak decays by using the constituent quark model. We assume the short range spin-dependent correlations between quarks together with the confinement force, and calculate the baryon masses and other static properties. In order to deal with the spin-dependent correlation correctly, we must rigorously solve the three body problem. For this purpose, we adopt the coupled-rearrangement-channel variational method with the Gaussian basis functions which has been developed by Hiyama and Kamimura[5].

We also focus on the SU(6) breaking effects on baryon properties. Introduction of the spin-dependent correlation naturally spoils the SU(6) spin-flavor symmetry which is known to work well for e.g. the baryon magnetic moments. We shall calculate the magnetic moments to estimate the SU(6) breaking effects clearly.

2 Calculation of Weak Decay Matrix Elements

Let us write the low energy effective weak interaction Hamiltonian[6];

$$\mathcal{H}_W = \frac{G_F \sin\theta \cos\theta}{\sqrt{2}} \sum_i c_i(\mu^2)\, O_i + \text{h.c.} \tag{1}$$

where O_i are the quark 4-Fermi operators, and O_1, O_2 give dominant contributions in our case. We take values of c_i given in ref. [6] at 1GeV^2.

Our task here is to evaluate the matrix element $\langle B_f\ \pi^a|\mathcal{H}_W|B_i\rangle$ for the strangeness changing process $B_i \to B_f + \pi^a$. The PCAC relation and soft pion theorem are suitable to deal with the strong interacting pion-nucleon system. Using them, one can find the baryon pole formula to the parity conserving amplitudes. For example, $\Lambda^0 \to n + \pi^0$ pole amplitude is given by,

$$\frac{M_N + M_\Lambda}{f_\pi}\left[G^{\pi 0}_{nn}\frac{1}{M_\Lambda - M_N}\langle n\,|\mathcal{H}_W|\,\Lambda\rangle + \langle n\,|\mathcal{H}_W|\,\Sigma^0\rangle\frac{1}{M_N - M_\Sigma}G^{\pi 0}_{\Lambda\Sigma}\right] \tag{2}$$

where $\langle n\,|\mathcal{H}_W|\,\Lambda\rangle$ and $\langle n\,|\mathcal{H}_W|\,\Sigma^0\rangle$ are the matrix elements of eq. (1) with appropriate baryon states, and $G^{\pi a}_{B\ B'}$ denote the axial vector coupling constants which gives probabilities for the pion emission $B \to B' + \pi^a$ and are constrained by experiments.

We come to determine the matrix elements of $\mathcal{H}_W$, $\langle n|\mathcal{H}_W|\Lambda\rangle$ and $\langle p|\mathcal{H}_W|\Sigma^+\rangle$. We recall the quark models such as Harmonic Oscillator model or MIT bag model give much smaller values for these matrix elements than the data[2]. It is instructive to rewrite the $V-A$ operators O_1, O_2 in the non-relativistic limit in the coordinate space as

$$O_1,\, O_2 \to a_d^\dagger\, a_u^\dagger (1 - \boldsymbol{\sigma}_u \cdot \boldsymbol{\sigma}_s)\; \delta^{(3)}(\boldsymbol{r}_{us})\; a_u a_s \tag{3}$$

where a_i, $a_i^\dagger$ are annihilation and creation operators of quarks with flavor i. Presence of the spin-projection operator $(1 - \boldsymbol{\sigma}_u \cdot \boldsymbol{\sigma}_s)$ tells us that the weak transition is generated by the two body process between spin-0 quark pairs; $(us)^0 \to (ud)^0$, which guarantees the $\Delta I = 1/2$ dominance on the non-leptonic hyperon decays due to the antisymmetrization of the quark-pairs. Now it is clear that this decay amplitude is sensitive to the correlation of the spin-0 quark pair in the baryons. The standard constituent quark model never incorporates such a quark-quark correlation properly.

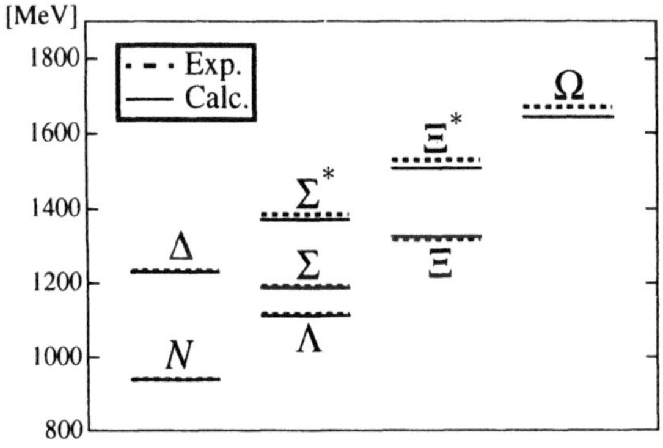

Figure 1. SU(3) baryon mass spectrum

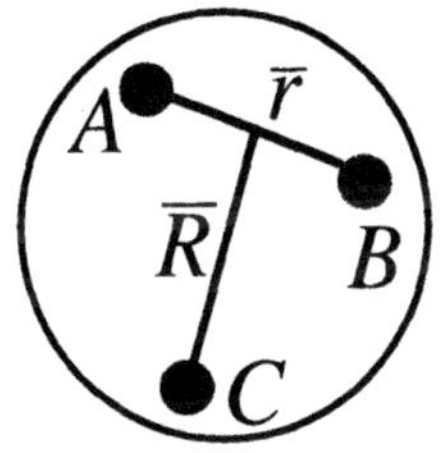

Figure 2. Configuration of three quarks in nucleon

3 Constituent Quark Model with the Spin-Dependent Correlations

We phenomenologically introduce the effective Hamiltonian which includes the confinement force and the spin-dependent part as;

$$\mathcal{H} = \sum_i \frac{p_i^2}{2m_i} + \sum_{i<j} \frac{1}{2} K \left(\boldsymbol{r}_i - \boldsymbol{r}_j\right)^2 + \sum_{i<j} V_S(ij) + V_0 \tag{4}$$

$$V_S(ij) = \begin{cases} 0 \quad (s = 1 \text{ pair}) & \\ \frac{C_{SS}}{m_i m_j} \mathrm{Exp}\left[-\left(\boldsymbol{r}_i - \boldsymbol{r}_j\right)^2 / \beta^2\right] & (s = 0 \text{ pair}) \end{cases} \tag{5}$$

where m_i are the constituent quark masses, and K, C_{SS}, β are the model parameters. V_0 contributes to the over all shift of the resulting spectrum and is chosen to adjust the energy of the lowest state to the nucleon mass. Constituent quark masses are taken to be $m_u = m_d = 330\mathrm{MeV}$ and $m_s = 510\mathrm{MeV}$.

Using this Hamiltonian, we shall solve non-relativistic three body problem rigorously. We use the coupled-rearrangement-channel variational method with Gaussian basis functions[5]. We assume the isospin symmetry between u and d quarks, and solve the three body problem without further approximations or assumptions. The strange quark is explicitly distinguished from light u, d quarks.

We shall fix the model parameters so as to reproduce the nucleon and Δ masses, proton radius and the $\Sigma^+ \to n\pi^+$ amplitude, since this decay is completely dominated by the baryon pole diagrams (see Table 1). We obtain the parameters $K = 0.005\text{GeV}^3$, $\beta = 0.5\text{fm}$ and $C_{SS}/m_u^2 = 1.4\text{GeV}$.

Resulting mass spectrum is shown in Fig.1. It could be possible to obtain a better agreement by modifying the potential or parameters, but the present results are enough for our purpose. The effect of the attractive correlation can be seen clearly by introducing the average distances of the Jacobi coordinate $\bar{r}$ and $\bar{R}$ defined in Fig.2, where all possible rearrangement channels are transformed to the configuration of Fig.2. We find, for the nucleon, $\bar{r} = 0.92\text{fm}$ and $\bar{R} = 0.97\text{fm}$ when the total spin s of the quark pair A and B is zero, while $\bar{r} = 1.1\text{fm}$ and $\bar{R} = 0.81\text{fm}$ in the $s = 1$ case. Apparently, the quark correlation modifies quark distribution in the nucleon. The value of the wave function at origin $\int d^3R\, d\Omega_r \Psi(r = 0, R)$ in the $s = t = 0$ case is about three times as large as that of the $s = t = 1$ case, which provides an huge enhancement for the $\Delta I = 1/2$ weak decay amplitudes.

The matrix elements of the weak Hamiltonian are calculated in terms of our wave functions. We find $\langle n|\mathcal{H}_W|\Lambda\rangle = -0.946(\times 10^{-2}\text{GeV}^3)$ as the full result and -0.310 when we omit the correlation. Similarly, $\langle p|\mathcal{H}_W|\Sigma^+\rangle = 2.86$ with V_S, while it becomes 0.762 without V_S. In the absence of the correlation V_S, a ratio $\langle p|H_W|\Sigma^+\rangle/\langle n|H_W|\Lambda\rangle = -2.45$ shows a perfect agreement with the SU(6) expectation $\sqrt{6} \simeq -2.4494\cdots$. In the realistic case with V_S, one can observe the substantial enhancement and the SU(6) breaking effect.

With these values, calculated weak transition amplitudes are tabulated in Table 1. We show the pole contributions only in the second column, and the sum of the pole, factorization and penguin contributions in the third column to be compared with the experiments. We find a good agreement for $\Sigma \to N\pi$ decays, while the $\Lambda \to N\pi$ amplitude is not enough. We note that the small value of the $\Lambda \to N\pi$ pole contribution is caused by the strong cancellation of the two terms in eq. (2). If we vary the axial-vector coupling G within the experimental errors ($\sim 10\%$), we find about 40% increase of the $\Lambda \to N\pi$ pole contribution with the Σ decay amplitudes almost unchanged.

4 SU(6) Symmetry Breaking Effects on the Magnetic Moment

Our wave functions clearly violate the naive SU(6) spin-flavor symmetry for the baryon. In fact, the ratio $\langle p|H_W|\Sigma^+\rangle/\langle n|H_W|\Lambda\rangle$ becomes -3.02, instead of the SU(6) value -2.45. Thus, we estimate the size of the SU(6) breaking effect to be about 20%, which is significant. On the other hand, it is historically known that the light baryon magnetic moments are well reproduced in the naive quark model by virtue of the SU(6) spin-flavor symmetry. Hence, it is important to examine the SU(6) breaking by calculating the magnetic moments.

Our results are shown in Table 2. It is manifest that the results are almost unchanged even after introducing the spin-dependent correlations. The differences are of order of a few % in any cases. It seems that the global baryon properties such as magnetic moments obtained by integrating the wave function over space are insensitve to the quark correlation, although the local structure of the quark wave function is modified substantially.

Table 1 Parity conserving weak transition amplitude (in 10^{-7} unit)

	Pole	total	Exp.
Σ_0^+	24.0	26.1	26.24
Σ_+^+	43.3	43.3	41.83
Λ_0^0	−3.82	−8.84	−15.61

Table 2 Magnetic Moments

	full	no V_S	Exp.
μ_p	2.75	2.84	2.79
μ_n	−1.78	−1.90	−1.91
μ_Λ	−0.60	−0.61	−0.61
μ_{Σ^+}	2.67	2.73	2.46
μ_{Σ^-}	−1.05	−1.06	−1.16
μ_{Ξ^0}	−1.40	−1.45	−1.25

5 Summary

In conclusion, we have studied the roles of the spin-dependent quark correlations in the baryon structure. We have emphasized that the non-leptonic weak transition of the hyperon is unique quantity to investigate the quark-quark correlation in the spin-0 channel. We have solved three body problem explicitly using the coupled-rearrangement-channel variational method. Results for static baryon properties as well as the transition amplitudes of the non-leptonic hyperon decay reasonably agree with the empirical values. We have also discussed the SU(6) breaking effects on the baryon properties, and pointed out that this symmetry is sill useful for the static baryon properties, although local behavior of the quark wave function considerably departs from the SU(6) expectation.

References

1. J.F. Donoghue, E. Golowich, B. Holstein: Phys. Reports **131**, 319 (1986)
2. D. Tadic and J. Trampetic: Phys. Rev. **D23**, 144 (1981)
3. K. Suzuki and H. Toki: Mod. Phys. Lett. **A9**, 1059 (1994)
4. See e.g., *Diquarks* I, II, ed. M. Anselmino, E. Predazzi, World Scientific
5. E. Hiyama and M. Kamimura: Nucl. Phys. **A588**, 35c (1995)
6. G. Buchalla, A.J. Buras, M.E. Lautenbacher: Rev. Mod. Phys. **68**, 1125 (1996)

Few-Body Systems Suppl. 12, 214–222 (2000)

Few-Body Systems

Spectroscopy and Liftime of Mesons with Bottom and Charm Quarks

M.A.K. Lodhi[1], A.A. El-Hady[2], J.P. Vary[3]

[1] Department of Physics, Texas Tech University, Lubbock Texas 79409, USA

[2] Department of Physics, Zagazig University, Zagazig, Egypt

[3] Department of Physics and Astronomy, Iowa State University , Ames, Iowa 50011, USA

Abstract. The understanding of hadronic matter properties in high density and temperature environment is necessary for interpreting relativistic heavy ion collision data. Accordingly the bound state of a bottom and acharm quark B_c is of great theoretical and experimental importance. We have calculated the spectrum of this meson in Bethe-Salpeter model using two reductions. The ground state mass and decay lifetime of B_c predicted by this model are found in excellent agreement with the experimental values observed at Fermi Lab. We evaluate the mass of the pseudo-scalar B_c meson as 6.356 GeV/c^2 and 6.380 GeV/c^2 and the lifetime as 0.47 ps and 0.46 ps respectively in the two reductions , in good agreement with the recently reported mass of 6.40 $\pm$ 0.39 (stat.) $\pm$ 0.13 (syst.) GeV/c^2 and lifetime of $0.46^{+0.18}_{-0.16}$ (stat.) $\pm$ 0.03 (syst.) ps by the CDF Collaboration. We evaluate the decay constant of the B_c meson and compare different contributions to its decay width.

1 Introduction

At present it is widely believed that the standard model can describe particles and their interactions upto a few hundred GeV successfully [1]. The fermion degrees of freedom in the standard model are quarks and leptons and the interactions are mediated by gauge bosons and the putative Higgs boson. Gluons are the gauge bosons of quantum chromodynamics (QCD), and photons, W and Z are the gauge bosons of electroweak interaction (EW). The gauge interactions have their own strength parameters. These parameters are not predicted by standard model but are measured experimentally. Attempts to relate these parameters and reduce their number are widely investigated in the context of of physics beyond the standard model. one of the most inspiring facts is that the physical spectrym of the standard model does not include quarks or gluons, i.e they do not appear as asymptotic states.

In the case of of EW interaction. one can perform perturbative calculations for small coupling through which impressive agreement has been achieved with the experiments. For strong coupling in the case QCD at large distances, perturbation theory fails, and one has to rely on other methods [1]. In this paper we are using that "other method," namely, the reduction of Bethe Salpeter Equation (BSE) to incorporate the large distance QCD features and apply in investigating B_c meson spectra and decay lifetime.

2 Bethe Salpeter Equation and Reduction Technique

Just as the Dirac Equation provides a relativistic description of a fermion in an external field, BSE [2] provides an appealing starting point to describe hadrons as relativistic bound states of quarks and gluons [3]. The BSE has four elements, the interaction kernel (V) and the propagator (G) which we provide as an input with energy, and the amplitude (ψ). We solve in the momentum space for the amplitude and the energy, which is contained in the propagator. The covariance of the BSE intoduces an extra relative time in the problem which is difficult to interpret physically. Different approaches have been developed to make the problem more tractable and physically appealing. In Quasi Potential equation (QPE) approach the two particle propagator is modified in a way which keeps covariance and reduces the 4-dimensional BSE to a 3-dimensional equation. This approach has been successfully used in different nuclear and particle physics applications [3, 4, 5].

Eyre and Vary [6] intoduced a method to include confinement in the interaction kernel in a nonrelativistic problem in momentum space. Spence and Vary [7] extended this method to the relativistic problem of quarkonium without and with the inclusion of spin in the Instantaneous Approximation (IA) and in one of the QPE, the Blankenbecler-Sugar (BbS) reduction [8]. Sommerer, Spence and Vary [9] investigated different reductions as special cases of general QPE written in terms of three parameters. They tested varios reductions in applications to heavy quarkonium systems.

In this paper, we use the model [9, 10] based on the Bethe-Salpeter Equation to include the bottom-charmed mesons. The BSE for a bound state may be written in momentum space in the form

$$G^{-1}(P,p)\psi(P,p) = \int \frac{1}{(2\pi)^4} V(P,p-p')\psi(P,p')d^4p' \tag{1}$$

Where P is the four-momentum of the bound state, p is the relative four-momentum of the constituents.

We have used two reductions of the QPE to study the meson spectrum [9, 10]. These reductions correspond to different choices of the two particle propagator used to reduce the problem into three dimensions. We refer to these reductions as A and B. Reduction A corresponds to a spinor form of the Thompson equation [11] and reduction B corresponds to a new QPE introduced in Ref. [12]. These two reductions are chosen because they are shown to give good fits to the meson spectrum [9].

We assume the interaction kernel to consists of a one gluon exchange interaction, V_{OGE}, in the ladder approximation, and a phenomenological, long range scalar confinement potential, V_{CON} given in the form

$$V_{OGE} + V_{CON} = -\frac{4}{3}\alpha_s \frac{\gamma_\mu \otimes \gamma_\mu}{(p-p')^2} + \sigma \lim_{\mu\to 0} \frac{\partial^2}{\partial\mu^2} \frac{\mathbf{1}\otimes\mathbf{1}}{-(p-p')^2+\mu^2} \tag{2}$$

Here, α_s is the strong coupling, which is weighted by the meson color factor of $\frac{4}{3}$, and the string tension σ is the strength of the confining part of the interaction.

The strong coupling, α_s is assumed to be

$$\alpha_s(Q^2) = \frac{4\pi\alpha_s(\mu^2)}{4\pi + \beta_1\alpha_s(\mu^2)\ln(Q^2/\mu^2)} \tag{3}$$

where $\beta_1 = 11 - 2n_f/3$ and $n_f = 4$ is the number of quark flavors taken to be fixed. At the scale of the Z-boson, $\alpha_s(\mu^2 = M_Z^2) \simeq 0.12$ and Q^2 is related to the meson mass scale through,

$$Q^2 = \gamma^2 M^2_{meson} + \beta^2, \tag{4}$$

where γ and β are parameters determined by a fit to the meson spectrum. In this formulation of BSE there are seven parameters : four masses, m_u=m_d, m_s, m_c, m_b; the string tension σ, and the parameters γ and β used given in Table 1.

Table 1. Values of the parameters used in reductions A, B

	Reduction A	Reduction B
m_b (GeV)	4.65	4.68
m_c (GeV)	1.37	1.39
m_s (GeV)	0.397	0.405
m_u (GeV)	0.339	0.346
σ (GeV^2)	0.233	0.211
γ	0.616	0.444
β (GeV)	0.198	0.187

In this formalism the mesons are taken as bound states of a quark and an anti-quark. The wavefunctions for the mesons are calculated by solving reductions of Bethe-Salpeter equation [9, 10]. We construct the meson states as [13]

$$|M(\mathbf{P_M}, J, m_J)\rangle = \sqrt{2M}\int d^3\mathbf{p}\langle Lm_L S m_S|Jm_J\rangle\ \langle l m_l \bar{l} m_{\bar{l}}|Lm_L\rangle$$
$$\Phi_{Lm_L}(\mathbf{p})|\bar{q}(\frac{m_{\bar{q}}}{M_{q\bar{q}}}\mathbf{P_M} - \mathbf{p}, m_{\bar{s}})\rangle|q(\frac{m_q}{M_{q\bar{q}}}\mathbf{P_M} + \mathbf{p}, m_s)\rangle \tag{5}$$

where the quark states are given by

$$
\begin{aligned}
|q(\mathbf{p}, m_s)\rangle &= \sqrt{\frac{(E_q + m_q)}{2m_q}} \begin{pmatrix} \chi^{m_s} \\ \frac{\sigma \cdot \mathbf{p}}{(E_q+m_q)} \chi^{m_s} \end{pmatrix} \\
M_{q\bar{q}} &= m_q + m_{\bar{q}} \\
E_q &= \sqrt{m_q^2 + \mathbf{p}^2}
\end{aligned} \tag{6}
$$

In Eq. 5 M is the meson mass.

3 B_c Meson Spectrum

Recently the CDFCollaboration reported the observation of the bottom-charmed mesons B_c in 1.8 TeV $p\bar{p}$ collisions using the CDF detector at the Fermilab Tevatron [14]. This pseudo-scalar state is the lowest energy state of the family of mesons composed of a $\bar{b}$ anti-quark and a c quark. It decays only through weak interaction. This eliminates uncertainties encountered in strong decays and renders the decay width of B_c more theoretically tractable.

Different approaches have been used to evaluate the spectrum of B_c mesons. Non-relativistic potential models have been used by Eichten and Quigg [15] where they discussed four potentials and gave mass values for the B_c meson in the range 6.248–6.266 GeV/c^2. Gershtein *et al.* [16] used two potentials and reported predictions of 6.253 and 6.264. QCD sum rules have been used by Chabab [17] where he predicted a prediction of 6.25 GeV/c^2.

To study the meson spectrum we use the aforementioned reductions for solving the BSE to calculate the B_c states within a reasonable range of a root mean square (rms) deviation from the observed values, say 40 to 50 MeV. We have thus calculated the spectrum of B_c meson in reductionS A and B , and compare In the Table 2 with the work of Eichten and Quigg [15], Gershtein at al [16] using Martin potential and Gershtein *et al.* [16] using Buchmuller-Tye (BT) potential. The first row compares with the experimental results [14] of 6.40 $\pm$ 0.39 (stat.) $\pm$ 0.13 (syst.) GeV/c^2. Both reduction A and B compare reasonably well with the experimental results though the experimental uncertainties are large.

4 Decay Modes and Lifetime of B_c Meson

We shall next apply this model to examine some of the decays which have attracted a great deal of attention paricularly of heavy ion of scientists. These decays are weak decays of B mesons. The interest in B decays grew especially after the heavy quark symmetry (HQS)[18, 19, 20] and heavy quark effective theory (HQET) [21]. Although the strength of the weak interaction is small, and perturbative calculations could be justifiable, the problem arises from the fact that the decaying of b-quark in B meson is not a free quark decay but is in a bound system with all its strong interaction complexities which cannot be

Table 2. spectrum of B_c mesons in different channels (GeV/c^2).

State	This work Reduction A	This work Reduction B	Eichten and Quigg Ref. [15]	Gershtein *et al.*[16] Martin potential	Gershtein *et al.*[16] BT potential
1^1S_0	6.356	6.380	6.264	6.253	6.246
1^3S_1	6.397	6.415	6.337	6.317	6.337
1^3P_0	6.673	6.692	6.700	6.683	6.700
1^3P_2	6.751	6.773	6.747	6.743	6.747
1^1P_1	6.752	6.777		6.729	6.736
2^1S_0	6.888	6.874	6.856	6.867	6.856
2^3S_1	6.910	6.891	6.899	6.902	6.899
1^3D_1	6.984	6.955	7.012	7.008	7.012

treated with perturbative techniques. Better understanding of strong interaction responsible for bound state would help us understand the weak decays. Similarly, the understanding of weak decays may help us better understand the strong interaction dynamics. The semi-leptonic form factors for B to D and D^* mesons, and the leptonic decay constants were calculated in previous works [22, 23, 24], using the similar wavefunctions . Following the same approach for this work we define the weak decay constants for the pseudo-scalar and vector mesons by

$$
\begin{aligned}
< 0|J_\mu|P(p) > &= i f_P p_\mu \\
< 0|J_\mu|V(p) > &= M_V f_V \varepsilon_\mu \\
J_\mu &= V_\mu - A_\mu
\end{aligned}
\tag{7}
$$

where P and V are pseudo-scalar and vector states and V_μ and A_μ are the vector and axial vector currents.

Taking into account the relativistic effects, the expressions of the decay constants in terms of the wavefunctions are given by [25]

$$
f_i = \sqrt{\frac{12}{M}} \int_0^\infty \frac{p^2 dp}{2\pi^3} \sqrt{\frac{(m_q + E_q)(m_{\bar{q}} + E_{\bar{q}})}{4E_q E_{\bar{q}}}} F_i(p) \tag{8}
$$

(9)

where the subscript i represents P or V and

$$
F_P(p) = \left[1 - \frac{p^2}{(m_q + E_q)(m_{\bar{q}} + E_{\bar{q}})}\right] \psi_P(p) \tag{10}
$$

$$
F_V(p) = \left[1 - \frac{p^2}{3(m_q + E_q)(m_{\bar{q}} + E_{\bar{q}})}\right] \psi_V(p) \tag{11}
$$

where ψ_P, ψ_V are the wavefunctions of the pseudo-scalar and vector states respectively. The non-relativistic limit of these expressions yields a relation between f_i and the wavefunction at the origin in coordinate space , $R(0)$,

$$f_i = \sqrt{\frac{3}{\pi M}} R(0). \tag{12}$$

The leptonic decay constant (f_{B_c}) is relevant for the annihilation channel of the B_c pseudo-scalar meson. In Table 3 we compare different predictions for this quantity.

Table 3. Leptonic decay constant of B_c (f_{B_c}) in MeV.

This work A	This work B	Eichten and Quigg Ref. [15]	Gershtein *et al.*[16] Martin potential	Gershtein *et al.*[16] BT potential
578	490	500	512	500

The lifetime of B_c is a very important quantity which may help us understand the basic properties of the weak interaction at a fundamental level especially since the strong interaction effects can be estimated reliably. The total width can be approximated by the sum of the widths of $\bar{b}$-quark decay with the spectator c-quark, the c-quark decay with the spectator $\bar{b}$-quark, and the annihilation channel $B_c^+ \to l^+\nu_l(c\bar{s}, u\bar{s})$, $l = e,\ \mu,\ \tau$. Since all these decays lead to different final states, we have no interference between different amplitudes. The total width is then given by

$$\Gamma(B_c \to X) = \Gamma(b \to X) + \Gamma(c \to X) + \Gamma(\text{ann}) \ . \tag{13}$$

Neglecting the quark binding effects, we obtain for the b and c inclusive widths in the spectator approximation,

$$\begin{aligned} \Gamma(b \to X) &= \frac{G_F^2 |V_{cb}|^2 m_b^5}{192\pi^3} \cdot 9 \ , \\ \Gamma(c \to X) &= \frac{G_F^2 |V_{cs}|^2 m_c^5}{192\pi^3} \cdot 5 \ , \end{aligned} \tag{14}$$

The width of the annihilation channel is given by

$$\Gamma(\text{ann}) = \sum_i \frac{G_F^2}{8\pi} |V_{bc}|^2 f_{B_c}^2 M_{bc} m_i^2 \left(1 - \frac{m_i^2}{m_{Bc}^2}\right)^2 \cdot C_i \ , \tag{15}$$

where $C_i = 1$ for the $\tau\nu_\tau$ channel and $C_i = 3|V_{cs}|^2$ for the $\bar{c}s$ channel, and m_i is the mass of the heaviest fermion (τ or c). Table 4 shows various contributions to the width of B_c in our model.

Table 4. Various contributions to the decay width of B_c in 10^{-12} GeV.

	$\Gamma(b \to X)$	$\Gamma(c \to X)$	Γ (ann)
Reduction A	0.75	0.51	0.14
Reduction B	0.78	0.55	0.11

We have used $V_{cb} = 0.041$, and $V_{cs} = 0.96$. From Table 4 we see that both reductions predict that the b decay dominates c decay in B_c meson.

In Table 5, we compare the lifetime of B_c in different models with the CDF experimental result. The experimental result indicates that the binding effects may not be very important as suggested by Quigg [26].

Table 5. Comparison of the lifetime of B_c meson (in ps) in different models.

Experiment [14]	Reduction A	Reduction B	Quigg [26]	Gershtein *et al.* [16]
$0.46^{+0.18}_{-0.16}$ (stat.) $\pm$ 0.03 (syst.)	0.47	0.46	$1.1 - 1.4$	0.55 ± 0.15

5 Concluding Remarks

We have evaluated the meson spectrum of the B_c meson in two reductions of BSE. The parameters used in this work have been taken from the previous work in which they were adjusted to fit the observed masses within a range of rms deviation of about 50 MeV. Although the B_c meson is heavy enough to be treated nonrelativistically; we insisted to calculate its prperties from the wave function derived relativistically, in order to stay in the same formalism,and to be consisten with the light mesons treatment. The results for properties of B_c agree with the recent measurement of the CDF Collaboration of B_c mass in the ground state.

One can imagine that the strength of the long range confinement interaction would decrease in a medium similar to Debye screening in quqntum electrodynamics. There the Coulomb interaction is exponentially screened in a medium with the screening length decreasing with the density. In a hot and/or dense medium quarks and gluons may become deconfined to form a quark-gluon plasma (QGP). Such a situation should help us understand the confinement mechanism [27]. The leading effect of the medium would be on the long range confinement strength. However, the medium effects may also change the effective masses of quarks. For these calculations we have used the confinement parameter given in Table 1 and predicted the leptonic decay constant and evaluated various contributions to the decay width of B_c. The partial width of B_c

due to b-quark decay dominates that due to the c-quark decay. Our result for the B_c lifetime is again in good agreement with the CDF measurement.

This work was supported in part by the US Department of Energy, Grant No. DE-FG02-87ER40371, Division of High Energy and Nuclear Physics and by the International Institute of Theoretical and Applied Physics, Ames, Iowa.

References

1. J.F. Donoghue, E. Golowich, Holstein: *Dynamics of the Standard Model.* Cambridge University Press 1992
2. E.E. Salpeter and H.A. Bethe: Phys. Rev. **84**, 1232 (1951)
3. C. Itzykson and J.B. Zuber: *Quantum Field theory.* New York: McGraw-Hill 1980 (Chapter 10 gives a review of Bethe-Salpeter equation)
4. G.E. Brown and A.D. Jackson: *The Nucleon-Nucleon Interaction.* New York: North-Holland 1976 and references therein
5. B. Silvestre-Brac, A. Bilal, C. Gignoux, P. Schuck: Phys. Rev. **D29**, 2275 (1984)
6. D. Eyre and J.P. Vary: Phys. Rev. **D34**, 3467 (1986)
7. J.R. Spence and J.P. Vary: Phys. Rev. **D35**, 2191 (1987); J.R. Spence and J.P. Vary: Phys. Rev. **C47**, 1282 (1993)
8. R. Blankenbecler and R. Sugar: Phys. Rev. **142**, 1051 (1966)
9. A.J. Sommerer, J.R. Spence, J.P. Vary: Mod. Phys. Lett. **A8**, 3537 (1993); A.J. Sommerer, J.R. Spence, J.P. Vary: Phys. Rev. **C49**, 513 (1994)
10. Alan J. Sommerer, A. Abd El-Hady, John R. Spence, James P. Vary: Phys. Lett. **B348**, 277 (1995)
11. R.H. Thompson: Phys. Rev. **D1**, 110 (1970)
12. J.R. Spence and J.P. Vary: Phys. Rev. **C47**, 1282 (1993); A.J. Sommerer, J.R. Spence, J.P. Vary: Mod. Phys. Lett. **A8**, 3537 (1993)
13. Nathan Isgur, Daryl Scora, Benjamin Grinstein, Mark B. Wise: Phys. Rev. **D39**, 799 (1989)
14. F. Abe et al.: The CDF Collaboration, hep-ex/9804014; W. Hav and V. Papadimitriou: private communication
15. E. Eichten and C. Quigg: Phys. Rev. **D49**, 5845 (1994)
16. S.S. Gershtein, V.V. Kiselev, A.K. Likhoded, A.V. Tkabladze, A.V. Berezhnoi, A.I. Onishchenko: hep-ph/9803433; V.V. Kiselev, A.K. Likhoded, A.V. Tkabladze: Phys. Rev. **D51**, 3613 (1995); S.S. Gershtein, V.V. Kiselev, A.K. Likhoded, A.V. Tkabladze: Phys. Usp. **38**, 1 (1995)

17. M. Chabab: LMPHE/97-04; Phys. Lett. **B325**, 205 (1994)

18. N. Isgur and M.B.Wise: Phys. Lett. **B232**, 113 (1989); **B237**, 527 (1990)

19. M.B. Voloshin and M.A. Shifman: Yad. Fiz. **47**, 801 [Sov. J.N.P. **47**, 511 (1988)]

20. H. Georgi: Phys. Lett. **B24**

21. M. Neubert: Phys. Rep. **245**, 259 (1994) and references therein

22. A. Abd El-Hady, K.S. Gupta, A.J. Sommerer, J. Spence, J.P. Vary: Phys. Rev. **D51**, 5245 (1995)

23. A. Abd El-Hady, A. Datta, K.S. Gupta, J.P. Vary: Phys. Rev. **D55**, 6780 (1997)

24. A. Abd El-Hady, A. Datta, J.P. Vary: Phys. Rev. **D58**, 6780 (1998)

25. S. Veseli and I. Dunietz: Phys. Rev. **D54**, 6803 (1996)

26. C. Quigg: *Proceedings of the Workshop on B Physics at Hadron Accelerators*, ed. by P. McBride and C. Shekhar Mishra, Fermilab-CONF-93/267 (SSCL-SR-1225) 1994

27. A.M. Poskanzer, J.W. Harris, L.S. Schroeder (eds.): Nucl. Phys. **A590**, 1c-653c (1995)

Few-Body Systems Suppl. 12, 223–228 (2000)

Few-Body Systems

Implementing PCAC in Nonperturbative Models of Pion Production

B. Blankleider, A.N. Kvinikhidze*

Department of Physics, Flinders University of South Australia, Bedford Park, SA 5042, Australia

Abstract. Traditional few-body descriptions of pion production use integral equations to sum the strong interactions nonperturbatively. Although much physics is thereby included, there has not been a practical way of incorporating the constraints of chiral symmetry into such approaches. Thus the traditional few-body descriptions fail to reflect the underlying theory of strong interactions, QCD, which is largely chirally symmetric. In addition, the lack of chiral symmetry in the few-body approaches means that their predictions of pion production are in principle not consistent with the partial conservation of axial current (PCAC), a fact that has especially large consequences at low energies. We discuss how the recent introduction of the "gauging of equations method" can be used to include PCAC into traditional few-body descriptions and thereby solve this long standing problem

1 Introduction

Approximate chiral symmetry is an important property of quantum chromodynamics (QCD) and should therefore be an attribute of any effective description of strong interaction processes. Yet most nonperturbative effective descriptions of pion production in few-body processes ($NN \to \pi NN$, $\pi N \to \pi\pi N$, $\gamma N \to \pi N$, etc.) have been based on traditional few-body approaches which are not consistent with chiral symmetry – even if they are based on chiral Lagrangians (i.e. the input functions are chirally invariant). The essential idea of a traditional few-body approach is to take as input two-body t matrices and pion production vertices, and then use integral equations to sum the multiple-scattering series nonperturbatively. A feature of such an approach is that the input can be constructed phenomenologically without the need to specify an explicit Lagrangian for the strong interactions. Despite the large amount of physics taken into account (e.g. all possible pair-like interactions are included),

*On leave from the Mathematical Institute of Georgian Academy of Sciences, Tbilisi, Georgia

the lack of approximate chiral symmetry in this approach results in a pion production amplitude that does not obey PCAC. Although this is a particularly serious problem at low energies where low-energy theorems apply, the absence of PCAC is undesirable at any energy because of its inconsistency with QCD. In this contribution we show how the gauging of equations method [1, 2] can be used to construct a nonperturbative few-body approach that is consistent with chiral symmetry and whose pion production amplitude obeys PCAC.

2 Traditional Few-Body Model of NN → πNN without PCAC

We shall base our discussion on the example of the relativistic πNN system for which a traditional few-body approach has recently been developed [3, 4]. For simplicity of presentation we shall treat the nucleons as distinguishable. The few-body approach to the πNN system provides a nonperturbative simultaneous description of the processes $NN \to \pi NN$, $NN \to NN$, and $\pi NN \to \pi NN$ within the context of relativistic quantum field theory. In this case the integral equations are four-dimensional and can be expressed symbolically as[1]

$$\mathcal{T} = \mathcal{V} + \mathcal{V}\mathcal{G}_t\mathcal{T} \tag{1}$$

where, for distinguishable nucleons, $\mathcal{T}$, $\mathcal{V}$, and $\mathcal{G}_t$ are 4×4 matrices written as

$$\mathcal{T} = \begin{pmatrix} T_{NN} & \bar{T}_N \\ T_N & T \end{pmatrix}; \quad \mathcal{V} = \begin{pmatrix} V_{NN} & \bar{\mathcal{F}} \\ \mathcal{F} & G_0^{-1}\mathcal{I} \end{pmatrix}; \quad \mathcal{G}_t = \begin{pmatrix} D_0 & 0 \\ 0 & G_0 w^0 G_0 \end{pmatrix}. \tag{2}$$

The elements of matrix $\mathcal{T}$ are defined as follows. T is a 3×3 matrix whose elements $T_{\lambda\mu}$ are Alt-Grassberger-Sandhas (AGS) amplitudes (generalised to four dimensions) describing the process $\pi NN \to \pi NN$. Note that the following "subsystem-spectator" labelling convention is used: $\lambda = 1$ or 2 labels the channel where nucleon λ forms a subsystem with the pion, the other nucleon being a spectator, while $\lambda = 3$ labels the channel where the two nucleons form the subsystem with the pion being the spectator. In a similar way, T_{NN} is the amplitude for NN scattering while T_N and $\bar{T}_N$ are 3×1 and 1×3 matrices whose elements $T_{\lambda N}$ and $T_{N\mu}$ describe $NN \to \pi NN$ and $\pi NN \to NN$, respectively. For simplicity of presentation we shall neglect connected diagrams that are simultaneously NN- and πNN- irreducible. Then the elements making up the kernel matrix $\mathcal{V}$ specified in Eq. (2) take the following form:

$$V_{NN} = V_{NN}^{\mathrm{OPE}} - \Delta \tag{3}$$

where V_{NN}^{OPE} is the nucleon-nucleon one pion exchange potential and Δ is a subtraction term that eliminates overcounting. $\mathcal{F}$ is a 3×1 matrix with

$$\mathcal{F}_\lambda = \sum_{i=1}^{2} \bar{\delta}_{\lambda i} F_i - B \tag{4}$$

[1]In this paper it should be understood that all amplitudes and currents with external nucleon legs are actually operators in Dirac space and need to be sandwiched between appropriate Dirac spinors $\bar{u}$ and u to obtain the corresponding physical quantities.

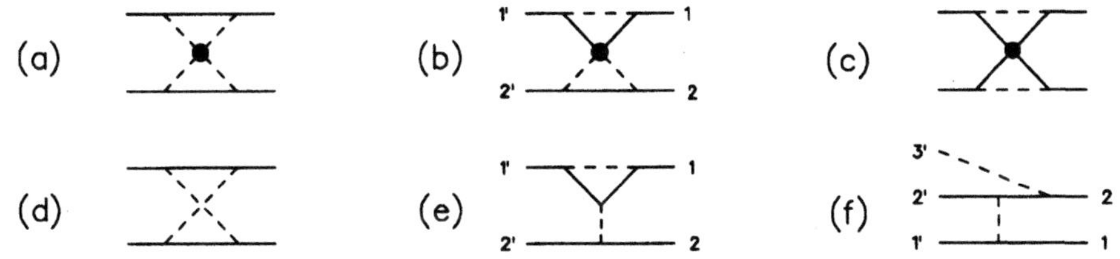

Figure 1. Diagrams making up the subtraction terms. (a) $W_{\pi\pi}$, (b) $W_{\pi N}$, (c) W_{NN}, (d) X, (e) Y, and (f) B. The dark circles represent the following two-body amplitudes: (a) full $\pi\pi$ t-matrix, (b) one-nucleon irreducible πN t-matrix, and (c) full NN t-matrix minus the NN one-pion-exchange potential.

where $F_i = f_i d_j^{-1}$ consists of f_i, the vertex for $N_i \to \pi N_i$, and d_j, the Feynman propagator of nucleon $j \neq i$. The subtraction term B in Eq. (4) likewise eliminates overcounting. $\bar{\mathcal{F}}$ is the 1×3 matrix that is the time reversed version of $\mathcal{F}$, G_0 is the πNN propagator, and $\mathcal{I}$ is the 3×3 matrix whose (λ, μ)'th element is $\bar{\delta}_{\lambda,\mu}$. Finally the propagator term $\mathcal{G}_t$ is a diagonal matrix consisting of the NN propagator D_0, and the 3×3 diagonal matrix w^0 whose diagonal elements are $t_1 d_2^{-1}$, $t_2 d_1^{-1}$, and $t_3 d_3^{-1}$, with t_λ being the two-body t matrix for the subsystem particles in channel λ (for $\lambda = 1$ or 2, t_λ is defined to be the πN t matrix with the nucleon pole term removed). The subtraction terms Δ and B are defined with the help of Fig. 1 as follows:

$$\Delta = W_{\pi\pi} + W'_{\pi N} + W_{NN} + X + Y' - \bar{B}' G_0 B' \tag{5}$$

where $W'_{\pi N} = W_{\pi N} + P W_{\pi N} P$, $Y' = Y + PYP$, and $B' = B + PBP$, P being the nucleon exchange operator.

Despite the rich amount of physics incorporated into the $NN \to \pi NN$ amplitude within this model, it becomes evident from the following discussion that this amplitude cannot satisfy PCAC.

3 New Few-Body Model of NN $\to$ πNN with PCAC

3.1 *Choosing Appropriate Degrees of Freedom*

Although we do not specify the exact Lagrangian behind our new few-body approach, we do assume that this underlying Lagrangian is chirally invariant (up to a small explicit chiral symmetry breaking term). In turn, the chiral invariance of the Lagrangian puts a strong constraint on the nature of the fields that can be used to construct a practical few-body model. For example, if one would like to follow the πNN model above and have only pions and nucleons as the degrees of freedom, then the underlying chiral Lagrangian would necessarily involve an isovector pion field $\boldsymbol{\xi}$ that transforms into functions of itself under a chiral transformation. Unfortunately, this can happen only in a nonlinear way [5]:

$$\boldsymbol{\xi} \to \boldsymbol{\xi} + \frac{1}{2}\boldsymbol{\theta}(1 - \boldsymbol{\xi}^2) + \boldsymbol{\xi}\boldsymbol{\theta}\cdot\boldsymbol{\xi} \tag{6}$$

where $\boldsymbol{\theta}$ is the vector of three (infinitesimally small) rotation angles. Thus one pion transforms into two pions, two pions transform into four pions, etc., a situation that would make it difficult to formulate a chirally invariant few-body description (i.e. a description whose exposed states have a restricted number of pions – in our case 0 or 1). Alternatively, we can follow the example of the linear sigma model and choose an underlying Lagrangian that involves, in addition to pions and nucleons, the isoscalar field σ. In this case the σ, together with the pion field $\boldsymbol{\phi}$, transform under the chiral transformation in a linear way:

$$\boldsymbol{\phi} \to \boldsymbol{\phi} + \boldsymbol{\theta}\sigma, \qquad \sigma \to \sigma - \boldsymbol{\theta} \cdot \boldsymbol{\phi}. \tag{7}$$

Thus the four-component field $\phi \equiv (\sigma, \boldsymbol{\phi})$ transforms into itself - a situation that is ideal for a few-body approach. We shall therefore adopt the latter approach and treat pions and sigma particles on an equal footing. For the few-body πNN model of Sec. 2, this means mostly a formal change where the usual (isospin) three-component pion is replaced by a four-component one. There is, however, one new aspect in that terms with a three-meson vertex ($\pi\pi\sigma$) now need to be included. For example, one will need the new subtraction term illustrated in Fig. 2. Nevertheless, the equations of Sec. 2 retain their structure, the only essential change being an increase in the size of the matrices like T, T_N and $\bar{T}_N$ to take into account the introduction of a σNN channel. With these modifications we obtain few-body πNN equations that are consistent with an underlying chiral Lagrangian.

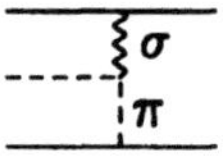

Figure 2. New subtraction term in the few-body πNN model where pions and sigmas are treated on an equal footing.

3.2 Coupling the Axial Vector Field Everywhere

Now that we have achieved consistency with an underlying strong interaction chiral Lagrangian, our next goal is to construct a few-body πNN model whose axial current is partially conserved. The problem at hand is analogous to the one of constructing a conserved electromagnetic current for a few-body system whose strong interactions are consistent with an underlying Lagrangian that conserves charge. In this case it is well known that exact current conservation is obtained by coupling an external electromagnetic field to all possible places in the strong interaction model. In a similar way, one would expect to obtain a partially conserved axial current by coupling an external axial vector field to all possible places in the strong interaction model. Until recently, however, what has not been known is the way to achieve this complete coupling for a few-body system whose strong interactions are described nonperturbatively by integral equations. Fortunately, the solution to this problem presented recently in the context of electromagnetic interactions [1, 2], is based on a topological argument and therefore applies equally well to the present case of an axial vector field.

The basic idea is to add a vector index (indicating an external axial vector field) to all possible terms in the *integral equations* describing the strong interaction model. Thus, in the case of the πNN system where the structure of the integral equations is given as in Eq. (1), coupling an external axial isovector field gives

$$\mathcal{T}^\mu = \mathcal{V}^\mu + \mathcal{V}^\mu \mathcal{G}_t \mathcal{T} + \mathcal{V}\mathcal{G}_t^\mu \mathcal{T} + \mathcal{V}\mathcal{G}_t \mathcal{T}^\mu \tag{8}$$

which can easily be solved to give a closed expression for $\mathcal{T}^\mu$:

$$\mathcal{T}^\mu = (1 + \mathcal{T}\mathcal{G}_t)\mathcal{V}^\mu(1 + \mathcal{G}_t\mathcal{T}) + \mathcal{T}\mathcal{G}_t^\mu \mathcal{T}. \tag{9}$$

$\mathcal{T}^\mu$ is a matrix of transition amplitudes T^μ_{NN}, $T^\mu_{N\Delta}$, T^μ_{Nd}, etc. (for the moment, we suppress the isovector index in these amplitudes). Here we are particularly interested in the transition amplitude T^μ_{NN} as it is closely related to the pion production amplitude we seek.

It is easy to see that the transition amplitude T^μ_{NN} given by Eq. (9) has the axial vector field attached everywhere in T_{NN} *except* on the external nucleon legs [1]. Including these external leg contributions then gives the complete axial vector transition current for $NN \to NN$:

$$j^\mu_{NN} = (\Gamma_1^\mu d_1 + \Gamma_2^\mu d_2)T_{NN} + T_{NN}(d_1\Gamma_1^\mu + d_2\Gamma_2^\mu) + T^\mu_{NN} \tag{10}$$

where Γ_i^μ is the axial vertex function of nucleon i. The input to the πNN equations consists of the two-body t matrices t_i, the pion production vertices f_i, and the single particle propagators d_i. Thus the input to Eq. (9) also includes the axial transition currents t_i^μ, f_i^μ, and axial vertex function Γ_i^μ. In order for these input quantities to be consistent with an external axial vector field being attached everywhere, they must be constructed to satisfy the Axial Ward-Takahashi (AWT) identities [6]. This is easily achieved by restricting the form of the corresponding bare quantities [1]. Thus, for example, $\Gamma_i^{a\mu}$ needs to satisfy the AWT identity

$$q_\mu \Gamma_i^{a\mu}(k,p) = i\left[d_i^{-1}(k)\gamma_5 + \gamma_5 d_i^{-1}(p)\right] t^a - i f_\pi m_\pi^2 f_i^a(k,p) d_\pi(q) \tag{11}$$

where p and k are the initial and final momenta of the nucleon, $q = k - p$, $f_i^a(k,p)$ is the πNN vertex for a pion of isospin component a, t^a is an isospin $1/2$ matrix, m_π is the mass of the pion, and f_π is the pion decay constant. With the other input quantities constructed to satisfy similar AWT identities, it can be shown that the two-nucleon axial current given by Eq. (10) satisfies the AWT corresponding to exact PCAC:

$$\begin{aligned}
& q_\mu j^{a\mu}_{NN}(k_1k_2, p_1p_2) \\
&= i\left[(\gamma_5 t^a)_1 T_{NN}(k_1 - q, k_2; p_1p_2) + T_{NN}(k_1k_2; p_1 + q, p_2)(\gamma_5 t^a)_1\right] \\
&+ i\left[(\gamma_5 t^a)_2 T_{NN}(k_1, k_2 - q; p_1p_2) + T_{NN}(k_1k_2; p_1, p_2 + q)(\gamma_5 t^a)_2\right] \\
&- i f_\pi m_\pi^2 T^a_{N0}(k_1k_2, p_1p_2) d_\pi(q)
\end{aligned} \tag{12}$$

where T_{N0}^a is the amplitude for $\pi NN \to NN$ and in its time reversed form, is just the the pion production amplitude that we are seeking. Note that the axial current $j_{NN}^{a\mu}$ contains a pion pole [5, 6]:

$$j_{NN}^{a\mu}(k_1k_2,p_1p_2) = \bar{j}_{NN}^{a\mu}(k_1k_2,p_1p_2) + d_\pi(q)F_\pi^\mu(q)T_{N0}^a(k_1k_2,p_1p_2), \tag{13}$$

where $\bar{j}_{NN}^{a\mu}$ has no pion pole and F_π^μ is the pion decay vertex function. Thus an alternative way of obtaining the pion production amplitude T_{N0}, is to take the residue of Eq. (10) at the pion pole.

3.3 Physical Content of the New Model

Not only does the pion production amplitude T_{N0} obey exact PCAC, but it contains a very rich amount of physics that goes beyond what is included in the traditional few-body approach of Sec. 2. A few of the infinite number of new contributions are illustrated in Fig. 3.

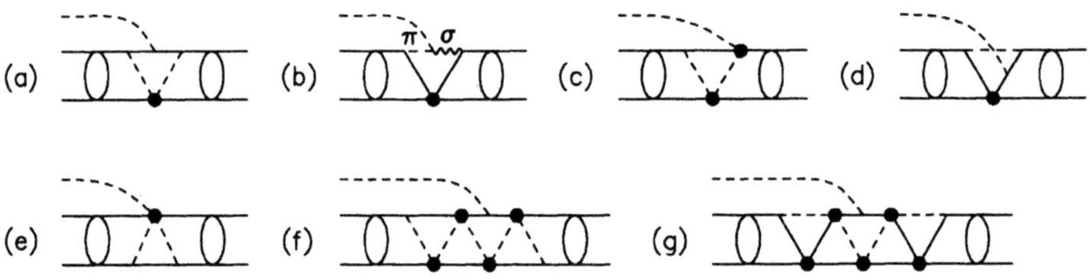

Figure 3. Examples of new pion production mechanisms included in the few-body model with PCAC.

Finally we would like to stress the practicality of our approach: PCAC and inclusion of an infinite number of new pion production mechanisms has been achieved through a simple closed expression, Eq. (9), involving scattering amplitudes $\mathcal{T}$ obtained from a traditional few-body model, Eq. (1). Our method also does not depend on the model used for kernel $\mathcal{V}$ – rather than Eq. (2), we could equally have used the simpler case of an NN one-pion-exchange potential.

Acknowledgement. This work was supported by a grant from the Flinders University Research Committee.

References

1. A. Kvinikhidze and B. Blankleider: Phys. Rev. **C60**, 044003 (1999)
2. A. Kvinikhidze and B. Blankleider: Phys. Rev. **C60**, 044004 (1999)
3. A.N. Kvinikhidze and B. Blankleider: Nucl. Phys. **A574**, 788 (1994)
4. D.R. Phillips and I.R. Afnan: Ann. Phys. (N.Y.) **247**, 19 (1996)
5. S. Weinberg: *The Quantum Theory of Fields II.* Cambridge: Cambridge University Press 1966
6. W. Bentz: Nucl. Phys. **A446**, 678 (1985)

Few-Body Systems Suppl. 12, 229–234 (2000)

Few-Body Systems

Electron-Deuteron Scattering in a Three-Dimensional Relativistic Framework*

D.R. Phillips

Department of Physics, University of Washington, Seattle, WA 98195-1560
Department of Physics, University of Maryland, College Park, MD 20742-4111

Abstract. I discuss a three-dimensional relativistic formalism that includes the effects of negative-energy states, approximate boosts of the two-body system, and current conservation. I also display the electromagnetic form factors of the deuteron predicted by this approach—both in impulse approximation and with a $\rho\pi\gamma$ current included—for the range of Q^2 explored in JLab experiments.

Recently experiments in Halls A and C at the Jefferson Laboratory (JLab) have probed the electromagnetic structure functions of the deuteron at large space-like momentum transfers. Results for T_{20} at Q^2 up to almost 2 GeV^2, for B for $Q^2 < 1.3$ GeV^2, and for A out to $Q^2 = 6$ GeV^2 have been obtained. At these momentum transfers relativistic kinematics and dynamics would appear to be a necessary part of any theoretical description. Hence there has been much effort invested in formulating the NN bound-state problem in a relativistic fashion [1]–[7]. If nucleonic degrees of freedom are employed then this is a logical extension of the standard nonrelativistic treatment of the NN system. Of course, it is also crucial that any formalism attempting to tackle this problem results in a conserved deuteron electromagnetic current.

A, B, and T_{20} can be calculated using a nonrelativistic NN interaction which is fit to data. This approach has met with a significant amount of success (see, e.g. Ref. [8]). The work I report on here attempts to imitate the nonrelativistic approach, but within a relativistic formalism. To this end we construct an NN interaction, place it in a relativistic scattering equation, and fit the parameters of our interaction to NN data. The electromagnetic form factors of the deuteron predicted by this model are then calculated. Thus our three-dimensional (3D) framework includes relativistic kinematics, negative-energy states, boost effects, and relativistic pieces of the electromagnetic current explicitly at all stages of the calculation. It was developed and applied in Refs. [3]–[6]. Here I will focus on our most recent results [6] for the electron-deuteron

* *E-mail address:* phillips@phys.washington.edu

scattering experiments at JLab [9, 10]. First, I outline the formalism, displaying both the bound-state equation and the impulse approximation current matrix element in the case of an instantaneous two-body interaction. I then specify the ingredients of our calculation: the NN interaction and the single-nucleon form factors, as well as the $\rho\pi\gamma$ meson-exchange-current (MEC). Finally, I present and discuss our results for A, B, and T_{20}.

1 Formalism

The Bethe-Salpeter equation (BSE) for a bound-state vertex function, Γ, is

$$\Gamma = K G_0 \Gamma, \tag{1}$$

where G_0 is the NN propagator, $G_0 = i d_1 d_2$. In studies for the deuteron bound-state [11] the kernel K included a set of single-boson exchanges, yielding the "ladder" approximation: $K \equiv K^{(2)}$. However, in such an approximation the BSE does not give the correct one-body limit. This limit is only properly treated in Eq. (1) if the full set of ladder and crossed-ladder graphs is taken for K [12]. In Ref. [3] we remedied this, showing that the pieces of the graphs which appear in K and are responsible for the one-body limit can be resummed to obtain another 4D equation, which, even if only the lowest-order kernel is employed, has the correct one-body limit:

$$\Gamma = K^{(2)}(G_0 + G_C)\Gamma. \tag{2}$$

Equation (2) is still four-dimensional. A straightforward way to reduce it to three dimensions is to imitate Salpeter [13] and make the replacement:

$$K^{(2)}(q) = \frac{1}{q^2 - \mu^2} \quad \longrightarrow \quad K^{(2)}_{\text{inst}}(\mathbf{q}) = -\frac{1}{\mathbf{q}^2 + \mu^2}, \tag{3}$$

where $q = (q_0, \mathbf{q})$ is the four-momentum of the meson. $K^{(2)}_{\text{inst}}$ depends only on the three-vector $\mathbf{q}$ and so Eq. (2) becomes a 3D equation [14]

$$\Gamma_{\text{inst}} = K^{(2)}_{\text{inst}} \langle G_0 + G_C \rangle \Gamma_{\text{inst}}. \tag{4}$$

Here the 3D propagator $\langle G_0 + G_C \rangle$ is obtained by integrating over the time-component of relative four-momentum, p_0. Throughout the rest of this work such integration over p_0 is denoted by angled brackets:

$$\langle G_0 + G_C \rangle = \int \frac{dp_0}{2\pi} \left[G_0(p;P) + G_C(p;P)\right] \tag{5}$$

$$= \frac{\Lambda_1^+ \Lambda_2^+}{E^+ - \epsilon_1 - \epsilon_2} - \frac{\Lambda_1^+ \Lambda_2^-}{2\epsilon_2} - \frac{\Lambda_1^- \Lambda_2^+}{2\epsilon_1} - \frac{\Lambda_1^- \Lambda_2^-}{E^- + \epsilon_1 + \epsilon_2}. \tag{6}$$

$\Lambda^{\pm}$ are related to projection operators onto positive and negative-energy states of the nucleon. E is the total energy, and $\epsilon_i = \sqrt{\mathbf{p}_i^2 + m_i^2}$. Comparing the

$++ \to ++$ piece of $K^{(2)}_{\text{inst}}\langle G_0 + G_C\rangle K^{(2)}_{\text{inst}}$ with the amplitude obtained at fourth order in the 4D field theory we see that the contribution of negative-energy states agrees at leading order in $1/M$ [5]. Effects due to "Z-graphs" are included in a bound-state calculation that employs Eq. (4).

Of course, the instant approximation which led to Eq. (4) is uncontrolled. However, Refs. [15, 3] show how to implement similar 3D reductions in a systematically-improvable way. The numerical work of Ref. [5] then showed that the replacement of Eq. (3) by the systematic 3D reductions developed in Ref. [3] had little impact on the deuteron electromagnetic form factors. Thus here I will report only on results obtained using Eq. (4).

Next we construct a conserved electromagnetic deuteron current. The impulse-approximation current for the ladder BSE is:

$$\mathcal{A}^{(0)}_{\mu} = i\,\bar{\Gamma}(P+Q)\, d_1(p_1)\, d_2(p_2+Q) j^{(2)}_{\mu} d_2(p_2)\,\Gamma(P) + \ (1 \leftrightarrow 2), \qquad (7)$$

where Γ is the solution of Eq. (1), $P = p_1 + p_2$, and $P+Q$ are the four-momenta of the initial and final states, and $j^{(2)}_{\mu}$ is the usual one-nucleon current. Using a Ward-Takahashi identity it is easy to show that $\mathcal{A}^{(0)}_{\mu}$ is conserved.

However, here we do not begin with the ladder BSE, but instead with the 4D equation (2). Constructing a conserved impulse approximation current for the vertex function of Eq. (2) is a little more involved. A piece must be added to the current (7) if a four-dimensional conserved current is to be obtained [5]. A reduction of this current to a 3D form can then be made in an analogous fashion to the reduction used for the bound-state equation itself. This reduction can also be performed in a systematic fashion, but here we need only the "instant" current, constructed from objects in the 3D theory:

$$\mathcal{A}_{\text{inst},\mu} = \bar{\Gamma}_{\text{inst}}\, \mathcal{G}^{\gamma}_{\text{inst},\mu}\, \Gamma_{\text{inst}}. \qquad (8)$$

This current is conserved, provided that Γ_{inst} is the solution of Eq. (4). An explicit form for the 3D operator $\mathcal{G}^{\gamma}_{\text{inst},\mu}$ is displayed in Ref. [5]. Eq. (8) includes the effects of both time-ordered graphs like that on the left-hand side of Fig. 1, and those due to photons coupling to Z-graphs.

2 Input to the Calculation

Equation (8) defines our impulse-approximation current. Detailed results for this current were presented in Ref. [5]. The results are very similar to those of Ref. [1]. Significantly, in our results negative-energy states played only a small role in the deuteron structure functions if $Q^2 < 4$ GeV^2. In this paper I will move beyond impulse approximation and discuss the inclusion of the $\rho\pi\gamma$ MEC, which is shown on the right-hand side of Fig. 1. The form factors employed at the πNN and ρNN vertices in this diagram are the same as those in the NN potential.

The vertex functions Γ_{inst} were obtained with all positive and negative-energy states included in the calculation. Without the negative-energy states

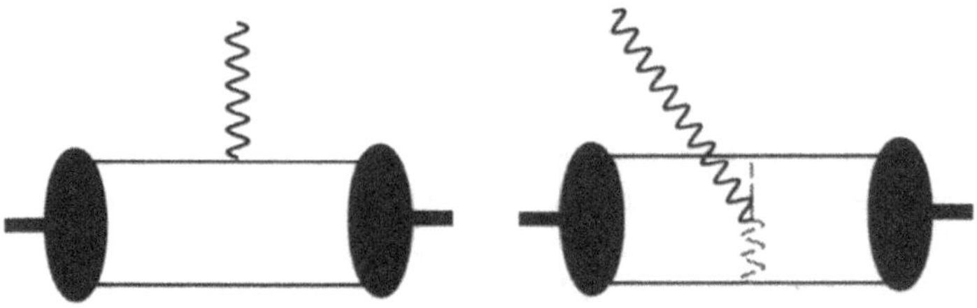

Figure 1. On the left we depict the positive-energy-state impulse approximation mechanism for electron-deuteron scattering. On the right the $\rho\pi\gamma$ MEC diagram is shown.

the interaction is exactly the Bonn-B potential [16]. To compensate for slight changes in the deuteron mass when the negative-energy states are included we adjust the σ-meson coupling from the value of Ref. [16], $\frac{g_\sigma^2}{4\pi} = 8.08$, to 8.55.

The single-nucleon form factors chosen were those of Mergell *et al.* [17]. The $\rho\pi\gamma$ coupling was set equal to $g_{\rho\pi\gamma} = 0.563$. Lastly, we followed Ref. [1] and used vector-meson dominance to get a $\rho\pi\gamma$ form factor given by the ω-meson pole: $F_{\rho\pi\gamma}(Q) = (Q^2 + m_\omega^2)^{-1}$.

3 Results and Discussion

The left panel of Fig. 2 shows our results up to $Q^2 \approx 2$ GeV^2 for A, while the right panel gives results and data over the full range of experimental Q^2. The two JLab experiments with data for A [9] are denoted by triangles and squares.

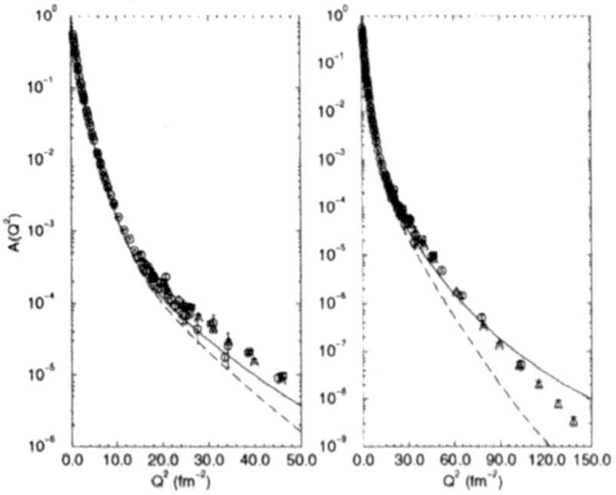

Figure 2. The deuteron structure function A. The left panel is an enlarged version of the right one. In both the dashed line is the impulse approximation result, and the solid line is the result with the $\rho\pi\gamma$ MEC included. The experimental data of Refs. [18, 19] are denoted by the open circles, while those of Refs. [9] are represented by triangles and squares.

Our impulse approximation underpredicts the A data for $Q^2 = 1 - 2\,\text{GeV}^2$. Some of this lack of strength is compensated for by the $\rho\pi\gamma$ MEC, and the data is then reproduced in the region $Q^2 = 2 - 4\,\text{GeV}^2$. However, A is then *over*predicted at large Q^2, indicating that the $\rho\pi\gamma$ MEC is too strong at large Q in the present model.

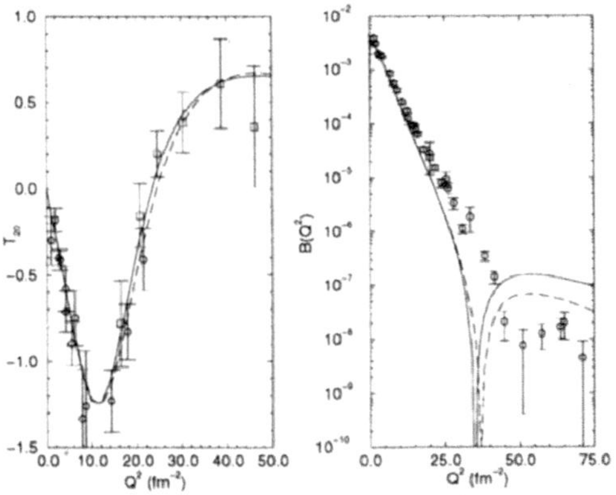

Figure 3. The deuteron structure functions B and T_{20}. Older experimental data are given by the open circles, the recent NIKHEF data of Bouwhuis *et al.* by diamonds, and the preliminary JLab data of Ref. [10] by squares. The dashed and solid line represent impulse approximation and $\rho\pi\gamma$ MEC calculations.

As seen in the left half of Fig. 3, the extant experimental data for B [19, 20] (which does not yet include JLab data) are even less well-described than the data for A. At $Q^2 < 1$ GeV^2 there is significant discrepancy between our calculation and the data. This discrepancy worsens slightly when the relativistic version of the $\rho\pi\gamma$ MEC is included.

On the other hand, the T_{20} data of Refs. [10, 21] are well-described by our approach out to $Q^2 \approx 2\,\text{GeV}^2$. T_{20} is quite sensitive to relativistic effects [7, 8], so it is pleasing that our calculation reproduces the data at large Q^2 well.

Acknowledgements

I thank Steve Wallace for a most enjoyable and productive collaboration. I also thank Neal Devine for his computational help and codes, and Betsy Beise for a number of discussions on the experimental situation. Finally, I am grateful to the U. S. Department of Energy, Nuclear Physics Division, for its support under grants DE-FG02-93ER-40762 and DE-FG03-97ER41014.

References

1. E. Hummel and J.A. Tjon: Phys. Rev. Lett. **63**, 1788 (1989); Phys. Rev. **C42**, 423 (1990); *ibid.* **C49**, 21 (1994)
2. J.W. van Orden, N.K. Devine, F. Gross: Phys. Rev. Lett. **75**, 4369 (1995)

3. D.R. Phillips and S.J. Wallace: Phys. Rev. **C54**, 507 (1996)

4. D.R. Phillips and S.J. Wallace: Few-body Syst. **24**, 175 (1997)

5. D.R. Phillips, S.J. Wallace, N.K. Devine: Phys. Rev. **C58**, 2261 (1998)

6. D.R. Phillips, S.J. Wallace, N.K. Devine: `nucl-th/9906091`

7. J. Carbonell and V.A. Karmanov: `nucl-th/9902053`

8. R.B. Wiringa, V.G.J. Stoks, R. Schiavilla: Phys. Rev. **C51**, 38 (1995)

9. L.C. Alexa et al.: Phys. Rev. Lett. **82**, 1374 (1999); D. Abbott et al.: Phys. Rev. Lett. **82**, 1379 (1999)

10. E.J. Beise et al.: Few-Body Syst. Suppl. **10**, 315 (1999)

11. J. Fleischer and J.A. Tjon: Nucl. Phys. **B84**, 375 (1975); Phys. Rev. **D15**, 2537 (1977); *ibid.* **D21**, 87 (1980)

12. F. Gross: Phys. Rev. **C26**, 2203 (1982)

13. E.E. Salpeter: Phys. Rev. **87**, 328 (1952)

14. V.B. Mandelzweig and S.J. Wallace: Phys. Lett. **B197**, 469 (1987); S.J. Wallace and V.B. Mandelzweig: Nucl. Phys. **A503**, 673 (1989)

15. A. Klein: Phys. Rev. **90**, 1101 (1953); *ibid.* **94**, 1052 (1954)

16. R. Machleidt: Adv. Nucl. Phys. **19**, 189 (1989)

17. P. Mergell, U.-G. Meissner, D. Drechsel: Nucl. Phys. **A596**, 367 (1996)

18. J.E. Elias et al.: Phys. Rev. **177**, 2075 (1969); R.G. Arnold et al.: Phys. Rev. Lett. **35**, 776 (1975); S. Platchkov et al.: Nucl. Phys. **A510**, 740 (1990)

19. G.G. Simon, C. Schmitt, V.H. Walther: Nucl. Phys. **A364**, 285 (1981); R. Cramer et al.: Z. Phys. **C29**, 513 (1985)

20. S. Auffret et al.: Phys. Rev. Lett. **54**, 649 (1985); P. Bosted et al.: Phys. Rev. **C42**, 38 (1990)

21. M.E. Schulze et al.: Phys. Rev. Lett. **52**, 597 (1984); V.F. Dmitriev et al.: Phys. Lett. **B157**, 143 (1985); B.B. Voitsekhovskii et al.: JETP Lett. **43**, 733 (1986); R. Gilman et al.: Phys. Rev. Lett. **65**, 1733 (1990); M. Ferro-Luzzi et al.: Phys. Rev. Lett. **77**, 2630 (1996); M. Bouwhuis et al.: Phys. Rev. Lett. **82**, 3755 (1999); I. The et al.: Phys. Rev. Lett. **67**, 173 (1991)

Few-Body Systems Suppl. 12, 235–239 (2000)

A Free-of-Ambiguity Electromagnetic Current for the Deuteron within the Front-Form Hamiltonian Dynamics

G. Salmè[1], F.M. Lev[2], E. Pace[3]

[1] Istituto Nazionale di Fisica Nucleare, Sezione di Roma, P.le A. Moro 2, I-00185 Rome, Italy

[2] Laboratory of Nuclear Problems, Joint Institute for Nuclear Research, Dubna, Moscow region 141980, Russia

[3] Dipartimento di Fisica, Università di Roma "Tor Vergata", and Istituto Nazionale di Fisica Nucleare, Sezione Tor Vergata, Via della Ricerca Scientifica 1, I-00133, Rome, Italy

Abstract. The deuteron electromagnetic properties are investigated in the few-(Gev/c)^2 region by using a novel current within the front-form Hamiltonian dynamics. This current is not affected by the usual ambiguities related to the choice of the subset of independent matrix elements to be treated in one-body approximation.

The deuteron electromagnetic (em) form factor, $A(Q^2)$, recently measured at TJNAF [1, 2] over a wide range of momentum transfer, Q^2, represents a good candidate, together with $B(Q^2)$ and $T_{20}(Q^2)$, for testing different, relativistically coherent descriptions of the deuteron em properties. In order to obtain [3], for an *interacting system*, an em current operator, $J^\mu(x)$, properly commuting with the Poincaré generators and satisfying: i) Hermiticity, ii) parity and time reversal covariance, and iii) current conservation, we have adopted the front form of the relativistic Hamiltonian dynamics (FFHD) [4]. If one considers only the two-nucleon Hilbert space, the Hamiltonian dynamics is very convenient for implementing a relativistic calculation, since it is possible to construct fully relativistic, but non local, models retaining a large part of the successfull phenomenology developed within the non relativistic approach. This means that one can explicitly construct the generators of the Poincaré group, even in presence of the interaction, and then investigate the behaviour of relevant states and operators. In particular, once an ansatz for the interacting mass operator M is given, the generators of the Poincaré group can be obtained through the Bakamjian-Thomas procedure (see, e.g., [5]). It is easy to show that the ansatz

$M^2 = M_0^2 + U$, where $M_0^2 = 4(m^2 + k^2)$ and $U = 4m\mathcal{R_M}V\mathcal{R_M}^{-1}$, with V the NN interaction in the usual potential approach and $\mathcal{R_M}$ the generalized Melosh rotations [5], leads to the Schrödinger equation. Therefore, the intrinsic wave functions obtained within the usual non relativistic approach can be safely exploited in a genuine relativistic framework.

In [3, 6], we have shown that a current with the proper behavior under the Poincaré transformations and satisfying all the above mentioned properties can be constructed starting from auxiliary operators which act on the internal variables of the system. These operators are the matrix elements (between the CM momentum eigenstates) of the components of the current operator, defined in a Breit frame where the three-momentum transfer is along the z axis. For instance, these auxiliary operators can be obtained from the familiar one-body current, vid. [3, 6]

$$j^\mu(Ke_z; m, m) = \frac{1}{2}\{\mathcal{J}^\mu(Ke_z; m, m) + L^\mu_\nu[r_x(-\pi)]\ exp(\imath\pi S_x)\ [\mathcal{J}^\nu(Ke_z; m, m)]^*\ exp(-\imath\pi S_x)\}, \qquad (1)$$

where $K = Q/2$, $L[r_x(-\pi)]$ is the element of the Lorentz group corresponding to a rotation of $-\pi$ around the x axis, S_x is the x component of the front-form spin operator, * means Hermitian conjugation in internal space and $\mathcal{J}^\mu(Ke_z; m, m)$ is given by

$$\begin{aligned} \mathcal{J}^+(Ke_z; m, m) &= \langle 0, P^+|\Pi J^+_{free}(0)\Pi|0, P^{'+}\rangle \\ \mathcal{J}_\perp(Ke_z; m, m) &= \langle 0, P^+|\Pi J_{\perp free}(0)\Pi|0, P^{'+}\rangle \\ \mathcal{J}^-(Ke_z; m, m) &= \mathcal{J}^+(Ke_z; m, m). \end{aligned} \qquad (2)$$

In Eq. (2), Π is the projector onto the subspace of bound states with mass m and spin S, $|\vec{P}_\perp, P^+\rangle$ is an eigenstate of the total deuteron momentum, $\vec{P}_\perp \equiv (P_x, P_y)$ and $P^+ \equiv (P_0 + P_z)/\sqrt{2}$. Eq. (1) represents the basic ingredient of our approach (the auxiliary operators are unitarily equivalent to the current $J^\mu(x)$) and allows one to calculate all the independent matrix elements from the same formula. The second term in Eq. (1) implicitely introduces two-body currents, because of the presence of the spin operator S_x, and automatically produces a current fulfilling current conservation. Differently, in previous models, developed within FFHD (see, e.g., [8, 9]) in the so-called q^+ reference frame, the one-body approximation was applied only to a subset of independent matrix elements of the current operator. This procedure clearly contains a degree of arbitrariness, as shown, by the many different proposals on the choice of the subset (only three independent matrix elements, out of four, are involved in the calculations of the em properties, since the deuteron form factors are three). As a first application of the current defined in Eq. (1), the magnetic and quadrupole moments of the deuteron have been calculated [6], obtaining a remarkable improvement in the description of these quantities. Then, detailed calculations of $A(Q^2)$, $B(Q^2)$ and $T_{20}(Q^2)$ for different nucleon form factors and NN interactions have been carried out. In Figs. 1-3 preliminary

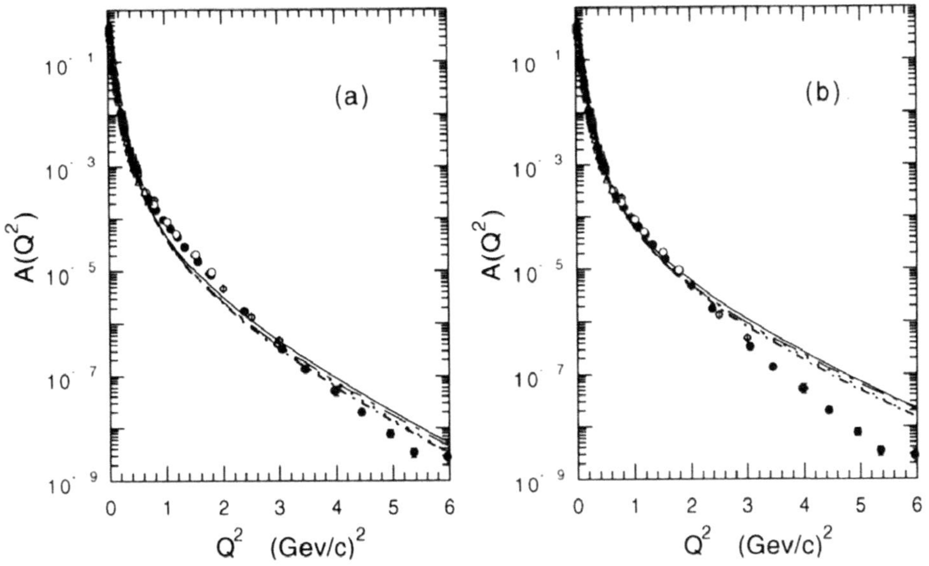

Figure 1. (a) The deuteron form factor $A(Q^2)$ vs Q^2, obtained using the Höhler nucleon form factors [11] and different NN interactions: RSC [10](a) (solid line); AV14 [10](b) (dot-dashed line); AV18 [10](c) (long-dashed line); Paris [10](d) (dotted line). Full dots: [1]; open dots: [2]; for the other experimental data see [12]. - (b) The same as in Fig. 1(a), but using Gari-Krümpelmann nucleon form factors [13]. (After [12])

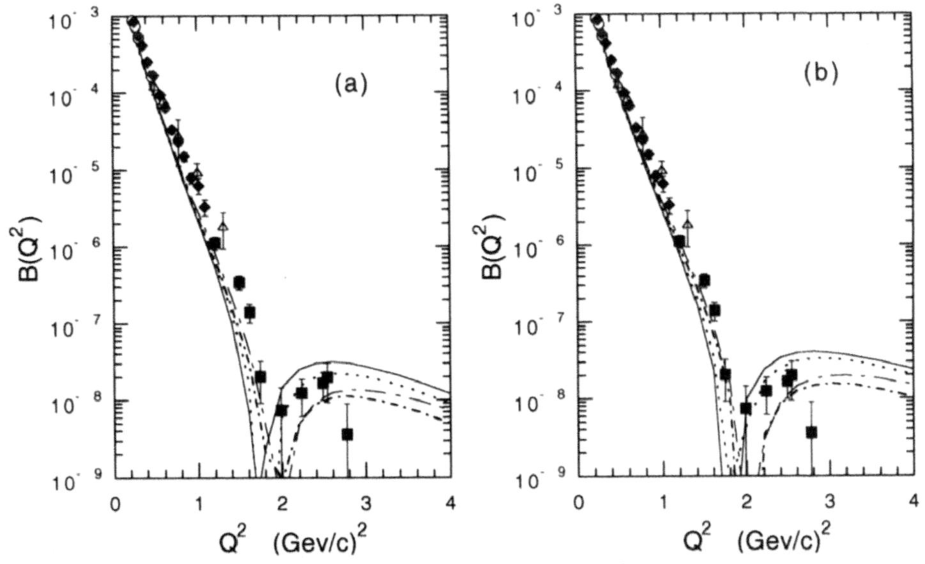

Figure 2. (a) The deuteron form factor $B(Q^2)$ vs Q^2, obtained as in Fig. 1(a). For the experimental data see [12]. - (b) The same as in Fig. 2(a), but using Gari-Krümpelmann nucleon form factors [13]. (After [12])

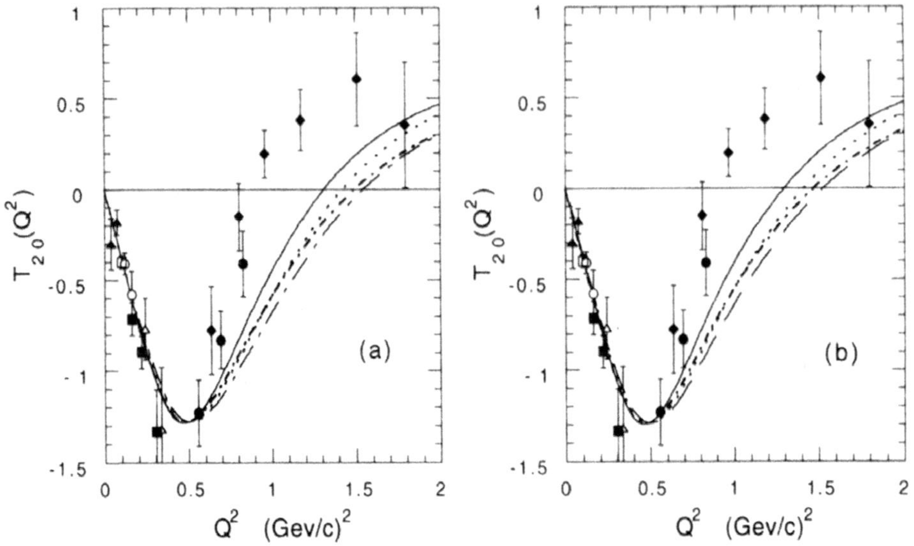

Figure 3. (a) The tensor polarization $T_{20}(Q^2)$ vs Q^2, obtained as in Fig. 1 (a). Diamonds: [14]; for the other experimental data see [12]. - (b) The same as in Fig. 3(a), but using Gari-Krümpelmann nucleon form factors [13].

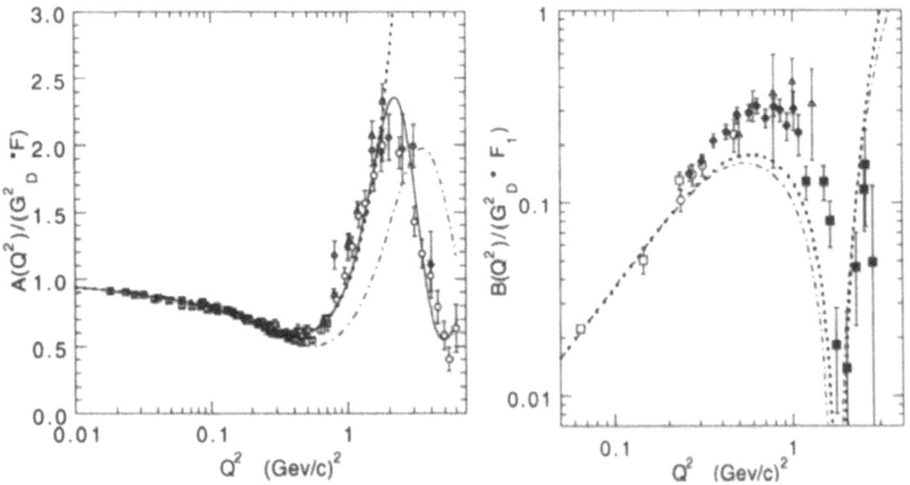

Figure 4. (a) The ratio $A(Q^2)/(G_D^2\ F)$ (with $F = 1/(1 + Q^2/0.1)^{2.5}$ and $G_D = 1/(1 + Q^2/0.71)^2)$ vs Q^2. Solid line: FFHD calculation with the RSC interaction, $G^p_{E(M)}$ and G^n_M from the Gari-Krümpelmann model and a fitted G^n_E, see text; dotted line: the same as the solid line, but with the full Gari-Krümpelmann model [13] for the nucleon form factors; dash-dotted line: the same as the solid line, but with the full Höhler model [11]. - (b) The ratio $B(Q^2)/(G_D^2\ F_1)$ (with $F_1 = 1/(1 + Q^2/0.1)^3$) vs Q^2. The solid curve is not represented, since it overlaps the dotted one.

results of $A(Q^2)$, $B(Q^2)$ and $T_{20}(Q^2)$ are presented. The dependence upon the choice of NN interaction is remarkable, not only for the NN interactions shown in this contribution, but also in the case of modern $\chi^2 \approx 1$ NN interactions (see [7]). From Figs. 1 and 3 respectively, one can see that $A(Q^2)$ shows a great sensitivity to the poorly-known $G_E^n(Q^2)$, while, as well known, the deuteron tensor polarization $T_{20}(Q^2)$ is not sizeably affected by changes of the nucleon form factors. In Fig. 4, in view of the very accurate data measured at TJNAF, for $A(Q^2)$ and, in the near future, for $B(Q^2)$, the deuteron form factors are presented in more detailed plots. Preliminary calculations obtained by keeping fixed the proton form factors and the neutron magnetic one to the corresponding quantities of the Gari-Krümpelmann model and by fitting $G_E^n(Q^2)$ are also shown. It turns out that a parametrization for G_E^n, somewhat higher than the Gari-Krümpelmann model around 0.8 $(GeV/c)^2$ and smaller beyond, allows one to nicely reproduce $A(Q^2)$, leaving basically unchanged $B(Q^2)$ and $T_{20}(Q^2)$.

References

1. L.C. Alexa et al.: Phys. Rev. Lett. **82**, 1374 (1999)
2. D. Abbott et al.: Phys. Rev. Lett. **82**, 1379 (1999)
3. F. Lev, E. Pace, G. Salmè: Nucl. Phys. **A 641**, 299 (1998)
4. P.A.M. Dirac: Rev. Mod. Phys. **21**, 392 (1949)
5. B.D. Keister and W. Polyzou: Adv. Nucl. Phys. **21**, 225 (1991)
6. F. Lev, E. Pace, G. Salmè: Phys. Rev. Lett. in press and nucl-th/9910049
7. F. Lev, E. Pace, G. Salmè: to be published
8. I.L. Grach and L.A. Kondratyuk: Yad. Fiz.**39**, 316 (1984)
9. P.L. Chung, F. Coester, B.D. Keister, W.N. Polyzou: Phys. Rev. **C37**, 2000 (1988)
10. (a) R.V. Reid: Ann. of Phys. **50**, 411 (1968); (b) R.B. Wiringa, R.A. Smith, T.A. Ainsworth: Phys. Rev. **C29**, 1207 (1984); (c) R.B. Wiringa, V.G.J. Stoks, R. Schiavilla: Phys. Rev. **C51**, 38 (1995); (d) M. Lacombe et al.: Phys. Rev. **C21**, 861 (1980)
11. G. Hoehler et al.: Nucl. Phys. **B114**, 505 (1975)
12. F. Lev, E. Pace, G. Salmè: In *Proceedings of the Workshop on "Electron-nucleus scattering"*, Eds. O. Benhar, A. Fabrocini, R. Schiavilla, ETS Pisa 1999, p.401
13. M. Gari and W. Krümpelmann: Z. Phys. **322**, 689 (1985)
14. E.J. Beise: Few-body Syst. Suppl. **10**, 315 (1999)

Few-Body Systems Suppl. 12, 240–244 (2000)

New Data on A_{yy} in Deuteron Inclusive Breakup at Large Transverse Momenta and Relativistic Description of the Deuteron at Short Distances

V.P. Ladygin[1,*], L.S. Azhgirey[1], S.V. Afanasiev[1], V.V. Arkhipov[1], V.K. Bondarev[1,2], Yu.T. Borzounov[1], G.Filipov [1,3], L.B. Golovanov[1], A.Yu. Isupov[1], A.A. Kartamyshev[4], V.A. Kashirin[1], A.N. Khrenov[1], V.I. Kolesnikov[1], V.A. Kuznezov[1], N.B. Ladygina[1], A.G. Litvinenko[1], S.G. Reznikov[1], P.A. Rukoyatkin[1], A.Yu. Semenov[1], I.A. Semenova[1], G.D. Stoletov[1], A.P. Tzvinev[1], N.P. Yudin[5], V.N. Zhmyrov[1], L.S. Zolin[1]

[1] Joint Institute for Nuclear Researches, 141980 Dubna, Russia

[2] St.-Petersburg State University, 198350 St.-Petersburg, Russia

[3] Institute of Nuclear Research and Nuclear Eenergy, 1784 Sofia, Bulgaria

[4] Russian Scientific Center "Kurchatov Institute", 123182 Moscow, Russia

[5] Moscow State University, 117234 Moscow, Russia

Abstract. Preliminary data on the tensor analyzing power A_{yy} in breakup of 9 GeV/c deuterons on hydrogen and carbon nuclei at secondary proton transverse momenta up to $\sim$ 900 MeV/c are presented. The results suggest that a deuteron structure function at short distances can depend on more than one independent variable unlike an ordinary wave function.

1 Introduction

The breakup of fast deuterons on nuclei is the traditional tool in studying the deuteron structure at short distances between the constituents. Relativistic hard scattering (RHS) model [1] gives the following qualitative picture of breakup process: at zero angle the main process is the direct fragmentation: as the proton detection angle increases the hard scattering of the deuteron nucleon on the target nucleon plays an ever increasing role. Such a simple model allows to describe the data on the differential cross section using the standard deuteron wave functions (DWFs) [2]. Taking account of the rescattering and

* *E-mail address:* ladygin@sunhe.jinr.ru

virtual pion production [3] improves the description at small detection angles of protons.

However, RHS model (even taking into account additional mechanisms) fails to reproduce the behaviour of polarization observables, especially at large momenta of protons [4]. The results of our recent measurements of the tensor analyzing power A_{yy} in the breakup of 9 GeV/c deuterons with the emission of protons at angle of 85 mr [5] do not agree with the calculations performed within RHS using conventional DWFs [6].

In this report we present new preliminary data on A_{yy} in breakup of 9 GeV/c deuterons on hydrogen and carbon nuclei with transverse momenta of secondary protons up to $\sim$ 900 MeV/c.

2 Experiment

The experiment has been performed using a tensorially polarized deuteron beam of the Synchrophasotron of the Laboratory for High Energies of JINR. The tensor polarization of the beam has been measured from the asymmetry of protons from the deuteron breakup on nuclear targets, $d + A \to p + X$ at zero angle and the momenta $p_p \sim 2/3p_d$ [7]. The vector polarization of the beam has been measured from the asymmetry of quasi-elastic pp scattering on CH_2 target [8]. The tensor and vector polarizations, p_{zz} and p_z, were $p^+_{zz} = 0.798 \pm 0.002(stat) \pm 0.040(sys)$, $p^-_{zz} = -0.803 \pm 0.002(stat) \pm 0.040(sys)$, $p^+_z = 0.231 \pm 0.014(stat) \pm 0.012(sys)$ and $p^-_z = 0.242 \pm 0.014(stat) \pm 0.012(sys)$, respectively.

A slowly extracted 9 GeV/c deuteron beam with a typical intensity of $\sim 5 \cdot 10^8 \div 10^9\ \vec{d}$/spill was directed onto a liquid hydrogen target of 30 cm in length or onto carbon targets with varied length. The data were obtained at secondary proton emission angles of 85, 130 and 160 mr and secondary proton momenta between 4.5 and 7.0 GeV/c. The separation of the protons and inelastically scattered deuterons was done by the measurements of their time-of-flight (TOF) over a base line of $\sim$ 34 m. The residual background was completely eliminated by the requirement that particles are detected at least in two prompt TOF windows.

3 Results

The data on A_{yy} at four different fixed longitudinal momentum fractions $\alpha = (E_p + P_p)/(E_d + P_d)$, where E_d and E_p are the energies, and p_d and p_p the momenta of the incoming deuteron and detected proton, respectively, are presented in Fig.1 versus transverse momentum of the proton, P_T. The full and empty symbols are the data obtained on hydrogen and carbon targets, respectively. The empty triangles are the data from ref. [4]. One can see that the A_{yy} data do not depend on the atomic number of the target, and hence (i) multiple scattering processes plays a minor role and (ii) the obtained information reflects the internal deuteron structure.

The A_{yy} data obtained at different fixed longitudinal momentum fractions, α, demonstrate a linear dependence upon the transverse momentum of the protons, P_T. A_{yy} has a positive value at small P_T and changes the sign at P_T higher than $\sim$ 600 MeV/c independently on the longitudinal momentum fraction α. Such a behaviour contradicts the predictions of the RHS model with conventional deuteron wave functions [6].

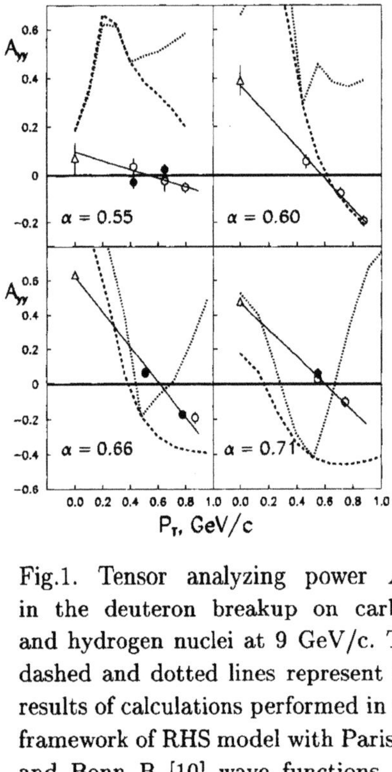

Fig.1. Tensor analyzing power A_{yy} in the deuteron breakup on carbon and hydrogen nuclei at 9 GeV/c. The dashed and dotted lines represent the results of calculations performed in the framework of RHS model with Paris [9] and Bonn B [10] wave functions, respectively. The solid lines are the result of the linear fit.

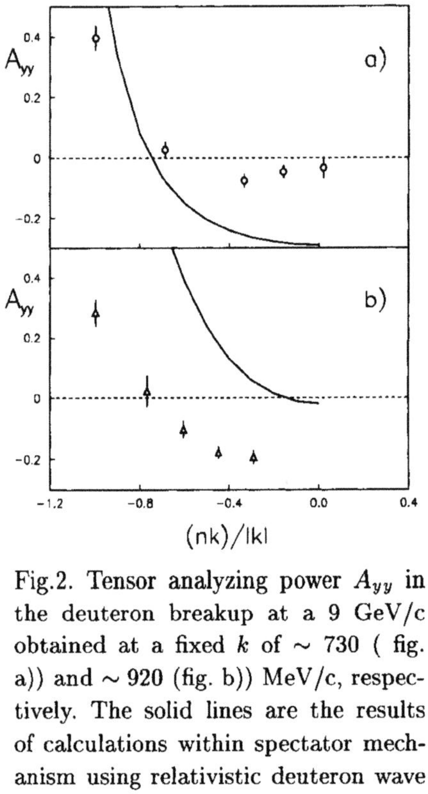

Fig.2. Tensor analyzing power A_{yy} in the deuteron breakup at a 9 GeV/c obtained at a fixed k of $\sim$ 730 (fig. a)) and $\sim$ 920 (fig. b)) MeV/c, respectively. The solid lines are the results of calculations within spectator mechanism using relativistic deuteron wave function of the covariant approach [11].

The calculations performed in the framework of RHS model [6] with Paris [9] and Bonn B [10] deuteron wave functions (dashed and dotted lines, respectively) can not reproduce the data.

The observed features of the A_{yy} data suggest that an adequate description of the data may be achieved by using a deuteron structure function that depends at short distances on more than one variable. In the covariant approach [11] the internal structure of the deuteron is described by six functions depending on two variables.

The A_{yy} data plotted for the internal nucleon momenta k of $\sim$ 730 (circles) and $\sim$ 920 MeV (triangles) are shown in Fig.2. They slope down fastly as the variable $(\vec{n}\vec{k})/|k|$ increases. This variable is defined as

$$k^2 = \frac{m_p^2 + p_T^2}{4\alpha(1-\alpha)} - m_p^2, \qquad (\vec{n}\vec{k}) = \sqrt{\frac{m_p^2 + p_T^2}{\alpha(1-\alpha)}} \cdot \left(\frac{1}{2} - \alpha\right).$$

The results of calculations in the framework of the spectator mechanism using deuteron wave function from ref. [11] are shown in Fig.2 by the solid lines. Only two (f_1 and f_2) from six components of this wave function were taken into account. One can see that the calculations qualitatively reproduce the angular dependence of the A_{yy} data.

In Fig.3 the data on A_{yy} in breakup of 9 GeV/c deuterons on carbon with emission of protons at an angle of 85 mr [5] are plotted versus proton momentum p in comparison with the calculations based on different approaches. The dashed and dotted lines are the results of the RHS model with Paris [9] and Bonn [10] DWFs, respectively. The results of calculations within spectator mechanism with DWF of ref. [11] are given by the dash-dotted and solid lines taking into account $f_1 - f_2$ and $f_1 - f_6$ functions, respectively. It is seen that these calculations fail to reproduce the behaviour of the experimental data, especially at high momenta of protons.

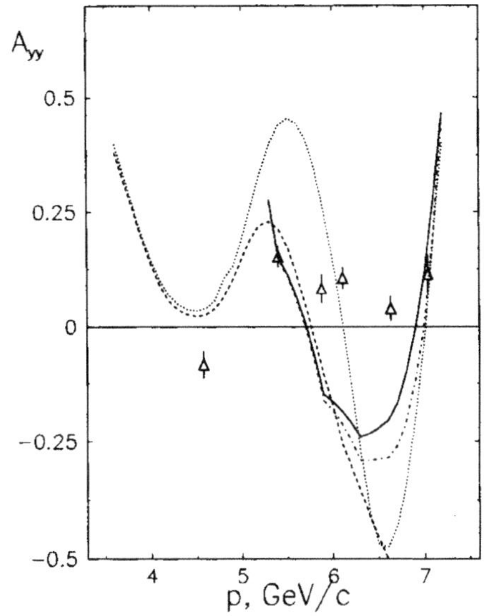

Fig.3. Tensor analyzing power A_{yy} in breakup of 9 GeV/c deuterons with emission of protons at angle of 85 mr versus proton momentum in lab. [5]. The curves are explained in the text.

4 Conclusions

New data on the tensor analyzing power A_{yy} in breakup of 9 GeV/c deuterons on hydrogen and carbon nuclei have been obtained at proton transverse momenta up to $\sim$ 900 MeV/c. These data are in good agreement with the previous data [5] on A_{yy} for the same reaction on carbon with proton emission at an angle of 85 mr.

The independence of A_{yy} for the breakup process with protons emitted at large transverse momenta on the atomic number of the target is observed.

The A_{yy} data obtained at different fixed longitudinal momentum fractions, α, demonstrate a linear dependence upon the transverse momentum of the protons, P_T. A_{yy} is positive at small P_T and changes the sign at P_T higher than $\sim$ 600 MeV/c independently on the longitudinal momentum fraction α. Such a behaviour definitely contradicts the predictions of the relativistic hard scattering model with conventional deuteron wave functions.

The observed features of the A_{yy} data suggest that a deuteron structure function at short distances, where relativistic effects are significant, depends on more than one variable, as it is the case, for example, in some relativistic models [11]. The measurements of A_{yy} and polarization transfer coefficient at different initial deuteron energies and various α and P_T will provide necessary experimental base for development of such models.

Acknowledgement. Authors thank the Director of LHE A.I.Malakhov and vice-Director of LHE V.N.Penev for their permanent help. Authors would like to express their thanks to the accelerator crew and POLARIS staff. They are grateful to L.V.Budkin, V.P.Ershov, V.V.Fimushkin, V.I.Ivanov, A.S.Nikiforov, Yu.K.Pilipenko, V.G.Perevozchikov, E.V.Ryzhov, A.I.Shirokov and O.A.Titov for their assistance during the experiment. Authors thank A.N.Prokofiev and A.A.Zhdanov for the lending of the scintillation counters for the polarimeter. They thank V.A.Karmanov for helpful comments. One of authors (V.P.L.) is grateful to the Organizing Committee of this Conference for the financial support and kind hospitality, to Prof. S.Oryu for his help and to Russian Academy of Science for the travel grant.

References

1. R. Blankenbecler et al.: Phys. Rev. **D15**, 3321 (1977); Phys. Rev. **C22**, 2433 (1980); L.S. Azhgirey et al.: Yad. Fiz. **48** 87 (1988)
2. L.S. Azhgirey et al.: Nucl. Phys. **A528**, 621 (1991)
3. G.I. Lykasov: Part. and Nucl. **24**, 140 (1993)
4. V.G. Ableev et al.: JINR Rapid Comm. **4[43]-90**, 5 (1990); T. Aono et al.: Phys.Rev.Lett. **74**, 4997 (1995); L.S. Azhgirey et al.: Phys. Lett. **B387**, 37 (1996)
5. S.V. Afanasiev et al.: Phys. Lett. **B434**, 21 (1998); V.P. Ladygin et al.: Few Body Systems Suppl. **10**, 451 (1999); L.S. Azhgirey et al.: Yad. Fiz. **62** (1999) in press
6. L.S. Azhgirey and N.P. Yudin: Yad. Fiz. **57**, 160 (1994)
7. L.S. Zolin et al.: JINR Rapid Comm. **2[88]-98**, 27 (1998) 27
8. L.S. Azhgirey et al.: PTE **1**, 51 (1997)
9. M. Lacombe et al.: Phys. Lett. **B101**, 139 (1981)
10. R. Machleidt et al.: Phys. Reports **149**, 1 (1987)
11. V.A. Karmanov and A.V. Smirnov: Nucl. Phys. **A575**, 520 (1994); J. Carbonell and V.A. Karmanov : Nucl. Phys. **A581**, 625 (1994)

Few-Body Systems Suppl. 12, 245–248 (2000)

Few-Body Systems

Relativistic Multichannel Scattering Theory with Particle Creation

C. Chandler[1,*], A.G. Gibson[2,†]

[1] Department of Physics and Astronomy, University of New Mexico, Albuquerque, NM 87101, USA

[2] Department of Mathematics and Statistics, University of New Mexico, Albuquerque, NM 87101, USA

The creation of a mathematically rigorous multichannel relativistic quantum scattering theory that includes particle creation and annihilation has been an important goal in theoretical physics for many decades. Even in such a thoroughly worked-over field as pion-nucleon physics the goal has remained unrealized [1]. This basic problem is here reexamined from a perspective that is different in some important respects from the main traditional approaches. In particular, it is not a local quantum field theory [2] nor is it based on the Bakamjian-Thomas construction [3].

The fundamental interaction considered here is a traditional Yukawa-type interaction in which a "meson" is emitted or absorbed by a "nucleon." Included in this interaction is a vertex function. Nucleon number is conserved, and antinucleons do not appear. On the one hand, this interaction has the disadvantage that the theory is not a local relativistic quantum field theory, and fundamental properties (such as relativistic covariance) previously proved on the basis of locality must be proved anew. On the other hand, interactions of this type have been in use for a long time [4], and the presence of vertex functions permits a rigorous proof, indicated here, that the Hamiltonian is self-adjoint.

The fundamental Hilbert space has the usual Fock-space structure that accommodates a variable number of mesons. The Yukawa nature of the fundamental interaction implies that the nucleons of the Hamiltonian and the fundamental Hilbert space must be "bare" particles. The asymptotically observed nucleons are "dressed," on the other hand, implying that the Fock space of asymptotic states is distinct from the fundamental Hilbert space. This theory is, then, necessarily a two-Hilbert-space theory.

* *Email address:* chandler@unm.edu

† *Email address:* archie@math.unm.edu

It is shown here how to construct the appropriate two-Hilbert-space framework for a simplified system consisting of a single bare nucleon and an indeterminate number of a single type of meson. The spin and charge of the nucleon is ignored. The meson is its own antiparticle, and it has zero charge and spin.

The bare nucleon Hilbert space $\mathcal{H}$. The fundamental Hilbert space $\mathcal{H}$ is the Fock space of vectors of the form $\Phi = \bigoplus_{m=0}^{\infty} \varphi_m$, where $\varphi_m(\mathbf{p}; \mathbf{k}_1, \ldots, \mathbf{k}_m)$ is the square integrable component that specifies the probability density for a bare nucleon (momentum $\mathbf{p}$) and m mesons (momenta $\mathbf{k}_1, \ldots, \mathbf{k}_m$).

The bare nucleon Hamiltonian H. The fundamental Hamiltonian H has the form $H = K + \Gamma + \Gamma^{\dagger}$. For suitable $\Phi = \bigoplus_{m=0}^{\infty} \varphi_m \in \mathcal{H}$, K is defined by $K\Phi = \bigoplus_{m=0}^{\infty} (K\Phi)_m$,

$$(K\Phi)_m \equiv \left[\sqrt{M_0^2 + |\mathbf{p}|^2} + \sum_{j=1}^{m} \sqrt{\mu^2 + |\mathbf{k}_j|^2} \right] \varphi_m(\mathbf{p}; \mathbf{k}_1, \ldots, \mathbf{k}_m). \tag{1}$$

With similar notation Γ and $\Gamma^{\dagger}$ are defined by

$$(\Gamma\Phi)_m \equiv \sqrt{m+1} \int d\mathbf{q}\, g^*(\mathbf{q}, \mathbf{p} - \mathbf{q})\, \varphi_{m+1}(\mathbf{p} - \mathbf{q}; \mathbf{k}_1, \ldots, \mathbf{k}_m, \mathbf{q}), \tag{2}$$

$$(\Gamma^{\dagger}\Phi)_m \equiv \begin{cases} 0, & m = 0, \\ \sqrt{m}\, \mathsf{S}_m\, g(\mathbf{k}_m, \mathbf{p}) \varphi_{m-1}(\mathbf{p} + \mathbf{k}_m; \mathbf{k}_1, \ldots, \mathbf{k}_{m-1}), & m \geq 1. \end{cases} \tag{3}$$

Here, M_0 is the (finite) bare nucleon mass, μ is the meson mass, and S_m is the operator that symmetrizes the meson variables. The square root factors are of combinatorial origin. The form factor g is presumed to satisfy

$$|||g|||^2 \equiv \sup_{\mathbf{p}} \int d\mathbf{q}\, |g(\mathbf{q}, \mathbf{p} - \mathbf{q})|^2 < \infty. \tag{4}$$

This Hamiltonian is self-adjoint and bounded below. This is proved by establishing with straightforward estimates the inequalities

$$||\Gamma^{\dagger}\Phi|| \leq |||g|||(||N^{1/2}\Phi|| + ||\Phi||), \qquad ||\Gamma\Phi|| \leq |||g|||\,||N^{1/2}\Phi||, \tag{5}$$

$$\mu||N^{1/2}\Phi|| \leq \mu||N\Phi|| \leq ||K\Phi||, \tag{6}$$

where N is the meson number operator. These inequalities imply that the Kato-Rellich Theorem [5] is applicable, establishing that H is self-adjoint and bounded below on the domain of K (which is dense in $\mathcal{H}$).

It is noteworthy that if antiparticles are incorporated into H in the obvious way, the Hamiltonian becomes that of a nonlocal field theory of the type considered years ago by Pais and Uhlenbeck [6, 7]. Should the nonlocality be removed by insisting that g be a constant, the above proof of self-adjointness would fail. Indeed, H would not even be essentially self-adjoint, in a manner analogous to the situation for the three-dimensional delta function potential in nonrelativistic quantum mechanics [8]. In that case, the Hamiltonian H as written would not completely describe the dynamics and additional information would be required.

The asymptotic Hilbert space $\mathcal{F}$. The Hilbert space $\mathcal{F}$ is the Fock space of vectors of the form $\Phi = \bigoplus_{m=0}^{\infty} \varphi_m$, where $\varphi_m(\mathbf{p}; \mathbf{k}_1, \ldots, \mathbf{k}_m)$ is the square integrable component that specifies the probability density for a physical dressed nucleon (momentum $\mathbf{p}$) and m mesons (momenta $\mathbf{k}_1, \ldots, \mathbf{k}_m$).

The asymptotic Hamiltonian F. The asymptotic Hamiltonian F is just the kinetic energy operator defined in Eq. (1), with the bare nucleon mass M_0 replaced by the physical dressed nucleon mass M.

The identification operator J. The identification operator J is a linear operator that maps states in $\mathcal{F}$ into $\mathcal{H}$. For suitable $\Psi = \bigoplus_m \psi_m \in \mathcal{F}$, $J\Psi = \bigoplus_m (J\Psi)_m$ is defined by

$$\begin{aligned}(J\Psi)_m \quad &\equiv \quad \mathsf{S}_m \sum_{m_1, m_2 = 0}^{\infty} \delta_{m, m_1+m_2} (m!/m_1! m_2!)^{1/2} \chi_{m_2}(\mathbf{p}; \mathbf{k}_{m_1+1}, \ldots, \mathbf{k}_{m_1+m_2}) \\ &\qquad \times \psi_{m_1}(\mathbf{p} + \mathbf{k}_{m_1+1} + \cdots + \mathbf{k}_{m_1+m_2}; \mathbf{k}_1, \ldots, \mathbf{k}_{m_1}). \qquad (7)\end{aligned}$$

The square root factor is combinatorial in nature. The functions χ_{m_2} are the Fock space components of the eigenstate of H that represents a dressed nucleon of momentum $\mathbf{P}$ and energy $(M^2 + \mathbf{P}^2)^{1/2}$, with $\mathbf{P} = \mathbf{p} + \mathbf{k}_{m_1+1} + \cdots + \mathbf{k}_{m_1+m_2}$ in Eq. (7).

For the operator J to have a nontrivial domain, we have to assume that $\sum_{m=0}^{\infty} 2^m ||\chi_m||_{\mathbf{P}}^2 < \infty$ uniformly in $\mathbf{P}$, where $||\cdot||_{\mathbf{P}}$ is the square integrable norm over all the momentum variables with the total momentum held fixed at $\mathbf{P}$. Since χ_m for large M and fixed $\mathbf{P}$ (hence, fixed energy) would be describing virtual mesons, it seems reasonable that the sum would converge. An actual proof, however, remains for the future.

The wave operators $\Omega^{(\pm)}$. The scattering operator for the system is $S \equiv \Omega^{(+)\dagger}\Omega^{(-)}$, with

$$\Omega^{(\pm)} \equiv \mathrm{s} - \lim_{t \to \pm\infty} e^{iHt} J e^{-iFt}. \qquad (8)$$

It is well known that these limits exist if for any suitable $\Psi \in \mathcal{F}$ the quantity $||(HJ - JF)e^{-iFt}\Psi||$ is integrable in the time t from minus to plus infinity. If Ψ has only one nonzero component $\psi_{\bar{m}}$, then

$$\begin{aligned}([HJ - JF]e^{-iFt}\Psi)_m &= \bar{m}[(m!/\bar{m}!(m - \bar{m} + 1)!]^{1/2} \\ &\times \mathsf{S}_m \int d\mathbf{q}\, g^*(\mathbf{q}, \mathbf{p} - \mathbf{q}) \chi_{m-\bar{m}+1}(\mathbf{p} - \mathbf{q}; \mathbf{k}_1, \ldots, \mathbf{k}_{m-\bar{m}+1}) \\ &\times (e^{-iFt}\Psi)_{\bar{m}}(\mathbf{p} - \mathbf{q} + \mathbf{k}_1 + \cdots + \mathbf{k}_{m-\bar{m}+1}; \mathbf{q}, \mathbf{k}_{m-\bar{m}+2}, \ldots, \mathbf{k}_m). \quad (9)\end{aligned}$$

The proof of the desired integrability in t is now obtained by transforming Eq. (9) to coordinate space and employing estimates similar to those employed by Reed and Simon [9] in the analogous nonrelativistic proof. The wave operators $\Omega^{(\pm)}$ therefore exist.

It is straightforward to prove that

$$\text{s} - \lim_{t\to\pm\infty} e^{iFt} J^\dagger J e^{-iFt} = I_{\mathcal{F}}, \tag{10}$$

where $I_{\mathcal{F}}$ is the identity operator on $\mathcal{F}$. From this it follows that $\Omega^{(\pm)\dagger}\Omega^{(\pm)} = I_{\mathcal{F}}$.

The relativistic invariance. A theory is deemed relativistic if the generators of the relativistic transformations satisfy the Poincaré algebra [2, 3]. Because the asymptotic space $\mathcal{F}$ represents noninteracting particles, the Poincaré generators G_{0i}, $i = 1, \ldots, 10$, are well known. It is straightforward to verify that the operators $G_i^{(\pm)} \equiv \Omega^{(\pm)} G_{0i} \Omega^{(\pm)\dagger}$ also satisfy the Poincaré algebra on $\mathcal{H}$. Moreover, the generator of the time translations is $HE^{(\pm)} = E^{(\pm)}H$, where the $E^{(\pm)}$ are the orthogonal projections onto the ranges of $\Omega^{(\pm)}$. For these two sets of generators to be the same (i.e., $G_i^{(+)} = G_i^{(-)}$), so that there is a unique set of generators for the theory, it is sufficient that the scattering operator S be unitary and commute with the generators G_{0i}. Establishing these two conditions with mathematical rigor is a daunting task that remains for the future.

References

1. J. Carlson and R. Schiavilla: Rev. Mod. Phys. **70**, 743 (1998)
2. S. Weinberg: *The Quantum Theory of Fields, Vol. I.* Cambridge: Cambridge University Press, 1995
3. B. D. Keister and W. N. Polyzou: In: J. W. Negele and E. W. Vogt (eds.): Advances in Nuclear Physics, Vol. 20, p.225. New York: Plenum 1991
4. H. Garcilazo and T. Mitzutani: *πNN Systems*, p.76. Singapore: World Scientific 1990
5. M. Reed and B. Simon: *Methods of Modern Mathematical Physics, Vol. II*, p.162. New York: Academic 1975
6. A. Pais and G.E. Uhlenbeck: Phys. Rev. **79** 145, (1950)
7. K. Namsrei: *Nonlocal Quantum Field Theory and Stochastic Quantum Mechanics.* Dordrecht: Reidel, 1986
8. S. Albeverio: *Solvable Models in Quantum Mechanics.* New York: Springer 1988
9. M. Reed and B. Simon: *Methods of Modern Mathematical Physics, Vol. III*, p.83. New York: Academic 1979

Few-Body Systems Suppl. 12, 249–255 (2000)

Few-Body Systems

Spin Observables for Electromagnetic Nuclear Physics with Few-Body ystems

J.F.J. van den Brand

Department of Physics, Vrije Universiteit, de Boelelaan 1081, 1081 HV Amsterdam, The Netherlands

Abstract. We discuss aspects of the physics program with polarized electrons and polarized few-body systems in Amsterdam. Polarized electrons were injected into the AmPS storage ring at a beam energy of 720 MeV, while a Siberian snake was employed to preserve longitudinal polarization at the interaction point. Vector-polarized deuterium was produced by an atomic beam source and injected into an open-ended cylindrical cell, internal to the electron storage ring. The spin correlation parameter A^V_{ed} was measured for the $(e, e'p)$ and $(e, e'n)$ reactions at a four-momentum transfer squared of 0.21 $(\text{GeV}/c)^2$. The data are sensitive to the spin structure of the deuteron and allow to extract a value for the charge form factor of the neutron.

1 Introduction

Although the neutron has no net electric charge, it does have a charge distribution. Precise measurements [1] where thermal neutrons ¿from a nuclear reactor are scattered from atomic electrons indicate that the neutron has a positive core surrounded by a region of negative charge. The actual distribution is described by the charge form factor G^n_E, which enters the cross section for elastic electron scattering. It is related to the Fourier transform of the charge distribution and is generally expressed as a function of Q^2, the square of the four-momentum transfer. Data on G^n_E are important for our understanding of the nucleon and are essential for the interpretation of electromagnetic multipoles of nuclei, e.g. the deuteron.

Since a practical target of free neutrons is not available, experimentalists mostly resorted to (quasi)elastic scattering of electrons from unpolarized deuterium [2, 3] to determine this form factor. The shape of G^n_E as function of Q^2 is relatively well known from high precision elastic electron-deuteron scattering [3]. However, in this case the cross section is dominated by scattering ¿from the proton and, moreover, is sensitive to nuclear-structure uncertainties and reaction-mechanism effects. Consequently, the absolute scale of G_E^n still contains a systematic uncertainty of about 50%.

Many of the aforementioned uncertainties can be significantly reduced through the measurement of electronuclear spin observables. The scattering cross section with both longitudinal polarized electrons and a polarized target for the $^2\mathrm{H}(e,e'N)$ reaction, can be written as [4]

$$S = S_0 \left\{1 + P_1^d A_d^V + P_2^d A_d^T + h(A_e + P_1^d A_{ed}^V + P_2^d A_{ed}^T)\right\}, \qquad (1)$$

where S_0 is the unpolarized cross section, h the polarization of the electrons, and P_1^d (P_2^d) the vector (tensor) polarization of the target. A_e is the beam analyzing power. The vector and tensor target analyzing powers $A_d^{V/T}$ and spin-correlation parameters $A_{ed}^{V/T}$, depend on the orientation of the target spin. The polarization direction of the deuteron is defined by the angles Θ_d and Φ_d in the frame where the z-axis is along the direction of the three-momentum transfer ($\mathbf{q}$) and the y-axis is defined by the vector product of the incoming and outgoing electron momenta. $A_{ed}^V(\Theta_d = 90°, \Phi_d = 0°)$ contains an interference term, where the effect of the small charge form factor is amplified by the dominant magnetic form factor. Experiments with external beams have been carried out at Mainz and MIT. Here, we describe a measurement performed at NIKHEF (Amsterdam), which uses a stored polarized electron beam and a vector-polarized deuterium target.

2 Experimental Aspects

The experiment was performed with a polarized gas target internal to the AmPS electron storage ring. An atomic beam source (ABS) was used to inject a flux of 3×10^{16} deuterium atoms/s (in two hyperfine states) into the feed tube of a cylindrical storage cell cooled to 75 K. The cell had a diameter of 15 mm and was 60 cm long. An electromagnet was used to provide a guide field of 40 mT over the storage cell which oriented the deuteron polarization axis perpendicular to $\mathbf{q}$ in the scattering plane. A doublet of steering magnets around the target region compensated for the deflection of the electron beam by the guide field. In addition, two sets of four beam scrapers preceding the internal target were used to reduce events that originated from beam halo scattering from the cell. By alternating two high-frequency transitions in the ABS, the vector polarization of the target ($P_1^d = \sqrt{\frac{3}{2}}(n_+ - n_-)$), with $n_\pm$ the fraction of deuterons with spin projection ± 1) was varied every 10 seconds. Compared to our previous experiments with tensor-polarized deuterium [5, 6], this target setup resulted in an increase of the figure of merit by more than one order of magnitude, with a typical target thickness of 1×10^{14} deuterons/cm^2.

Polarized electrons were produced by photo-emission from a strained-layer semiconductor cathode (InGaAsP). After linear acceleration to 720 MeV the electrons were injected and stacked in the AmPS storage ring. The polarization of the stored electrons was maintained by setting the spin tune to 0.5 with a strong solenoidal field (using the well-known Siberian snake principle). Optimization of the longitudinal polarization at the interaction point was achieved by varying the orientation of the spin axis at the source and by measuring the polarization of the stored electrons with a polarimeter based on spin-dependent Compton backscattering. Scattered electrons were detected in the

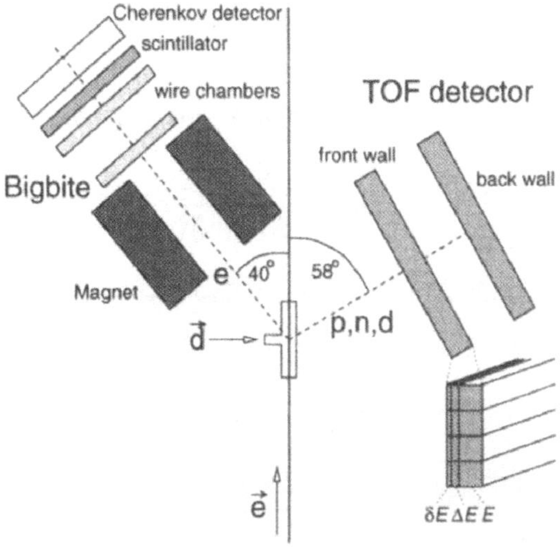

Figure 1. Layout of the detector setup. The electron spectrometer consists of a 1 Tm magnet, two drift chambers of four planes each, a scintillator and a Čerenkov detector. The time of flight system consists of two identical walls of four E-scintillators preceded by two veto scintillators. The second wall was used only for neutron detection.

large-acceptance magnetic spectrometer Bigbite with a momentum acceptance from 250 to 720 MeV/c and a solid angle of 96 msr (see Fig. 1). Kinematics were chosen such that G_E^n was probed near its maximum (as determined from Ref. [3]), resulting in the most sensitive measurement of G_E^n for a given statistical accuracy. Consequently, the electron detector was positioned at a central angle of 40°, with an acceptance for the electron scattering angle of $35° \leq \theta_e \leq 45°$, resulting in a central value of $Q^2 = 0.21(\mathrm{GeV}/c)^2$.

Neutrons and protons were detected in a time-of-flight (TOF) system made of two subsequent and identical scintillator arrays. Each array consisted of four 160 cm long, 20 cm high, and 20 cm thick plastic scintillator bars stacked vertically. Each bar was preceded by two (δE and ΔE) plastic scintillators (3 and 10 mm thick, respectively) of equal length and height, used to identify and/or veto charged particles. Each of the 24 scintillators was read out at both ends to obtain position information along the bars (resolution $\sim$ 4 cm) nd good coincidence timing resolution ($\sim$ 0.5 ns). The TOF detector was positioned at a central angle of 58° and covered a solid angle of about 250 msr. Protons with kinetic energies in excess of 40 MeV were detected with an energy resolution of about 10%. The $e'N$ trigger was formed by a coincidence between the electron arm trigger and a hit in any one of the eight TOF bars. By simultaneously detecting protons and neutrons in the same detector, one can construct asymmetry ratios for the two reaction channels ^{2}H$(e,e'p)n$ and ^{2}H$(e,e'n)p$, in this way minimizing systematic uncertainties associated with the deuteron ground-state wave function, absolute beam and target polarizations, and possible dilution by cell-wall background events.

3 Results

An experimental asymmetry (A_{exp}) can be constructed, via

$$A_{exp} = \frac{N_+ - N_-}{N_+ + N_-} \tag{2}$$

where $N_\pm$ is the number of events that pass the selection criteria, with hP_1^d either positive or negative. Data on A_{exp} were compared to the predictions of the model of Arenhövel *et al.* [4]. This advanced, non-relativistic model includes the effects from final-state interaction, meson-exchange currents, isobar configurations, and relativistic corrections, and has shown to provide good descriptions for quasifree proton knockout from tensor-polarized deuterium [6]. Finite acceptance effects were taken into account with a Monte Carlo code that interpolated the model predictions between a dense grid of calculations over the full kinematical range and detector acceptance.

Protons are selected by a valid hit in two PMT's of at least one E-bar and a valid hit in both PMT's of one of the preceding ΔE bars.

This requirement allowed us to use particle identification in order to discriminate between protons and either deuterons or pions. To select two-body breakup, the electron energy is required to be larger than 450 MeV, and the reconstructed missing energy between -50 MeV and 50 MeV. This requirements resulted in a clean sample of two-body breakup events, with some dilution of cell wall events.

This sample has been used (after the background correction, described earlier), to extract $A_V^{ed}(90°, 0°)$. Fig. 2 shows the experimental results, and compares it to the predictions [4] of the model of Arenhoevel et al.

At small missing momentum, the spin correlation function is equal to what is expected for elastic electron-proton scattering, while for increasing momenta, the predictions and the data show that the asymmetry reverses sign, as expected from the increasing D wave in the wave function of the deuteron.

It can also be observed, that inclusion of reaction-mechanism corrections, mainly for Isobar Configurations (IC), is required for a better description of the data. We have investigated the dependence of the predictions on the nucleon-nucleon potential for the Bonn, Paris, Argonne and Nijmegen-V14 potential. The effect is negligible for p_m < 200 MeV/c, and increases to 0.02 for p_m = 400 MeV/c, much smaller than the experimental accuracy or the reaction mechanism effects. The data suggests that the model overestimates A_V^{ed} around p_m = 200 MeV/c. This could be contributed to an underestimate of the D wave or the effects of IC. This is in agreement with earlier measurements of

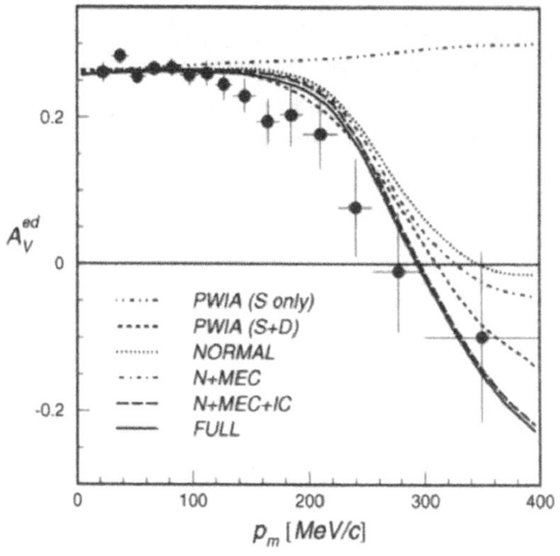

Figure 2. Beam-vector spin correlation parameter as function of missing momentum for the two-body breakup reaction at a four-momentum transfer of 0.21 (GeV/c)2. The dot-dot-dashed curve and dashed curve represent PWIA predictions with and without inclusion of the *D*-wave, respectively. The other curves are predictions of the model of Arenhoevel et al., for normal (dotted), normal+mec (dashed-dotted), normal+mec+ic (long-dashed) and total (full) calculations, as indicated in Ref. 4. The predictions are folded over the detector acceptance using Monte Carlo methods.

the deuteron quadrupole moment, unpolarized quasi-elastic electron-deuteron scattering, and measurements of T_{20}.

Neutrons were identified by a valid hit in one E-scintillator or two neighboring E-scintillators (to allow for events that deposit energy in two neighboring E-scintillators) and no hits in the preceding thin scintillators, which resulted in an 8- to 12-fold veto requirement. Minimum-ionizing particles and photons were rejected by a cut on the time of flight, resulting in a clean sample of neutrons, with only a small proton contamination. The spin-correlation parameter was obtained from the experimental asymmetry by correcting for the contribution of protons misidentified as neutrons (less than 1 %, as determined from a calibration with hydrogen), and for the product of beam and target polarization, as determined from the $(e, e'p)$ channel.

The data are compared to the predictions of the full model of Arenhoevel *et al.*, assuming the dipole parameterization for the magnetic form factor

of the neutron and the Paris nucleon-nucleon (NN) potential, folded over the detector acceptance with our Monte Carlo code for various values of G_E^n. Full model calculations are required for a reliable extraction of G_E^n. We extract $G_E^n(Q^2 = 0.21(\mathrm{GeV/c})2) = 0.066 \pm 0.015 \pm 0.004$, where the first (second) error indicates the statistical (systematic) uncertainty. The systematic error is mainly due to the uncertainty in the correction for misidentified protons and the orientation of the holding field (thus the contribution of the spin-correlation parameter $A_V^{ed}(0°, 0°)$ to our experimental asymmetry).

We have investigated the influence of the NN potential on the calculated spin-correlation parameters using Arenhoevel's full treatment. The results for $A_V^{ed}(90°, 0°)$ using the Paris, Bonn, Nijmegen, and Argonne V14 NN potential differ by less than 5 % for missing momenta below 200 MeV/c.

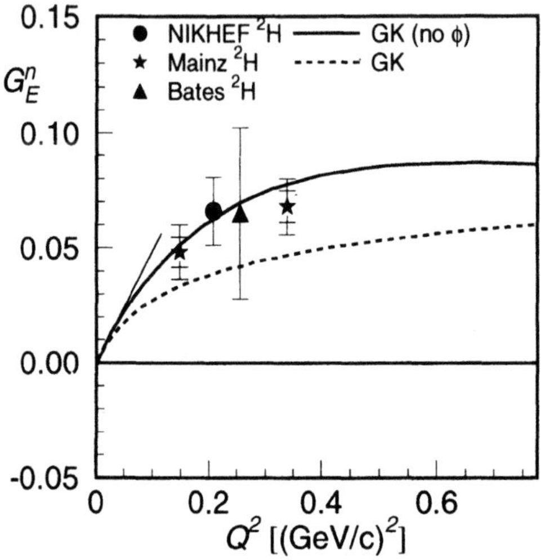

Figure 3. Data on the charge form factor of the neutron from double polarization experiments as function of squared four-momentum transfer. The solid circle shows our result[7], the triangle corresponds to the Bates result [8], the stars to the results obtained at Mainz [9]. Both data sets have been obtained on deuterium and have been analyzed with the full model of Arenhoevel et al. The curves represent theoretical predictions according to the VMD model of Gari and Krumpelmann, with and without the inclusion of the ϕ meson. The line at the origin represents the data on the slope of G_E^n.

In Fig. 3 we compare our experimental result [7] to other data btained with spin-dependent electron scattering from deuterium [8, 9]. The igure also shows the precize data on the charge radius of the neutron [1].

By comparison to the predictions of the QCD-VM model by Gari and Krumpelmann[10], with and without the inclusion of the coupling of the ϕ-meson to the nucleon (which these authors identify with the effect of strangeness in the neutron), our datum favors the prediction without strangeness in the neutron included.

4 Conclusion

In summary, we presented the first measurement of the sideways spin-correlation parameters $A_{ed}^V(90°, 0°)$ in quasifree electron-deuteron scattering.

We obtained information on the spin structure of the deuteron. In addition, we extract the neutron charge form factor at $Q^2 = 0.21$ (GeV/c)^2. When combined with the known value and slope [1] at $Q^2 = 0$ (GeV/c)^2 and the elastic electron-deuteron scattering data from Ref. [3], this result puts strong constraints on G_E^n up to $Q^2 = 0.7$ (GeV/c)^2.

Acknowledgement. I would like to thank the the 97-01 collaboration, and the NIKHEF and Vrije Universiteit technical groups for their outstanding support. This work was supported in part by the Stichting voor Fundamenteel Onderzoek der Materie (FOM), which is financially supported by the Nederlandse Organisatie voor Wetenschappelijk Onderzoek (NWO).

References

1. S. Kopecki et al.: Phys. Rev. Lett. **74** 2427 (1995); Phys. Rev. **C56**, 2229 (1997)
2. A. Lung et al.: Phys. Rev. Lett. **70**, 718 (1993)
3. S. Platchkov et al.: Nucl. Phys. **A510**, 740 (1990)
4. H. Arenhövel, W. Leidemann, E.L. Tomusiak: Z. Phys. **A331**, 123 (1988); Erratum, Z. Phys. **A334**, 363 (1989)
5. M. Ferro-Luzzi et al.: Phys. Rev. Lett. **77**, 2630 (1996); M. Bouwhuis et al.: Phys. Rev. Lett. **82**, 3755 (1999)
6. Z.-L. Zhou et al.: Phys. Rev. Lett. **82**, 687 (1999)
7. I. Passchier et al.: Phys. Rev. Lett. **82**, 4988 (1999)
8. T. Eden et al.: Phys. Rev. **C50**, R1749 (1994)
9. C. Herberg et al.: Eur. Phys. J. **A5**, 131 (1999)
10. M.F. Gari and W. Krumpelmann: Phys. Lett. **B274**, 159 (1992)

Few-Body Systems Suppl. 12, 256–262 (2000)

Few-Body Systems

Precise Measurement of the Transverse Asymmetry in Quasielastic $^3\vec{\mathrm{He}}(\vec{e}, e')$ and the Neutron Magnetic Form Factor

H. Gao*†

Laboratory for Nuclear Science and Department of Physics, Massachusetts Institute of Technology, Cambridge, MA 02139, U.S.A.

Abstract. We have measured the transverse asymmetry $A_{T'}$ in $^3\vec{\mathrm{He}}(\vec{e}, e')$ quasielastic scattering with high statistical precision for Q^2-values from 0.1 to 0.6 $(\mathrm{GeV/c})^2$. The data are expected to allow extraction of the neutron magnetic form factor G_M^n with an uncertainty similar to that of recent experiments on deuterium. Data analysis is currently in progress.

1 Introduction

Electromagnetic form factors are of fundamental importance for an understanding of the underlying structure of nucleons. Knowledge of the distribution of charge, magnetization within the nucleons provides a sensitive test of models based on Quantum Chromodynamics (QCD), as well as a basis for calculations of processes involving the electromagnetic interaction with complex nuclei. The understanding of the nucleon structure in terms of quark and gluon degrees of freedom of QCD will provide a basis to understand more complex strongly interacting matter at the level of quarks and gluons. While the proton form factors are known with excellent precision over a large range of four-momentum transfer Q^2, the corresponding data for the neutron are of inferior quality due to the lack of free neutron targets. Over the past decade, with the advent of improved experimental techniques, the precise determination of both the neutron electric form factor, G_E^n, and the magnetic form factor, G_M^n, has become a focus of experimental activity. While improving the precision of G_M^n is interesting in itself, it also benefits experiments designed to determine G_E^n, which

*For the Thomas Jefferson National Accelerator Facility (JLab) experiment E95-001 collaboration

†Supported in part by the U.S. Department of Energy, the National Science Foundation, and the French Atomic Energy Commission and National Center for Scientific Research.

usually measure the ratio G_E^n/G_M^n. Furthermore, precise data for the nucleon electromagnetic form factors are essential for the analysis of parity violation experiments designed to measure the strangeness content of the nucleon.

Until recently, most data on G_M^n had been deduced from elastic and quasielastic electron-deuteron scattering. For inclusive measurements, this procedure requires the subtraction of a large proton contribution and suffers from large theoretical uncertainties due to the deuteron model employed and corrections for final-state interactions (FSI) and meson-exchange currents (MEC), limiting the precision of G_M^n to ~20% at low Q^2. The proton subtraction is avoided in coincidence $d(e,e'n)$ experiments [1], and the sensitivity to nuclear structure can be greatly reduced by measuring the cross section ratio $d(e,e'n)/d(e,e'p)$ in quasielastic kinematics. Several recent experiments [2, 3, 4] have employed the latter technique to extract G_M^n with uncertainties of <2% in the momentum transfer range $Q^2 = 0.1$ to 0.8 (GeV/c)^2. While this precision is excellent, the results of the experiments [1, 2, 3, 4] are not fully consistent (cf. Figure 1). Furthermore, the two most precise data sets [3, 4] of G_M^n seem to suggest a very different Q^2 dependence. In addition to the existing data, several theoretical calculations of G_M^n are shown in Figure 1. The solid curve is an improved quark model calculation by Lu, Thomas and Williams [6] with a bag constant of 0.9 fm. The dotted curve is the minimal vector dominance model by Meissner [7]. The short and long dashed curves are the non-relativistic and relativistic quark model calculations by Eich [8] and Schlumpf [9], respectively. The dash-dotted line is a recent calculation by Mergell *et al.* [10] based on a fit of the proton data using dispersion theoretical arguments. These calculations clearly show very a different behavior in the low Q^2 region. Thus, further data are desirable to clarify the situation with respect to the discrepancies among different measurements and theoretical calculations.

An alternative approach to a precision measurement of G_M^n is the inclusive reaction $^3\vec{\mathrm{He}}(\vec{e},e')$. In comparison to deuterium experiments, this technique employs a different target and relies on polarization degrees of freedom. It is thus subject to completely different systematics. A first such experiment was done at MIT-Bates [5], and a result for G_M^n was extracted as shown in Figure 1. Here, we report on a continuation of these measurements recently performed at the Thomas Jefferson National Accelerator Facility (JLab), experiment E95-001.

Polarized ^{3}He is a good candidate for an effective neutron target because its ground state wave function is dominated by the S-state in which the proton spins cancel and the nuclear spin is entirely due to the neutron. The cross section for the process $^3\vec{\mathrm{He}}(\vec{e},e')$ can be written in terms of four nuclear response functions $R_K(Q^2,\nu)$ [11] as

$$\frac{d^2\sigma}{d\Omega dE'} = \sigma_{Mott}\left[v_L R_L + v_T R_T - h(\cos\theta^* v_{T'} R_{T'} + 2\sin\theta^* \cos\phi^* v_{TL'} R_{TL'})\right], \tag{1}$$

where θ^* and ϕ^* are the polar and azimuthal angles defining the direction of the target spin with respect to the momentum transfer vector $\vec{q}$ the v_K are

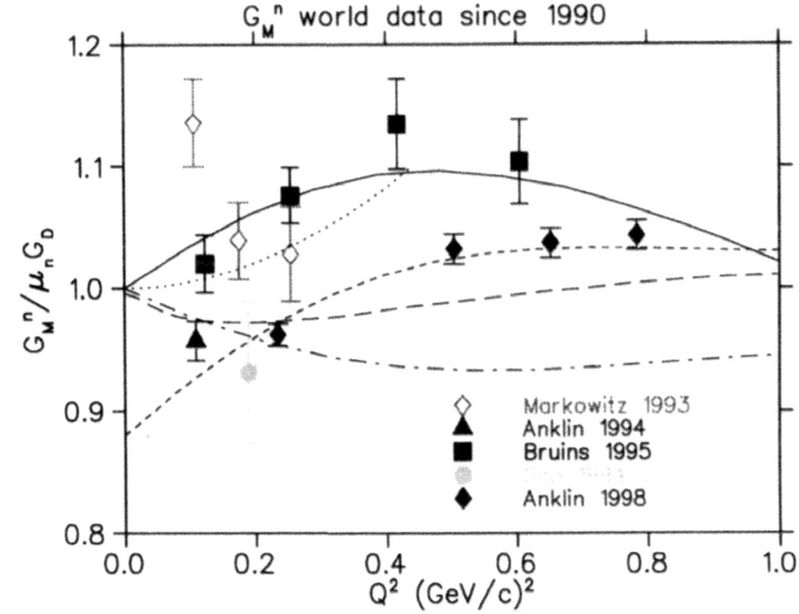

Figure 1. The neutron magnetic form factor G_M^n in units of the standard dipole parameterization, $\mu_n G_D$, in the low Q^2 region, as determined in several recent measurements: Markowitz *et al.* [1] (open diamonds) using $d(e, e'n)$; Anklin *et al.* [2] (triangle), Bruins *et al.* [3] (squares), and Anklin *et al.* [4] (solid diamonds) using the ratio $d(e, e'n)/d(e, e'p)$; and Gao *et al.* [5] (circle) using $^3\vec{\mathrm{He}}(\vec{e}, e')$.

kinematic factors, ν is the electron energy loss, h is the helicity of the incident electron, and $Q^2 \equiv \vec{q}^{\,2} - \nu^2$. $R_{T'}$ and $R_{TL'}$ are two response functions arising from the polarization degrees of freedom. An experimentally clean signature of these observables is the spin-dependent asymmetry, defined as

$$A = \frac{\sigma_+ - \sigma_-}{\sigma_+ + \sigma_-} = -\frac{\cos\theta^* v_{T'} R_{T'} + 2\sin\theta^* \cos\phi^* v_{TL'} R_{TL'}}{v_T R_T + v_L R_L}, \tag{2}$$

where the subscript + (−) refers to the electron helicity h. Orienting the target spin at $\theta^* = 0°$ selects the transverse asymmetry $A_{T'}$ proportional to $R_{T'}$. In plane wave impulse approximation (PWIA) the cross section for $^3\vec{\mathrm{He}}(\vec{e}, e')$ at the center of the quasielastic peak is (roughly) proportional to the sum of the *en* plus twice the *ep* elastic cross section. The cross section for polarized electron-nucleon scattering is

$$\frac{d\sigma_{eN}}{d\Omega} = \sigma_{Mott}\frac{E'}{E}\left[v_L(1+\tau)G_E^2 + v_{TL'}2\tau G_M^2 - h\,\bar{p}_N(\cos\theta^* v_{T'}2\tau G_M^2 \right.$$
$$\left. -2\sin\theta^*\cos\phi^* v_{TL'}\sqrt{2\tau(1+\tau)}G_M G_E)\right], \quad (3)$$

where $\tau = Q^2/(4M^2)$, and $\bar{p}_N$ is the effective nucleon polarization. As a consequence of the S-state dominance, the neutron in ^{3}He is almost fully polarized, $\bar{p}_n \approx 1$, at the quasielastic peak while the remaining small components of the ^{3}He ground state, the D state (~8%) and the mixed-symmetry S' state (~1%), give rise to a small net proton polarization of $\bar{p}_p \approx -0.03$ [12]. Combining the above equations, the transverse asymmetry can be written

$$A_{T'} \propto \frac{\sigma_n}{(\sigma_n + 2\sigma_p)}\bar{p}_n(G_M^n)^2 + \frac{2\sigma_p}{(\sigma_n + 2\sigma_p)}\bar{p}_p(G_M^p)^2, \quad (4)$$

where σ_n and σ_p are the electron-neutron and electron-proton elastic scattering cross sections, respectively. Since G_M^n and G_M^p are comparable in magnitude, but $|\bar{p}_p| \ll |\bar{p}_n|$, $A_{T'}$ is dominated by the neutron contribution and so is essentially proportional to $(G_M^n)^2$. Note that it is mostly the proton contribution that is sensitive to details of the ^{3}He ground state wave function.

The strong sensitivity of $A_{T'}$ to $(G_M^n)^2$ in quasielastic kinematics has been verified in a number of recent calculations [13, 14, 15, 16]. The most advanced of these include corrections for FSI [15] and FSI and MEC [16], which are relatively small at the quasielastic peak. One may conclude that $A_{T'}$ depends only weakly on the details of the ^{3}He nuclear ground state and the reaction mechanism. Thus, a precision measurement of $A_{T'}$ is suitable to extract precise information on G_M^n.

2 Experiment

The experiment was carried out in Hall A at JLab in early 1999 using a longitudinally polarized CW electron beam at energies of 0.778 and 1.727 GeV and 10 μA current. A high pressure ^{3}He gas target was polarized via spin-exchange optical pumping at a density of 2.5×10^{20} nuclei/cm^3. The effective target length seen by the spectrometers was 10 cm/sin θ_e, where θ_e is the electron scattering angle. To facilitate optical pumping, the target contained small admixtures of nitrogen (~10^{18} cm^{-3}) and rubidium (~10^{14} cm^{-3}). Background from the nitrogen was determined in calibration measurements using a reference cell which has the same dimensions as those of the ^{3}He target cell and is corrected for in the analysis; the contribution from rubidium is negligible. The beam and target polarizations were approximately 70% and 30%, respectively. The beam helicity was reversed randomly at a rate of 1 Hz. A total beam charge of approximately 22 C was accumulated, resulting in a total data set of 1.3×10^9 quasielastic events after background subtraction.

Six kinematic points were measured corresponding to $Q^2 = 0.1$ to 0.6 (GeV/c)^2 in steps of 0.1 (GeV/c)^2. To maximize sensitivity to $A_{T'}$, the target spin was oriented at $-62.5°$ with respect to the beam direction, resulting in a contribution to the asymmetry due to $R_{TL'}$ of less than 2% for all kinematic points. The target spin direction was rotated by 180° every 24-48 hours for systematic checks, causing the asymmetry to change sign. The beam helicity definition was also reversed every 24 hours for systematic checks associated with the electron beam helicity. The scattered electrons were observed in the two Hall A High Resolution Spectrometers, HRSe and HRSh. Both spectrometers were configured to detect electrons in single-arm mode using nearly identical detector packages consisting of two dual-plane vertical drift chambers for tracking, two planes of segmented plastic scintillators for trigger formation, and a CO_2 gas Cherenkov detector and Pb-glass total-absorption shower counter for particle identification. Pion background was rejected using the Cherenkov and shower counter information. The spectrometer momentum and angular acceptances were approximately ±4.5% and 5.5 msr, respectively. The level of background from the walls of the glass target was measured at regular intervals with the target cell empty and was less than about 5% of the full target yield. The HRSe was set for quasielastic kinematics while the HRSh detected elastically scattered electrons. The elastic asymmetry can be calculated to better than 2% using the well-known elastic form factors of ^{3}He [17], and so the elastic measurement allows precise monitoring of the product of beam and target polarizations. Standard Møller and NMR polarimetry was also performed and served as a cross-check.

3 Expected Results

A statistical precision in $A_{T'}$ of $\sim$ 2.0% was achieved for each Q^2 point in a ±15 MeV bin around the center of the quasielastic peak. This precision is better by about a factor of five that that of our previous experiment on ^{3}He [5] at $Q^2 = 0.19$ (GeV/c)^2. Extraction of G_M^n from the data requires the use of theoretical calculations. Currently the Bochum-Krakow group [16] is carrying out extensive calculations of $A_{T'}$ as a function of G_M^n for the kinematics of this experiment which include the FSI and MEC effects. This is the state-of-the-art three-body calculation. In addition, an independent calculation of the ^{3}He quasielastic asymmetry which will include FSI and MEC effects is currently in progress [18]. The Bochum-Krakow calculation will be convoluted with the experimental acceptances, and G_M^n will be determined using a best fit of $A_{T'}(G_M^n)$ to the data in the vicinity of the quasielastic peak. Fig. 2 shows the expected precision for G_M^n from this experiment. The data also allow a detailed analysis of the dependence of $A_{T'}$ on the electron energy transfer ν. The regions away from the quasielastic peak are expected to be sensitive to details of the reaction mechanism. Thus, the shape of $A_{T'}(\nu)$ can be used to constrain calculations that include FSI and MEC corrections. Data analysis is currently in progress and results are expected in late 1999.

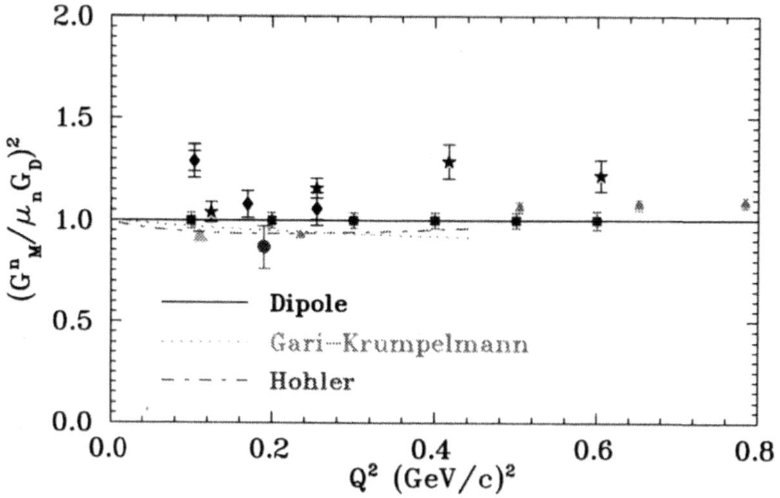

Figure 2. The neutron magnetic form factor ${G_M^n}^2$ in units of the standard dipole parameterization, $\mu_n {G_D}^2$, in the low Q^2 region, as determined in several recent measurements. The projected results from this experiment (E95-001) are shown as solid squares using the $^3\vec{\mathrm{He}}(\vec{e}, e')$ process.

References

1. P. Markowitz et al.: Phys. Rev. **C48**, R5 (1993)
2. H. Anklin et al.: Phys. Lett. **B336**, 313 (1994)
3. E.E.W. Bruins et al.: Phys. Rev. Lett. **75**, 21 (1995)
4. H. Anklin et al.: Phys. Lett. **B428**, 248 (1998)
5. H. Gao et al.: Phys. Rev. **C50**, R546 (1994)
6. D.H. Lu, A.W. Thomas, A.G. Williams: Phys. Rev. **C57**, 2628 (1998)
7. U.-G. Meissner: Phys. Rep. **161**, 213 (1988)
8. E. Eich: Z. Phys. **C45**, 627 (1988)

9. F. Schlumpf: J. Phys. **G20**, 237 (1994)

10. P. Mergell, U.-G. Meissner, D. Dreshel: Nucl. Phys. **A596**, 367 (1996)

11. T.W. Donnelly and A.S. Raskin: Ann. Phys. (N.Y.) **169**, 247 (1986)

12. J.L. Friar et al.: Phys. Rev. **C42**, 2310 (1990)

13. R.-W. Schulze and P.U. Sauer: Phys. Rev. **C48**, 38 (1993)

14. C. Ciofi degli Atti, E. Pace, G. Salmè: Phys. Rev. **C51**, 1108 (1995)

15. S. Ishikawa et al.: Phys. Rev. **C57**, 39 (1998); and private communication.

16. J. Golak: private communication

17. A. Amroun et al.: Nucl. Phys. **A579**, 596 (1994)

18. G. Salmè: private communication

Few-Body Systems Suppl. 12, 263–268 (2000)

Few-Body Systems

The Off-Shell Nucleon-Nucleon Amplitude: Why it is Unmeasurable in Nucleon-Nucleon Bremsstrahlung

H.W. Fearing

TRIUMF, 4004 Wesbrook Mall, Vancouver, B.C., Canada V6T 2A3

Abstract. Nucleon-nucleon bremsstrahlung has long been considered a way of getting information about the off-shell nucleon-nucleon amplitude which would allow one to distinguish among nucleon-nucleon potentials based on their off-shell properties. There have been many calculations and many experiments devoted to this aim. We show here, in contrast to this standard view, that such off-shell amplitudes are not measurable as a matter of principle. This follows formally from the invariance of the S-matrix under transformations of the fields. This result is discussed here and illustrated via two simple models, one applying to spin zero, and one to spin one half, processes. The latter model is very closely related to phenomenological models which have been used to study off-shell effects at electromagnetic vertices.

Almost exactly fifty years ago Ashkin and Marshak [1] proposed using nucleon-nucleon bremsstrahlung to get information about the off-shell parts of the nucleon-nucleon amplitude. Since then there have been many calculations, mostly in non-relativistic potential models, of this process and a number of experiments. More recently a number of new experiments have been designed with the express purpose of getting further off shell and thus providing a more sensitive test of this picture.

The central point of this talk is to show that this historical motivation for nucleon-nucleon bremsstrahlung is incorrect. In actual fact the off-shell amplitude is not a physically well defined quantity and *as a matter of principle* cannot be measured in nucleon-nucleon bremsstrahlung, or any other process.

Let us begin with some general background. The process of interest is $p + p \to p + p + \gamma$. Since this process involves an intermediate virtual particle, it in principle provides access to an off-shell amplitude. It is interesting because the nucleon-nucleon interaction is perhaps the most fundamental and certainly most accessible of the strong interactions. Modern potentials describe the elastic scattering well but in principle produce different off-shell matrix elements. One can not measure these directly, but must have some other process to put the particle back on shell. The well known electromagnetic interaction is the obvious choice. Thus in this normal, but incorrect, scenario, measurement of $p + p \to p + p + \gamma$ in kinematic regions far enough off shell will determine the half off-shell T-matrix and distinguish among potentials.

To understand the problem with this approach we first review the standard potential model description of bremsstrahlung. First observe that in an abstract mathematical sense the off-shell amplitude is well defined as the solution of the Lippman-Schwinger equation for off-shell momenta. It is only part of the full physical bremsstrahlung process however. To get the full process standard potential models include first the so called external radiation graphs in which a photon is emitted from an external leg. These involve off-shell effects both at the strong vertex and at the electromagnetic vertex. Usually the double scattering term in which the photon is emitted from a line in between two full scattering T-matrices is also included. Finally a few 'contact' graphs are included in an ad hoc fashion. These may include diagrams with Δ's or the one with a $\pi\rho\gamma$ or $\pi\omega\gamma$ vertex. Modern nucleon-nucleon potentials are built from diagrams with multi meson exchanges, and in principle diagrams with photons attached to any interior charged line should also be included. There are many such 'contact' diagrams which are irreducible and thus can not be included in the double scattering or external radiation graphs. In potential models, these are just dropped.

At the very simplest, qualitative level one can begin to understand the ambiguities inherent in the off-shell amplitude as follows. The amplitude for a typical bremsstrahlung external radiation graph can be written as

$$\{T_{on} + (p^2 - m^2)T_{off}\} \quad \frac{i}{p^2 - m^2} \quad \Gamma_{em}$$

corresponding to an on-shell part and an off-shell part of the nucleon-nucleon interaction, together with propagator and electromagnetic interaction. However, since the factor $p^2 - m^2$ appearing in the off-shell part exactly cancels the same factor appearing in the propagator, the amplitude can also be written as

$$T_{on} \quad \frac{i}{p^2 - m^2} \quad \Gamma_{em} \quad + \quad T_{off} i \Gamma_{em}$$

which has the form of an external radiation graph with on-shell amplitude plus a contact term. Thus the off-shell terms can be written alternatively as contact terms, and there will always be an ambiguity in how one separates from the full measurable process the part associated with an off-shell amplitude.

At a more fundamental level this ambiguity can be related to a theorem [2] which tells us that a transformation can be made on the fields appearing in a Lagrangian without changing the S-matrix elements, which correspond to the physical, measurable quantities. Such transformations generally change the 'equation of motion' terms in the Lagrangian, which are those leading to the $p^2 - m^2$ factor which gives the off-shell part of the vertex functions.[3] This means that such transformations can be used to change the off-shell T-matrix for the elastic process in an arbitrary fashion, without changing the full bremsstrahlung amplitude. Thus one can never measure an off-shell amplitude since the measurable quantity, the bremsstrahlung cross section, corresponds to an infinite number of different off-shell amplitudes.

We can summarize what we have learned so far as follows. There will always be some ambiguity as to what is called off-shell and what is called contact. At a more formal level field transformations can change the coefficient of the off-shell amplitude without changing the measurable bremsstrahlung. Thus the off-shell amplitude is not measurable as there are an infinite number of different such amplitudes which give the same bremsstrahlung amplitude.

Why wasn't this noticed in the fifty years of work on bremsstrahlung? Perhaps it was because potential model calculations tend to drop the contact terms from the beginning and because field transformation concepts, though well known in some areas of physics, are quite different from the techniques used in most non relativistic potential model approaches.

We turn now to a couple of simple field theory models which will illustrate the concepts so far described only in a qualitative way. Consider first a model of the spin zero bremsstrahlung process, e.g. $\pi^+ + \pi^0 \to \pi^+ + \pi^0 + \gamma$, described in detail in ref. [4]. We take as the Lagrangian for this process, $\mathcal{L} = \mathcal{L}_2 + \mathcal{L}_4^{GL} + \Delta\mathcal{L}_4$, where $\mathcal{L}_2 + \mathcal{L}_4^{GL}$ is the usual chiral perturbation theory Lagrangian to $\mathcal{O}(p^4)$ [5] written in terms of χ, U, which contain the fields, and of the covariant derivative D_μ. The last term involves the equation of motion and is given by the expression [6]

$$\Delta\mathcal{L}_4 = \beta_1 Tr\left(\mathcal{O}\mathcal{O}^\dagger\right) + \beta_2 Tr\left[(\chi U^\dagger - U\chi^\dagger)\mathcal{O}\right].$$

Here $\mathcal{O} = 0$ is the equation of motion, with $\mathcal{O}$ a somewhat complicated function also of χ, U, and D_μ

This Lagrangian generates a simple effective field theory which allows exact calculations to a given order. Though it was originally motivated by chiral perturbation theory, the result to be obtained from it has absolutely nothing to do with chiral perturbation theory.

Using this Lagrangian we can calculate the elastic amplitude for the process $\pi^+(p_1) + \pi^0(p_2) \to \pi^+(p_3) + \pi^0(p_4)$, obtaining a result given (schematically) as

$$\Gamma_{4\pi} = T_{on} + (....)(\Lambda_1 + \Lambda_3) + (....)\beta_2(\Lambda_1 + \Lambda_3) + (....)\beta_1\Lambda_1 + (....)\beta_1\Lambda_3$$

where (....) are known factors and $\Lambda_i = p_i^2 - m_\pi^2$. This is the exact analog of the off-shell elastic amplitude one would calculate from a potential by solving the Lippman-Schwinger equation. The off-shell part is proportional to Λ_1 or Λ_3 and for the most part to the parameters β_1 and β_2. The on-shell amplitude is obtained by taking the limit $\Lambda_1, \Lambda_3 \to 0$

Now what happens when we apply a field transformation to the fields in this Lagrangian? To that end consider the transformation $U' \to \exp(iS)U$, for arbitrary α_1, α_2, where S is given by

$$S = \frac{4i}{F_0^2}\left[\alpha_1\mathcal{O} - \alpha_2\left(\chi U^\dagger - U\chi^\dagger - \frac{1}{2}Tr(\chi U^\dagger - U\chi^\dagger)\right)\right].$$

Under this transformation $\mathcal{L}_2(U') \to \mathcal{L}_2(U) + \delta\mathcal{L}_2(U)$ with

$$\delta\mathcal{L}_2 = \alpha_1 Tr(\mathcal{O}\mathcal{O}^\dagger) + \alpha_2 Tr((\chi U^\dagger - U\chi^\dagger)\mathcal{O})$$

Clearly $\delta\mathcal{L}_2(U) \sim \Delta\mathcal{L}_4$. Thus what the transformation does, since α_1, α_2 are arbitrary, is to arbitrarily change the values of the coefficients β_1, β_2 in the Lagrangian, i.e. to change the coefficients which give the off-shell elastic amplitude. By the general result [2] such transformation does not change the full measurable amplitude, though it clearly changes the off-shell part of the elastic amplitude.

Now let us apply this model to bremsstrahlung. The external radiation diagrams can be obtained from the off-shell elastic amplitude, a propagator, and the electromagnetic $\pi\pi\gamma$ vertex, all consistently calculated using the model Lagrangian. The result is [4] in abbreviated form,

$$\begin{aligned} M_3 + M_1 &= T_{on}\left(\frac{\epsilon \cdot p_3}{k \cdot p_3} - \frac{\epsilon \cdot p_1}{k \cdot p_1}\right) + \text{constant} \\ &+ \beta_1(\ldots\text{A}\ldots) + \beta_2(\ldots\text{B}\ldots) \end{aligned}$$

where the A and B pieces are known kinematic dependent factors.

$M_3 + M_1$ is the analogue of what would be calculated in a potential model from the half off-shell T-matrix, a non-relativistic propagator, and an electromagnetic vertex. In a potential model it would be compared with data to extract the off-shell information 'contained' in β_1 and β_2. But, this approach has to be wrong because field transformations change the value of β_1 and β_2 arbitrarily without changing the value of the bremsstrahlung amplitude. In fact the bremsstrahlung amplitude must be independent of β_1 and β_2 since 0 is one possible value.

To understand what is happening, consider the contact term, which can be calculated explicitly in this model, unlike in potential model calculations.

$$\Gamma_{4\pi\gamma} = \text{const.} - \beta_1(\ldots\text{A}\ldots) - \beta_2(\ldots\text{B}\ldots)$$

Comparison of the contact term with $M_3 + M_1$ shows that all terms involving the parameters β_1, β_2, which govern the strength of the off-shell amplitude, cancel in the full result for bremsstrahlung which is thus independent of β_1 and β_2 as it must be. A similar result is obtained when this Lagrangian is used for Compton scattering.[7]

The fragment of the amplitude, $M_1 + M_3$, which is analogous to the usual potential model bremsstrahlung result, does however depend on β_1, β_2 and if one considers it alone, one is led to the (spurious) conclusion that off-shell amplitudes can be measured by measuring bremsstrahlung.

Let us now consider a different model [8], one applicable to spin one-half particles. This model corresponds very closely to the types of phenomenological models used in calculations investigating off-shell effects at electromagnetic vertices, e.g. [9]. The prototype reaction will be $p + n \to p + n + \gamma$, i.e. proton-neutron bremsstrahlung. We consider the p-n system, rather than the p-p system just to reduce the number of diagrams to be considered and to avoid the extra algebra required for identical particles.

We thus start with the Lagrangian,

$$\mathcal{L}_0 = \overline{\Psi}(i\not{D} - m)\Psi - \frac{e\kappa}{4m}\overline{\Psi}\sigma_{\mu\nu}F^{\mu\nu}\Psi + g\overline{\Psi}\Psi\overline{\Phi}\Phi. \tag{1}$$

Here D is the covariant derivative, e and κ are the proton charge and anomalous magnetic moment, A_μ, $F_{\mu\nu}$ are photon field and field strength tensor, and $\Psi \sim$ proton, $\Phi \sim$ neutron. This Lagrangian leads to the standard electromagnetic coupling to a spin one-half particle, the standard free equation for the proton and to a standard, but simplified, scalar boson exchange form for the strong interaction. However it generates no off-shell effects to lowest order.

Now consider a field transformation on this Lagrangian of the form

$$\Psi' \to \Psi + \tilde{a}g\overline{\Phi}\Phi\Psi + \tilde{b}e\sigma_{\mu\nu}F^{\mu\nu}\Psi$$

where $\tilde{a}$ and $\tilde{b}$ are arbitrary constants This transformation generates a new Lagrangian

$$\mathcal{L}_0(\Psi') = \mathcal{L}_0(\Psi) + \Delta_1\mathcal{L}(\Psi) + \Delta_2\mathcal{L}(\Psi)$$

The new piece of the Lagrangian $\Delta_1\mathcal{L}(\Psi)$ generates an off-shell contribution to the strong amplitude proportional to $\tilde{a}$ and an off-shell part of the electromagnetic vertex proportional to $\tilde{b}$ as well as a contact term necessary for gauge invariance. $\Delta_2\mathcal{L}(\Psi)$ generates only contact terms.

If we now take $\mathcal{L}_0(\Psi)+\Delta_1\mathcal{L}(\Psi)$ as our Lagrangian we have something corresponding very closely to the Lagrangian used in phenomenological calculations such as those of [9, 10]. It produces off-shell contributions at both strong and electromagnetic vertices. One could compare the predictions of this Lagrangian for bremsstrahlung with data and in principle extract values of the parameters $\tilde{a}$ and $\tilde{b}$. However this does not give any information about off-shell nucleon-nucleon amplitudes. Since the change in the Lagrangian, $\Delta_1\mathcal{L}(\Psi) + \Delta_2\mathcal{L}(\Psi)$, originates in a field transformation it can not affect the bremsstrahlung amplitude. This means that the amplitude using the full Lagrangian must be independent of the parameters $\tilde{a}$ and $\tilde{b}$, which can be verified explicitly. As a corollary, the Lagrangian $\mathcal{L}_0(\Psi) - \Delta_2\mathcal{L}(\Psi)$ must give exactly the same bremsstrahlung result as $\mathcal{L}_0(\Psi) + \Delta_1\mathcal{L}(\Psi)$. But the former Lagrangian contains only contact terms and no off-shell effects while the latter has mainly off-shell effects. One could also use any linear combination of these two Lagrangians. Thus there are an infinite number of Lagrangians, with different proportions of off-shell effects and contact terms which give the same bremsstrahlung amplitude. So again we see that the concept of an off-shell amplitude as a part of a bremsstrahlung amplitude is not well defined and such off-shell amplitudes are not measurable.

Let us now summarize what has been learned. Field transformations change the coefficients of the off-shell part of the elastic amplitude without changing the bremsstrahlung amplitude. In effect such transformations simply move terms back and forth between the bremsstrahlung diagrams containing off-shell amplitudes and the contact terms. This means that the off-shell contributions are not uniquely defined and therefore that the 'off-shell amplitude' is not measurable in bremsstrahlung reactions, contrary to historical expectations. Hence the claim that measuring bremsstrahlung will distinguish among potentials on the basis of their off-shell behavior is just not valid.

This result means that most previous calculations and experiments dealing with nucleon-nucleon bremsstrahlung were driven by an aim – to measure off-shell effects – which is just not possible. Does this make nucleon-nucleon bremsstrahlung any less interesting? The answer is clearly no. What emerges from these results is the importance of the contact terms. These contact terms originate from the photon probe of the currents internal to the strong interaction. To understand them in detail one must understand these interactions at a microscopic level, which is probably much more interesting and gives us much more insight about the physical mechanisms involved, than does understanding gross properties of a phenomenological potential.

This means however that for the future we need both comprehensive experiments with enough data over a wide enough kinematic range to distinguish details of the process and also microscopic calculations which include specifically details of the contact terms.

Finally nothing here depends specifically on the bremsstrahlung process, and so it is probable that off-shell amplitudes are unmeasurable in any process. Thus calculations purporting to show sensitivity to off-shell amplitudes should be viewed with suspicion.

The author would like to thank Stefan Scherer who has collaborated on much of what is reported here. The work was also supported in part by a grant from the Natural Sciences and Engineering Research Council of Canada.

References

1. J. Ashkin and R. Marshak: Phys. Rev. **76**, 58 (1949); Erratum, *ibid*, 989
2. R. Haag: Phys. Rev. **112**, 669 (1958); J.S.R. Chisholm: Nucl. Phys. **26**, 469 (1961); S. Kamefuchi, L. O'Raifeartaigh, A. Salam: Nucl. Phys. **28**, 529 (1961)
3. S. Weinberg: *Quantum Theory of Fields*, vol.1, p.331. Cambridge Univ. Press 1995
4. H.W. Fearing: Phys. Rev. Lett. **81**, 758 (1998)
5. J. Gasser and H. Leutwyler: Ann. Phys. **158**, 142 (1984)
6. T.E. Rudy, H.W. Fearing, S. Scherer: Phys. Rev. **C50**, 447 (1994)
7. S. Scherer and H. W. Fearing: Phys. Rev. **C51**, 359 (1995)
8. H.W. Fearing and S. Scherer: TRI-PP-99-30, nucl-th/9909076
9. E.M. Nyman: Nucl. Phys. **A154**, 97 (1970); **A160**, 517 (1971)
10. S. Kondratyuk, G. Martinus, O. Scholten: Phys. Lett. **B418**, 20 (1998)

Few-Body Systems Suppl. 12, 269–274 (2000)

Few-Body Systems

Probing Few-Body Systems with Bremsstrahlung

N. Kalantar-Nayestanaki*

Kernfysisch Versneller Instituut (KVI), Zernikelaan 25, 9747 AA Groningen, The Netherlands

Abstract. A series of bremsstrahlung measurements have been performed with the superconducting cyclotron, AGOR, at KVI. These measurements, on the proton-proton and proton-deuteron systems, aim to investigate the nucleon-nucleon interaction as the nucleons go off their mass shell. Cross sections and analyzing powers have been measured for all possible exit channels of both systems with an incident proton energy of 190 MeV, which is below the particle-production threshold. Some results of the real-photon production are presented for both systems and compared to theoretical predictions.

1 Introduction

Nucleon-nucleon bremsstrahlung is the most fundamental reaction used in studying the off-shell behavior of the nucleon-nucleon interaction as it involves the strong interaction between two nucleons, and the well-known electromagnetic interaction. In 1958 Low proved the soft-photon theorem, which states, that in a series expansion in the photon momentum, the first two terms of the scattering amplitude can be expressed exactly in terms of the elastic amplitude for the same system [1]. Current potential models describe the elastic data very accurately and therefore also soft-photon production. To test the off-shell behaviour of the potentials, one has to go beyond the validity regime of the Low expansion and thus investigate high-energy bremsstrahlung.

Recently, the study of nucleon-nucleon bremsstrahlung below the particle-production threshold has been receiving increasing attention both theoretically and experimentally. The reason for this is the success of modern potential-model calculations on the one hand and the advent of high-precision detection systems capable of processing high count rates on the other. Polarized-proton beams and high-flux neutron beams have also delivered a boost to the experimental efforts. Cross-section and analyzing-power data from TRIUMF [2] clearly show the need for the inclusion of off-shell effects in the calculations. With even

* *E-mail address:* nasser@kvi.nl

higher accuracies achievable nowadays, one can hope to study, in detail, the properties of the off-shell nucleon as it propagates in an intermediate state as well as the non-nucleonic degrees of freedom (such as the Δ–excitation) which can even admix at energies below the pion-production threshold. In this contribution, after a brief review of nucleon-nucleon bremsstrahlung in section 2, the experimental setup used in the KVI measurements is sketched in section 3. The analysis method along with the first results are discussed in section 4. A summary with some outlook are presented in the last section.

2 Nucleon-nucleon Bremsstrahlung and the KVI Experiment

The dominant ingredients in nucleon-nucleon bremsstrahlung are radiation from external and internal nucleon lines. The higher-order effects which have been investigated in the past few years include meson-exchange currents (MEC), the delta contributions and relativistic contributions. The Δ-isobar contribution in the intermediate state and the MEC effects have been studied in detail by several people [3, 4, 5, 6]. Even though all these contributions are relatively small, precise measurements and selected kinematics should distinguish between models with or without them. As a result of the absence of the electric dipole radiation and electric MECs in the proton-proton system, one is more sensitive to higher-order effects in this system.

The experimental situation on the proton-neutron bremsstrahlung is quite different than that of the proton-proton system. Generally speaking, the neutron beams lack the high quality and intensity required for precise measurements. Because of this, the cross sections presented in the literature have been inclusive. In order to investigate the proton-neutron bremsstrahlung, the proton-deuteron system was chosen in the past for which no good calculations exist. Any calculation attempting to describe the inclusive results which have been published in the literature [7] should treat all the possible outgoing channels of this reaction properly. These are: radiative capture, three- and four-body final states.

The KVI bremsstrahlung program with AGOR on few-body systems, set out to study and measure the cross sections and analyzing powers of all the above-mentioned reactions in a series of measurements specifically tuned for nucleon-nucleon bremsstrahlung. The measurements started with the proton-proton system and extended to the proton-deuteron system in which all the different final states were studied. In addition to coplanar geometry, non-coplanar geometries have been measured for a large number of angle combinations. Angular coverage of protons is from 6° to 26° and photons were measured from 60° to 170°. All the measurements are exclusive in the sense that all the outgoing particles were measured. Besides, all the measurements were carried out using polarized protons. In this paper, some experimental results on the proton-proton system will be presented along with some recent theoretical calculations. In addition, first results on proton-deuteron system will be presented.

3 Experimental Apparatus

To detect the outgoing hadrons at forward angles, the Small-Angle Large-Acceptance Detector, SALAD, was designed and built. This detector which was built for a precise measurement of the absolute cross sections and analyzing powers possesses an almost cylindrical symmetry and detects particles between 6° and 19° for the whole azimuthal angular range and proton energies of up to 135 MeV with a resolution of about 10%. The upper limit of the range of the polar angles increases to about 26° for a smaller region of the azimuthal angles. The large solid angle (about 400 msr) is mandated by the very small cross sections of the nucleon-nucleon bremsstrahlung process. To allow the beam free passage, our detector was designed with a central hole. The detector and the performance of various parts of it have been discussed in detail in several publications [8, 9, 10]. We used a liquid hydrogen/deuterium target with very thin synthetic foils for these measurements eliminating the need for a good target-window study [11].

For the measurement of the position and the energy of the photons, we used the Two-Arm Photon Spectrometer, TAPS [12], which was placed primarily at backward angles.

4 Results

For the analysis of the experiments, the fact that most measured channels were kinematically overdetermined, was used for background rejection. For instance, in the proton-proton case, five measured angles, namely the polar and azimuthal angles, θ and ϕ, of the two outgoing protons and the polar angle of the photon, θ_γ, and the incident proton energy were used to reconstruct the other parameters. By comparing these parameters to their measured counterparts most of the background left was rejected. In this section, some results are presented for the $p+p$ and $p+d$ systems.

For all the measurements performed, a fraction of the elastically-scattered particles were recorded to be used in the determination of the luminosity and the polarization of the beam as well as for calibration purposes.

As was mentioned earlier, our detection system is capable of measuring a very large number of angle combinations of the outgoing particles. Here, we only show a few percent of the available real-bremsstrahlung phase space. In figure 1, the typical results of the coplanar cross sections and analyzing powers are presented as a function of the polar angle of one of the protons, while the polar angle of the other proton and the polar angle of the photon are fixed [13]. Systematic errors on the cross sections are around 5% for the present measurements. For comparison, a data point from an earlier TRIUMF experiment at 200 MeV is also shown in the figure [14].

Three different types of calculations were performed for our kinematics. The first one uses the Soft-Photon Approximation (SPA) of Liou, Timmermans and Gibson [15]. This calculation is consistent with the low-energy theorem and is relativistic and gauge invariant. The second set of calculations is done by

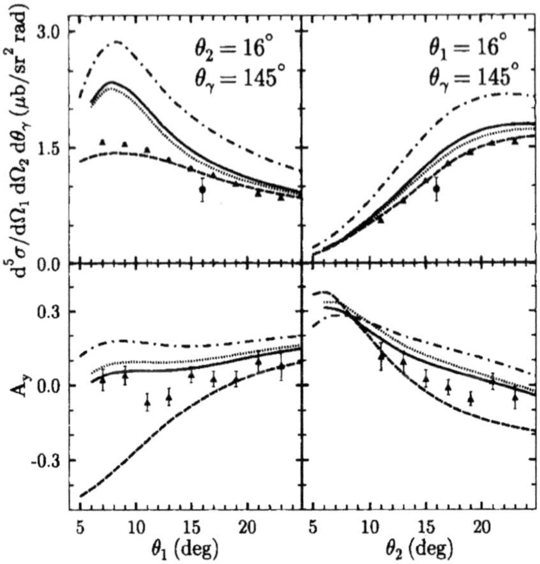

Figure 1. Cross sections and analyzing powers as a function of first (left) and second (right) proton polar angles for $\theta_\gamma = 145°$. The second (first) proton angle is fixed at 16°, respectively. The calculations are SPA (dash), the "nucleonic calculation" (dotted), the full calculation (solid), and another full calculation by Eden and Gari (dash-dotted). The circle data point is from TRIUMF for E_p=200 MeV [14].

Martinus, Scholten and Tjon [6]. These relativistic, microscopic calculations are covariant, thereby including the negative-energy states. To see the higher order effects, a calculation called the "nucleonic calculation" is shown in which only the internal and external bremsstrahlung (rescattering) contributions are considered, see the dotted curve, and one which includes MEC, $N\bar{N}$ and Δ contributions as well (solid curve). As can be seen in the figure, there are sizable differences between the data and the calculations for the cross sections and analyzing powers (especially the ones shown in the upper-left panel). A third semi-relativistic, microscopic calculation is done by Eden and Gari which is based on the non-relativistic Ruhrpot potential (dash-dotted curve in Figure 1) [3]. The big differences between the two microscopic calculations are not yet understood. In should be noted that Coulomb corrections are expected to be negligible for the kinematics considered here. Aside from our measurements, results from other laboratories have emerged on this system [16, 17]. The results for the virtual-photon production which, for the first time, was measured in parallel have already been published [18].

Finally, the preliminary results for one of the proton-deuteron channels, namely the "coherent" channel in which the outgoing channel consists of a proton, a deuteron and a photon, is presented. Figure 2 shows the cross sections as a function of photon angle for a number of proton-deuteron angle combinations, indicated in the figure. Note that to this date, there are no calculations available in the literature for this channel.

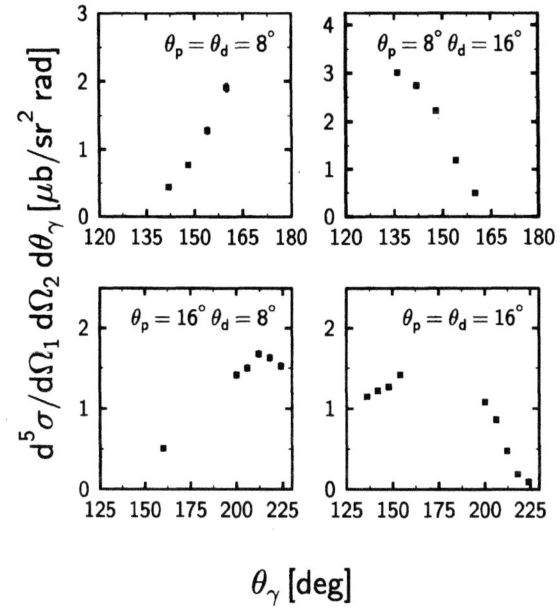

Figure 2. *Preliminary* cross sections for the "coherent" proton-deuteron bremsstrahlung as a function of θ_γ for four different angle combinations (given in the figures).

5 Summary and Outlook

In this contribution, the most complete bremsstrahlung measurements on the $p+p$ and $p+d$ systems have been outlined. All the measurements were done with the new superconducting cyclotron, AGOR, at KVI with a polarized proton beam with an energy of 190 MeV. The detection systems SALAD (specially built for these measurements) and TAPS were utilized for these investigations in which all the outgoing particles were measured. For the first time, large-area liquid targets contained within very thin foils were used for these systems eliminating the need for a good target-window study. Since all the outgoing channels of these reactions are kinematically overdetermined, the background levels were reduced to a minimum allowing the extraction of cross sections and analyzing powers which are extremely clean. The elastic cross sections serve as a monitor of the integrity of the system and as a luminosity monitor.

Very high-accuracy cross sections are obtained for the real-photon production which do not agree with the most recent calculations which attempt to include all the possible effects. For the proton-deuteron system, data with high accuracy have been obtained. These data should guide the theorists in their calculations for this simple system.

There is still a large part of the phase space which has been measured and for which the results are emerging. Data with high statistical accuracies from non-coplanar geometries are being looked at which will certainly produce interesting physics and new observables, never measured before.

I would like to acknowledge the assistance of all those who helped build and commission SALAD, took part in all the preparations and data taking and are involved in the analysis. Here, I should thank my graduate students H. Huisman and M. Volkerts who are responsible for the analysis of the data presented here.

References

1. F.E. Low: Phys. Rev. **110**, 974 (1958) and E.M. Nyman: Phys. Lett. **B25**, 135 (1967); Phys. Rev. **170**, 1628 (1968)
2. K. Michaelian et al.: Phys. Rev. **D41**, 2689 (1990)
3. J.A. Eden and M.F. Gari: Phys. Lett. **B347**, 187 (1995); Phys. Rev. **C53**, 1102 (1996)
4. F. de Jong et al.: Phys. Lett. **B333**, 1 (1994); F. de Jong et al.: Phys. Rev. **C51**, 2334 (1995)
5. M. Jetter and H.W. Fearing: Phys. Rev. **C51**, 1666 (1995)
6. G. Martinus et al.: Phys. Lett. **B402**, 7 (1997); Phys. Rev. **C56**, 2945 (1997); Phys. Rev. **C58**, 686 (1998)
7. J.A. Pinston et al.: Phys. Lett. **B249**, 402 (1990) and J. Clayton et al.: Phys. Rev. **C45**, 1810 (1992)
8. N. Kalantar-Nayestanaki et al.: Nucl. Inst. and Meth., in press.
9. M. Volkerts et al.: Nucl. Inst. and Meth. **A428**, 432 (1999)
10. S. Schadmand et al.: Nucl. Inst. and Meth. **A423**, 174 (1999)
11. N. Kalantar-Nayestanaki et al.: Nucl. Inst. and Meth. **A417**, 215 (1998)
12. O. Schwalb et al.: Nucl. Inst. and Meth. **A295**, 191 (1990); H. Ströher: Nucl. Phys. News Int. Vol. **6**, No. 1, 7 (1996)
13. H. Huisman et al.: Phys. Rev. Lett., in press.
14. J.G. Rogers et al.: Phys. Rev. **C22**, 2512 (1980)
15. M.K. Liou et al.: Phys. Lett. **B345**, 372 (1995); Phys. Rev. **C54**, 1574 (1996); R. Timmermans: private communication.
16. R. Bilger et al.: Phys. Lett. **B429**, 195 (1998)
17. K. Yasuda et al.: Phys. Rev. Lett. **82**, 4775 (1999)
18. J.G. Messchendorp et al.: Phys. Rev. Lett. **82**, 2649 (1999); Phys. Rev. Lett. **83**, 2530 (1999)

Few-Body Systems Suppl. 12, 275–280 (2000)

Few-Body Systems

Radiative Muon Capture on ^{3}He

D.H. Wright[1], D.S. Armstrong[2], P.C. Bergbusch[3], M. Blecher[4], B.C. Doyle[5], T.P. Gorringe[5], P. Gumplinger[1], M.D. Hasinoff[3], J. A. Macdonald[1], J-M. Poutissou[1], R. Poutissou[1], C.M. Sigler[4]

[1] TRIUMF, Vancouver, British Columbia, V6T 2A3 Canada

[2] College of William and Mary, Williamsburg, Virginia 23187 USA

[3] University of British Columbia, Vancouver, British Columbia, V6T 1Z5 Canada

[4] Virginia Polytechnic Institute and State University, Blacksburg, Virginia 24061 USA

[5] University of Kentucky, Lexington, Kentucky 40506 USA

Abstract. The radiative muon capture process has been observed in ^{3}He by detecting a high-energy photon in coincidence with the recoil triton. A cylindrical pair spectrometer and a scintillating liquid ^{3}He target were used to detect the photon and triton, respectively. The resulting photon spectrum is compared to impulse approximation calculations both with and without meson exchange corrections. The calculations over-estimate the measured spectrum by 30% and 20%, respectively. The recoil energy spectrum indicates that only 11% of the total radiative capture rate goes to final states which include deuterons and protons, with the remainder of the rate going to a triton final state.

1 Introduction

In the presence of hadrons the $V - A$ form of the weak interaction is modified to include at most four additional couplings

$$V_\lambda = g_v(q^2)\gamma_\lambda + ig_m(q^2)\sigma_{\lambda\nu}\frac{q^\nu}{2M} + g_s(q^2)\frac{q_\lambda}{m_l} \tag{1}$$

$$A_\lambda = g_a(q^2)\gamma_\lambda\gamma_5 + g_p(q^2)\frac{q_\lambda}{m_l}\gamma_5 + ig_t(q^2)\sigma_{\lambda\nu}\gamma_5\frac{q^\nu}{2M}, \tag{2}$$

where q^2 is the momentum transfer squared, m_l is the lepton mass and M is the hadron mass.

Experimentally, the least well-known of these couplings is the pseudoscalar form factor g_p, although theoretically it seems to be well understood. It has been shown using PCAC [1, 2] and more recently using QCD and heavy baryon chiral perturbation theory [3] that the form of g_p is given by

$$g_p(q^2) = \frac{2m_\mu g_{\pi NN} F_\pi}{m_\pi^2 - q^2} - \frac{1}{3} g_a(0) m_\mu M r_a^2. \tag{3}$$

Here m_μ and m_π are the muon and pion masses, $g_{\pi NN}$ is the pion-nucleon coupling constant, F_π is the pion decay constant, M is the nucleon mass and r_a is the axial radius of the nucleon. In the ordinary muon capture reaction (OMC)

$$\mu^- + p \rightarrow n + \nu_\mu \tag{4}$$

the momentum transfer is fixed at $q^2 = -0.88 m_\mu^2$, leading to a g_p value of 8.4 or about $6.7 g_a(0)$. In the radiative process (RMC),

$$\mu^- + p \rightarrow n + \nu_\mu + \gamma \tag{5}$$

the momentum transfer is no longer fixed and for photon energies near 100 MeV q^2 can approach $0.9 m_\mu^2$ and g_p can approach 27.3. This increased sensitivity to the pseudoscalar form factor makes radiative muon capture an attractive choice for measuring g_p despite typical branching ratios which are factors of 10^4 or 10^5 smaller than ordinary muon capture. While the elementary process (Eq. 5) has been measured [4, 5] the next most interesting target is ^{3}He. Experimentally ^{3}He has two advantages over ^{1}H: an expected branching ratio which is 20 times larger and a charged particle in the final state. Theoretically, excellent three-body wave functions are available so that nuclear effects should not cloud the experimental results. The most recent calculation of RMC on ^{3}He is that of Klieb and Rood [6] who use the impulse approximation with exchange effects. Their calculation is of the exclusive RMC channel

$$\mu^- +^3 He \rightarrow t + \nu_\mu + \gamma. \tag{6}$$

Currently no RMC calculations for the "breakup" channels

$$\mu^- +^3 He \rightarrow d + n + \nu_\mu + \gamma \tag{7}$$

$$\mu^- +^3 He \rightarrow p + n + n + \nu_\mu + \gamma \tag{8}$$

exist so that any measurement of the branching ratio and hence g_p will need to remove these channels before comparison with theory.

2 Experiment

The experiment was performed using the radiative muon capture spectrometer at TRIUMF (shown in Fig. 1) [7] and a specially constructed liquid helium target. An incident beam of 59 MeV/c negative muons stopped in liquid ^{3}He where photons, recoil tritons, deuterons and protons were produced. Photons were converted to $e^+ - e^-$ pairs which were then momentum analyzed in a cylindrical drift chamber. The recoiling particles deposited energy in the liquid

^3He target, some of which was converted to scintillation light, which in turn was collected in a UV-sensitive phototube. The pulse from the phototube was sent to GaAs CCD transient digitizers with 2 ns resolving time [8]. Selecting events in which the digitized recoil pulse was in coincidence with the detected photon allowed the exclusive RMC channel (Eq. 6) to be separated from the breakup modes (Eqs. 7,8).

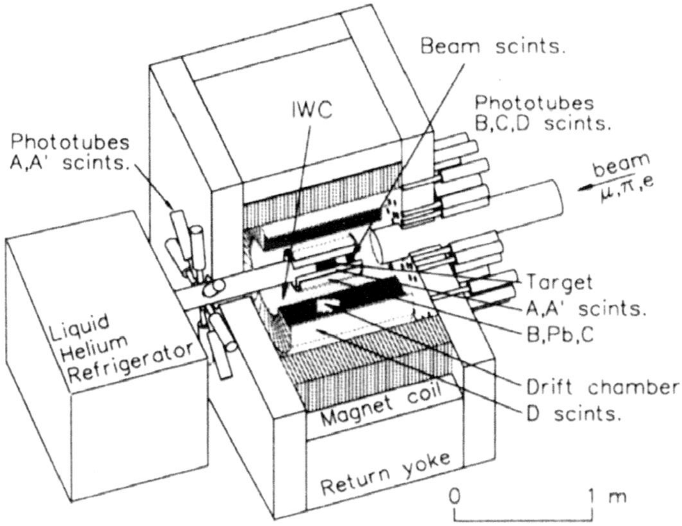

Figure 1. Radiative muon capture spectrometer with liquid helium target.

As a check of the target detection efficiency two analyses were carried out, one using photon-recoil coincidence events and one using events which required only the detection of a photon. In the photon-recoil coincidence analysis, pion-induced events were removed by rejecting all events with prompt timing. In the photon-only analysis a 440 ns blank window was imposed in order to remove the much more frequent RMC events which occurred after a muon stopped in the target's silver walls. After correcting the number of events for the differing widths of the prompt and blank windows the resulting photon spectra were compared and found to be nearly identical. The photon-recoil spectrum contained about 2% fewer events than the photon-only spectrum, indicating a target efficiency of 98%.

The photon acceptance of the pair spectrometer was checked by stopping negative pions in the ^3He target and observing the pion capture reactions

$$\pi^- +^3 He \rightarrow t + \gamma \tag{9}$$

$$\pi^- +^3 He \rightarrow t + \pi^0 \rightarrow t + \gamma\gamma. \tag{10}$$

The resulting photon spectrum was compared to a previous measurement of the above reactions [9] and excellent agreement in both spectrum shape and branching ratio was observed.

3 Preliminary Results

A total of 589 photon-recoil coincidence events were collected during our second experimental run and represents about 60% of our total data sample. The recoil pulse height spectrum of these events is shown in Fig. 2. The large peak at low energy was identified as recoil tritons while the higher energy events were identified as recoil deuterons and protons. This identification was confirmed by

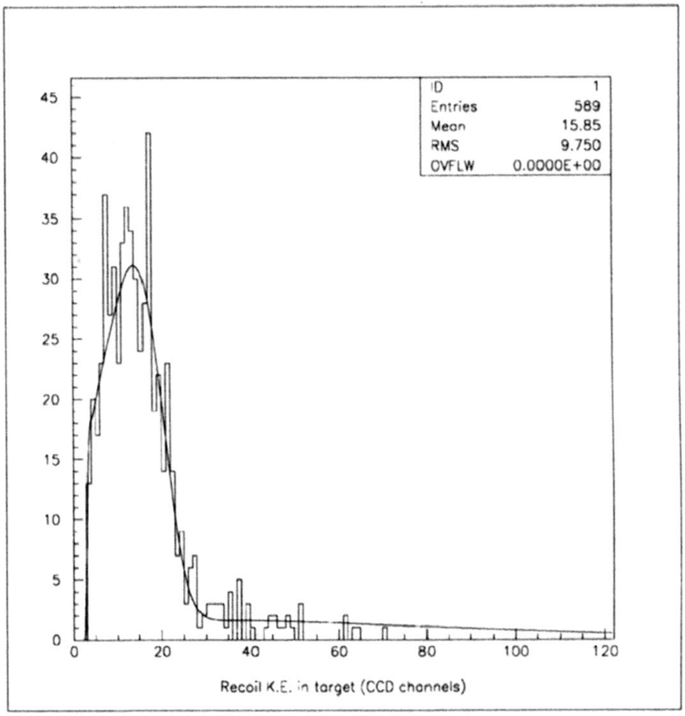

Figure 2. Pulse height spectrum of scintillation light from recoiling tritons, deuterons and protons in coincidence with high-energy photons.

a determination of the energy scale: measuring the centroid of the pulse height spectrum of the 4.1 MeV muon from stopped pion decay gave a value of 0.095 MeV/channel. Hence the sharp upper edge of the triton recoil spectrum at 1.9 MeV should occur around channel 20 in the pulse height spectrum.

The small broad peak of the recoil deuterons and protons as well as their relative scarcity made it easy to subtract these events from the total recoil spectrum in order to obtain the exclusive triton-photon events. Most of the photons corresponding to deuteron and proton recoils were unambiguously removed by using a scatterplot of photon energy vs. recoil pulse height and noting that there are no triton pulse heights above channel 30 in Fig. 2. In order to remove deuteron and proton recoil events with pulse heights below channel 30, it was necessary to assume that the recoiling particle energies were determined completely by phase space. The phase space shape was then used to subtract a small number of photons from the corresponding energy bins. The estimated error introduced by this procedure was small and the total number of events subtracted amounted to 11 ± 3% of the inclusive spectrum.

The photon spectrum resulting from this subtraction is shown in Fig. 3. Here only the events above 58 MeV are retained because the photon acceptance of the pair-spectrometer begins to fall off below that value and because the rate of accidental coincidences between bremsstrahlung photons and recoil particles becomes non-negligible.

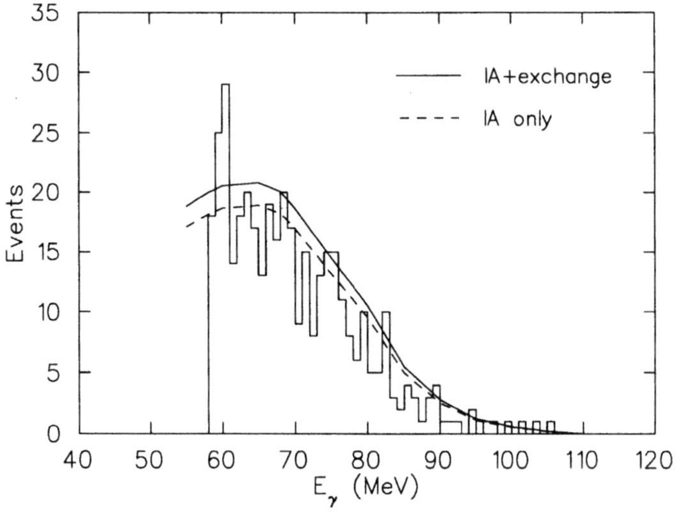

Figure 3. Energy spectrum of photons in coincidence with recoil tritons compared to the prediction of the impulse approximation (dashed curve) and the impulse approximation with meson exchange effects (solid curve).

This spectrum was compared to the impulse approximation calculation of Klieb and Rood [6] which is shown in Fig. 3. The solid and dashed curves are the predicted photon spectra with and without exchange effects which have been convoluted with the spectrometer response function. In both cases there is good agreement between the spectral shapes but the theory over-estimates the magnitude of the measured spectrum by about 30% when exchange effects are included and by about 20% with exchange omitted.

4 Conclusions

The most recent impulse approximation calculation of radiative muon capture on ^{3}He was found to over-estimate the measured exclusive photon spectrum by 30% if exchange effects are included and by 20% if they are not. Agreement between experiment and theory can be obtained only if $g_p(q^2 = -0.88m_\mu^2)$

is assumed to be 3.4, instead of the predicted value of 8.4. Given the recent success of both impulse approximation and elementary particle calculations [10] in describing a very precise measurement of ordinary muon capture in ^{3}He [11], this seems unlikely. When the remaining data are analyzed it is possible that the experimental spectrum may increase somewhat. However, it is useful to point out that no calculations of RMC on ^{3}He have been performed since that of Klieb and Rood in 1983. Their treatment of RMC involved multiplying certain amplitudes by an empircal factor of 1.1 which brought their OMC calculations into agreement with the OMC measurements at that time. A more modern calculation of RMC in ^{3}He, based on the recent OMC calculations, is required.

Another preliminary result of the present experiment is that the RMC final states which involve a deuteron or a proton represent only 11% of the total RMC rate, well below the 30% measured in the case of ordinary muon capture [12]. It is possible that the extra body (the photon) in the RMC final state suppresses the breakup modes relative to the OMC process by reducing the available phase space. A resolution of this question will require that calculations of RMC on ^{3}He be done which treat the final states containing deuterons, protons and neutrons.

Acknowledgement. This work was supported by the National Research Council and the Natural Sciences and Engineering Council of Canada and by the US National Science Foundation. The authors would like to thank Lawrence Livermore National Laboratory for the loan of the ^{3}He gas.

References

1. S.L. Adler and Y. Dothan: Phys. Rev. **151**, 1267 (1966)
2. L. Wolfenstein: *High-Energy Physics and Nuclear Structure*, edited by S. Devons, p.661. New York: Plenum 1970
3. V. Bernard, N. Kaiser, Ulf-G. Meissner: Phys. Rev. **D50**, 6899 (1994)
4. G. Jonkmans et al.: Phys. Rev. Lett. **77**, 4512 (1996)
5. D.H. Wright et al.: Phys. Rev. **C57**, 373 (1998)
6. L. Klieb and H.P.C Rood: Phys. Rev. **C29**, 223 (1984)
7. D.H. Wright et al.: Nucl. Instrum. Methods Phys. Res. **A320**, 249 (1992)
8. D.A. Bryman et al.: Nucl. Instrum. Methods Phys. Res. **A396**, 394 (1997)
9. F. Corriveau, M.D. Hasinoff, D.F. Measday, J.-M. Poutissou, M. Salomon: Nucl. Phys. **A473**, 747 (1987)
10. J.G. Congleton and E. Truhlik: Phys. Rev. **C53**, 956 (1996)
11. P. Ackerbauer et al.: Phys. Lett. **B417**, 224 (1998)
12. A.C. Phillips and F. Roig: Nucl. Phys. **A234**, 378 (1974)

Few-Body Systems Suppl. 12, 281–286 (2000)

Few-Body Systems

Measurements of Q^2 Evolution of the GDH Sum Rule and the Spin Structure of the ^{3}He and the Neutron

J. P. Chen*

Jefferson Lab, Newport News, VA 23606, USA

Abstract. We have recently completed experiment E94-010 at Jefferson Lab. The experiment used a polarized ^{3}He target and polarized electron beam to measure asymmetries and cross sections for $^3\vec{He}(\vec{e}, e')$ from threshold to deep inelastic region, covering the quasielastic and the resonance region.The Q^2 range covered is from 0.03 to 1 GeV^2. From these data, the ^{3}He and the neutron spin structure and the Q^2 evolution of the Gerasimov-Drell-Hearn (GDH) sum rule will be studied This is the first measurement of the Q^2 evolution of the GDH sum rule. Preliminary results are presented. Plan for future study on this subject at Jefferson Lab is discussed.

1 Introduction

The nucleon spin has been of central interest ever since the EMC experiment found that at small distances the quarks carry only a fraction of the nucleon spin. Going from shorter to larger distances the quarks are dressed with gluons and $q\bar{q}$ pairs and acquire more and more of the nucleon spin. How is this process evolving with the distance scale? At the two extreme kinematic regions we have two fundamental sum rules: the Bjorken sum rule (Bj-SR)[1] which holds for the proton-neutron difference in the asymptotic limit, and the Gerasimov-Drell-Hearn sum rule (GDH-SR)[2, 3] at $Q^2 = 0$. What is the connection between the Bj-SR and the GDH-SR? Study the evolution of the GDH-SR at low Q^2 is crucial to answer these questions.

2 Bjorken and GDH Sum Rules and the Connection

2.1 Bjorken Sum Rule

From the 20 years of spin structure experiments in the deep inelastic region, one of the most important outcomes is the experimental test of the Bjorken Sum

For the Jefferson Lab E94-010 Collaboration

Rule to about 10% level. The Bj-SR relates the spin structure over the entire energy range to a static property of the nucleon, the axial coupling constant g_A:

$$\int_0^1 (g_1^p - g_1^n)dx = \frac{1}{6}g_A \tag{1}$$

The Bj-SR is based on very general principle (QCD) and is valid at the Bjorken limit: $Q^2 \to \infty$, $\nu \to \infty$ while $x = Q^2/2m\nu$ is finite. For finite Q^2, there is a QCD correction factor.

2.2 GDH Sum Rule

Another fundamental sum rule, the Gerasimov Drell-Hearn sum rule (GDH-SR) [2, 3], holds at the other extreme of the kinematics, $Q^2 = 0$:

$$\int_{thr}^{\infty} (\frac{\sigma_{3/2} - \sigma_{1/2}}{\nu})d\nu = \frac{2\pi^2\alpha}{M^2}\kappa^2 \tag{2}$$

here again the sum rule relates the helicity (doubly polarized) cross section difference over the whole energy range to a static property of the nucleon, the anomalous magnetic moment of the nucleon, κ.

The GDH-SR is derived using the dispersion relation (without subtraction) on the forward Compton scattering amplitude, combined with the Optical and the Low Energy Theorems. The input assumptions are very general. The only asummption could be open to "reasonable" questions is the no-subtraction condition in the dispersion relation. Partial wave analysis based on unpolarized and singly polarized data, gave an estimate of the double polarization cross sections and therefore the indirect 'experimental' sum, showed a discrepancy from the GDH-SR. The direct experimental test becomes possible only now with the availability of the high luminosity polarized beam and polarized targets and the modern advanced detection technology.

2.3 GDH Sum Rule for ^{3}He

The GDH-SR is also valid for any nucleus:

$$\int_{thr}^{\infty} (\frac{\sigma_P - \sigma_A}{\nu})d\nu = \frac{4\pi^2\mu_A^2}{J} \tag{3}$$

where the subscription P (A) stand for the target polarization parallel (anti-parallel) to the beam polarization. J and μ_A are the angular momentum and the magnetic moment of the target nucleus.

In the case of polarized ^{3}He target, since it is approximately a polarized neutron target, it can be used to study both the GDH-SR for ^{3}He and for the neutron. $GDH(^3He) = 496\mu b$ and $GDH(n) = 233\mu b$. The difference of the two sum rules gives us an elegant method to study the 'nuclear' contribution to the GDH-SR for ^{3}He. The difference is $263\mu b$ and should be approximately equal to the GDH sum (^{3}He) from break-up threshold to the π threshold. One can learn how good a polarized ^{3}He target is a polarized neutron target by experimentally checking the GDH(^{3}He) sum below the π threshold.

2.4 *Generalized GDH Sum Rule: Connecting GDH and Bjorken Sum Rules*

Phenomenological models have been proposed to extend the GDH-SR integral for the proton and neutron to finite Q^2 and connect it to the Bj-SR in deep inelastic regime [4, 5]. Recently Ji and Osborne[6] made a rigorous extension of the GDH-SR to the entire region of Q^2. With the same assumptions as the GDH-SR, Ji and Osborne derived the generalized GDH-SR:

$$4\int_{el}^{\infty}\frac{G_1(\nu,Q^2)}{\nu}d\nu = S_1(Q^2)$$

where $S_1(Q^2)$ is the forward virtual Compton Scattering amplitude. The GDH-SR and Bj-SR are the two limiting cases ($Q^2 = 0$ and $Q^2 = \infty$) of the generalized GDH-SR. Other than the two limiting cases, the $S_1(Q^2)$ can also be calculated at small Q^2, where hadrons are the relevant degree of freedom, with Chiral Perturbation theory and at large Q^2, where quarks and gluons(partons) are the relevant degree of freedom, with higher order QCD expansion (twist expansion). At small Q^2, it was calculated to the leading order[6] using the Chiral Perturbation theory with the Heavy Baryon approximation. At large Q^2, twist-2 and twist-4 terms have been calculated. Efforts are underway to calculate to higher orders for both small and large Q^2.

A crucial question in this connection is how low in Q^2 the Bj-SR can be evolved using the high twist expansion? Recent estimates [6] suggest a Q^2 value as low as 0.5 GeV^2. Also at the other end, chiral perturbation theory is applicable at small Q^2, and may allow evolution of the GDH-SR to $Q^2 = 0.02$ GeV^2 or higher. Theoretical efforts (such as lattice calculations) are needed to bridge the remaining gap.

The importance of such efforts cannot be overemphasized as it would mark the first time that hadronic structure is described by a fundamental theory in the entire kinematics regime, from short to large distances.

3 JLab E94-010 Experiment

Experiment JLab E94-010[7], has recently been completed with a polarized ^{3}He targets to study the Q^2 evolution of the GDH integral for ^{3}He and the neutron. The experiment covers a range of $Q^2 = 0.03 - 1.1\ GeV^2$ and from the elastic to the deep inelastic regime, covering the quasielastic and the resonance region. Figure 1 shows the kinematic coverage of the experiment.

3.1 *The Experimental Setup*

The experiment was carried out in Hall A of the Jefferson Lab. Highly polarized (70%) electron beam with current up to 15 μA scattered off a high density

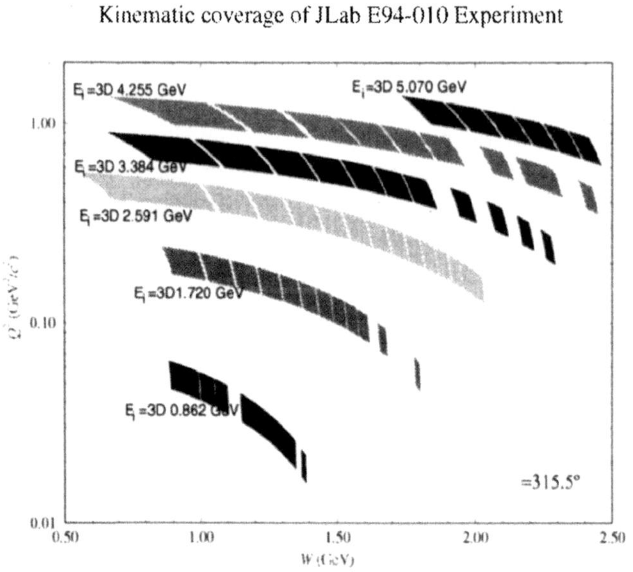

Figure 1. E94-010 Kinematics

(10 atm gaseous in a 40 cm long glass cell) polarized ^{3}He target (30-40% polarization in-beam). The recently built polarized ^{3}He target was based on the spin exchange principle. Rubidium atoms were polarized by optical pumping, and the polarization was transfered to ^{3}He nuclei via spin exchange collisions. The target was polarized either parallel or perpendicular to the beam direction and the polarization was measured with two independent methods: NMR with Adiabatic Fast Passage, and EPR (Electron Paramagnetic Resonance).

Two nearly identical spectrometers, setting at same scattering angle of 15.5°, were used to detect the scattered electrons. Both spectrometers were equipped with standard electron detector packages. The comparison of the data from the two spectrometers gives us a check of the data quality and minimizes some systematic uncertainties.

3.2 Preliminary Results

The data analysis is still underway. The raw ^{3}He elastic asymmetries were checked against the world elastic data weighted by the beam and target polarizations. Figure 2 shows the spectra of the raw parallel and perpendicular asymmetries for four energies. The data are still preliminary since radiative corrections, acceptance corrections, dilution corrections and some other correction factors have not been applied. The error bars are statistical only. Final results on the ^{3}He asymmetries are expected to be available in the next a few months and the results on the generalized ^{3}He GDH sum will follow afterwards.

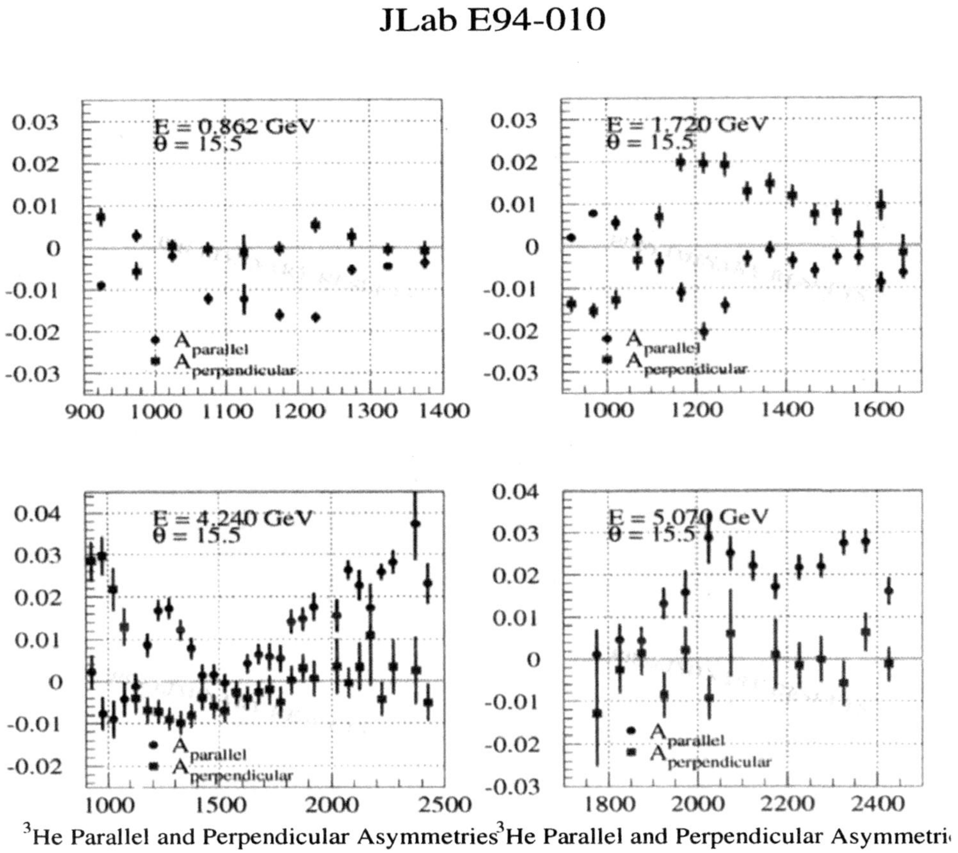

Figure 2. Preliminary Raw ^{3}He Asymmetries

4 Extraction of the Neutron Results

Due to the small components of S' and D waves in the ^{3}He ground state, the polarized ^{3}He target is only approximately a polarized neutron target. J. Friar et al. [8] calculated the effective polarization of the neutron and proton in a polarized ^{3}He target. The results are that the effective neutron polarization is about 87% and the effective proton polarization is about -3%. Recently, C. degli Atti and S. Scopetta[9] studied the extraction of the neutron asymmetries and the generalized GDH sum from a polarized ^{3}He target. Although at the resonance region at low Q^2, the extracted neutron spin structure function g_1

could differ significantly from g_1 of the free neutron, the extracted neutron GDH sum does not differ much from the free neutron GDH sum. The reason is that the Fermi motion, being the main effect, does not affect the integrated results. Other theory groups[10] will also study this problem in the future.

5 Outlook and Summary

The spin structure and the GDH sum rule study are one of the major efforts at Jefferson Lab. Experiments on the proton[11] and the deuteron[12] have been partially completed. An extension of the ^{3}He experiment to very low Q^2[13] and higher energy (6 GeV) is planned. It will enable us to extrapolate to the real photon point and also study the convergence of the sum rule. JLab is planning to upgrade the accelerator to 12 GeV. In the future, extension to 12 GeV can test the Regge behavior of the GDH sum rule.

In summary, the spin structure and the generalized GDH sum rule of the nucleon and few-body nuclei are one important area of research at Jefferson Lab. JLab E94-010 is the first completed experiment measuring the spin structure and the generalized GDH sum rule for ^{3}He and the neutron at low Q^2 region. Preliminary results were shown. Future extension to nearly real photon kinematics and higher energies are planned. Experiments on the proton and deuteron are also being carried out at JLab. Fruitful results from the GDH sum rule and spin structure study are expected in the next few years.

References

1. J.D. Bjorken: Phys. Rev. **148**, 1467 (1966)
2. S.B. Gerasimov: Sov. J. Nucl. Phys. **2**, 430 (1966)
3. S.D. Drell and A.C. Hearn: Phys. Rev. Lett. **16**, 908 (1966)
4. M. Anselmino, B. Ioffe, E. Leader: Sov. J. Nucl. Phys. **49**, 136 (1989)
5. V. Burkert and B. Ioffe: JETP **105**, 1153 (1994)
6. X. Ji and J. Osborne: hep-ph/9905010 (1999)
7. JLab E94-010, Spokespersons, Z. E. Meziani, G. Cates, J. P. Chen
8. J. Friar et al.: Phys. Rev. **C42**, 2310 (1990)
9. C. degli Atti and S. Scopetta: Phys. Lett. **B44**, 223 (1997)
10. R. Schiavilla, W. Glöckle: private communication
11. JLab E91-023, Spokespersons, V. Burkert, D. Crabb, R. Minehart
12. JLab E93-009, Spokespersons, S. Kuhn and M. Taiuti
13. JLab E97-110, Spokespersons, J.P. Chen, G. Cates, F. Garibaldi

Few-Body Systems Suppl. 12, 287–292 (2000)

Threshold Contribution to the Gerasimov-Drell-Hearn Sum Rule in ^{3}He

H.R. Weller*

Department of Physics and Triangle University Nuclear Laboratory, Box 90308, Duke University, Durham, NC 27708

Abstract. Measurements of the cross section, vector and tensor analyzing powers, and gamma-ray polarization in the radiative capture reactions $D(p,\gamma)^3$He and $p(d,\gamma)^3$He at c.m. energies in the range 0–53 keV allow the determination of the reduced matrix elements (RMEs) relevant for these transitions. The RMEs provide values of the doublet and quartet M1 cross sections, as well as the asymptotic D-to-S state ratio. These quantities are in good agreement with theoretical predictions. The RMEs can also be used to compute the value of the integral which determines the Gerasimov-Drell-Hearn sum rule for ^{3}He in the energy region of our experiment. The results are compared with the predictions of an *ab initio* microscopic three-body model calculation. The theoretical predictions for the value of this integral based on a "nucleons-only" assumption are an order-of-magnitude smaller than experiment. The discrepancy is reduced to about a factor of two when meson-exchange currents are taken into account.

In the present work, we have been able to evaluate the integral which determines the Gerasimov-Drell-Hearn (GDH) sum rule [1, 2] in the threshold photodisintegration region of ^{3}He. The GDH sum rule connects the helicity structure of the photo-absorption cross section to the anomalous magnetic moment of the nuclear target. It is derived using Lorentz and gauge invariance, crossing symmetry, causality and unitarity of the forward Compton scattering amplitude, and is explicitly given by

$$I_T = \int_{\omega_{\rm th}}^{\infty} d\omega \frac{\sigma_P(\omega) - \sigma_A(\omega)}{\omega} = 4\pi^2 \alpha s_T \left(\frac{\kappa_T}{m_T}\right)^2 , \tag{1}$$

where σ_P and σ_A are the cross sections for absorption of polarized photons of energy ω and helicities parallel and antiparallel to the target spin s_T (in its maximum state), $\omega_{\rm th}$ is the threshold photon energy for inelastic processes, α is

*Partially supported by the U.S. Department of Energy contract numbers DE-FG02-97ER41033.

the fine-structure constant, and m_T and κ_T are the target mass and anomalous magnetic moment, respectively.

In ^{3}He, the GDH sum rule is $I_{^3\mathrm{He}} = 498$ μb, using the experimental value $\kappa_{^3\mathrm{He}} = -8.366$. It is useful to divide the integral of Eq.(1) into the part up to pion threshold, and the part above this threshold. For the part above pion threshold, the ^{3}He nucleus should have roughly the same strength as the neutron, i.e. $\simeq 230$ μb. Such an expectation is based on the fact that the ^{3}He ground state consists predominantly of a spherically symmetric S-wave component, in which the proton spin projections are opposite and the net polarization is therefore due entirely to the neutron. Ignoring corrections to this naive estimate, it is expected that the ω-region from the photodisintegration threshold ($\omega_{\mathrm{th}} = 5.4949$ MeV) up to the pion threshold should contribute about 266 μb to the ^{3}He GDH sum rule.

We have not made a direct measurement of the σ_P and σ_A photo-absorption cross sections to determine the contribution to the integral of Eq.(1) in the threshold regon. Instead we have determined its value from measurements of the cross sections, vector and tensor analyzing powers for the radiative capture reactions $D(p,\gamma)^3$He and $p(d,\gamma)^3$He. These measurements allow the determination of the complex reduced matrix elements (RMEs) of these reactions. This has been performed at TUNL for proton and deuteron incident energies in the range 0–80 keV, corresponding to values $\omega_{\mathrm{th}} \leq \omega \leq 5.548$ MeV.

In the energy region under consideration here (*pd* relative energies less than 53 keV), there are only six dominant RMEs corresponding to magnetic and electric dipole transitions (M1 and E1, respectively) resulting in emission of *pd* pairs in relative S- and P-wave states. Thus, ignoring the contribution of higher order multipoles, we find Ref. [6]

$$\begin{aligned} \Delta\sigma &\equiv \sigma_P - \sigma_A \\ &= \frac{16\pi^2\alpha\mu p}{\omega}\left[-|s_2|^2 + \frac{|s_4|^2}{2}\right. \\ &\qquad \left. -|p_2|^2 + \frac{|p_4|^2}{2} - |q_2|^2 + \frac{|q_4|^2}{2}\right], \end{aligned} \tag{2}$$

where for ease of presentation we have introduced the notation $s_{2J+1} \equiv M_1^{0JJ}$, $p_{2J+1} \equiv E_1^{1\frac{1}{2}J}$, and $q_{2J+1} \equiv E_1^{1\frac{3}{2}J}$ for J=1/2 and 3/2 (the superscripts in X_ℓ^{LSJ}, $X = M, E$, refer respectively to the quantum numbers LSJ of the continuum state). The energy dependence of the RMEs and $\Delta\sigma$ is understood.

The data of this experiment were obtained using frozen H_2O and D_2O targets and large HPGe detectors. This, as discussed in Ref. [4], allowed us to obtain the cross section and analyzing powers as a function of energy. This was accomplished by de-convoluting the full-energy peak in the spectrum, which was broadened when the 80 keV proton beam, for example, stopped in the ice target. Ofcourse it was the high resolution (about 4.2 keV) of the detector which made this possible. A typical spectrum is shown in Fig. 1.

An alternative method of analyzing the data was to simply bin the data into 10 keV wide bins. The angular distribution data on the cross section and

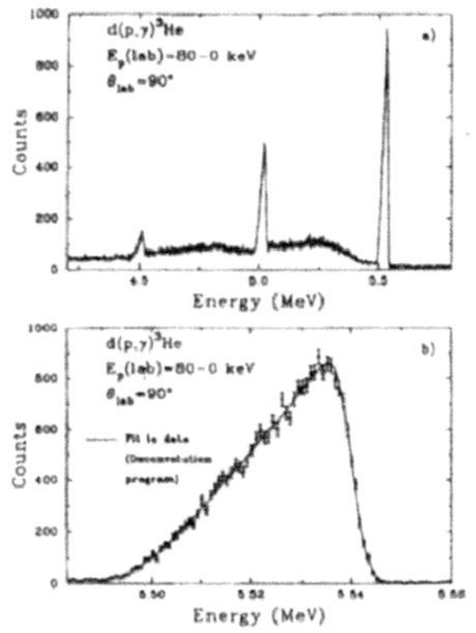

Figure 1. A HPGe spectrum for the $^2H(p,\gamma)^3He$ reaction(a). The spectrum in (b) is a blowup of the full energy peak seen in (a).

analyzing powers obtained from binning the data were analyzed to obtain the fraction of the cross section which was M1 vs. E1. The de-convolution analysis was extrapolated to produce angular distribution data at E=0, and an M1/E1 ratio was obtained there as well. The resulting M1 fraction as a function of energy is shown in Fig. 2.

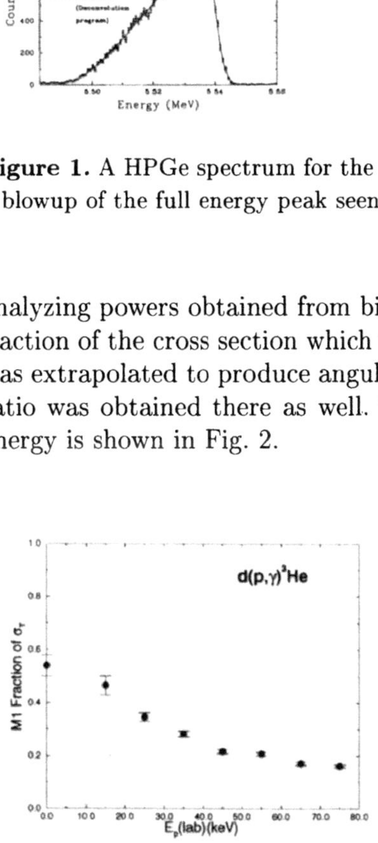

Figure 2. The M1 fraction of σ_T vs beam energy deduced from the data.

The current data set are sufficient to determine the six contributing matrix elements and their five relative phases in an unconstrained fit as shown in Fig. 3. Knowledge of these matrix elements has provided us with a number of new insights into the physics near threshold in ^{3}He. For example, these results determine the doublet and quartet M1 cross sections at E_{cm} = 27 keV [7]. The ratio of these cross sections is very sensitive to explicit MEC effects, changing from a value of 1.95 to a value of 0.48 when MEC effects are not and are included, respectively. Our experimentally determined value for the ratio of the quartet to the doublet strength is 0.49 ± 0.04, in excellent agreement with

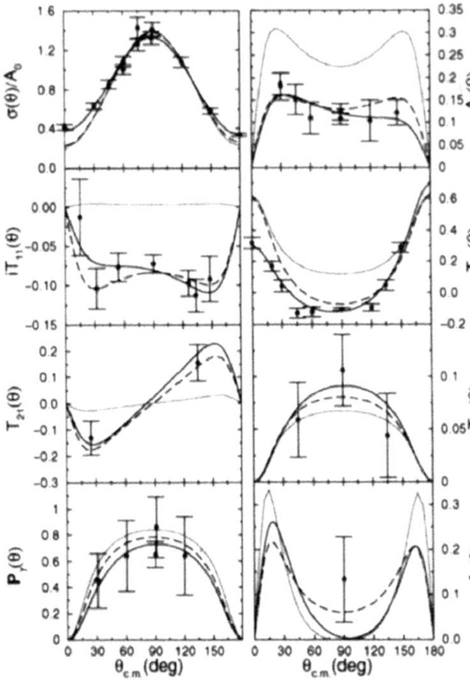

Figure 3. *pd* capture observables (full dots with error bars) and their fit (solid lines) used to determine the six leading RMEs. Theoretical predictions including both one- and two-body currents are shown by the dashed lines.

the theory which includes MEC effects. Even more impressive is the comparison of the absolute cross sections for these two terms. Our experimental value for the doublet M1 cross section at $E_{cm} = 27$ keV is 15.8 ± 1.0 nb; theory predicts 15.8 nb. And for the quartet we obtain 7.75 ± 0.62 nb, while theory gives 7.5 nb. This excellent agreement is clearly a triumph for few body nuclear physics.

Another physical quantity which we have extracted from these data is the value of the asymptotic D-to-S state ratio η. The value of η was deduced from the value of $T_{20}(90°)$ [8]. In order to do this, it was necessary to remove the M1 contribution from the measured value of $T_{20}(90°)$. This was done using the values of the six RMEs obtained from the fit to all of the data and setting the M1 strengths equal to zero in the expression for T_{20}. The resulting value of $T_{20}(90°)$ was -0.0654 ± 0.0165. This E1 determined value of $T_{20}(90°)$ arises from the interference of the S=3/2 and the S=1/2 E1 strengths. Since at this low energy the reaction occurs at relatively large distances, the S=3/2 E1 strength can only arise via E1 capture to the D-state of ^{3}He. This allowed us to calculate the E1 RMEs using asymptotic Whittaker functions to describe the bound S- and D-state wavefunctions of ^{3}He, as well as the scattering state wavefunctions. The integral was cut-off at the small distance of about 3 fm. Once again, the low energy of the reaction resulted in a very low sensitivity

Table 1. The contributions $I(\overline{\omega})$ (in nb) for two energies $\overline{\omega}$. In the third column, the values obtained from a fit of the experimental *pd* capture data are reported. The results of the theoretical calculations obtained with one-body only and both one- and two-body currents are listed in the fourth and fifth columns, labeled IA and FULL respectively. The lines denoted by M1, E1 S=1/2 and E1 S=3/2 report the partial contributions to $I(\overline{\omega}$=5.522 MeV) of the corresponding RMEs.

$\overline{\omega}$ (MeV)	$I(\overline{\omega})$	FIT	IA	FULL
5.522	M1	-0.0524 ± 0.0077	-0.0016	-0.0485
5.522	E1 S =1/2	-0.0361 ± 0.0095	-0.0030	$+0.0027$
5.522	E1 S =3/2	-0.0026 ± 0.0007	-0.0050	-0.0001
5.522	Total	-0.0911 ± 0.0123	-0.0096	-0.0459
5.548	Total	-1.120 ± 0.218	-0.149	-0.435

to this cutoff value, thereby making the extracted value of η essentially model independent. The value we obtained for η was -0.0340 $\pm$ 0.0102 $\pm$ 0.013, which compares well with the value previously deduced from transfer reactions (η = -0.0386 $\pm$ 0.0046 $\pm$ 0.0012) as well as the theoretical value [5] of η = -0.043 $\pm$ 0.001.

If we assume that the ratio of the doublet-to-quartet M1 strengths and the $J = 3/2$ to $1/2$ E1 strengths are equal to the values we obtained at $E_{\rm c.m.} = 26.6$ keV (the uncertainty in our results due to this assumption has been shown to be quite small), we can perform the integral of Eq. (1). The results corresponding to integrating up to $\overline{\omega}_1 = 5.522$ MeV ($E_p = 40$ keV) or to $\overline{\omega}_2 = 5.548$ MeV ($E_p = 80$ keV) are presented in Table 1 (I($\overline{\omega}$). While this is an insignificant (negative) contribution to the total strength expected below pion threshold, it is an interesting result which can be compared directly with theory.

A detailed description of the theoretical calculations which provide the basis for a comparison and interpretation of the experimental capture results obtained above has been previously published [3]. With respect to this earlier reference, however, we note that in the present work the variational treatment of the *pd* continuum states has been improved. As a result, the present calculations are in better agreement than reported in 1996 [3] with the many experimental capture data obtained below 80 keV. In particular, the discrepancy between the calculated and measured A_y observable has been reduced substantially. Theoretical predictions for this and a number of other observables are compared with data in Fig. 3 where the results for both one-body (nucleons only) and one and two-body (MEC) currents are presented.

The value of the GDH sum rule integral $I(\overline{\omega})$ (where $\overline{\omega}$ is the upper limit on the integral) predicted with inclusion of one-body only and both one- and two-body currents (columns labeled IA and FULL, respectively) are reported in Table 1 for $\overline{\omega}_1$=5.522 MeV and $\overline{\omega}_2$=5.548 MeV. Table 1 also lists the individual contributions to $I(\overline{\omega}_1)$ from the $-|s_2|^2+|s_4|^2/2$, $-|p_2|^2+|p_4|^2/2$, $-|q_2|^2+|q_4|^2/2$ RME combinations (rows labeled M1, E1 S=1/2, and E1 S=3/2, respectively).

The total contribution to $I(\overline{\omega})$ is mostly due to M1 strength. While the results in the "full"calculation are in much better agreement with data than those in IA, a factor of two discrepancy persists between theory and experiment. This discrepancy is mostly due to the difference between the calculated and measured S=1/2 E1 strength, as the second row in Table 1 makes clear. The difference in the strengths of $|p_2|^2$ and $|p_4|^2/2$ observed in the experimental results is presently unaccounted for. Perhaps it is due to a spin-dependent 3-body force recently proposed to account for the so-called A_y problem [9]. Further theoretical work is needed here.

The factor of 10 discrepancy between the IA calculation and the integral of Eq.1 observed here can be observed as discrepancies with several of the observables of Fig. 3. Note, for example the large discrepancies in $A_y(\theta)$, $iT_{11}(\theta)$, and $T_{20}(\theta)$. While these are largely removed when two-body currents are included (see Fig. 3), the remaining discrepancy of a factor of 2 for the GDH integral observable show up in Fig. 3 as discrepancies in $\sigma(\theta)$ and $T_{20}(\theta)$, especially in the vicinity of 0°.

In summary, polarized capture data have made it possible to determine the value of the integral which defines the GDH sum rule in region just above threshold in ^{3}He. While the results indicate that an extremely tiny piece of the total sum-rule strength is located in this region, we have found, by direct comparison with theory, that this integral observable is very sensitive to the effects of two-body currents. The inclusion of these currents reduces the discrepancy between theory and experiment from a factor of 10 to a factor of 2. Further theoretical progress is needed to understand the physical origin of the difference in the p-wave E1 RMEs responsible for the remaining discrepancy. This first glimpse of the GDH integral for ^{3}He, although insignificant to the Sum Rule value, demonstrates the new knowledge contained in the quantity and emphasizes the need for measurements at higher energies.

References

1. S.B. Gerasimov: Sov. J. Nucl. Phys. **2**, 430 (1966)
2. S.D. Drell and A.C. Hearn: Phys. Rev. Lett. **16**, 908 (1996)
3. M. Viviani, R. Schiavilla, A. Kievsky: Phys. Rev. **C54**, 534 (1996)
4. G.J. Schmid et al.: Phys. Rev. Lett. **76**, 3088 (1996);
   G.J. Schmid et al.: Phys. Rev. **C56**, 2565 (1997)
5. J.L. Friar et al.: Phys. Rev. **C37**, 2859 (1988)
6. H.R. Weller et al.: Atomic Data and Nuclear Data Tables **58**, 219 (1994)
7. B.J. Rice et al.: Phys. Rev. **C56**, R2917 (1997).
8. B.J. Rice and H.R. Weller: Phys. Rev. **C55**, 2700 (1997)
9. D. Hüber and J. L. Friar: Phys. Rev. **C58**, 674 (1998)

Few-Body Systems Suppl. 12, 293–296 (2000)

Few-Body Systems

Measurement of Proton-Proton Bremsstrahlung at 389 MeV

K. Yasuda*, RCNP E73/E103 collaborations

Research Center for Nuclear Physics, Osaka University, Ibaraki, Japan

Abstract. Differential cross sections and analyzing powers for the proton-proton bremsstrahlung have been measured at 389 MeV incident energy. At the present kinematical conditions, an enhancement of the cross sections due to the Δ current contribution is predicted. The measured cross sections are larger than the theoretical predictions including the Δ current contribution.

1 Introduction

Proton-proton bremsstrahlung ($pp\gamma$) is one of the most fundamental nucleon-nucleon (NN) inelastic scattering reactions. In some theoretical studies on the $pp\gamma$ reaction at the intermediate energy, it is predicted that the non-nucleonic degrees of freedom, such as meson-exchange currents, negative energy states and the Δ current, are important [1, 2]. Some of these calculations predict that the Δ current contribution may increase the cross section about 100% even at about 400 MeV incident proton energy, which is far below the Δ resonance region [1, 2]. Recently, measurements of the $pp\gamma$ reaction were performed at several facilities, such as TRIUMF [3], IUCF [4], COSY [5] and KVI [6]. In these measurements, the incident energies were around or below the pion production threshold and not enough high to investigate the Δ current contribution. Precise experimental data were desired at a higher incident energy to investigate the Δ effect on the $pp\gamma$ process.

In this contribution we report the cross sections and analyzing powers for the $pp\gamma$ reaction at a proton incident energy of 389 MeV. We measured the two outgoing protons with two magnetic spectrometers. The spectrometers were placed at the most forward angles, which were 26°, on each side of the beam so as to be close to the kinematical condition that gives maximum Δ contribution. In this kinematical condition, the Δ current contribution is predicted to be large, especially at forward photon emission angles [1, 2]. The cross section data were published in [7].

* *E-mail address:*yasuda@rcnp.osaka-u.ac.jp

2 Experiment

The experiment was performed at the Research Center for Nuclear Physics (RCNP), Osaka University. A 389 MeV proton beam from the RCNP ring cyclotron was delivered to a liquid hydrogen target, which was developed by the Kyushu University group [8] The target thickness was about 9 mm and the container windows were made of 12.5 μm thick aramid foil. Two outgoing protons were detected with the two-arm spectrometer, Grand Raiden (GR) [9] and Large Acceptance Spectrometer (LAS) [10], both at 26.0°. The four-momentum and invariant mass of the third outgoing particle were calculated with the measured proton momenta. The beam intensity and polarization were monitored with beam line polarimeters placed at upstream of the target, which counted the *pp* elastic scattering events from an aramid and CH_2 foil targets. The beam intensity and polarization were about 15 nA and 70%, respectively. The luminosity was measured by observing the *pp* elastic scattering on the liquid hydrogen target.

The squared missing mass spectra reconstructed from the detected two protons are shown in Fig. 1. These are obtained from one magnetic field settings which correspond to about 120° photon emission angle. A clear peak around 0 MeV^2 which corresponds to the $pp\gamma$ process is obtained. The background events due to the accidental coincidence and from the target container foils are also shown in the left figure. The difference spectra, given by open circles with statistical errors, are shown in the right figure. It shows that the background has been estimated correctly and has been subtracted from the data. The experimental data are compared to the result of a Monte Carlo simulation. In this simulation, the events were generated with the phase space distribution, and the energy straggling and multiple scattering in the target and the momentum and angular resolutions of the spectrometer were taken into account. It is clear that the missing mass spectra were well accounted by the simulation.

3 Results and Discussion

Figure 2 shows the coplanar $pp\gamma$ differential cross sections and analyzing powers as a function of the reconstructed photon emission angle θ_γ in the laboratory system. The error bars include statistical errors and systematic errors due to the error in the magnetic field setting of the spectrometer and the uncertainty of the missing mass gate. In addition to these errors, there are systematic errors for the absolute normalization of 5.9% in the cross section data.

The $pp\gamma$ differential cross sections and analyzing powers of the present work were compared to the theoretical calculations of de Jong *et al.* [1]. In Fig. 2, the solid line corresponds to the calculation including the contribution from the Δ current, and the dashed line is the result of the calculation that only incorporates the nucleonic current contribution. These calculations predict that the Δ effect is seen at forward photon emission angles and the differential cross section increases by a factor two at about $\theta_\gamma = 70°$. The present data favor the calculations including the Δ current, but especially at about $\theta_\gamma = 70°$,

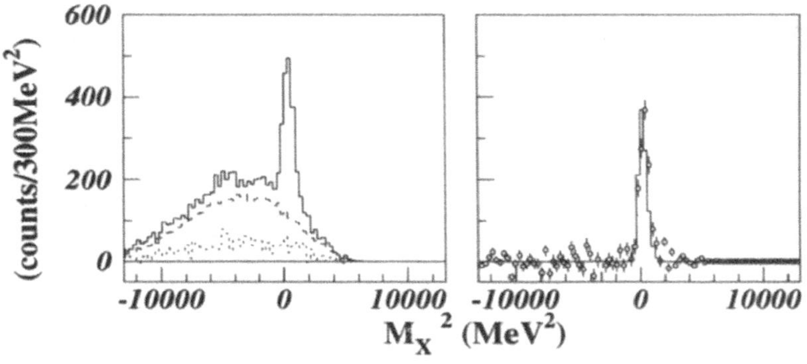

Figure 1. The squared missing mass ($M_X{}^2$) spectra reconstructed from the two observed proton momenta. left: The spectrum as taken with the liquid hydrogen target runs (solid), contribution from the accidental coincidence events (dashed), and the "empty target" runs (dotted), right: The missing mass spectrum after subtracting the accidental coincidence and the background events (open circles with error bars). The solid lines show the result of a Monte Carlo simulation.

where the effect of the Δ current seems to be large, the present result is about 70% larger than the theoretical prediction including the contribution of the Δ current. On the other hand, the present result is not much different from the theoretical predictions at the backward angles. For the analyzing powers, there are discrepancies between the data and the theoretical calculations at about $\theta_\gamma = 70°$. Another result of the calculation including the meson-exchange currents and the Δ current at our kinematical conditions [2] also predict smaller differential cross section than that of the present data. This suggests that there might be contributions of some other mechanisms to the $pp\gamma$ cross section and a further theoretical study is desired.

Acknowledgement. We wish to acknowledge K. Sagara for our use of their liquid hydrogen target system. We also thank K. Nakayama, O. Scholten and H. Toki for useful discussions and theoretical calculations.

References

1. F. de Jong, K. Nakayama, T.-S. Lee: Phys. Rev. **C51** 2334 (1995)

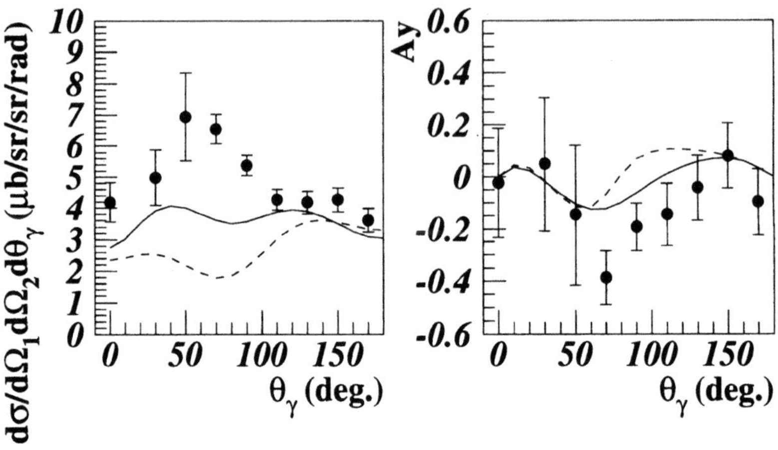

Figure 2. The differential cross sections for the $pp \to pp\gamma$ in the laboratory system. The lines indicate the results of the theoretical calculations including the Δ current (solid) and only the nucleonic current (dashed).

2. G.H. Martinus, O. Scholten, J.A. Tjon: Phys. Rev. **C56,** 2945 (1997); **58** 686 (1998)

3. K. Michaelian et al.: Phys. Rev. **D41,** 2689 (1990)

4. B.V. Przewoski et al.: Phys. Rev. **C45,** 2001 (1992)

5. R. Bilger et al.: Phys. Lett. **B429,** 195 (1998)

6. N. Kalantar-Nayestanaki et al.: submitted to Phys. Rev. Lett.

7. K. Yasuda et al.: Phys. Rev. Lett. **82,** 4775 (1999)

8. K. Sagara et al.: RCNP Annual Report 1995, p. 158

9. M. Fujiwara et al.: Nucl. Instrum. Methods Phys. Res. Sect. A **422**, 484 (1999)

10. N. Matsuoka et al.: RCNP Annual Report 1990, p. 235

Few-Body Systems Suppl. 12, 297–301 (2000)

Few-Body Systems

The $\gamma^* N \leftrightarrow \Delta$ Transition Form Factors

S.S. Kamalov*, Shin Nan Yang

Department of Physics, National Taiwan University, Taipei 10617, Taiwan

Abstract. We show that recent experiment data for the ratios E_{1+}/M_{1+} and S_{1+}/M_{1+} can be explained in a dynamical model for electromagnetic production of pions, together with a simple scaling assumption for the bare $\gamma^* N\Delta$ form factors. Within our model we find that the bare Δ is almost spherical and the electric E2 and Coulomb C2 quardrupole excitations of the physical Δ are nearly saturated by pion cloud contribution in $Q^2 \leq 4.0\ GeV^2$.

It has been well recognized that the study of the excitations of the hadrons can shed light on the nonperturbative aspects of QCD. One case which has recently been under intensive study is the electromagnetic excitation of the $\Delta(1232)$ resonance. At low four-momentum transfer squared Q^2, the interest is motivated by the possibility of observing a D−state in the Δ. The existence of a D−state in the Δ has the consequence that the Δ is deformed and the photon can excite a nucleon through electric $E2$ and Coulomb $C2$ quardrupole transitions. In a symmetric SU(6) quark model, the electromagnetic excitation of the Δ could proceed only via $M1$ transition. In pion electroproduction, $E2$ and $C2$ excitations would give rise to nonvanishing $E_{1+}^{(3/2)}$ and $S_{1+}^{(3/2)}$ multipole amplitudes. Recent experiments give nonvanishing ratio $R_{EM} = E_{1+}^{(3/2)}/M_{1+}^{(3/2)} \sim -0.03$ [1] at $Q^2 = 0$ which has been widely taken as an indication of the Delta deformation.

At sufficiently large Q^2, the perturbative QCD (pQCD) is expected to work. It predicts that only helicity-conserving amplitudes contribute at high Q^2, leading to $R_{EM} = E_{1+}^{(3/2)}/M_{1+}^{(3/2)} \to 1$ and $R_{SM} = S_{1+}^{(3/2)}/M_{1+}^{(3/2)} \to const.$ This behavior in the perturbative domain is very different from that in the nonperturbative one. It is an intriguing question to find the region of Q^2 which signals the onset of the pQCD.

In a recent measurement [2], the electromagnetic excitation of the Δ was studied at $Q^2 = 2.8$ and $4.0\ GeV^2$ via the reaction $p(e, e'p)\pi^0$. The extracted

*Permanent address: Laboratory of Theoretical Physics, JINR Dubna,141980 Moscow region, Russia

ratios R_{EM} and R_{SM} remain small and *negative.* This disagrees with the previous analysis [3] of the earlier DESY data which gave small but *positive* R_{EM} and R_{SM} at $Q^2 = 3.2\ GeV^2$, though both analyses indicate that pQCD is still not applicable in this region of Q^2. In this talk, we want to show that the recent data of Ref. [2] can be understood from the dominance of the pion cloud contribution at low Q^2, in both $E_{1+}^{(3/2)}$ and $S_{1+}^{(3/2)}$, as predicted by a dynamical model [4, 5] for electromagnetic production of pion, together with a simple scaling assumption for the bare $\gamma^* N\Delta$ form factors.

The main feature of the dynamical approach to the pion photo- and electroproduction [4, 5] is that the unitarity is built in by explicitly including the final state πN interaction in the theory, namely, t-matrix is expressed as

$$t_{\gamma\pi}(E) = v_{\gamma\pi} + v_{\gamma\pi}\, g_0(E)\, t_{\pi N}(E)\,, \tag{1}$$

where $v_{\gamma\pi}$ is the transition potential operator for $\gamma^* N \to \pi N$ and, $t_{\pi N}$ and g_0 denote the πN t-matrix and free propagator, respectively, with $E \equiv W$ the total energy in the CM frame.

In the (3,3) channel the transition potential $v_{\gamma\pi}$ consists of two terms

$$v_{\gamma\pi}(E) = v^B_{\gamma\pi} + v^\Delta_{\gamma\pi}(E)\,, \tag{2}$$

where $v^B_{\gamma\pi}$ is the background transition potential which includes Born terms and vector mesons exchange contributions, as described in Ref. [6]. The second term of Eq. (2) corresponds to the contribution of bare Δ.

We decompose Eq. (1) in the following way,

$$t_{\gamma\pi}(E) = t^B_{\gamma\pi} + t^\Delta_{\gamma\pi}(E)\,, \tag{3}$$

where

$$t^B_{\gamma\pi}(E) = v^B_{\gamma\pi} + v^B_{\gamma\pi}\, g_0(E)\, t_{\pi N}(E)\,, \qquad t^\Delta_{\gamma\pi}(E) = v^\Delta_{\gamma\pi} + v^\Delta_{\gamma\pi}\, g_0(E)\, t_{\pi N}(E)\,. \tag{4}$$

The advantage of such a decomposition is that all the processes which start with the electromagnetic excitation of the bare Δ are summed up in $t^\Delta_{\gamma\pi}$.

We evaluate $t^B_{\gamma\pi}$ with pion scattering matrix $t_{\pi N}$ obtained in a meson-exchange model [7]. Note that to make principal value integration associated with $v^B_{\gamma\pi}$ convergent, we introduce an off-shell dipole form factor with cut-off parameter Λ=440 MeV. The gauge invariance, violated due to the off-shell rescattering effects, is restored by the substitution $J^B_\mu \to J^B_\mu - k_\mu k \cdot J^B/k^2$, where J^B_μ is the electromagnetic current corresponding to the background contribution $v^B_{\gamma\pi}$ and k is the photon momentum.

For the the Δ resonance contribution $t^\Delta_{\gamma\pi}$, the corresponding multipole amplitudes A^Δ is evaluated using a Breit-Wigner form, as was done in the isobar model of Ref. [6],

$$A^\Delta(W,Q^2) \;=\; \bar{A}^\Delta(Q^2)\, \frac{f_{\gamma\Delta}\, \Gamma_\Delta\, M_\Delta\, f_{\pi\Delta}}{M^2_\Delta - W^2 - iM_\Delta \Gamma_\Delta}\, e^{i\phi}\,, \tag{5}$$

where $f_{\pi\Delta}(W)$ is the usual Breit-Wigner factor describing the decay of the Δ resonance with total width $\Gamma_\Delta(W)$ and physical mass M_Δ=1232 MeV. The expressions for $f_{\gamma\Delta}$, $f_{\pi\Delta}$ and Γ_Δ are taken from Ref. [6]. The phase $\phi(W)$ in Eq. (5) is to adjust the phase of A^Δ to be equal to the corresponding pion-nucleon scattering phase $\delta^{(33)}$. Note that at the resonance energy, $\phi(M_\Delta) = 0$.

The main parameters in the bare $\gamma^* N\Delta$ vertex are the $\bar{A}^\Delta$'s in Eq. (5). For the magnetic dipole $\bar{\mathcal{M}}^\Delta$ and electric quadrupole $\bar{\mathcal{E}}^\Delta$ transitions they are related to the conventional electromagnetic helicity amplitudes $A^\Delta_{1/2}$ and $A^\Delta_{3/2}$ by

$$\bar{\mathcal{M}}^\Delta(Q^2) = -\frac{1}{2}(A^\Delta_{1/2} + \sqrt{3}A^\Delta_{3/2}), \qquad \bar{\mathcal{E}}^\Delta(Q^2) = \frac{1}{2}(-A^\Delta_{1/2} + \frac{1}{\sqrt{3}}A^\Delta_{3/2}). \quad (6)$$

At the photon point $Q^2 = 0$, the bare amplitudes $\bar{\mathcal{M}}^\Delta(0)$ and $\bar{\mathcal{E}}^\Delta(0)$ of Eq. (5) were extracted from the best fit to the results of the recent analyses of Mainz [8] and VPI group [9] as shown in Fig. 1 by solid curves. The dashed curves denote the contribution from $t^B_{\gamma\pi}$ only. The dotted curves represented the K-matrix approximation to $t^B_{\gamma\pi}$, namely, without the principal value integral term included.

The numerical values for $\bar{\mathcal{M}}^\Delta$ and $\bar{\mathcal{E}}^\Delta$ and the helicity amplitudes, at $Q^2 = 0$, are given in Table. 1. Here we also give "dressed" values obtained using K-matrix approximation for $t^B_{\gamma\pi}$. One notices that the determined above bare values for the helicity amplitudes, which amount to only about 60% of the corresponding dressed values, are close to the predictions of the constituent quark model (CQM), as was pointed out by Sato and Lee [10]. The large reduction of the helicity amplitudes from the dressed to the bares ones results from the fact that the principal value integral part of $t^B_{\gamma\pi}$, which represents the effects of the off-shell pion rescattering, contributes approximately for half of the M_{1+} as indicated by the dashed curves in Fig. 1.

Table 1. Comparison of the "bare" and "dressed" values for the amplitudes $\bar{A}^\Delta$, $A^\Delta_{1/2}$ and $A^\Delta_{3/2}$ (in $10^{-3}\, GeV^{-1/2}$).

Amplitudes	"bare"	"dressed"	PDG
$\bar{\mathcal{M}}^\Delta$	158 ± 2	289 ± 2	293 ± 8
$\bar{\mathcal{E}}^\Delta$	0.4 ± 0.3	-7 ± 0.4	-4.5 ± 4.2
$A^\Delta_{1/2}$	-80 ± 2	-134 ± 2	-140 ± 5
$A^\Delta_{3/2}$	-136 ± 3	-256 ± 2	-258 ± 6

We now turn to the Q^2 evolution of the multipoles in the (3,3) channel. In the present work, we parametrize the Q^2 dependence of the dominant $\bar{\mathcal{M}}^\Delta$ amplitude by

$$\bar{\mathcal{M}}^\Delta(Q^2) = \bar{\mathcal{M}}(0)\frac{|\,\mathbf{k}\,|}{k_\Delta}(1 + \beta Q^2)\, e^{-\gamma Q^2}\, G_D(Q^2)\,, \quad (7)$$

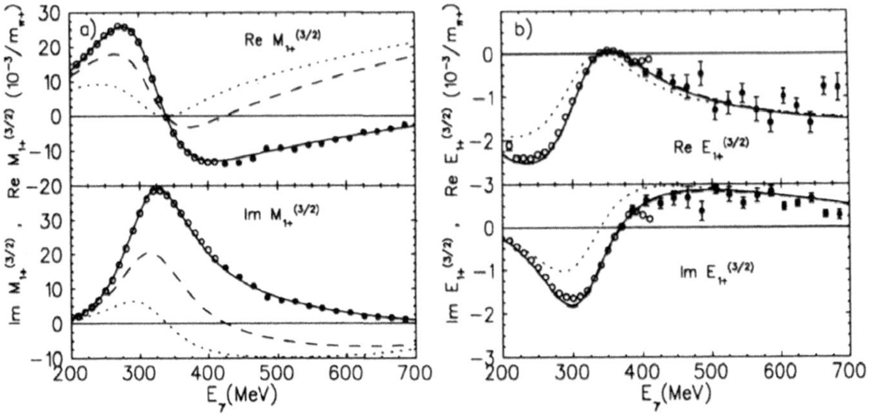

Figure 1. Real and imaginary parts of the, **(a)** $M_{1+}^{(3/2)}$, and **(b)** $E_{1+}^{(3/2)}$ multipoles. The open and full circles are the results from the Mainz dispersion analysis [8] and from the VPI analysis [9]. Notations for the curves are given in the text.

where G_D is the nucleon dipole form factor. For the small $\bar{\mathcal{E}}^\Delta$ and $\bar{\mathcal{S}}^\Delta$ amplitudes, following Ref. [6], we assume that they have the same Q^2 dependence as $\bar{\mathcal{M}}^\Delta$. This is motivated by the scaling law which has been observed for the nucleon form factors.

Using the β and γ in Eq. (7) as free parameters, we fit the recent experimental data [2] as well as old one quoted in Ref. [6] on the Q^2 dependence of the $M_{1+}^{(3/2)}$ multipole or equivalently, the G_M^* form factor defined by Eq. (24) from Ref. [6]. Our result is shown in Fig. 2a. The obtained values for the β and γ parameters are: $\beta = 0.44\ GeV^{-2}$ and $\gamma = 0.38\ GeV^{-2}$. Here the dashed curve corresponds to contribution from the bare Δ, i.e., $t_{\gamma\pi}^{\Delta}$ of Eq. (4). The results for the ratios $R_{EM} = E_{1+}^{(3/2)}/M_{1+}^{(3/2)}$ and $R_{SM} = S_{1+}^{(3/2)}/M_{1+}^{(3/2)}$ are shown in Fig. 2b. It is seen that they are in good agreement with the results of the model independent analysis of Ref. [2] up to Q^2 as high as 4.0 GeV^2. Note that since the bare values for the electric and Coulomb excitations are small, the absolute values and shape of these ratios are determined, to a large extent, by the pion rescattering contribution. The bare Δ excitation contributes mostly to the $M_{1+}^{(3/2)}$ multipole.

In summary, we calculate the Q^2 dependence of the ratios E_{1+}/M_{1+} and S_{1+}/M_{1+} in the $\gamma^* N \to \Delta$ transition, with the use of a dynamical model for electromagnetic production of pions. We find that both ratios E_{1+}/M_{1+} and S_{1+}/M_{1+} remain small and negative for $Q^2 \leq 4.0\ GeV^2$. Our results agree well with the recent measurement of Frolov et al. [2], but deviate strongly from the predictions of pQCD. Our results indicate that the bare Δ is almost spherical and hence very difficult to be directly excited via electric E2 and Coulomb C2 quardrupole excitations. The experimentally observed $E_{1+}^{(3/2)}$ and $S_{1+}^{(3/2)}$ multipoles are, to a very large extent, saturated by the contribution from pion cloud, i.e., pion rescattering effects.

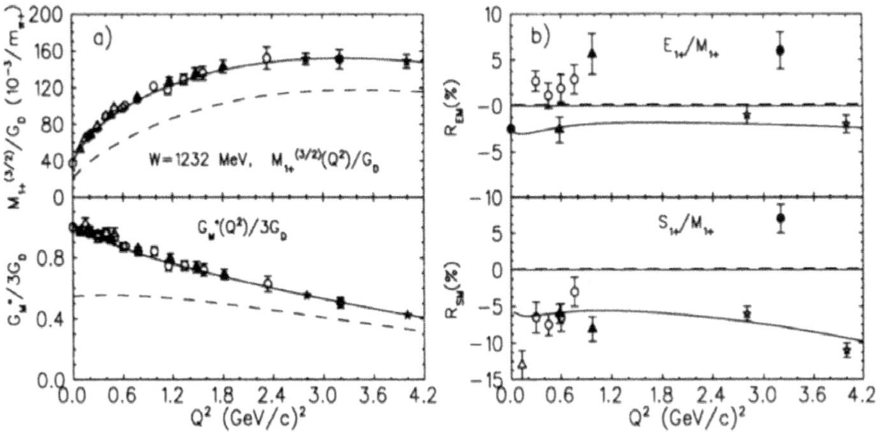

Figure 2. The Q^2 dependence of **(a)** $Im\,M_{1+}^{(3/2)}$ with corresponding G_M^* form factor, and **(b)** ratios $E_{1+}^{(3/2)}/M_{1+}^{(3/2)}$ and $S_{1+}^{(3/2)}/M_{1+}^{(3/2)}$ at W=1232 MeV. The full and dashed curves are the results for the "dressed" and "bare" $\gamma N\Delta$ vertexes, respectively. Experimental data as quoted in Ref. [6]. The new data at Q^2=2.8 and 4.0 $(GeV/c)^2$ are from Ref. [2].

References

1. R. Beck et al.: Phys. Rev. Lett. **78**, 606 (1997); G. Blanpied et al.: Phys. Rev. Lett. **79**, 4337 (1997)
2. V.V. Frolov et al.: Phys. Rev. Lett. **82**, 45 (1999)
3. V. Burkert and L. Elouadrhiri: Phys. Rev. Lett. **75**, 3614 (1995)
4. S.N. Yang: J. Phys. **G11**, L205 (1985)
5. H. Tanabe and K. Ohta: Phys. Rev. **C31**, 1876 (1985)
6. D. Drechsel, O. Hanstein, S.S. Kamalov, L. Tiator: Nucl. Phys. **A645**, 145 (1999)
7. C.T. Hung, S.N. Yang, T.-S.H. Lee: J. Phys. **G20**, 1531 (1994); C. Lee, S.N. Yang, T.-S.H. Lee: *ibid.* **G17**, L131 (1991)
8. O. Hanstein, D. Drechsel, L. Tiator: Nucl. Phys. **A632**, 561 (1998)
9. R.A. Arndt, I.I. Strakovsky, R.L. Workman: Phys. Rev. **C53**, 430 (1996) (SP97 solution of the VPI analysis)
10. T. Sato and T.-S.H. Lee: Phys. Rev. **C54**, 2660 (1996); T.-S. H. Lee: in N^* *Physics*, eds. T.-S.H. Lee and W. Roberts, p.19. Singapore: World Scientific 1997

Few-Body Systems Suppl. 12, 302–310 (2000)

Few-Body Systems

Measurement of Elastic Electron-Deuteron Scattering at Large Momentum Transfers

A. Saha*

Thomas Jefferson National Accelerator Facility
12000 Jefferson Avenue, Newport News, VA 23606, USA
E-mail: saha@jlab.org

Abstract. Coincidence measurements of elastic electron-deuteron cross sections have been performed in the range of Q^2 between 0.7 and 6.0 $(\mathrm{GeV/c})^2$. The extracted values of the deuteron electric structure function $[A(Q^2)]$ in this kinematic range are presented. Values of the deuteron magnetic structure function $[B(Q^2)]$ in the range $0.7 \leq Q^2 \leq 1.3$ $(\mathrm{GeV/c})^2$ were also extracted and preliminary values are presented. These data are compared with theoretical predictions of both nucleon-meson and quark-gluon based models.

1 Introduction

The unpolarized elastic electron-deuteron cross section is given by:

$$\frac{d\sigma}{d\Omega} = \frac{\alpha^2 \; E' \; \cos^2\left(\frac{\theta}{2}\right)}{4 \; E^3 \; \sin^4\left(\frac{\theta}{2}\right)} \left[A(Q^2) + B(Q^2) \; \tan^2\left(\frac{\theta}{2}\right)\right], \qquad (1)$$

where α is the fine structure constant, E and E' are the incident and scattered electron energies, θ is the electron scattering angle, and $Q^2 \sim 4EE' \sin^2(\theta/2)$ is the invariant four-momentum transfer. The exchange of a single virtual photon is assumed.

The electric $[A(Q^2)]$ and magnetic $[B(Q^2)]$ deuteron structure functions are given by:

$$A(Q^2) = F_c^2(Q^2) + \frac{8}{9}\tau^2 F_q^2(Q^2) + \frac{2}{3}\tau F_m^2(Q^2), \qquad (2)$$

$$B(Q^2) = \frac{4}{3}\tau(1+\tau) F_m^2(Q^2), \qquad (3)$$

where $\tau = Q^2/4M^2$ and M is the deuteron rest mass. $[F_c(Q^2)]$ is the charge monopole, $[F_q(Q^2)]$ is the charge quadrupole and $[F_m(Q^2)]$ is the magnetic dipole form factors.

*Representing the JLab Hall A collaboration.

The structure functions $A(Q^2)$ and $B(Q^2)$ can be separated by the Rosenbluth separation technique of measuring two or more unpolarized cross section values at a constant value of Q^2 and different electron scattering angles θ. To separate the individual $F_c(Q^2)$ and $F_q(Q^2)$ contributions, an additional polarization observable is needed.

The experiment being reported here measured unpolarized cross sections in the range of $0.7 \leq Q^2 \leq 6.0$ (GeV/c)^2 and $15.5° \leq \theta \leq 40.6°$ to extract $A(Q^2)$ [forward data]. Cross sections were also measured in the range of $0.7 \leq Q^2 \leq 1.3$ (GeV/c)^2 at a fixed angle of $\theta = 144.5°$ [backward data] which allowed separation of $A(Q^2)$ and $B(Q^2)$ in this Q^2 range via the Rosenbluth technique.

2 Experiment

The experiment was conducted in Hall A at the Thomas Jefferson National Accelerator Facility (JLab). Incident electron beams with a 100% duty factor, energies from 0.5 to 4.4 GeV, and beam currents from 5 to 120 μA were used. Beam current and energy uncertainties were estimated to be ±2% and ±0.2%, respectively. Uncertainties due to beam position and angle at the target are negligible.

The target system consisted of three independent loops. Two of the loops were filled with liquid deuterium and the third with liquid hydrogen. Each loop consisted of a 15 cm long cylindrical target cell with its long axis of symmetry parallel to the beam path, a heat exchanger to remove the heat deposited by the beam in the liquid, and a fan system that circulated the liquid between the target cell and the heat exchanger. The heat exchange system maintained the liquid deuterium (hydrogen) at a temperature of 22 (19) K and a pressure of 22 (27) psia.

One of the deuterium targets was used exclusively for the forward-angle measurements while the other was used to acquire all the backward data. The latter target had a collimator attached to it on the electron spectrometer side. The collimator allowed the electron spectrometer to view a well-defined section of the target cell (~10cm) at 144.5° without interference from angle dependent obstructions due to the target cell design used by this experiment.

In order to reduce beam-induced density changes, the beam was rastered in both the horizontal and vertical direction in a ~2x2 mm^2 pattern at frequencies of 17.6 and 24.4 kHz respectively. Measured beam-induced deuterium density changes were ~2.5% at 120 μA. A 15 cm long "empty" target was used to measure possible contributions from the full cell end-caps to the measured cross sections. They were found to be negligible.

The Hall A experimental facility consists of two identical High Resolution Spectrometers (HRS) of the type QQDQ. The dipole has a built-in index which provides additional focusing. One of the HRSs was set to bend negatively charged particles into its focal plane (electron HRS) while the other for positive particles (recoil HRS). Each HRS was instrumented with two planes of plastic scintillators for triggering and timing as well as a pair of vertical drift chambers for track reconstruction. In addition, the electron HRS was equipped with a CO_2 gas Čerenkov and a lead-glass calorimeter for electron identification. The trigger logic was set to accept all electron-recoil spectrometer coincidences as well as samples of single-arm triggers for detector efficiency studies.

Coincidence elastic electron-proton (*e-p*) cross sections were measured in this experiment to check our understanding of spectrometer optics and double-arm acceptance. The *e-p* kinematics were selected such that the electron-recoil solid angle Jacobian for *e-p* was the same as for electron-deuteron (*e-d*) for all forward angle data. For the $\theta = 144.5°$ *e-d* data, the corresponding *e-p* data was taken at a fixed angle $\theta = 90°$ to avoid any partial occlusion of the hydrogen target cell. The Jacobian ratio of *e-p* to *e-d* was ~ 0.7 for these conditions. All the *e-p* and the low Q^2 *e-d* data were taken with and without solid-angle defining collimators in front of the spectrometers. The large Q^2 *e-d* data was taken without collimators due to the small cross sections measured.

In the data analysis, the electron spectrometer section of a raw coincidence event was first validated by requiring a pulse height in the Čerenkov counter above pedestal and an energy deposition in the calorimeter of at least 60% of the momentum determined from the drift chamber track. Raw coincidence events passing the above cuts were then required to have relative time-of-flight (TOF) between the electron and recoil triggers of ± 7.5 nanoseconds or less. Random coincidences were negligible.

The elastic *e-p* and *e-d* cross sections were calculated according to:

$$\frac{d\sigma}{d\Omega} = \frac{N_{ep(ed)}\ C}{N_i\ N_t\ F\ \Delta\Omega}, \tag{4}$$

where $N_{ep(ed)}$ is the number of *e-p* (*e-d*) coincidence events passing the cuts described above, N_i is the number of incident electrons, N_t is the number of target-nuclei/cm^2, $\Delta\Omega$ is the effective double-arm solid angle including the spectrometer-acceptance dependent part of the radiative corrections, F is the portion of the radiative corrections that depends only on Q^2 and target thickness. C is the product of corrections such as detector inefficiency (1%), trigger inefficiency including proton (2.5-2.8%) and deuteron (3.2-3.5%) absorption in the scintillators, computer dead time (typically 10% for *e-d*), and proton (forward data: 0.7-1.8%, backward data: 0.8%) and deuteron (forward data: 2.8-5.1%, backward data: 5.6-8.2%) absorption in the target and spectrometer windows.

The effective double-arm solid angle $\Delta\Omega$ was evaluated with a Monte Carlo computer program that simulated elastic *e-p* and *e-d* scattering under identical conditions as our measurements. The program traced scattered electrons and recoil nuclei from the target to the detectors through models representing the magnetic characteristics, physical apertures, and alignment of each HRS. The effects from ionization energy losses and multiple scattering in the target and vacuum windows were taken into account for both electrons and recoil nuclei. Bremsstrahlung radiation losses for both incident and scattered electrons in the target and vacuum windows as well as internal radiative effects were also taken into account. Details on this simulation method can be found in Katramatou *et al.* [1]

The elastic *e-p* cross sections measured with the acceptance-defining collimators were found to agree, on average, within 0.3% with the values calculated from a recent parametrization of the proton electromagnetic form factors by Bosted *et al.* [2] Elastic *e-p* cross sections measured without collimators were found to be higher by 2.6% than the values calculated from the same parametrization. Consequently, the elastic *e-d* cross sections taken without collimator have been normalized by this amount.

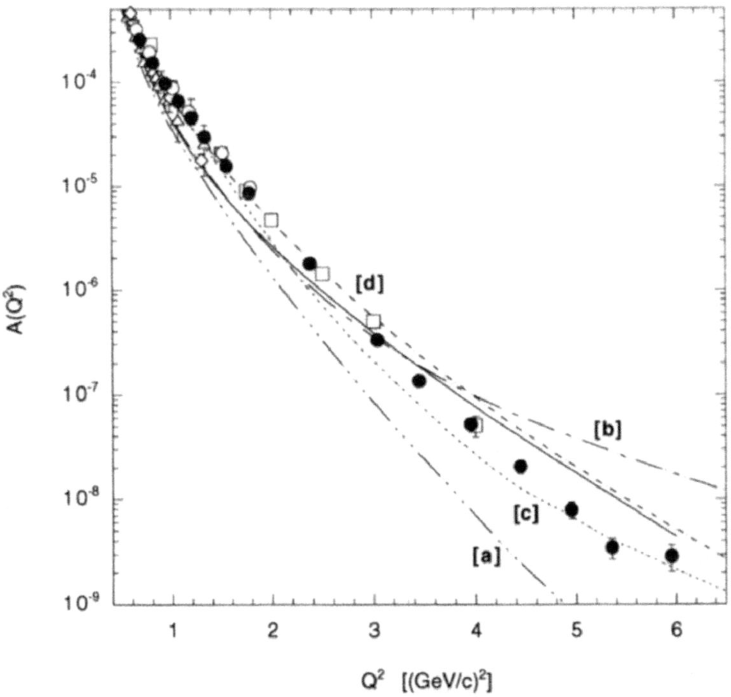

Figure 1. Shown are the values of $A(Q^2)$ obtained by this experiment [3] [•] together with existing data from CEA [4] [△], SLAC [5] [▫], Bonn [6] [⋄], Saclay [7] [▽] and, JLab-C [8] [∘]. Also shown are theoretical predictions from Hummel and Tjon [9] without [a] and with [b] meson exchange currents (MEC), Van Orden *et al.* [10] [c] and Carbonell and Karmanov [14] [d]. The solid line represents the model of Lev *et al.*[15]

3 Results

Values of $A(Q^2)$ were extracted from the measured forward-angle *e-d* cross sections under the assumption that $B(Q^2)$ does not contribute in any sizable way to the cross section (supported by the existing $B(Q^2)$ data). The extracted $A(Q^2)$ values are presented in Fig. 1. The error bars represent statistical and

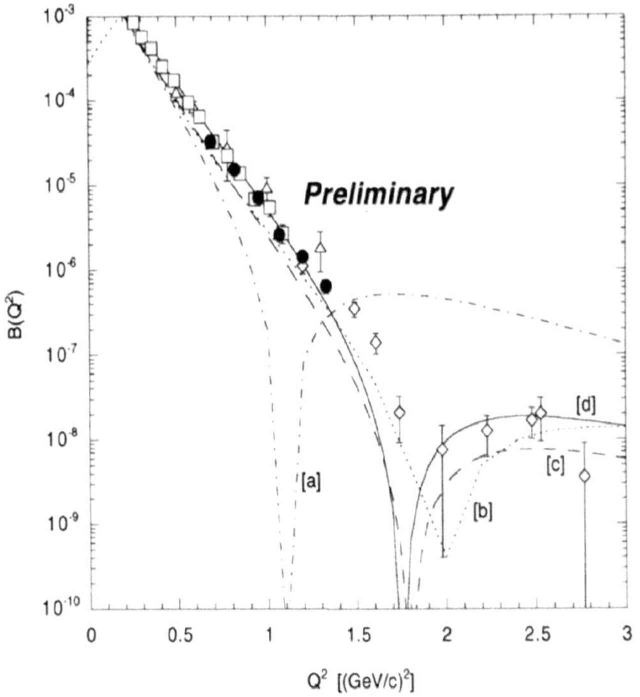

Figure 2. Preliminary values of $B(Q^2)$ obtained by this experiment [•] together with existing data from Bonn[6] [△], Saclay[18] [□] and, SLAC[19] [◇]. Also shown are theoretical predictions from Carbonell and Karmanov [14] [a], Lev *et al.* [15] [b], Hummel and Tjon [9] [c] and, Van Orden *et al.* [10] [d].

systematic uncertainties added in quadrature. The statistical error ranged from $\pm 1\%$ to $\pm 30\%$ at $Q^2 = 6$ (GeV/c)^2. The systematic error has been estimated to be $\sim \pm 6\%$. Our results agree very well with previous SLAC data and continue to follow the trend of a smooth fall-off versus momentum transfer. Figure 2 shows our preliminary values of $B(Q^2)$. The error bars include statistical and systematic uncertainties added in quadrature and range from 5% at 0.7 (GeV/c)^2 to 20% at 1.3 (GeV/c)^2. Our data agree well with previously existing data but do not reach far enough in momentum transfer to confirm the location of the $B(Q^2)$ minimum observed in the SLAC data.

Results from several theoretical calculations are also shown in Figs. 1 and 2. The predictions by Hummel and Tjon[9] (HT) are shown as curves [a] and [b] in Fig. 1 and [c] in Fig. 2. The HT calculations use the Blankenbecler-Sugar equation,[13] a "quasipotential" approximation of the Bethe-Salpeter (BS) equation[11] obtained by allowing both nucleons to be equally off-shell so that their relative energy is set to zero. The nucleon form factor parametrization of Höhler *et al.*[16] is used. Calculations of $A(Q^2)$ based on the relativistic impulse approximation (RIA) without meson exchange currents (MEC) typically lack enough strength at large Q^2 to account for the data (curve [a] of Fig. 1). Additional strength is obtained by including the effects of MEC and isobar contributions in the calculations. For elastic scattering from the deuteron, only isoscalar two-body exchange currents such as $\rho\gamma\pi$, $\omega\sigma\gamma$, $\omega\eta\gamma$, etc. contribute. However, the form factors for these exchange currents are neither experimentally known nor theoretically constrained by an invariance. Their behavior needs to be modeled. In the HT case, the $\rho\pi\gamma$ and $\omega\sigma\gamma$ MEC have been included with form factors given by the vector dominance model (VDM) (curve [b] of Fig. 1 and curve [c] of Fig. 2). In this model, the $\rho\pi\gamma$ MEC has a larger effect in $A(Q^2)$ than the $\omega\sigma\gamma$ MEC. The situation is reversed for $B(Q^2)$.

Curves [c] of Fig. 1 and [d] of Fig. 2 show the predictions of Van Orden *et al.*[10] This model uses the Spectator (or Gross) equation,[12] a "quasipotential" approximation of the BS equation obtained by restricting one of the nucleons to its positive energy mass shell. The model includes a $\rho\pi\gamma$ MEC evaluated in a covariant-separable-quark model[21] which gives a "softer" Q^2 dependence of the $\rho\pi\gamma$ form factor. The location of the $B(Q^2)$ diffraction minimum in the Van Orden *et al.* model is very sensitive to the interference between S and D components of the deuteron wave function with the two smaller P-state components.

The predictions by Carbonell and Karmanov[14] are shown as curves [d] and [a] of Figs. 1 and 2 respectively. The approach taken by these authors uses an explicitly covariant version of the Light-Front Dynamics (LFD) but with a deuteron wavefunction which is calculated perturbatively from the non-relativistic wave function.[22] For the one-boson exchange kernel found in the LFD calculation, the set of mesons, coupling constants, and form factors corresponding to the Bonn model[23] were used. The nucleon form factors used are those of Mergell *et al.*[24] The results of the calculation for $A(Q^2)$ agree well with the data up to $Q^2 \sim 2.5$ (GeV/c)^2 (see Fig. 1). Departure between the experiment and theory for larger Q^2 is attributed to the use of the non-relativistic deuteron wave function. The location of the $B(Q^2)$ diffraction minimum predicted by this model (see Fig. 2) is sensitive to the interference among various components of the deuteron wave function. The authors believe that the perturbative approach used to calculate the deuteron wave function does not provide enough accuracy in the calculation of these components.

The solid line of Fig. 1 and curve [b] of Fig. 2 show the predictions of Lev, Pace, and Salme[15] for $A(Q^2)$ and $B(Q^2)$. The authors use the front-form approach but select the Breit frame to build a current operator (obtained from the free current) which meets all the requirements of extended Poincare covariance. The predictions shown make use of the nucleon form factor parametrization of Höhler *et al.*[16] and the Argonne AV18[17] nucleon-nucleon potential.

Deuteron models based on dimensional scaling[26, 27] and perturbative QCD[28] predict the deuteron form factor $F_d(Q^2) \equiv \sqrt{A(Q^2)}$ to fall as $(Q^2)^{-5}$. Consequently, the quantity $A(Q^2) \times (Q^2)^{10}$ should scale. Our data exhibits a scaling behavior compatible with those expectations (top panel of Fig. 3). The bottom panel of Fig. 3 shows the reduced deuteron form factor $f_d(Q^2) \equiv F_d(Q^2)/F_N^2(Q^2/4)$ where the two powers of the nucleon form factor

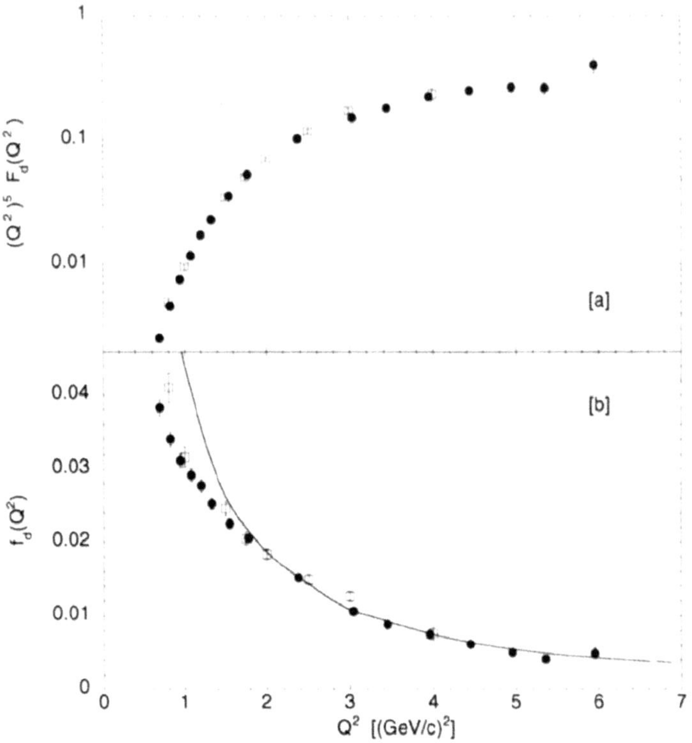

Figure 3. The deuteron form factor $F_d(Q^2)$ times $(Q^2)^5$ [a] and the reduced deuteron form factor $f_d(Q^2)$ [b] from this experiment [•] and from SLAC[5] [□]. The solid curve represents the asymptotic pQCD prediction of Brodsky *et al.*[25] for $\Lambda = 100$ MeV, arbitrarily normalized to the data at $Q^2 = 4$ (GeV/c)^2.

$F_N(Q^2) = (1 + Q^2/0.71)^{-2}$ remove in a minimal way the effects of nucleon structure.[29] Our $f_d(Q^2)$ data appear to follow, for $Q^2 > 2$ (GeV/c)^2, the asymptotic Q^2 prediction of pQCD.[25] Although several authors have questioned the validity of dimensional scaling and pQCD at the momentum transfers of this experiment,[30, 31] similar scaling-like behavior has been reported in deuteron photodisintegration cross sections at moderate photon energies.[32]

In summary, we have extended the measured range of $A(Q^2)$ up to 6 (GeV/c)^2 and clarified inconsistencies in previous low Q^2 data. Our $B(Q^2)$ results confirm previous measurements[19] and improve their precision. The measured $A(Q^2)$ data have a dependence on momentum transfer which is qualitatively reproduced by the conventional nuclear physics models presented above. The results are also consistent with predictions of dimensional quark scaling and perturbative QCD. On the theoretical side, the location of the $B(Q^2)$ diffraction minimum is very sensitive to the details of the model used. It is therefore essential to extend the measurements of $B(Q^2)$ to higher momentum transfers to test the validity of the various models proposed. Future measurements at higher Q^2 of $A(Q^2)$ and $B(Q^2)$ as well as of the elastic form factors of the helium isotopes are also essential to test the validity of the apparent scaling behavior observed.

We gratefully acknowledge the outstanding support of the staff of the Accelerator and Physics Divisions of JLab. This work is supported in part by the U.S. Department of Energy and the National Science Foundation.

References

1. A.T. Katramatou et al.: Nucl. Instr. Meth. **A267**, 448 (1988); A.T. Katramatou: SLAC-NPAS-TN-86-08, 1986 (unpublished)
2. P.E. Bosted et al.: Phys. Rev. **C51**, 409 (1995)
3. L.C. Alexa et al.: Phys. Rev. Lett. **82**, 1374 (1999)
4. J.E. Elias et al.: Phys. Rev. **177**, 2075 (1969)
5. R.G. Arnold et al.: Phys. Rev. Lett. **35**, 776 (1975)
6. R. Cramer et al.: Z. Phys. **C29**, 513 (1985)
7. S. Platchkov et al.: Nucl. Phys. **A510**, 740 (1990)
8. D. Abbott et al.: Phys. Rev. Lett. **82**, 1379 (1999)
9. E. Hummel and J.A. Tjon: Phys. Rev. Lett. **63**, 1788 (1989); E. Hummel and J.A. Tjon: Phys. Rev. **C42**, 423 (1990)
10. J.W. Van Orden, N. Devine, F. Gross: Phys. Rev. Lett. **75**, 4369 (1995) and references therein
11. E.E. Salpeter and H.A. Bethe: Phys. Rev. **84**, 1232 (1951)
12. F. Gross: Phys. Rev. **186**, 1448 (1969); Phys. Rev. **D10**, 223 (1974); Phys. Rev. **C26**, 2203 (1982)
13. R. Blankenbecler and R. Sugar: Phys. Rev. **142**, 1051 (1966)
14. J. Carbonell and V.A. Karmanov: Los Alamos preprint nucl-th/9902053
15. F.M. Lev, E. Pace, G. Salme: Los Alamos preprint nucl-th/9812030
16. G. Höhler et al.: Nucl. Phys. **B114**, 505 (1975)
17. R.B. Wiringa, V.G.J. Stoks, R. Schiavilla: Phys. Rev. **C51**, 38 (1995)
18. S. Auffret et al.: Phys. Rev. Lett. **54**, 649 (1985)
19. R.G. Arnold et al.: Phys. Rev. Lett **58**, 1723 (1987): P.E. Bosted et al.: Phys. Rev. **C42**, 38 (1990)

20. H. Ito and F. Gross: Phys. Rev. Lett. **71**, 2555 (1993)

21. K.L. Mitchell: Ph.D. Dissertation, Kent State University, 1995; P.C. Tandy: Prog. Part. Nucl. Phys. **36**, 97 (1996)

22. J. Carbonell and V.A. Karmanov: Nucl. Phys. **A581**, 625 (1995)

23. R. Machleidt, K. Holinde, Ch. Elster: Phys. Reports **149**, 1 (1987)

24. P. Mergell, U-G. Meissner, D. Drechsell: Nucl. Phys. **A596**, 367 (1996)

25. S.J. Brodsky, C-R. Ji, G.P. Lepage: Phys. Rev. Lett. **51**, 83 (1983)

26. S.J. Brodsky and G.R. Farrar: Phys. Rev. Lett. **31**, 1153 (1973)

27. V Matveev et al.: Nuovo Cimento Lett. **7**, 719 (1973)

28. S.J. Brodsky et al.: Phys. Rev. Lett. **51**, 83 (1983)

29. S.J. Brodsky and B.T. Chertok: Phys. Rev. Lett. **37**, 269 (1975)

30. N. Isgur and C.H. Llewellyn Smith: Phys. Rev. Lett. **52**, 1080 (1984); Phys. Lett. **217**, 535 (1989)

31. G.R. Farrar, K. Huleihel, H. Zhang: Phys. Rev. Lett. **74**, 650 (1995)

32. C. Bochna et al.: Phys. Rev. Lett. **81**, 4576 (1998)

Few-Body Systems Suppl. 12, 311–316 (2000)

Few-Body Systems

The Nucleon-Nucleon and Hyperon-Nucleon Interactions in the Quark Model

Y. Fujiwara[1,*], T. Fujita[1], M. Kohno[2], C. Nakamoto[3], Y. Suzuki[4]

[1] Department of Physics, Kyoto University, Kyoto 606-8502, Japan
[2] Physics Division, Kyushu Dental College, Kitakyushu 803-8580, Japan
[3] Suzuka National College of Technology, Suzuka 510-0294, Japan
[4] Department of Physics, Niigata University, Niigata 950-2181, Japan

Abstract. The SU_6 quark model for the NN and YN interactions, developed by Kyoto-Niigata group, has been upgraded to incorporate the momentum-dependent Bryan-Scott term and the vector-meson exchange potential acting between quarks. The phase-shift agreement in the NN sector at the non-relativistic energies up to $T_{lab.} = 350$ MeV is greatly improved. The single-particle potential calculated from the $t^{eff}\rho$ prescription shows the correct repulsive behavior in the high-momentum region. The essential feature of the ΛN-ΣN coupling is qualitatively similar to that obtained from the previous models. The G-matrix calculation of the ΛN-ΣN coupled-channel system in the model FSS shows that the Σ single-particle potential is repulsive in ordinary nuclear matter. The single-particle ℓs strength for Λ is found to be very small, in comparison with that of N.

1 Introduction

An important purpose of the quark-model (QM) study of the baryon-baryon interaction is to gain a natural and accurate understanding of the fundamental strong interaction, in which the quark-gluon degree of freedom is the most economical ingredient to describe the short-range part of the interaction, while the meson-exchange process is another physical entity dominating the medium- and long-range parts of the relative distance. The naive three-quark ($3q$) model for the nucleon (N) was applied to the ($3q$) - ($3q$) system in an attempt to gain a more fundamental understanding of the short-range repulsion of the NN interaction. We have recently achieved a simultaneous and realistic description of the NN and YN interactions in the RGM formalism of the spin-flavor SU_6

* *E-mail address:* fujiwara@ruby.scphys.kyoto-u.ac.jp

QM. [1, 2] In this model the effective quark-quark interaction is built by combining a phenomenological quark-confining potential and the colored version of the Fermi-Breit (FB) interaction with minimum effective meson-exchange potentials (EMEP) of scalar and pseudo-scalar meson nonets directly coupled to quarks. The flavor symmetry breaking for the YN system is explicitly introduced through the quark-mass dependence of the Hamiltonian. An advantage of introducing the EMEP at the quark level lies in the stringent relationship of the flavor dependence of the various NN and YN interaction pieces. In this way we can utilize our rich knowledge of the NN interaction to minimize the ambiguity of model parameters, which is crucial since the present experimental data for the YN interaction are very scarce.

In this report we upgrade the model FSS [2] to incorporate vector mesons and the momentum-dependent Bryan-Scott term, and compare the NN and YN observables with the existing experimental data. The agreement to the phase-shift parameters in the NN sector is greatly improved. The present

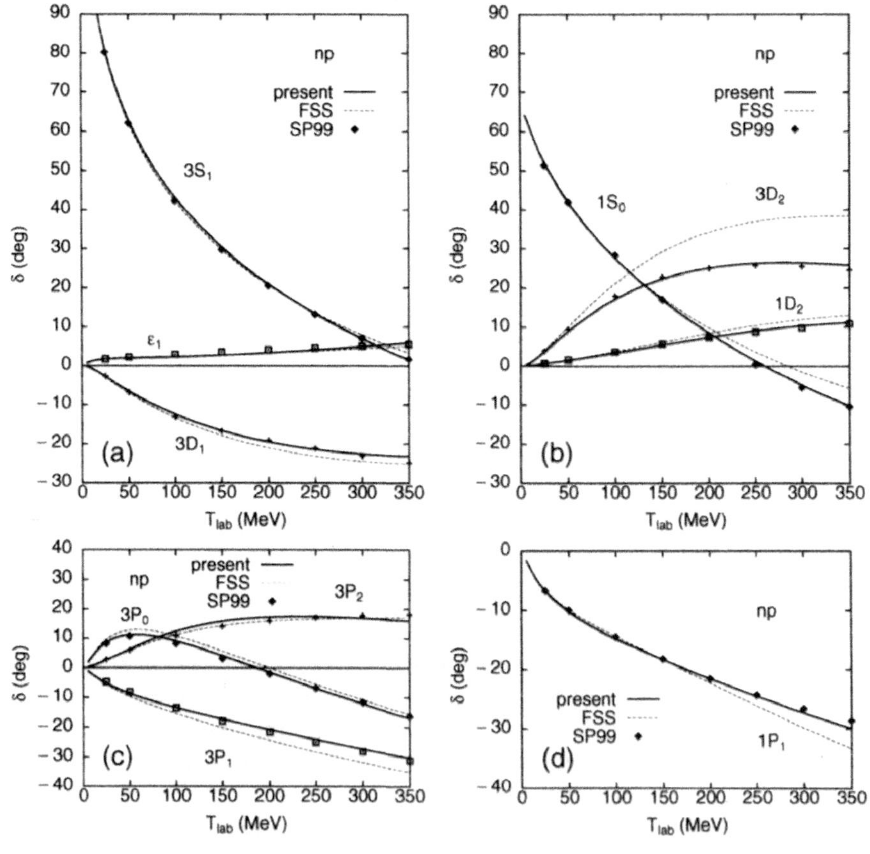

Figure 1. The *np* phase-shift parameters compared with the most recent phase-shift analysis SP99 [3]. The result by FSS is also shown by dashed curves.

model shares the good reproduction of the YN scattering data and the essential feature of the ΛN-ΣN coupling with our previous models. We also report the single-particle (s.p.) potentials of N, Λ and Σ, predicted by FSS through the G-matrix calculation in terms of the explicit quark-exchange kernel. The strength of the s.p. spin-orbit potential is also examined by using these G-matrices.

2 A New Model for NN and YN

A new version of our QM is generated from the Hamiltonian which consists of some extra pieces of interaction generated from the scalar- (S-), pseudoscalar- (PS-) and vector- (V-) meson-exchange potentials acting between quarks:

$$H = \sum_{i=1}^{6}\left(m_i c^2 + \frac{p_i^2}{2m_i}\right) + \sum_{i<j}^{6}\left(U_{ij}^{Cf} + U_{ij}^{FB} + \sum_{\beta} U_{ij}^{S\beta} + \sum_{\beta} U_{ij}^{PS\beta} + \sum_{\beta} U_{ij}^{V\beta}\right).$$

It is important to include the momentum-dependent Bryan-Scott term in the S- and V-meson contributions, in order to ensure the correct asymptotic behavior of the s.p. potentials in high-momentum region. We have calculated $U_N(q_1)$ by using the so-called $t^{eff}\rho$ prescription and made sure that it turns into repulsive

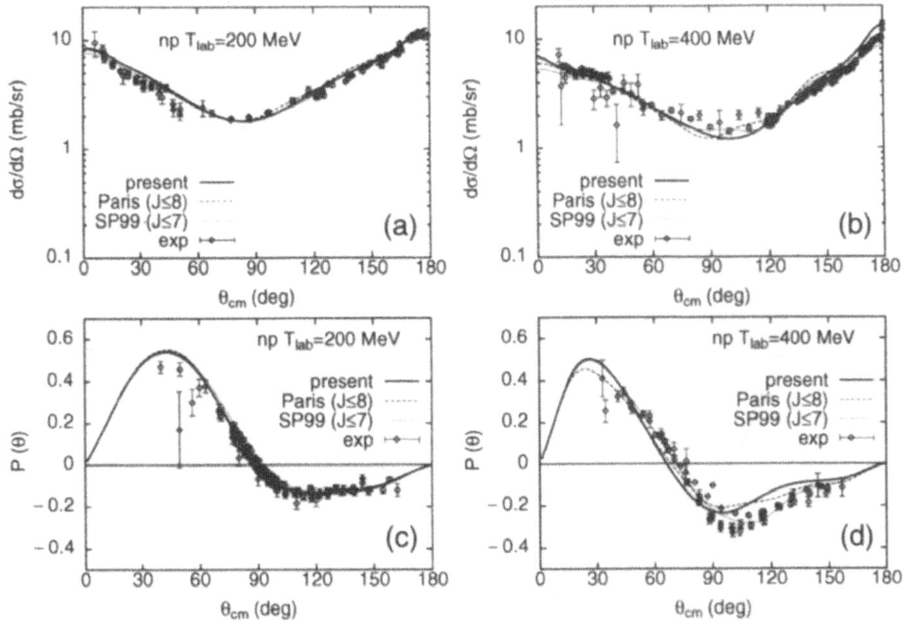

Figure 2. The differential cross sections and the polarization for the np scattering at T_{lab} = 200 and 400 MeV. The predictions by the Paris potential (Paris) and the phase-shift analysis (SP99) are also shown by the dashed curve and the dotted curve.

beyond the incident momentum $q_1 \sim 6\ \text{fm}^{-1}$, while our old model FSS is too attractive, having the minimum value ~ -70 MeV around $q_1 \sim 10\ \text{fm}^{-1}$. Another important feature of the present model is the introduction of vector mesons for improving the fit to the phase-shift parameters in the NN sector. Since the dominant effect of the ω-meson repulsion and the LS components of ρ, ω and K^* mesons are already accounted for by the FB interaction, only the quadratic LS (QLS) component of the octet mesons is expected to play an important role to cancel partially the strong one-pion tensor force. The FB tensor term is augmented by a factor of about 3, since the ε_1 parameter of the ${}^3S_1 + {}^3D_1$ coupling in the NN system is too small at higher energies. The delta-function contact term involved in the π-meson spin-spin part is reduced by a factor 0.381 as before, since the mass is exceptionally small. The spin-spin term from the QLS component is included with a factor 0.6352. The η and η' mesons are not included this time. The well-known unbalance of the ${}^3S$-${}^1S$ NN repulsion is resolved by taking a slightly smaller isovector S-meson mass $m_\delta =$ 720 MeV/c^2 only for NN. For the YN system, $m_\delta = 840$ MeV/c^2 is adopted. The model parameters including these are determined by fitting the most recent result of the phase shift analysis, SP99 [3], for the np scattering with the partial waves $J \leq 2$ and the incident energies $T_{lab} \leq 350$ MeV, under the constraint of the deuteron binding energy and the ${}^1S_0$ NN scattering length, as well as the low-energy YN total cross sections. The deuteron D-state probability is about 6 %, which is similar to FSS.

Figures 1 shows some important low-partial wave NN phase-shift parameters, compared with the experiment SP99. The previous result by FSS is also shown with the dotted curves. The ${}^3D_2$ phase shift is greatly improved by the QLS component. The good accuracy of the NN phase-shift parameters con-

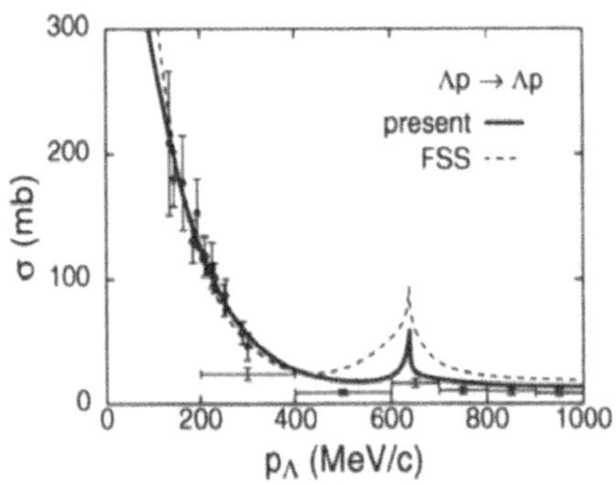

Figure 3. The Λp total cross sections compared with the experimental data [4]. The FSS result is also shown by the dashed curve.

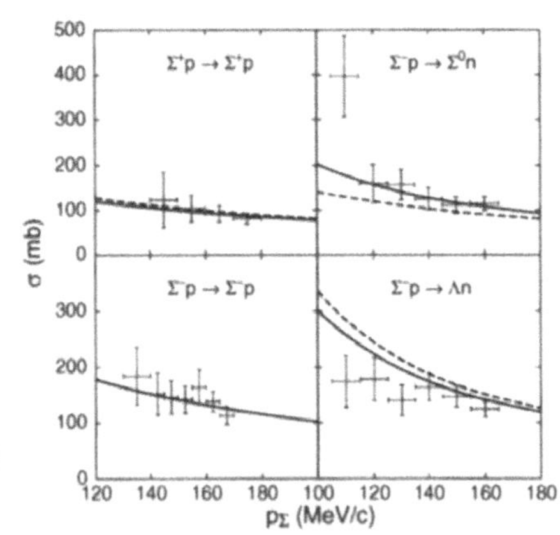

Figure 4. The low-energy Σ^+p and Σ^-p "total" cross sections compared with the experiment [4]. The dashed curves denote the original values without the Coulomb and Σ^-p-Σ^0n threshold corrections.

tinues up to $T_{lab} \sim 600$ MeV, where the inelasticity of the S-matrix becomes appreciable. The very good fit to the differential cross sections and the polarization is shown in Fig. 2 for $T_{lab} = 200$ and 400 MeV. Figures 3 and 4 show the Λp, and $\Sigma^+ p$, $\Sigma^- p$ total cross sections, respectively. We find that the previous enhancement of Λp total cross sections, found in FSS at the ΣN threshold, is hindered. In the present model, the $\Sigma N(I = 1/2)$ $^3P_1$ resonance still stays at the ΣN channel. For the ΛN-ΣN coupling, the present $^3S_1$-$^3D_1$ tensor coupling by the one-pion tensor force is rather close to our previous model RGM-F [1], while the $^1P_1$-$^3P_1$ coupling by the FB $LS^{(-)}$ force is similar to RGM-H [2]. For the $\Sigma^- p$ reactions, we have three outgoing channels; $\Sigma^- p$, $\Sigma^0 n$, and Λn. Figure 4 shows that our $\Sigma^- p \to \Lambda n$ reaction cross section is still somewhat too large. On the other hand, the $\Sigma^- p$ inelastic capture ratio at rest is given by $r_R = 0.458$, which reasonably agrees with the recent empirical values $r_R{}^{exp} = 0.474 \pm 0.016$ and 0.465 ± 0.011 [4].

3 G-Matrix Calculation by FSS

We have calculated the G-matrices for the NN, ΛN and ΣN systems by using the model FSS. The self-consistent procedure using the continuous prescription for intermediate spectra yields the central- and ℓs-type s.p. potentials for the nucleon and hyperons in symmetric nuclear matter. Partial-wave contributions of the s.p. potential $U_B(q_1 = 0)$ for the Λ and Σ at $k_F = 1.35$ fm^{-1} are reported in ref. [5]. The FSS produces a larger potential depth $U_\Lambda(0)$ than the Nijmegen soft-core potential NSC89 [6]. On the other hand, $U_\Sigma(0)$ has turned out to be repulsive, which is traced back to the characteristic Pauli repulsion in the $^3S_1 + {}^3D_1$ state of the $\Sigma N(I = 3/2)$ channel.

Table 1 shows the Scheerbaum factor $S_B(q_1 = 0)$ [7] calculated from the FSS G-matrices. In this calculation we have assumed spin-saturated nuclear matter with $N = Z$, in order to obtain the Thomas form of the s.p. ℓs potentials. The correct absolute value $S_N \sim -40$ MeV $\cdot$ fm^5 [7] is reproduced even in the Born approximation. The strong cancelation between the LS and $LS^{(-)}$ contributions takes place for S_Λ. The ratio S_Λ/S_N is 1/5 in the Born approximation [8], but is further reduced to less than 1/10 by the effect of the

Table 2.1. The Scheerbaum factors S_B (MeV$\cdot$fm^5) for N, Λ and Σ, predicted by FSS G-matrices for three different nuclear matter densities corresponding to $k_F = 1.20$, 1.35 and 1.50 fm^{-1}.

k_F (fm^{-1})	1.20	1.35		1.50
N	−42.1	−41.0	(1)	−39.8
Λ	−2.7	−3.5	$(\sim \frac{1}{12})$	−4.1
Σ	−21.9	−22.3	$(\sim \frac{1}{2})$	−22.2

short-range correlation. Table 1 shows that it becomes even smaller ($\sim 1/15$), if the density is lower ($\rho = 0.7\rho_0 \sim k_F = 1.20$ fm^{-1}). On the other hand, the N and Σ receive very little effect of the short-range correlation, resulting in the ratio $S_\Sigma/S_N \sim 1/2$ both in the Born approximation and in the realistic calculation.

4 Summary

We have upgraded our quark model FSS to incorporate the momentum-dependent Bryan-Scott term and the vector-meson exchange potential acting between quarks. With a few phenomenological ingredients, the accuracy of the model in the NN sector has now become almost comparable to that of the OBEP models. In spite of this improvement, the essential feature of the ΛN-ΣN coupling in the YN sector turns out to be almost the same as the previous models. [1, 2] We also calculated NN and YN G-matrices in ordinary nuclear matter, by solving the Bethe-Goldstone equations for the quark-exchange kernel generated from FSS. This model predicts a repulsive Σ s.p. potential, the strength of which is about 20 MeV. The repulsive isoscalar part of the Σ s.p. potential is reported to be compatible with recent BNL ($K^-, \pi^\pm$) data [9]. The relative ratio of the Scheerbaum factors S_Λ/S_N is found to be less than 1/10, which is consistent with recent experimental observations of very small single-particle spin-orbit potentials in ${}^{9}_{\Lambda}$Be and ${}^{13}_{\Lambda}$C [10].

References

1. Y. Fujiwara, C. Nakamoto, Y. Suzuki: Prog. Theor. Phys. **94**, 215 and 353 (1995); See also Prog. Theor. Phys. **94**, 65 (1995)
2. Y. Fujiwara, C. Nakamoto, Y. Suzuki: Phys. Rev. Lett. **76**, 2242 (1996); Phys. Rev. **C54**, 2180 (1996)
3. Scattering Analysis Interactive Dial-up (SAID), Virginia Polytechnic Institute, Blacksburg, Virginia; R.A. Arndt: Private Communication
4. For the YN experimental data, see M. M. Nagels, T.A. Rijken, J.J. de Swart: Ann. Phys. (N.Y.) **79**, 338 (1973), and references cited therin
5. M. Kohno, Y. Fujiwara, T. Fujita, C. Nakamoto, Y. Suzuki: In *Proceedings of the APCTP Workshop on Strangeness Nuclear Physics (SNP'99)* , Seoul, February 19-22, 1999. Singapore: World Scientific
6. H.-J. Schulze, M. Baldo, U. Lombardo, J. Cugnon, A. Lejeune: Phys. Rev. **C57**, 704 (1998)
7. R. R. Scheerbaum: Nucl. Phys. **A257**, 77 (1976)
8. O. Morimatsu, S. Ohta, K. Shimizu, K. Yazaki: Nucl. Phys. **A420**, 573 (1984)
9. J. Dabrowski: Phys. Rev. **C60**, 025205 (1999)
10. H. Tamura: Contribution to this conference APFB99

Few-Body Systems Suppl. 12, 317–323 (2000)

Few-Body Systems

The Soft-Core Nijmegen Hyperon-Nucleon and Hyperon-Hyperon Interactions

Th.A. Rijken

Institute for Theoretical Physics, University of Nijmegen, Nijmegen, The Netherlands

Abstract. In this talk we report on recent results obtained with the Extended-Soft-Core (ESC) interactions, where soft two-meson-exchange is taken into account, both uncorrelated and correlated. The most recent soft-core One-Boson-Exchange (OBE) models NSC97 [2, 3] have been reviewed in [4, 5]. We report on new results for the ESC-model for nucleon-nucleon (NN), hyperon-nucleon (YN), and hyperon-hyperon (YY). With this ESC-model, henceforth refered to ESC99, we succeeded in narrowing the gap between theory and experiment at low energies for NN considerably. For the energies $T_{lab} = 0 - 350$ MeV we reached a record low $\chi^2_{p.d.p.} = 1.15$, using the Nijmegen PWA. This NN-model allows a clearcut extension to YN and YY, in clear distinction to more phenomenological models. In this conference, we report for the first time on a succesfull ESC-description of the YN-scattering data. Several features of the OBE-models are improved with ESC.

1 Introduction

The Chiral-Quark-model (CQM) [1] explains the successes of the non-relativistic quark-model (NRQM) and is at the same time closely connected with the description of hadron dynamics through interactions involving mesons and baryons using effective chiral Lagrangians. Also, the coupling of mesons to quarks dressed by a pion cloud is in accordance with the ideas that the non-linear sigma-model is relevant for the description of hadronic interactions [1]. The picture that emerges from the chiral-quark-model, *i.e.* a quark-core surrounded by a meson cloud containing pions and other mesons, provides a natural basis for an approach to low energy baryon-baryon scattering using only mesonic degrees of freedom in the derivation of the baryon-baryon interactions. In [6, 7] we reported on an ESC-model for baryon-baryon scattering where many ingredients of the CQM picture were accounted for. Here we extended the soft-core OBE-models of the Nijmegen group [8, 9] to include uncorrelated two-meson-exchange and, for the first time in baryon-baryon models, meson-pair exchanges. The latter were inspired by the pion-pair interactions contained in the Lagrangians of the non-linear σ-models of the sixties, see e.g. [16].

In [4, 5] for example, we discussed the recently developed soft-core OBE model NSC97 [2, 3]. We examined in some detail the treatment of mesons, emphasizing and examining the important role of the scalar-meson nonet in the Nijmegen soft-core OBE-models. We found that the mixing was in conformity with a $Q\bar{Q}$-picture of the scalar-mesons with ideal mixing. We must reiterate that the YN-scattering data give essentially only information about s-wave interactions, and that these are best determined sofar. Indeed, it was found by Akaishi et al that some of the NSC97 solutions are very succesful in describing the neutron rich Λ-hypernuclei [10]. Although the recent NSC97 solutions [2, 3] provide the user with a variety of different s-wave interactions, the corresponding p-wave interactions are not checked by experimental scattering data. Indeed, the appearance of repulsive triplet-odd interactions in NSC97 solutions is rather unexpected, and could very well be doubtful. The ΛN-spin-orbit interaction seems to be too strong. This is in particularly the case for NSC97e,f, which are favored because of their succes in describing the s-shell Λ-hypernuclei. Furthermore, also the ΣN-sector is not fixed very well, since different Nijmegen models give rather different interactions in e.g. the $^3S_1(I = 3/2)$-wave. It was pointed out by Dabrowski that the NSC97 interactions are likely at variance with some experimental results [11]. Finally, we note that in general the mixing for the scalar mesons in the soft-core OBE-models was in conformity with a $Q\bar{Q}$-picture of the scalar-mesons with ideal mixing. The upshot of this is a rather weak attraction in the $\Lambda\Lambda$-channel.

All these remarks make it clear that many details of the YN-interactions are ill determined yet. Therefore, we continue our search for the baryon-baryon interactions along the lines of the Nijmegen approach, see [12]. In this report we describe the Nijmegen approach to the baryon-baryon interactions with the ESC-models. Stressed is, as always, the simultaneous description of all baryon-baryon channels. In the ESC99-model, the topic of this contribution, we include the soft-core: (i) OBE-interactions, like in the [8, 9, 2], (ii) uncorrelated two-pseudo-scalar-exchange (TPSE), and (iii) meson-pair-exchange (MPE). For NN this new model is very succesful, as will appear from the results in this paper. We note that we did not include uncorrelated two-meson-exchange potentials with vector- and scalar-mesons. Including these brings in a lot of exchanges with a mass > 1 GeV. Then, also heavy pseudo-scalar-, heavy scalar-, axial-, and tensor-mesons should be included. This we pospone to the future. In the MPE we included such exchanges, like e.g. $(\pi\rho)_1$, in order to account for quantum numbers not included in OBE.

The contents of this paper is as follows. In section 2 we describe briefly the physical contents of the ESC-models, indicate briefly how the ESC-model can be extended to all baryon-baryon channels, and report on the most recent results for nucleon-nucleon. In section 3 we give the first results for hyperon-nucleon, and in section 4 we discuss the situation of the ESC-model as compared to the OBE-models for the $\Lambda-\Lambda$-interactions. Finally, in section 5 we offer some conclusions and an outlook.

2 Extended Soft-Core Model for Baryon-Baryon

The potential of this new NN-model, henceforth referred to as the ESC-model, consists of the contributions of

(i) The OBE-potentials of [8, 9], which apart from the low lying pseudo-scalar-, vector-, and scalar-mesons includes also contributions of the Pomeron. The latter represents the multi-peripheral (soft)pion exchanges and multi-gluon exchanges.

(ii) The TPSE-potentials as given in [13, 14]. These are two-pseudo-scalar-exchange potentials based on a combination of pseudo-vector and pseudo-scalar coupling to the baryon octet. We include both the so-called BW-graphs and TMO-graphs (see [15] for the used nomenclature).

(iii) We extend the OBE-model of [8, 9] further through the inclusion of phenomenological baryon-baryon-meson-meson vertices, henceforth referred to as 'meson-pair-exchange' (MPE). The vertices are given in [6, 14].

The motivation for including these MPE-potentials is that similar interactions are required in chiral-lagrangians [16]. They can be viewed upon as the result of the out integration of the heavy-meson and resonance degrees of freedom. Moreover, they also represent two-meson exchange potentials. We are less radical than Weinberg, see *e.g.* [17], in that we do not integrate out the degrees of freedom of the mesons with masses below 1 GeV.

The $SU(3)$-extension of the ESC-model for NN to YN and YY is carried out by assigning the meson-pairs to the $SU(3)$-irreps. The proper $SU(3)$ combinations for respectively the singlet-, the octet-symmetric-, and octet-antisymmetric-irreps for the ps-ps-pairs can be found in [18].

Fitting this new model to the NN-data, using the 1993 Nijmegen single energy $pp+np$ phase shift analysis [19], leads to an excellent result. We reached for the energies in the range $0 \leq T_{lab} \leq 350$ MeV, which comprises 4233 data, a $\chi^2_{p.d.p.} = 1.15$. In Table 1 we give information about the quality of the NN-fit. Details about the couplings, form factors etc. will be given elsewhere [20]. Also, more detailed discussions of the ESc-model can be found in the given references. We notice that the ESC-model gives a great improvement with respect to the OBE-model [8]. In particular it appeared that the $^1P_1$-, the $^3D_2$-, and the $^3D_3$-waves have much improved.

3 Hyperon-Nucleon and Hyperon-Hyperon

In the NSC97-model [2], the form factors depend on the $SU_f(3)$-assignment of the mesons. In principle we introduce form factor masses Λ_8 and Λ_1 for each meson-nonet, for respectively the $\{8\}$- and $\{1\}$-members. In the application to YN and YY we allow for $SU_f(3)$-breaking, by using different cut-off's for the strange mesons K, K^*, κ. Also, we used for the first time an

T_{lab}	#data	χ_0^2	$\Delta\chi^2$	$\hat{\chi}_0^2$	$\Delta\hat{\chi}^2$
0.383	144	137.55	18.0	0.960	0.125
1	68	38.02	51.3	0.560	0.755
5	103	82.23	9.3	0.800	0.091
10	209	257.99	66.3	1.234	0.228
25	352	272.20	31.3	0.773	0.089
50	572	547.67	105.0	0.957	0.184
100	399	382.45	33.7	0.959	0.084
150	676	673.05	141.2	0.996	0.209
215	756	754.52	150.0	0.998	0.198
320	954	945.38	278.9	0.991	0.292
Total	4233	4091.12	875.8	0.948	0.204

Table 1. χ^2-distribution PSA93 and ESC99-model. χ_0^2: χ^2 PSA93, $\hat{\chi}_0^2$: χ^2_{pdpt} PSA93, The χ^2-access ESC99-model is denoted $\Delta\chi^2$ and $\hat{\chi}^2$, respectively.

$SU_f(3)$-symmetry breaking of the coupling constants, which is based on the $^3P_0$-mechanism of quark-anti-quark pair creation leading to the meson-baryon coupling. The $SU_f(3)$-breaking is generated by assuming that the $Q\bar{Q}$-pair creation constants γ are flavor-dependent, i.e. $\gamma_u = \gamma_d \neq \gamma_s$. The same ingredients have been used in the ESC99-model. In Table 2 we give a few results for the ESC99-model and compare these to those of earlier Nijmegen models. The results on U_Λ, U_Σ, and Γ_Σ have been obtained by Yamamoto [21], using the YNG-interactions of Bandōo and Yamamoto [22]. The different ESC99 solutions are obtained, like the NSC97 solutions, by varying α_V^m, and refitting the other free parameters pertinent to YN. The Λ well depth $U_\Lambda = -28$ MeV, obtained from analyses of the (π^+, K^+) and (K^-, π^-) cross sections on nuclear targets [23] is not produced by the ESC99-model using the YNG-interaction. This leaves for example room for repulsive three-body contributions in matter. Also interesting is that for NSC97 and ESC99 Γ_Σ is much smaller than is the case for the NSC89-model. This is more consistent with the recently confirmed $^4_\Sigma He$ hypernucleus [24], which has a conversion width $\Gamma \approx 7$ MeV. The ESC99 ΣN-interactions are such that the $^3S_1(I = 3/2)$-interaction is rather repulsive. Also, the p-wave interactions are attractive, as can be seen from Table 3. In the $S = -2$-systems, the experimental information is limited to the ground

	α_V^m	a_s	a_t	U_Λ	U_Σ	Γ_Σ	χ^2
NSC89	0.275	-2.73	-1.48	-30.8	-27.1	26.9	15.8
ESC99b	0.405	-2.70	-1.60	-39.6	20.0	17.4	17.4
ESC99d	0.505	-2.56	-1.61	-37.4	3.1	17.4	17.3
NSC97e	0.375	-2.10	-1.86	-34.3	-11.8	14.6	16.1
NSC97f	0.365	-2.51	-1.75	-31.1	-11.6	15.5	16.7
NHC(F)	0.588	-2.18	-1.93	-30.8	+5.8	9.6	27.1
NHC(D)	0.334	-1.77	-2.06	-40.1	-29.3	10.3	19.4

Table 2. Comparison ΛN-Scattering length's, χ^2 YN-data, etc. We note that for ESC99b and ESC99d for the pseudo-scalar mesons $\alpha_P = 0.434$ and 0.440 respectively.

states of ${}^{6}_{\Lambda\Lambda}He$, ${}^{10}_{\Lambda\Lambda}Be$, and ${}^{13}_{\Lambda\Lambda}B$. From the double-$\Lambda$ states, mentioned above, it is infered that $\Delta B_{\Lambda\Lambda} = 4 - 5$MeV, which indicates a rather strong $\Lambda\Lambda$-interaction. As mentioned before, the soft-core OBE-models have difficulty to produce strong attraction in $\Lambda\Lambda$. In the case of the ESC99-model the situation has changed. The mixing for the scalar mesons e.g. are $\theta_S = 1.3^o - 6.0^0$ for the solutions of ESC99, which is between ideal mixing for $Q\bar{Q}$ and that for $Q^2\bar{Q}^2$, So, in the case of ESC99 there is a sizeable attraction coming from both OBE as well as from TPSE.

4 Conclusions and Outlook

The multi-energy Nijmegen PWA poses a nice new challenge to the theory of the low momentum transfer baryon-baryon interactions. The success of our ESC-model indicates that the better quality of the multi-energy Nijmegen PWA with respect to other phase shift analyses, indeed opens the door to a more thorough understanding of the low energy NN-data.

As far as the YN and YY interactions are concerned, it seems that the ESC-approach is rather promising. The prospects for a satisfactory description of both scattering data and hyper-nuclei are good.

Model	$^1S_0$	$^3S_1$	$^1P_1$	$^3P_0$	$^3P_1$	$^3P_2$	Sum
NSC97f	-12.7	-22.0	2.3	0.1	4.5	-0.5	-28.3
ESC99a	-14.2	-23.1	-0.1	0.6	0.9	-1.9	-37.8
ESC99b	-14.6	-22.8	-1.0	0.6	0.4	-2.2	-39.6
ESC99c	-15.3	-21.9	-1.1	0.8	0.4	-2.2	-39.5
ESC99d	-13.9	-21.6	-1.0	0.7	0.2	-1.8	-37.4

Table 3. Partial wave contributions to the Λ potential energy $U_\Lambda(k_\Lambda = 0)$ at $k_F = 1.35$ fm^{-1} for various models. G-matrix calculations are performed with the QTQ prescription for the intermediate spectra. All entries are in MeV.

5 Acknowledgements

Part of this work was done while the author stayed at INPS, Tanashi-branch KEK, Tokyo. He is very grateful to Prof. Akaishi for the generous hospitality extended to him. It is great pleasure to thank Prof. Y. Yamamoto for his kind interest and very valuable G-matrix calculations, and Profs. T. Motoba, M. Wada, S. Shinmura, T. Harada, and T. Yamada for many stimulating meetings.

References

1. A. Manohar and H. Georgi: Nucl. Phys. **B234**, 189 (1984); S. Weinberg: Physica (Amsterdam) **96A**, 327 (1979); E. Witten: Nucl. Phys. **B160**, 57 (1979)
2. Th.A. Rijken, Y. Yamamoto, V.G.J. Stoks: Phys. Rev. **C59**, 21 (1999)
3. V.G.J. Stoks and Th.A. Rijken: Phys. Rev. **C59**, 3009 (1999)
4. Th.A. Rijken: Nucl. Phys. **A639**, 29c (1998)
5. Th.A. Rijken: Proceedings SNP99 Workshop, February 1999, Soul (to be published)
6. Th.A. Rijken: Few-Body Problems in Physics '93, Suppl. 7, 1-12, Springer-Verlag 1994
7. Th.A. Rijken: In *Nucl. and Part. Phys. with Meson Beams in the 1 GeV/c Region*, Proceedings of the 23th INS Int. Symp., Tokyo 1995, p.295. Universal Academy Press, Inc. 1995

8. M.M. Nagels, Th.A. Rijken, J.J. de Swart: Phys. Rev. **D17**, 768 (1978)

9. P.M.M. Maessen, Th.A. Rijken, J.J. de Swart: Phys. Rev. **C40**, 2226 (1989)

10. K.S. Myint, T. Harada, S. Shinmura, Y. Akaishi: *Neutron Rich Lambda Hypernuclei*; Proceedings of this Conference (to be published).

11. J. Dabrowski: Proceedings SNP99 Workshop, February 1999, Soul (to be published)

12. J.J. de Swart, Th.A. Rijken, P.M. Maessen, R.G. Timmermans: Nuov. Cim. **102A**, 203 (1988)

13. Th.A. Rijken and V.G.J. Stoks: Phys. Rev. **C54**, 2851 (1996)

14. Th.A. Rijken and V.G.J. Stoks: Phys. Rev. **C54**, 2869 (1996)

15. Th.A. Rijken: Ann. Phys. (N.Y.) **208**, 253 (1990)

16. S. Weinberg: Phys. Rev. **166**, 1568 (1968); Phys. Rev. **177**, 2604 (1969)

17. S. Weinberg: Phys. Lett. **B251**, 288 (1990); Nucl. Phys. **B363**, 3 (1991); C. Ordòñez and U. v. Kolck: Phys. Lett. **B291**, 459 (1992)

18. P.A. Carruthers: *Introduction to Unitary Symmetry*, New York: Intersciences Publishers, Wiley 1966

19. V.G.J. Stoks, R.A.M. Klomp, M.C.M. Rentmeester, J.J. de Swart: Phys. Rev. C**48**, 792 (1993)

20. Th.A. Rijken: (in preparation)

21. Y. Yamamoto: Private communication

22. Y. Yamamoto and H. Bandō: Progr. Theor. Phys. **83**, 254 (1990)

23. D.J. Millener, A. Gal, C.B. Dover, Phys. Rev. **C38**, 2700 (1988)

24. T. Nagae et al.: Phys. Rev. Lett. **80**, 1605 (1998); T. Harada: Phys. Rev. Lett. **81**, 5287 (1999)

Few-Body Systems Suppl. 12, 324–329 (2000)

Few-Body Systems

Coupled $\Lambda N - \Sigma N$ and $\Lambda NN - \Sigma NN$ Systems and Hyperon-Nucleon Interactions

K. Miyagawa[1], H.Kamada[2], W. Glöckle[2], H. Yamamura[1], T. Mart[3], C. Bennhold[4]

[1] Department of Applied Physics, Faculty of Science, Okayama University of Science, Okayama 700, Japan

[2] Institut für theoretische Physik II, Ruhr-Universität Bochum, D-44780 Bochum, Germany

[3] Jurusan, Fisika, FMIPA, Universitas Indonesia, Depok 16424, Indonesia

[4] Center for Nuclear Studies, Department of Physics, The George Washington University, Washington, D.C. 20052, USA

Abstract. This paper summarizes our studies of the hypertriton and describes our findings on the hyperon-nucleon interaction, especially on the role of the $\Lambda N - \Sigma N$ coupling and the strengths of the S-wave YN interactions. We briefly comment on our ongoing analyses with electromagnetic probes.

1 Introduction

One of the most prominent features of the ΛN interaction is the lack of the one-pion-exchange process[1] so that shorter range effects are more important than in the NN interaction. In addition, the presence of the $\Lambda N - \Sigma N$ coupling influences the nature of the interaction. The mass difference between Λ and Σ is about 80 MeV and this coupling can cause significant effects in hypernuclei. Unfortunately, these features of the YN interaction have not been well examined, mainly because YN scattering data are scarce. Even S-wave ΛN scattering lengths have yet to be determined.

Since 1993, we have analyzed the hypertriton[2], the lowest-mass hypernucleus, using various meson-theoretical YN interactions. These analyses have clarified properties of the S-wave YN interactions as well as confirmed the structure of the hypertriton as an extremely loosely bound system. Our findings are twofold: First, the binding energy is sensitive to the strengths of the $^3S_1$ and especially the $^1S_0$ YN forces. The analyses combined with the ΛN elastic total cross section data restrict the balance between the two force components to a small range. Second, the $\Lambda N - \Sigma N$ coupling plays the decisive role in the binding of the hypertriton. The expectation value of the $\Lambda N - \Sigma N$ coupling potential amounts to about half of the total YN potential.

Our another interest is in the behavior of the $\Lambda N - \Sigma N$ coupling around the ΣN threshold. Predictions of the ΛN elastic total cross sections based on various YN interaction models show an enhancement just around the ΣN threshold. Reference [3] demonstrates that this is not a simple threshold effect but is caused by a t-matrix pole around the threshold. This enhancement is being investigated by photo- and electroproduction processes of a K meson and a hyperon on the deuteron and ^{3}He targets. We are pursuing such studies [4], and report a part of them in this conference (See H.Yamamura). This paper briefly surveys our recent results.

2 Analyses of the Hypertriton

We solve the exact Faddeev equations for the coupled $\Lambda NN - \Sigma NN$ system. The formulation has appeared in several articles[2], to which we would like to refer the reader. For the YN system, various modern meson-theoretical interactions are used, the Jülich $\tilde{A}$[5] model, and the soft core models of the Nijmegen group, NSC89[6] and NSC97[7]. The model NSC97 has six different versions, namely a, b, c, d, e and f, which contain varying relative strengths between the $^1S_0$ and $^3S_1$ force components. Among these models, only NSC89 and NSC97f bind the hypertriton at the correct binding energy. (The experimental Λ separation energy is 0.13 ± 0.05 MeV.) The Jülich $\tilde{A}$ interaction cannot generate a bound state. This, as shown below, originates from a defect in their $^1S_0$ force. For the NN sector, we use the Paris, Bonn B and Nijmegen93 interactions, but the results above do not depend on the choice of these potentials. It is the $^1S_0$ and $^3S_1 -^3 D_1$ partial waves in the NN and YN subsystems that dominate the hypertriton binding energy. The convergence with increasing number of partial waves is demonstrated in Ref. [2].

Figure 1(a) illustrates baryon-baryon correlation functions in the hypertriton, the probability to find the two particles at a distance r. The NN correlation function is quite similar to the one for the deuteron which demonstrates that the NN state in the hypertriton is close to the deuteron. The correlation functions multiplied by r^2 are displayed in Fig. 1(b), showing that the ΛN correlation function has a large range. (The functions ρ are normalized as $\int r^2 \rho_{NN} = 1$ and $\int r^2 [\rho_{N\Lambda} + \rho_{N\Sigma}] = 1$.) Thus, as expected with the small Λ separation energy, the hypertriton is a loosely bound system where the Λ barely clings to a deuteron.

Table 1 shows expectation values for the kinetic and potential energies in the hypertriton. Using the Njimegen93 NN and NSC89 YN potentials we see that the NN subsystem is less bound than in the deuteron, and the relative kinetic energy of the Λ with respect to the two nucleons $< T_{\Lambda - NN} >$ is larger than its potential energy $< V_{\Lambda N, \Lambda N} >$. It is due to the $\Lambda - \Sigma$ conversion that the hyperon is bound to the two nucleons. The expectation value $< V_{\Lambda N, \Sigma N} >$

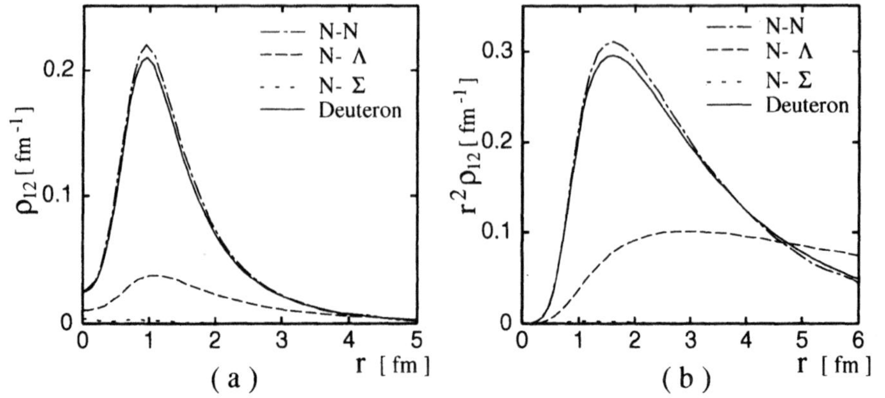

Figure 1. (a) Correlation functions for the NN, ΛN, and ΣN pairs in the hypertriton. The correlation function in the deuteron is also shown for comparison. (b) Correlation functions multiplied by r^2.

amounts to more than half of the total YN potential energy. In Table 2, we show the contributions from the $^1S_0$ and $^3S_1 - ^3D_1$ components to various YN potential energies. The $^1S_0$ component dominates the expectation value of the direct potential $< V_{\Lambda N,\Lambda N} >$, while the $^3S_1 - ^3D_1$ component dominates the coupling potential $< V_{\Lambda N,\Sigma N} >$.

Table 1. Expectation values for the various kinetic and potential energies in the hypertriton using the NSC89 YN and Nijmegen93 NN interactions.

$< T_{NN} >$	$< T_{\Lambda-NN} >$	$< T_{\Sigma-NN} >$	
20.57	2.28	0.83	
$< V_{NN} >$	$< V_{\Lambda N,\Lambda N} >$	$2 < V_{\Lambda N,\Sigma N} >$	$< V_{\Sigma N,\Sigma N} >$
−22.32	−1.65	−2.05	−0.02

Table 2. Contributions from the $^1S_0$ and $^3S_1 - ^3D_1$ force components to the expectation values of the YN potential energies

	$< V_{\Lambda N,\Lambda N} >$	$2 < V_{\Lambda N,\Sigma N} >$	$< V_{\Sigma N,\Sigma N} >$
$^1S_0$	−1.67	−0.40	0.03
$^3S_1 - ^3D_1$	0.02	−1.61	−0.06

Let us discuss the strengths of the $^1S_0$ and $^3S_1$ force components in relation to the restriction imposed by the binding of the hypertriton. In Fig. 2, the ΛN elastic total cross sections are shown for the various force models. All of the curves are adjusted to the sparse experimental data which have large error bars.

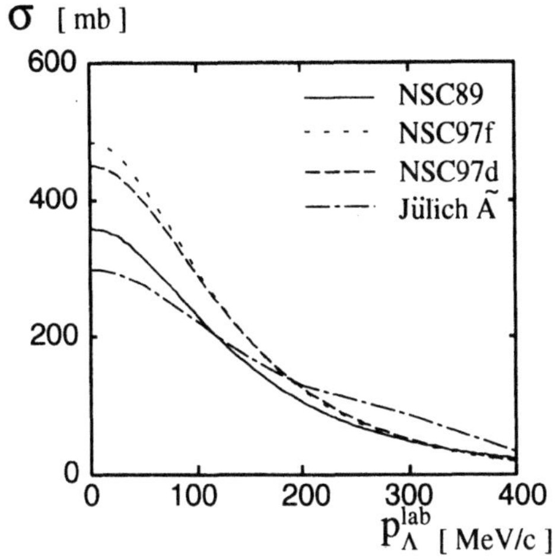

Figure 2. The ΛN elastic total cross sections σ predicted by the various YN interactions as a function of the Λ lab momentum.

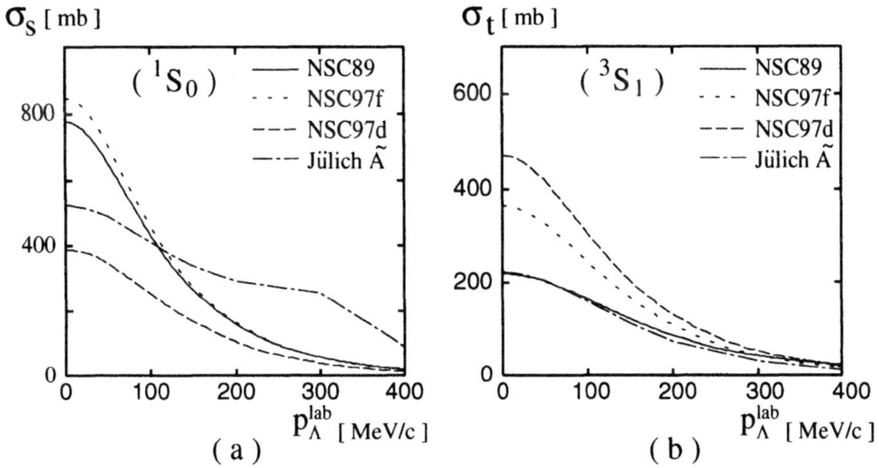

Figure 3. Partial ΛN elastic total cross sections (a) for $^1S_0$, and (b) for $^3S_1$, which are defined as $\sigma = \frac{1}{4}\sigma_s + \frac{3}{4}\sigma_t$.

Among these potentials, NSC89 and NSC97f bind the hypertriton. The NSC97 potential, as mentioned, has six versions with different magnitudes between the $^1S_0$ and $^3S_1$ force components. Here, we take NSC97d as an example which does not bind the hypertriton. In Figs. 3(a) and 3(b), the separate contributions from the $^1S_0$ and $^3S_1$ states to the ΛN elastic total cross sections are shown. They are defined as $\sigma = \frac{1}{4}\sigma_s + \frac{3}{4}\sigma_t$. As is seen in Figs. 3(a) and 3(b), the two potentials with the largest $^1S_0$ and small $^3S_1$ cross sections at low energies reproduce the hypertriton. From that we deduce that the scattering lengths are restricted to within $-2.7 \sim -2.4$ fm for $^1S_0$ and $-1.6 \sim -1.3$ fm for $^3S_1$.

3 Electromagnetic Probes for the YN and YNN Systems

YN scattering experiments would be the best way to clarify the properties of the YN interaction, but it is difficult to carry them out experimentally under present circumstances. Alternative and hopefully equally promising ways are experiments with electromagnetic probes such as $\gamma + d \to K^+ + Y + N$ and $e + d \to e' + K^+ + Y + N$, and similar experiments on ^{3}He. One can obtain the information on the YN interaction by analyzing the final state interactions among YN or YNN. An inclusive $d(e, e'K^+)YN$ experiment has already been performed at TJNAF at Hall C, while the data for $d(\gamma, K^+Y)N$ are being analyzed in Hall B.

What dominates the ΛN elastic total cross section curve is an enhancement around the ΣN threshold which is predicted by all force models. We have investigated the amplitude for the $\Lambda N - \Sigma N$ system and found that it has a pole close to that threshold in the complex momentum plane. This pole will have a significant influence on various observables in the reactions mentioned above.

Beginning with photoproduction, we recently studied the inclusive $d(\gamma, K^+)YN$ and exclusive $d(\gamma, K^+Y)N$ processes for $\theta_K = 0°$ and found sizeable effects of the YN final state interaction. This is reported by H.Yamamura in this conference. Even more pronounced are the final YN interaction effects on double polarization observables involving the recoil hyperon polarization along with a circularly polarized incoming photon. We plan to study the $e + d \to e' + K^+ + Y + N$ and $\gamma + {}^3\mathrm{He} \to K^+ + Y + N + N$ processes next. In the latter case, the $^3_\Lambda$H bound state just below the Λd threshold (0.13 MeV) is expected to yield significant effects on observables near threshold.

4 Summary

We have analyzed the hypertriton using various meson-theoretical YN interactions. Among them, only NSC89 and NSC97f give the correct binding energy. The hypertriton is dominated by the $^1S_0$ and $^3S_1 - {}^3D_1$ partial waves in the NN and YN subsystems, and has a structure such that a Λ clings to a deuteron as expected from the small Λ separation energy. The $\Lambda N - \Sigma N$ coupling plays the decisive role in this binding. The expectation value $< V_{\Lambda N,\Sigma N} >$ amounts to about half of the total potential energy $< V_{YN} >$. The strengths of the $^1S_0$ and $^3S_1$ YN forces can be restricted by the combined analyses of the hypertriton and the ΛN elastic cross section data. The scattering lengths have to lie within the intervals $-2.7 \sim -2.4$ fm for $^1S_0$ and $-1.6 \sim -1.3$ fm for $^3S_1$.

For further study of the YN interaction, we believe experiments with electromagnetic probes are promising. We have studied the $\gamma + d \to K^+ + Y + N$ processes and found sizeable effects of the YN final state interaction. The analyses of $e + d \to e' + K^+ + Y + N$ and $\gamma + {}^3\mathrm{He} \to K^+ + Y + N + N$ processes are underway.

References

1. See for instance, B.F. Gibson: Few-Body Systems, Suppl. **7**, 80 (1994)
2. W. Glöckle, K. Miyagawa, H. Kamada, J. Golak, H. Witała: Nucl. Phys. **A639**, 297c (1998); K. Miyagawa, H. Kamada, W. Glöckle, V. Stoks: Phys. Rev. **C51**, 2905 (1995); K. Miyagawa and W. Glöckle: Phys. Rev. **C48**, 2576 (1993)
3. K. Miyagawa and H. Yamamura: Phys. Rev. **C60**, 024003 (1999); nucl-th/9904002
4. H. Yamamura, K. Miyagawa, T. Mart, C. Bennhold, H. Haberzettl, W. Glöckle: Phys. Rev. C (in press); nucl-th/9907029
5. A.G. Reuber, K. Holinde, J. Speth: Czech. J. Phys. **42**, 1115 (1992)
6. P.M.M. Maessen, Th.A. Rijken, J.J. de Swart: Phys. Rev. **C40**, 2226 (1989)
7. Th.A. Rijken, V.G.J. Stoks, Y. Yamamoto: Phys. Rev. **C59**, 21 (1999)

Few-Body Systems Suppl. 12, 330–335 (2000)

Few-Body Systems

$\Xi^- d \to n\Lambda\Lambda$ as a Constraint on the $\Lambda\Lambda$ Amplitude !

I.R. Afnan

School of Chemistry, Physics and Earth Sciences, Flinders University, GPO Box 2100, Adelaide 5001, Australia

Abstract. With the recent success of the reaction $n(d,p)nn$ in determining the nn scattering length [1], we examine the sensitivity of the final state interaction peak in the reaction $d(\Xi, n)\Lambda\Lambda$ to the $\Lambda\Lambda$ scattering length.

1 Introduction

The interest in the strangeness $S \leq -2$ systems was originally motivated by Jaffe's [2] quark model prediction for a flavour singlet H dibaryon. More recently, the interest in the production of nuclei with $S \leq -2$ at RICH, and the possible formation of strange neutron stars has created the need for a baryon-baryon (BB) interaction in the $S \leq -2$ channels. At present the only data on $S = -2$ systems are the three $\Lambda\Lambda$ hypernuclei, ${}^{\,6}_{\Lambda\Lambda}$He, ${}^{10}_{\Lambda\Lambda}$Be and ${}^{13}_{\Lambda\Lambda}$B. Although these three nuclei may be used to constrain the $\Lambda\Lambda$ interaction via nuclear structure calculations, a more direct determination of the $\Lambda\Lambda$ amplitude in a process where the reaction mechanism is well understood is essential. The advent of tools to detect the $\Lambda\Lambda$ final states [3], may be used to reduce the background neutrons in a measurement of the neutron spectrum in $\Xi d \to n\Lambda\Lambda$, and thus put a constraint on the $\Lambda\Lambda$ scattering length.

In the present report we examine the sensitivity of the neutron differential energy spectrum (NDES) in the reaction $\Xi d \to n\Lambda\Lambda$ to the $\Lambda\Lambda$ scattering length. The amplitude for this reaction is calculated within the framework of the three-body equations [4]. For this reaction to proceed we need to know the BB interaction in the $S = -2$ channel, and in particular the potential in the coupled $\Lambda\Lambda - \Xi N$ channel that couples the initial and final states of this reaction. This potential is generated with the help of an $SU(3)$ flavour rotation of the Nijmegen model D [5] meson exchange potential, with the short range behaviour varied to generate different $\Lambda\Lambda$ scattering lengths.

2 The $\Lambda\Lambda - \Xi N$ Interaction

For Ξ^- capture on the deuteron, we restrict the interaction to $S-$waves only. In that case the potential for the exchange of the ith meson is given by

$$V_i(r) = V_c^{(i)}(r) + \boldsymbol{\sigma}_1 \cdot \boldsymbol{\sigma}_2 \, V_\sigma^{(i)}(r) \tag{1}$$

where the radial potential $V_\alpha^{(i)}(r)(\alpha = c, \sigma)$ for a meson with mass m_i is

$$V_\alpha^{(i)}(r) = V_0^{(i)} \left[\frac{e^{-m_i r}}{m_i r} - C \left(\frac{M}{m_i} \right) \right] \; ; \quad \alpha = c, \sigma \; . \tag{2}$$

Here we have replaced the hard core in the Nijmegen model D potential by a soft core, and to guarantee a one parameter short range repulsion, the mass $M = 2500$ MeV is chosen. The parameter C is then adjusted to change the $\Lambda\Lambda$ scattering length [6]. In Eq. (2), m_i is the physical masses of the exchanged meson, while $V_0^{(i)}$ is given in terms of the meson-baryon coupling constants which are determined by fits to the $S = 0, -1$ data. With a proper choice of the parameter C we have generated four potentials for the $\Lambda\Lambda - \Xi N$ system with different $\Lambda\Lambda$ scattering lengths. Two of the potentials (A and B) do not support a bound state, while the other two ($C1$ and $C2$) do support a bound state. In particular, we have chosen potential B to have a scattering length comparable to the nn scattering length which has an anti-bound state. This potential, we expect will give a large final state interaction peak.

To facilitate the three-body calculation for the reaction $\Xi d \rightarrow n\Lambda\Lambda$, we have approximated the meson exchange potentials by rank one Yamaguchi type separable potentials that gives the same $\Lambda\Lambda$ scattering lengths $a_{\Lambda\Lambda}$ as the original meson exchange potentials. In Table 1 we present the values of $a_{\Lambda\Lambda}$ for the meson exchange potentials and the equivalent separable potentials.

Table 1. The $\Lambda\Lambda$ scattering length for the meson exchange potentials A, B, $C1$ and $C2$ and the equivalent separable potentials SA, SB, $SC1$ and $SC2$.

Pot.	A	SA	B	SB	$C1$	$SC1$	$C2$	$SC2$
$a_{\Lambda\Lambda}$	-1.91	-1.90	-21.1	-21.0	7.82	7.84	3.37	3.36

3 The ${}_{\Lambda\Lambda}^{6}$He as a Constraint on $a_{\Lambda\Lambda}$

As stated in the Introduction, $\Lambda\Lambda$ hypernuclei can be used to put constraint on the $S = -2$ amplitude, and in particular on the $\Lambda\Lambda$ scattering length. To illustrate this, we consider the binding of ${}_{\Lambda\Lambda}^{6}$He. This hypernucleus can be modelled as a three-body problem ($\alpha\Lambda\Lambda - \alpha\Xi N$) including the important coupling between the $\Lambda\Lambda$ and the ΞN channels. In Table 2 we present the binding energy of ${}_{\Lambda\Lambda}^{6}$He for the four different potentials [6]. Comparing the calculated binding energy with the one experimental value [7], we may conclude that potential SB

gives the closest result to the experimental value. This suggests a $\Lambda\Lambda$ scattering length that is comparable to the nn scattering length, and possibly a final state interaction peak comparable to that observed in $d(n,p)nn$.

Table 2. The binding energy of ${}_{\Lambda\Lambda}^{6}\mathrm{He}$ for the four separable potentials considered

Model	SA	SB	$SC1$	$SC2$	Exp.
$a_{\Lambda\Lambda}$	-1.91	-21.2	7.82	3.37	?
B.E.$({}_{\Lambda\Lambda}^{6}\mathrm{He})$	9.738	12.268	15.912	19.836	10.9 ± 0.6

4 The Reaction $d(\Xi^-, n)\Lambda\Lambda$

The reaction $\Xi d \to n\Lambda\Lambda$ can be considered a three baryon problem ideally suited for treatment within the Faddeev framework. However, since the $\Lambda\Lambda$ and ΞN have thresholds that differ by $\approx$ 25 MeV[1], which is considerably larger than the binding energy of the deuteron, we need to include the mass difference between the baryons explicitly to get all the low energy thresholds for this reaction. As a result, we have two identical Fermions in both the initial (NN) and final ($\Lambda\Lambda$) states, and proper antisymmetry needs to be satisfied.

To satisfy the Pauli principle, we write the amplitude for breakup as a matrix element of the Alt-Grassberger-Sandhas (AGS) operator $U_{0\alpha}$ between antisymmetrized initial and final states, i.e.

$$T_{N\Lambda\Lambda\leftarrow\Xi d} = {}^{\mathrm{AS}}\langle N\Lambda\Lambda|\, U_{03}\, |\Xi d\rangle^{\mathrm{AS}} = {}^{\mathrm{AS}}\langle N\Lambda\Lambda|\, \sum_\alpha T_\alpha\, G_0\, U_{\alpha 3}\, |\Xi d\rangle^{\mathrm{AS}} \ , \quad (3)$$

where $U_{\alpha\beta}$ is a solution to the AGS equations [8]. Here the antisymmetrized initial state is given by

$$|\Xi d\rangle^{\mathrm{AS}} = \mathcal{A}_{12}\, |\Xi d\rangle \ , \quad \text{with} \quad {}^{\mathrm{AS}}\langle d\Xi|\Xi d\rangle^{\mathrm{AS}} = 1 \ , \quad (4)$$

where $\mathcal{A}_{ij}$ is the antisymetrization operator. Since the Ξ can convert on either nucleon in an isospin formulation, the final antisymmetric state is given by

$$|N\Lambda\Lambda\rangle^{\mathrm{AS}} = \mathcal{A}_{23}\, |N_1\Lambda_2\Lambda_3\rangle - \mathcal{A}_{13}\, |\Lambda_1 N_2 \Lambda_3\rangle \ , \quad \text{with} \quad {}^{\mathrm{AS}}\langle \Lambda\Lambda N|N\Lambda\Lambda\rangle^{\mathrm{AS}} = 1 \ . \quad (5)$$

With the help of the AGS equations and the definition of the antisymmetrization operators, we can write a set of coupled integral equations for the antisymmetrized amplitudes [4].

Although this reaction, in many ways, is similar to $d(n,p)nn$, and a detailed measurement should give a measure of the $\Lambda\Lambda$ scattering length, the fact that we have coupling between the $\Lambda\Lambda$ and ΞN channels gives a new dimension

[1] Ideally we should include also the coupling to the $\Sigma\Sigma$ channel. However, the mass difference between the Λ and Σ is such that the $\Sigma\Sigma$ threshold is some 150 MeV above the $\Lambda\Lambda$ threshold, and as a result we have neglected the coupling to the $\Sigma\Sigma$ channel.

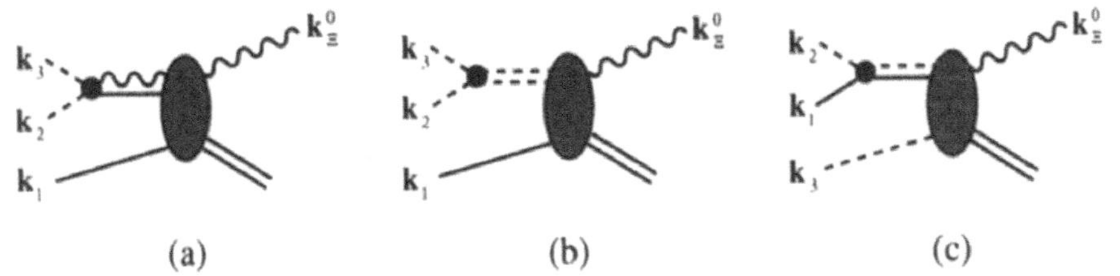

Figure 1. The three distinct contributions to the breakup amplitude, with (a) and (b) dominated by the final state interaction in the $\Lambda\Lambda - \Xi N$ channel and (c) playing the role of a background term.

to the final state interaction. In particular, there are three amplitudes that can contribute to the cross section. These are illustrated in Fig. 1. Two of the diagrams (a) and (b) are dominated by the final state interaction, while the third diagram (c) plays the role of a background amplitude. It is the coherent sum of these diagrams that will determine the sensitivity of the final state interaction peak to the $\Lambda\Lambda$ scattering length. This is to be compared with $nd \to pnn$ where there are only two diagrams, one dominated by the final state nn interaction, while the second is a background amplitude. The interference between these two diagrams results in a sharp final state interaction peak that can be used to extract the nn scattering length.

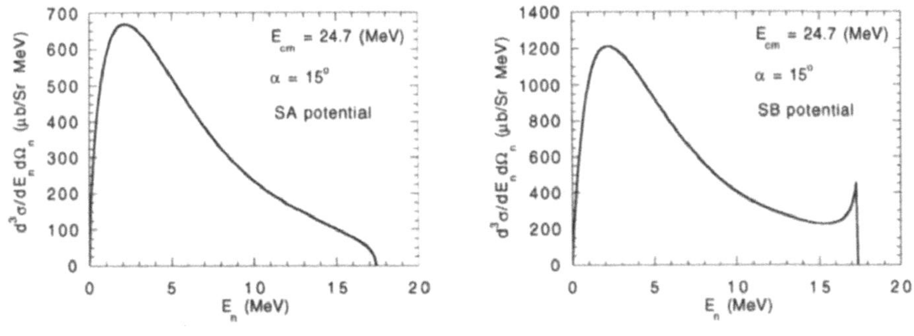

Figure 2. The NDES for the potentials SA, SB.

In Figs. 2 and 3 we present the NDES for the four separable potentials under consideration [4]. The total centre of mass energy E_{cm} is chosen to be above the Ξd threshold, but below the ΞNN threshold. In this way we get an estimate of the capture rate without having to introduce the Coulomb interaction in the initial state. The angle α is the centre of mass angle between the final neutron and the initial direction of the Ξ. The above results suggest that the final state interaction is considerably suppressed in this reaction. In fact with the exception of potential SB there is no final state interaction peak. Even for potential SB the magnitude of the final state interaction peak is considerably suppressed when compared with the cross sections for $nd \to pnn$.

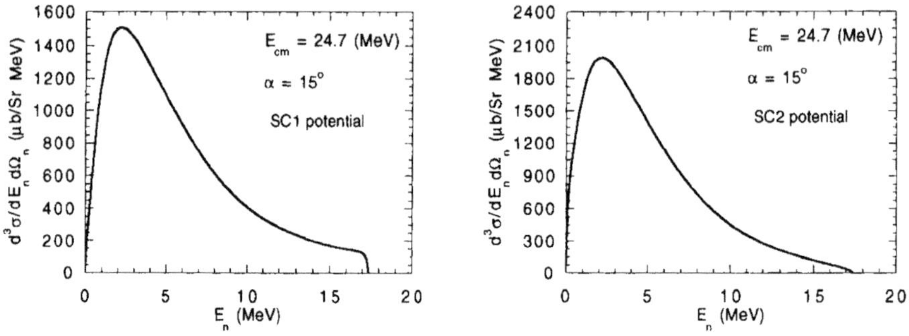

Figure 3. The NDES for the potentials $SC1$ and $SC2$.

To understand these results, we need to examine in more detail the interference pattern between the three amplitudes that contribute to the cross section. In fact, the results of Figs. 2 and 3 suggest that the two amplitudes dominated by the final state interaction, diagrams (a) and (b), are out of phase. To illustrate this we first examine the contribution of diagrams (a) and (b) to the cross section. Since the same amplitude (the $\Lambda\Lambda - \Xi N$) contributes to the final state interaction for these two diagrams, we expect the final state interaction peak to be identical for these two diagrams. This is illustrated in Fig. 4. It is clear that both of these diagrams give the same contribution to the final state interaction peak, and this contribution is considerably more than was observed in Figs. 2 and 3.

We next present in Fig. 5 the result of taking diagrams (a) plus (c) or (b) plus (c) for the total amplitude. Here we find that the final state interaction peak with diagrams (a) plus (c) is considerably less than is the case for diagrams (b) plus (c). This indicates that diagrams (b) and (c) are in phase, while diagrams (a) and (c) are out of phase. As a result, diagrams (a) and (b) are out of phase, which is the reason why the final state interaction is suppressed

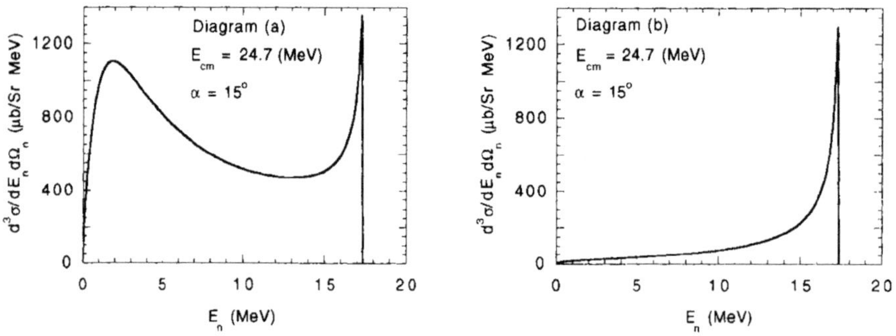

Figure 4. The NDES resulting from assuming that diagram (a) or diagram (b) only contribute to the total cross section for the potential SB.

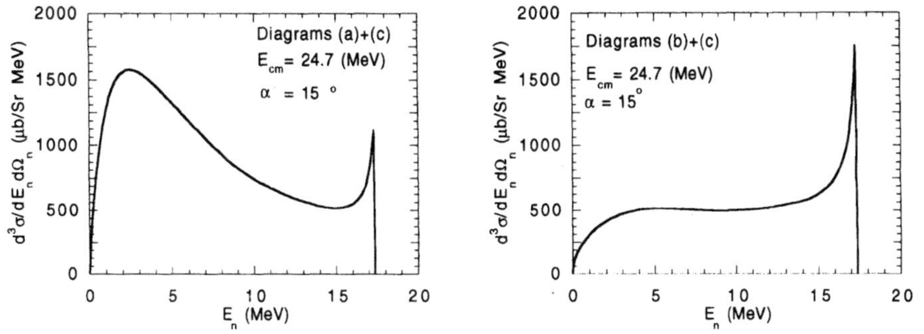

Figure 5. The NDES for diagrams (a) plus (c) and (b) plus (c) for potential *SB*.

in the cross sections presented in Figs. 2 and 3

5 Conclusion

From the above analysis we may conclude that: (i) If we fix the $\Lambda\Lambda$ scattering length to get the binding energy of ${}^{6}_{\Lambda\Lambda}$He, then the NDES for the reaction $\Xi d \to n\Lambda\Lambda$ has a final state interaction peak. However, this peak is suppressed. (ii) The fact that the final state interaction involves two coupled channels ($\Lambda\Lambda$ and ΞN) results in the cancellation between the two diagrams that are dominant in the final state interaction region. This makes this reaction very sensitive to the $\Lambda\Lambda$ scattering length, while at the same time making the cross section small and therefore difficult to measure.

Acknowledgement. The author would like to thank the Australian Research Council for their financial support. He is indebted to his collaborators Stephen Carr and Ben Gibson for many enlightening discussions and helpful suggestions.

References

1. C.R. Howell: This proceeding
2. A. Jaffe: Phys. Rev. Lett. **38**, 195 (1977)
3. T. Fukuda: Nucl. Phys. **A639**, 355c (1998)
4. S.B. Carr, I.R. Afnan, B.F. Gibson: Phys. Rev. **C57**, 2858 (1998)
5. M.M. Nagels, T.A. Rijken, J.J. de Swart: Phys. Rev. **D15**, 2547 (1977)
6. S.B. Carr, I.R. Afnan, B. F. Gibson: Nucl. Phys. **A625**, 143 (1997)
7. D.J. Prowse: Phys. Rev. Lett. **17**, 782 (1966)
8. E.O. Alt, P. Grassberger, W. Sandhas: Nucl. Phys. **B2**, 167 (1967)

Few-Body Systems Suppl. 12, 336–341 (2000)

Few-Body Systems

The Weak Decay of S Shell Hypernuclei and the ΔI=1/2 Rule*

E.V. Hungerford

Department of Physics, University of Houston, Houston, TX 77204, USA

Abstract. This paper summarizes the available information on S-shell hypernuclei, particularly with respect to their weak decays. It is also pointed out that a measurement of the branching ratios in the non-mesonic weak decay of ${}^4_\Lambda$H is crucial in order to test the ΔI=1/2 rule in the strangeness changing non-leptonic weak decay process. We emphasize that collectively, the S-shell hyperncului provide a unique system which can be used to test few body (A=1-5) calculations and extract new physics information.

1 Introduction

The structure of hadronic systems containing one or more hyperons (e.g. hypernuclei) has been under investigation for some time[1]. One motivation has been to sort out the hyperon-nucleon interaction in this many-body system. Although hypernuclear structure is now interpreted by conventional nuclear techniques, a understanding of the hyperon-nucleon interacton in terms of the underlying quark/gluon structure is also of interest. For example, it was suggested some time ago[2] that a hyperon in what would otherwise be an indistinguishable sea of up and down quarks, creates unique configurations of the strange quark when the Pauli principle is applied at the quark level. This may provide the key to interpret when (or if) a quark picture of the nucleus is appropriate.

In any event because the ΛN interaction is weak, one can with some confidence extract the ΛN potential from the effective Λ-Nucleus interaction by studying hypernuclear spectra. Of course this assumes that the concept of a Λ moving in a mean-field nuclear potential is valid, and that the spreading width of the Λ-N, particle-hole states is small. Although hypernuclear structure is still not completely understood, the application of mean-field calculations has been justified.

*Supported in part by the US DOE

However, in terms of the Λ-N interaction within the nucleus, the one pion exchange (OPE) between a Λ and a nucleon is not possible due to conservation of isospin. Thus the coupling of a lambda to an intermediate sigma, which leads to charge symmetry breaking and a sizeable three-body force, are important components of any hypernuclear potential[3]. In addition, the strangeness degree of freedom allows the nuclear particles to rearrange, in order to maximize the nuclear binding energy. Thus a hypernuclear system can emphasize rather different fundamental physics from conventional nuclei. As for example, the hyperon spin-orbit interaction is smaller, but charge symmetry breaking is larger than for nucleons.

Still while spectroscopy is important, much more information is potentially available. Indeed we emphasize here that the four S-shell hypernuclei collectively form a system where few body calculational methods can be tested, and utilized not only to extract structure information, but can be used to extract important new information on the weak interaction within the nuclear medium. Indeed the parity conserving amplitudes of the 4-fermion weak vertex describing $\Lambda N \rightarrow NN$, are only accessible via hypernuclear weak decays.

2 The $\Delta I=1/2$ Rule

Of particular interest is an investigation of the role that the $\Delta I=1/2$ rule plays in non-mesonic hypernuclear decay. Although a change of isospin of 1/2 or 3/2 is equally possible, in the non-leptonic, strangeness-changing weak decays, $\Delta S=1$, of Kaons and hyperons it is found that the I=1/2 amplitudes are significantly enhanced. For example the ratio ($\approx$ 1.9) of the decay probabilities of free Λ decay into a π^- and a π^0 is essentially given by the square of the Clebsch algebra for I=1/2 amplitudes[4]. In the hyperon decay this enhancement has been partially attributed to "penguin" diagrams, but in Kaon decays the effect is partially explained by final state interactions. In any event the experimental observation is generalized into the "$\Delta I=1/2$ rule", which states that the non-leptonic decays of all strange hadrons proceeds through $\Delta I=1/2$ amplitudes. However there is no universal explanation for this apparently universal rule, and most likely the effect is due to complicated dynamics in the decay process. In fact the rule may only be associated with pion decay, i.e. long range processes, as these are the only non-leptonic decays which have been studied in detail.

There is some theoretical prejudice, but no real experimental evidence that the $\Delta I=1/2$ rule is violated in the short ranged component of the ΛN weak interaction[5]. Here it has been argued that if quark effects can be observed at all in static nuclear properties, then they would be most evident in the high momentum transfer during the non-mesonic decay of the hypernuclear ground state. Non-mesonic hyperon decays occur because a Λ embedded in a nucleus finds that its mesonic decay channel is Pauli blocked, as the nucleon recoil in $\Lambda \rightarrow N + \pi$ decay has momentum much lower that the Fermi momentum of bound nucleons in the nucleus. Yet within the nuclear medium a Λ can weak decay through a non-mesonic process, $\Lambda N \rightarrow NN$, which provides more than sufficient momentum to eject nucleons from the nucleus. In hypernuclei with mass number greater than 5 non-mesonic decay dominates.

Without some other, unknown process the Δ I = 1/2 rule is not expected to be operative in quark dynamics[5]. We point out here that the key to an experimental test of the ΔI=1/2 rule is an investigation of the ratio of the non-mesonic neutron-stimulated to proton-stimulated weak decay of the ^{4}H hypernucleus, and this hypernucleus can now be produced using the Neutral Meson Spectrometer, NMS [7].

3 Weak Decay of S-Shell Hypernuclei

The status of the mesonic and non-mesonic weak decays of the S-shell hypernuclei are summarized it Table 1. Because there are spin-isospin correlations between the constituents of light hypernuclei, selective choices of S-shell hypernuclei can be used to weight various spin-isospin entrance channels for the ΛN weak decay. It can be demonstrated[8] that there are six possible transition amplitudes for S-shell hypernuclei decays when separated by spin and isospin. Measurements of the lifetimes, branching ratios, and asymmetry parameters can then be measured to extract these amplitudes. Indeed it can also be shown that not only the parity violating components, but the parity conserving weak amplitudes can be extracted, in contrast to other experiments involving hadrons, where the parity conserving amplitudes are masked by the much stronger hadronic force. Non-mesonic decay can be described by being either neutron stimulated, Λn →nn, or proton simulated, Λp →np. The ratio of the neutron to proton stimulated rate is experimentally found to be;

$$0.6 \leq \Gamma_{\rm nn}/\Gamma_{\rm np} \leq 0.9.$$

However the $^{13}S_1 \rightarrow ^{13}D_1$ tensor transition, which is only open to the proton channel, is expected to dominate the long range pion exchange component of the interaction. Therefore simple calculations based a meson exchange model exhibit a dominance of proton stimulated transitions, and a very small Λn→nn width, in disagreement with experiments. If the experiments are correct, perhaps the physics requires reliable quark calculations, but certainly better precision measurements of the partial decay rates of non-mesonic hypernuclear decays, as shown in Table 1, are needed.

A measure of the neutron to proton stimulated decay from three of the four S-shell hypernuclei, $^3_\Lambda$H, $^4_\Lambda$H, $^4_\Lambda$He, and $^5_\Lambda$He, can be used to verify the applicability of the ΔI=1/2 rule. The non-mesonic decays of $^4_\Lambda$He and $^5_\Lambda$He are reasonably well determined, and of the remaining two S-shell hypernuclei, the easiest and most sensitive additional measurement is observation of the non-mesonic decay of $^4_\Lambda$H. Non-mesonic decay of this hypernucleus has only previously been measured in emulsion, and with very poor statistics and uncertainties. It also should be pointed out that a measurement of the π^- to π^0 ratio for both $^4_\Lambda$H and $^4_\Lambda$He provides a direct test of the ΔI=1/2 rule in the mesonic decay of hypernuclei, although there is no reason to expect the rule would be violated

Table 1. Experimental Parameters of S-shell Hypernuclear Decay in Terms of the Λ Free Decay Width†

	$\Gamma_\Lambda/\Gamma_{free}$	Q^- (Γ_{nm}/π^-)	$\Gamma_{nm}/\Gamma_{free}$	α	$\Gamma_{\pi^-}/\Gamma_{free}$	$\Gamma_{\pi^0}/\Gamma_{free}$
${}^3_\Lambda$H	$1.13^{+0.19}_{-0.22}$(1)					
		0.4(2)				
	$2.05^{+0.42}_{-0.56}$(3)					
		0.0(4)				
${}^4_\Lambda$H	1.36±0.18(5)				$1.00^{+0.18}_{-0.15}$(5)	
		0.26±0.13(6)	0.29±0.14(6)			
${}^4_\Lambda$He	1.07±0.11(7)	1.13±0.22(7)	0.20±0.07(7)	0.27±0.14(7)	0.17±0.3(7)	
	1.15±0.48(6)	0.56±0.09(6)	0.14±0.03(6)	0.40±0.15(6)		
						0.55±0.07(5)
${}^5_\Lambda$He	1.03±0.08(8)	0.92±0.31(8)	0.41±0.14(8)	0.93±0.55(8)	0.44±.11(8)	0.18±0.20(8)

† Only most recent measurements are listed, for other values see; J. Cohen[6].
Prog. Part. Phys. 25(1990)139

1. G. Keyes, et al Nucl Phys B67(1973)269
2. V. Gorge, et al Nucl Phys 21(1960)599
3. G. Bohm, et al Nucl Phys B16(1970)46
4. B. Bhomik, et al, Phys Lett 3(1962)13
5. H. Outa, et al Nucl Phys A585(1955)109c
6. M.M. Block, et al Proc. Int. Conf. on Hyperfragments, St Cergue,1963, CERN Report 64-1
7. V. J. Zeps and G. B. Franklin, Proc. 23rd Int. Sypm. on Nucl. and Part. Phys., Universal Academy Press, Tokyo, 1995
8. J. J. Szymanski, et al, Phys Rev C43(1991)849

in this case.

4 Experiment E931

To undertake a measurement of the branching ratios of the decay of ${}^4_\Lambda$H, the NMS was mounted at the end of the C8 beamline at the Alternating Gradient Synchrotron, AGS, at the Brookhaven National Laboratory, BNL. It will be used to tag the production of ${}^4_\Lambda$H through the ^{4}He($K^-_{stopped}$, π^0)${}^4_\Lambda$H reaction. This reaction identifies when a ${}^4_\Lambda$H hypernucleus is the source of secondary particle emission from the target region. After the formation of this hypernucleus, one can determine the pion and neutron/proton stimulated decay spectra by observing, and tracking the secondary particles emitted from the target. For our purposes here we represent the experimental neutron to proton stimulated decay widths by the ratio, α, and use it in combination with previous measure-

Table 2. Expected counts for decay of $^4_\Lambda$H

Assuming 600 hours of production: Total $^4_\Lambda$H production: 4×10^4			
Particle Decay mode	$\eta \times \Omega/4\pi$	Expected Br. Frac.	Expected Counts
π^0	?	0.14	–
π^-	0.25	0.69	6900
Protons	0.27	0.02	215
pn-Coinc.	—	—	37
Neutrons	0.06	0.15	770
nn-Coinc.	—	—	61

ments of α in $^4_\Lambda$He and $^5_\Lambda$He, to extract the ratio $r=R_{n0}/R_{p0}$, the transition for the neutron stimulated decay from an initial spin zero state to a proton stimulated decay from an initial spin zero state. Theoretically, as argued above, one might expect that the value of α in $^4_\Lambda$H decay to be small, however using the value of α as measured in the other s-shell hypernuclei and the $\Delta I=1/2$ rule, one predicts a value of ≈ 7.0. We propose to measure α to at least 50 % accuracy, and so expect to easily test the validity of the $\Delta I=1/2$ rule by this experiment. Experiment rates are given in Table 2, which show that we can even obtain NN coincidence data. Although this will not be required to make a definitive test, coincidence data can help illuminate the neutron to proton stimulated decay puzzle by determining if multi-nucleon emission is important.

5 Summary

The investigation of strangeness in nuclear systems is not merely an extension of conventional nuclear physics. Certainly one cannot, nor would one want to, reproduce the wealth of information that has been accumulated on conventional nuclei. The hypernucleus, however, offers a selective probe of the hadronic many-body problem, providing insight in areas that cannot be easily addressed by conventional techniques. Such questions as isobar mixing, charge symmetry breaking, and quark confinement are more important, and thus more evident, in strange hadronic matter. What is now needed is a series of high precision studies with different reaction mechanisms and addressing a variety of topics.

As explained in the AGS proposal, E931 will determine the neutron to proton stimulated decay ratio, α, in $^4_\Lambda$H, and use this measurement in combination with previous measurements of α in $^4_\Lambda$He and $^5_\Lambda$He, to extract the ratio $r=R_{n0}/R_{p0}$. We believe that this measurement will settle the question about the validity of the $\Delta I=1/2$ rule in non-mesonic decay, and shed new light on the neutron to proton stiumlated decay puzzle.

References

1. B. F. Gibson and E. V. Hungerford: Phys. Rept. **257**, 350 (1995); *Proceedings of the International Conference on Hypernuclear and Strange Particle Physics*, D. J. Millener and R. E. Chrien, eds, Nucl. Phys. **A639**, 1c-596c (1998)
2. L. Biedenharn and E.V. Hungerford: Phys. Lett. **142B**, 232 (1984)
3. I. R. Afnan and B. F. Gibson: Phys. Rev. **C40**, R7 (1989); R. H. Dalitz, R. C. Herndon, Y. C. Tang: Nucl. Phys. **47B**, 109 (1972)
4. E. D. Commins and P. H. Bucksbaum: *W*eak Interactions of Leptons and Quarks, Cambridge Univ. Press, UK (1983)
5. T. Inoue, S. Takeuchi, M. Oka: Nucl. Phys. **A597**, 563 (1996); K. Maltman and M. Shmatikov: Nucl. Phys. **585**, 343c (1995)
6. J. Cohen: Progress in Particle and Nuclear Physics **25**, 139 (1990)
7. E907 and E931 proposals to the AGS, E. V. Hungerford, et al., spokespersons
8. R. H. Dalitz and L. Liu: Phys. Rev. **116**, 1312 (1959); H. H. Dalitz and G. Rajasekharan: Phys. Lett **1**, 96 (1962); M. M. Block and R. H. Dalitz: Phys. Rev. Lett. **11**, 96 (1963)

Few-Body Systems Suppl. 12, 342–346 (2000)

Few-Body Systems

High-Resolution Gamma-Spectroscopy of $^{7}_{\Lambda}$Li and $^{9}_{\Lambda}$Be

H. Tamura[1], D. Abe[1], S. Ajimura[2], H. Akikawa[3], K. Araki[1], H.C. Bhang[4], R.E. Chrien[5], T. Endo[1], P. Eugenio[6], G.B. Franklin[6], Y. Fujii[1], T. Fukuda[7], L. Gan[8], O. Hashimoto[1], H. Hotchi[9], K. Imai[3], Y. Kakiguchi[7], P. Khaustov[6], J.H. Kim[4], Y.D. Kim[10], H. Kohri[2], M. May[5], T. Miyoshi[1], T. Murakami[3], T. Nagae[7], J. Nakano[9], H. Noumi[7], H. Outa[7], K. Ozawa[1], P.H. Pile[5], B.P. Quinn[6], A. Rusek[5], T. Saito[11], J. Sasao[1], Y. Sato[1], S. Satoh[1], R.I. Sawafta[12], R.A. Schumacher[6], M. Sekimoto[7], T. Takahashi[1], T. Tamagawa[9], L. Tang[8], K. Tanida[9], H.H. Xia[13], L. Yuan[8], S.H. Zhou[13], L.H. Zhu[13,3], X.F. Zhu[13]

[1] Department of Physics, Tohoku University, Sendai 980-8578, Japan

[2] Department of Physics, Osaka University, Toyonaka 560-0043, Japan

[3] Department of Physics, Kyoto University, Kyoto 606-8502, Japan

[4] Department of Physics, Seoul National University, Seoul 151-742, Korea

[5] Brookhaven National Laboratory, Upton, NY 11973, USA

[6] Department of Physics, Carnegie Mellon University, Pittsburgh, PA 15213, USA

[7] Institute of Particle and Nuclear Physics, KEK, Tsukuba 305-0801, Japan

[8] Department of Physics, Hampton University, Hampton, VA 23668, USA

[9] Department of Physics, University of Tokyo, Tokyo 113-0033, Japan

[10] Department of Physics, Sejong University, Seoul 143-747, Korea

[11] Laboratory of Nuclear Science, Tohoku University, Sendai 980-0826, Japan

[12] North Carolina A&T State University, Greensboro, NC 27411, USA

[13] China Institute of Atomic Energy, Beijing 102413, China

Abstract. We report on the first two experiments of hypernuclear γ-ray spectroscopy with Hyperball, a large-acceptance germanium detector system. In a KEK experiment, we observed two γ transitions in $^{7}_{\Lambda}$Li, the spin-flip $M1(\frac{3}{2}^+\to\frac{1}{2}^+)$ at $691.7 \pm 0.6^{stat} \pm 1.0^{sys}$ keV and the $E2(\frac{5}{2}^+\to\frac{1}{2}^+)$ at $2050.4 \pm 0.4^{stat} \pm 0.7^{sys}$ keV. The former gives unambiguous information on the strength of the ΛN spin-spin interaction. $B(E2)$ of the latter transition was measured to be $3.6 \pm 0.5^{stat}\ {}^{+0.5}_{-0.4}{}^{\,sys} e^2\mathrm{fm}^4$. In the second experiment at BNL, we observed the $E2(\frac{5}{2}^+,\frac{3}{2}^+\to\frac{1}{2}^+)$ transitions of $^{9}_{\Lambda}$Be at around 3.05 MeV.

1 Introduction

High resolution γ spectroscopy with germanium (Ge) detectors is one of the most powerful means to investigate nuclear structure. However, it has been unavailable for hypernuclei because of technical problems caused by background from high energy meson beams. Recently we solved the problems and constructed "Hyperball", a large-acceptance Ge detector system dedicated to hypernuclear research. A series of experiments with Hyperball have started in 1998. Drastical improvement of energy resolution from a few MeV in the present (π^+,K^+) reaction spectroscopy into a few keV (FWHM) in our γ spectroscopy allows us to approach various interesting subjects in hypernuclear physics which were almost inaccessible before. One of the most important subjects is to investigate ΛN interactions, particularly spin dependent (spin-orbit, spin-spin, and tensor) parts, from fine structures of hypernuclei.

Hyperball consists of 14 coaxial Ge detectors equipped with fast electronics. It has a solid angle of about $15\% \times 4\pi$ sr and a photo-peak efficiency of about 3% at 1 MeV. Each Ge detector is surrounded by BGO counters which provide veto signals to suppress background in a Ge spectrum from π^0, high energy charged particles, as well as Compton scattering. A detailed description of the Hyperball project is found in Ref. [1].

2 Spectroscopy of ${}^{7}_{\Lambda}$Li (KEK E419)

The first experiment with Hyperball was performed at High Energy Accelerator Research Organization (KEK), employing the K6 beam line and Superconducting Kaon Spectrometer (SKS). Bound states of ${}^{7}_{\Lambda}$Li were produced by the (π^+,K^+) reaction at 1.05 GeV/c, and γ-rays were measured in coincidence. Figure 1 shows the γ-ray spectrum for the ${}^{7}_{\Lambda}$Li bound state events. The two peaks at 692 keV and at 2050 keV, observed only when the bound-state region in the (π^+,K^+) spectrum is gated, are attributed to transitions in ${}^{7}_{\Lambda}$Li.

2.1 Spin-Flip $M1$ Transition and the ΛN Spin-Spin Interaction

As shown in the left inset in Fig.1, the broad peak at 692 keV turns sharp by the Doppler shift correction which assumes γ-ray emission before slowing-down of the recoil ${}^{7}_{\Lambda}$Li. Therefore, the γ-ray emitting excited state has a lifetime less than 1 ps, which indicates it is a $M1$ transition of ${}^{7}_{\Lambda}$Li. This peak is assigned as the spin-flip $M1(\frac{3}{2}^+ \rightarrow \frac{1}{2}^+)$ transition but not the $M1(\frac{7}{2}^+ \rightarrow \frac{5}{2}^+)$ in the following reason. The $\frac{3}{2}^+$ and $\frac{7}{2}^+$ states have small production cross sections comparable with each other (see Fig.2) due to non-spin-flip property of the (π^+,K^+) reaction. However, the $\frac{3}{2}^+$ state can be much more populated through γ cascade from the $\frac{1}{2}^+$,T=1 state. Our observed 692 keV γ-ray yield agreed with the $\frac{3}{2}^+ \rightarrow \frac{1}{2}^+$ yield expected from the number of bound state events in the (π^+,K^+) spectrum and an estimated absolute efficiency of Hyperball.

Since the ^{6}Li(1^+) core has almost pure ${}^{3}S$ configuration, the ground-state doublet $(\frac{3}{2}^+, \frac{1}{2}^+)$ spacing is determined almost only by the ΛN spin-spin inter-

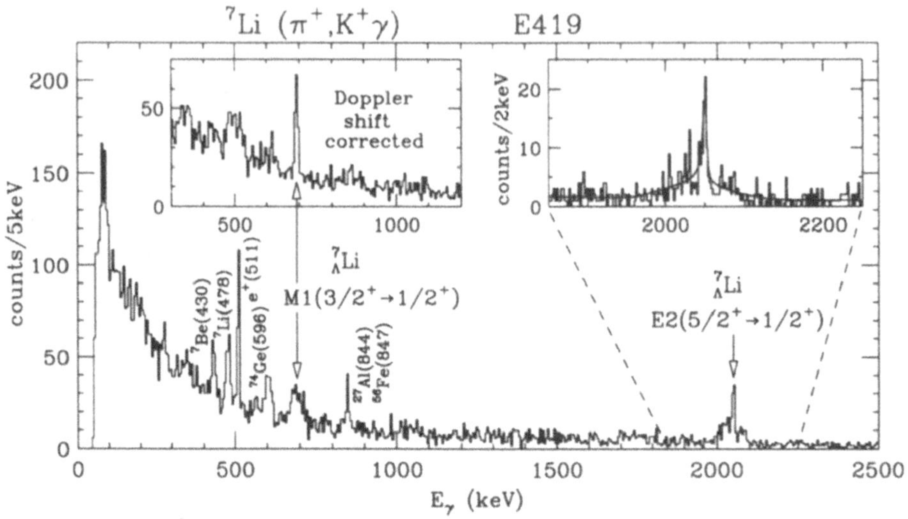

Figure 1. γ-ray spectrum for the bound-state region of ${}^7_\Lambda$Li. The two peaks at 692 keV and 2050 keV are attributed to $M1(\frac{3}{2}^+\rightarrow\frac{1}{2}^+)$ and $E2(\frac{5}{2}^+\rightarrow\frac{1}{2}^+)$ transitions of ${}^7_\Lambda$Li, respectively. See text for the inset figures.

action. A recent 4-body ($\alpha+p+n+\Lambda$) calculation by Hiyama *et al.* with the Nijmegen F interaction, where the strength of the spin-spin term was adjusted to reproduce the observed 1.1 MeV doublet spacing of the $A=4$ hypernuclei, predicted a ${}^7_\Lambda$Li ground-state doublet spacing of 650 keV without effect of the LS interaction [6]. A shell-model calculation by Millener *et al.* using a phenomenological spin-spin strength determined also from the $A=4$ hypernuclei predicted 610 keV [7]. Our observed spacing is thus consistent with the A=4 data and the strength of the ΛN spin-spin interaction was established.

2.2 B(E2) and Shrinking Effect

The peak at $2050.4\pm0.4\pm0.7$ keV is attributed to the $E2(\frac{5}{2}^+\rightarrow\frac{1}{2}^+)$ transition of ${}^7_\Lambda$Li (see Fig.2). It was previously observed at 2034±23 keV at Brookhaven National Laboratory (BNL) using NaI counters [2], but our high-resolution spectrum revealed the peak consisting of a sharp part and broad tails, indicating that the γ-ray energy is partly Doppler broadened. This fact supports the assignment of $E2$ and allows determination of the lifetime of the $\frac{5}{2}^+$ state using Doppler shift attenuation method. A Monte Carlo simulation was carried out to calculate the peak shape assuming various lifetimes, taking into account the recoil momentum of each peak event and the slowing-down process calculated from stopping power given by the SRIM code. The experimental spectrum was then fitted to the simulated peak shape with linear background as shown in the inset figure (right), and the lifetime was derived to be $5.8^{+0.9}_{-0.7}\pm0.7$ ps.

The lifetime was converted to $B(E2)$ of $3.6\pm0.5^{+0.5}_{-0.4}$ e^2fm^4, assuming that

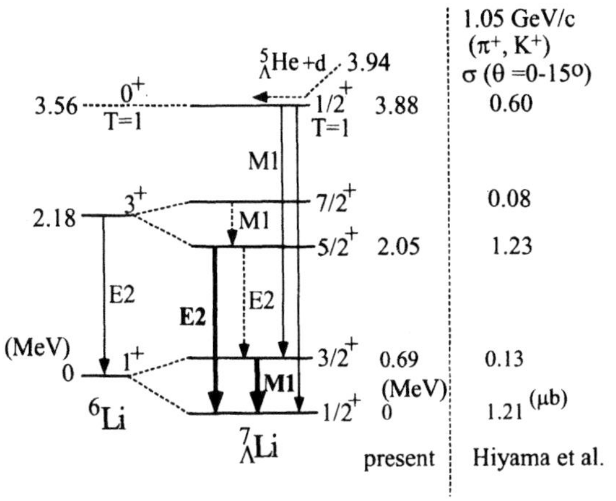

Figure 2. Level scheme and γ transitions of $^7_\Lambda$Li. Thick arrows and "present" indicate γ-transitions and level energies measured in the present experiment. $\sigma(\theta{=}0\text{–}15^0)$ shows calculated cross sections by (π^+,K^+) reaction at 1.05 GeV/c and integrated for the SKS acceptance (0^0–15^0) [3].

the branching ratio of $\frac{5}{2}^+ \to \frac{1}{2}^+$ is 94% (the branching ratio of $\frac{5}{2}^+ \to \frac{3}{2}^+$ is estimated to be 4% [3] and that of the weak decay is 2%). This is the first measurement of $B(E2)$ in hypernuclei. The measured $B(E2)$ is much smaller than that of the corresponding core transition, $B(E2;\, ^6\text{Li}\ 3^+ \to 1^+) = 10.9 \pm 0.9$ $e^2\text{fm}^4$ [4]. According to a cluster model calculation [3], this $B(E2)$ change is interpreted as shrinkage of the nuclear size of $^7_\Lambda$Li by 19 ± 4 % compared with ^{6}Li. It is the first observation of the phenomenon, predicted by Motoba, Bando and Ikeda in 1983, that a loosely-bound nucleus is contracted by a *Λ in the* 0*s orbit* due to attractive force between Λ and nucleons [5].

3 Spectroscopy of $^9_\Lambda$Be (BNL E930)

The second experiment (BNL E930) aims at determination of all the ΛN spin-dependent (spin-spin, spin-orbit, and tensor) interactions from γ spectroscopy of $^9_\Lambda$Be, $^{16}_\Lambda$O, $^7_\Lambda$Li, and $^{12}_\Lambda$C using the (K^-,π^-) reaction at the AGS D6 beam line. $^9_\Lambda$Be data were taken in 1998. The level spacing between the $\frac{5}{2}^+$ and $\frac{3}{2}^+$ states of $^9_\Lambda$Be provides crucial information on the spin-orbit interaction of Λ.

Figure 3 shows a preliminary γ-ray spectrum for the $^9_\Lambda$Be bound-state region. We observed a structure at 3.05 MeV, which is attributed to the $E2(\frac{5}{2}^+,\frac{3}{2}^+ \to \frac{1}{2}^+)$ transitions of $^9_\Lambda$Be. The energy is consistent with the previous BNL data with NaI counters where the two transitions were not separated [2]. In our high-resolution spectrum the 3.05 MeV structure seems to consist of closely-located two peak components, but further analysis is necessary to clearly identify each of the two transitions.

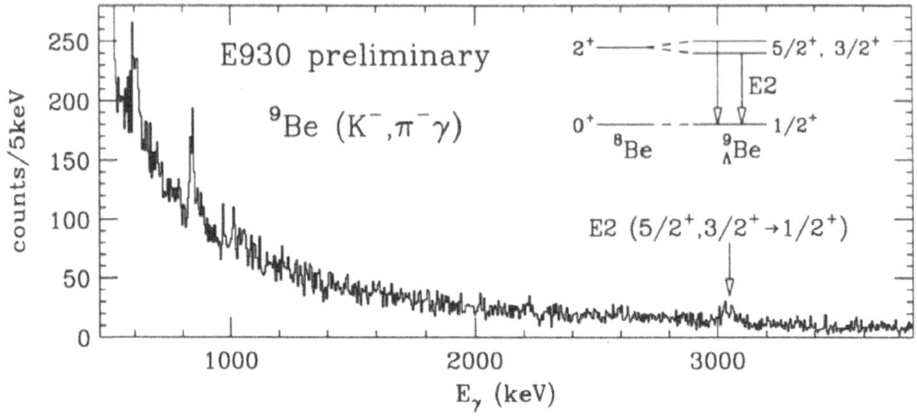

Figure 3. γ-ray spectrum for the bound-state region of ${}^{9}_{\Lambda}$Be (preliminary). Structure at 3.05 MeV is attributed to $E2(\frac{5}{2}^{+},\frac{3}{2}^{+}\rightarrow\frac{1}{2}^{+})$ transitions of ${}^{9}_{\Lambda}$Be.

4 Summary

We started experiments of high-resolution hypernuclear γ-ray spectroscopy using Hyperball. In the ${}^{7}_{\Lambda}$Li experiment, well-identified hypernuclear γ transitions were observed for the first time with Ge detectors; the $M1(\frac{3}{2}^{+}\rightarrow\frac{1}{2}^{+})$ and the $E2(\frac{5}{2}^{+}\rightarrow\frac{1}{2}^{+})$ transitions were observed at $691.7 \pm 0.6^{stat} \pm 1.0^{sys}$ keV and at $2050.4 \pm 0.4^{stat} \pm 0.7^{sys}$ keV, respectively. The strength of ΛN spin-spin interaction was experimentally established from the $M1$ energy. The $B(E2)$ was derived to be $3.6 \pm 0.5^{stat}\ {}^{+0.5}_{-0.4}{}^{\ sys} e^2\text{fm}^4$, which indicates a shrinkage of the nuclear size by Λ. In the ${}^{9}_{\Lambda}$Be experiment we observed the $E2(\frac{5}{2}^{+},\frac{3}{2}^{+}\rightarrow\frac{1}{2}^{+})$ transitions.

Acknowledgement. The authors are grateful to T. Motoba, E. Hiyama, and D.J. Millener, as well as the KEK-PS staff and BNL-AGS staff. This work is supported by Grant-In-Aid for Scientific Research on Priority Areas (Strangeness Nuclear Physics) from the Ministry of Education in Japan.

References

1. H. Tamura: Nucl. Phys. **A639**, 83c (1998)
2. M. May et al.: Phys. Rev. Lett. **51**, 2085 (1983)
3. E. Hiyama et al.: Phys. Rev. **C59**, 2351 (1999)
4. R.B. Firestone: *Table of Isotopes* (8th edition). A Wiley-Interscience Publication 1996
5. T. Motoba, H. Bandō and K. Ikeda: Prog. Theor. Phys. **80**, 189 (1983)
6. E. Hiyama et al.: Nucl. Phys. **A639**, 173c (1998)
7. D.J. Millener et al.: Phys. Rev. **C31**, 499 (1985)

Few-Body Systems Suppl. 12, 347–352 (2000)

Few-Body Systems

Four-Body Calculation of ${}^{4}_{\Lambda}$H and ${}^{4}_{\Lambda}$He with $\Sigma N - \Lambda N$ Conversion

E. Hiyama[1], M. Kamimura[2], T. Motoba[3], T. Yamada[4], Y. Yamamoto[5]

[1] Institute for Physical and Chemical Research (RIKEN), Wako, Saitama 351-0198, Japan

[2] Department of Physics, Kyushu University, Fukuoka 812-8581, Japan

[3] Laboratory of Physics, Osaka Electro-Communication University, Neyagawa 572-8530, Japan

[4] Laboratory of Physics, Kanto Gakuin University, Yokohama 236-8501, Japan

[5] Physics Section, Tsuru University, Tsuru, Yamanashi 402-8555, Japan

Abstract. We carried out four-body calculations of ${}^{4}_{\Lambda}$H and ${}^{4}_{\Lambda}$He taking both the $3N + \Lambda$ and $3N + \Sigma$ channels explicitly with the use of realistic NN and YN interactions. The Σ-channel component plays an important role in binding energies of the $A = 4$ hypernuclei though the admixture is approximately 1%. The $\Lambda N - \Sigma N$ coupling is found to be of central-force type in the Nijmegen model D and of tensor-force type in the model F.

1 Introduction

One of the main goals of the hypernuclear study is to obtain useful information on the hyperon-nucleon (YN) interaction. Although several types of YN interactions have been proposed, these interactions have a lot of ambiguity due to little information from the YN scattering experiments. Therefore, the study of hypernuclear structure should be helpful in elucidating the YN interaction properties. ${}^{3}_{\Lambda}$H is considered to be useful system to obtain information YN interaction. However, since this system is very weakly bound system, the off-energy shell effect in the YN interactions cannot be well examined. On the other hand, the study of ${}^{4}_{\Lambda}$H and ${}^{4}_{\Lambda}$He is much more useful to obtain information on the YN interactions because both of the spin-doublet states have been observed. Especially, it is expected to obtain information on the role of $\Lambda N - \Sigma N$ coupling in the binding mechanism of the $A = 4$ hypernuclei. In this work, we report four-body calculations of ${}^{4}_{\Lambda}$H and ${}^{4}_{\Lambda}$He taking the $\Lambda N - \Sigma N$

channel coupling into account explicitly.

2 Model and Method

We consider explicitly both the $3N+\Lambda$ and $3N+\Sigma$ channels explicitly and solve the four-body problem taking the $\Lambda N-\Sigma N$ conversion into account. The total Hamiltonian of $A=4$ hypernuclei is given by

$$H = T + V_{NN-NN} + V_{N\Lambda-N\Lambda} + V_{N\Sigma-N\Sigma} + V_{N\Lambda-N\Sigma} + V_{N\Sigma-N\Lambda} \quad (1)$$

where, T is the kinetic-energy operator. We employ the AV8 potential as the NN interaction, V_{NN-NN}. As for the YN interaction with $\Lambda N-\Sigma N$ coupling terms, we consider those of Nijmegen model D and model F, but we simulate them by using soft-core potentials with central and tensor terms [10] which reproduce the scattering phase shifts obtained by the original interactions. In the following, we refer to the phase-shift equivalent soft-core potentials as ND and NF. The total wavefunction is described as a sum of amplitudes of the rearrangement channels of Fig. 1 in the LS coupling scheme: [6, 8] as

$$\Psi_{JM}({}^4_\Lambda\mathrm{H},{}^4_\Lambda\mathrm{He}) = \sum_{Y=\Lambda,\Sigma}\sum_{c=1}^{4}\sum_{I,j}\sum_{l,L,\lambda}\sum_{s_0,s,S,t_0,t_0}\sum_{n,N,\nu} A^{(c)}_{Y,nlNL\nu\lambda Ijs_0sSt_0t}\mathcal{A}_N$$

$$\times\Big\{\Big[\Big[[\phi^{(c)}_{nl}(\mathbf{r}_c)\psi^{(c)}_{NL}(\mathbf{R}_c)]_I\xi^{(c)}_{\nu\lambda}(\boldsymbol{\rho}_c)\Big]_j\Big[[[\chi^{(\sigma)}_{1/2}(N_1)\chi^{(\sigma)}_{1/2}(N_2)]_{s_0}\chi^{(\sigma)}_{1/2}(N_3)]_s\chi^{(\sigma)}_{1/2}(Y)\Big]_S\Big]_{JM}$$

$$\times\Big[[[\chi^{(\tau)}_{1/2}(N_1)\chi^{(\tau)}_{1/2}(N_2)]_{t_0}\chi^{(\tau)}_{1/2}(N_3)]_t\chi^{(\tau)}_{t_Y}(Y)\Big]_{T=1/2}\Big\}, \quad (2.2)$$

where $\mathcal{A}_N$ are the three-nucleon antisymmetrization operator and the isospin $t_Y=0$ for $Y=\Lambda$ and $t_Y=1$ for $Y=\Sigma$. Here, n,N and ν denote the size of the Gaussian basis functions for four-body systems which are explicitly given in Ref. [6]. The basis functions were verified to be suited for describing both the short-range correlations and the long-range tail behaviour of few-body systems [1, 2, 3, 4, 5, 6, 7, 8]. The eigenenegies of Hamiltonian and coefficients A are determined by Rayleigh-Litz variational method. Calculation of the interaction matrix elements between the four-body basis functions can be performed rather easily with the use of the infinitesimally-shifted- Gaussian-Lobe representation of the basis functions [9, 4].

3 Results

All the calculations shown below have been done both for ${}^4_\Lambda$H and ${}^4_\Lambda$He. Calculated energies of the 0^+ ground state and the 1^+ excited state are illustrated in Fig. 2(a) and (b) together with the observed values. The energies are measured from the ${}^3\mathrm{H}+\Lambda$ threshold. In the case of only the $3N+\Lambda$ channel with ND included in the calculation, Fig. 2(a), the 0^+ state is barely bound and the 1^+ state is unbound. When the $(3N)_{t=1/2}+\Sigma$ channel with ND is further adopted,

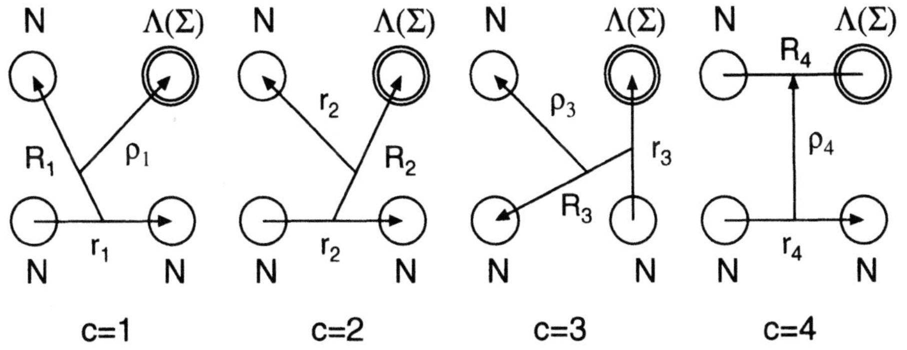

Figure 1. Jacobian coordinates for all the rearrangement channels of $3N + \Lambda(\Sigma)$ of ${}^4_\Lambda$H and ${}^4_\Lambda$He. The three nucleons are to be antisymmetrized.

both of the two states become bound evidently. Finally, further inclusion of the $(3N)_{t=3/2} + \Sigma$ channel makes the 1^+ state significantly more bound but with the 0^+ state almost unchanged. We understand that the Σ-channel component plays a very important role both in the energy gain and the relative splitting of the 0^+ and 1^+ states. However, the calculated binding energy of the 0^+ state is some 0.5 MeV smaller and hence the $0^+ - 1^+$ splitting is about a half of the observed value. In Fig. 2(b), the level structure calculated with NF is similar to the case of ND. Especially, we find that $(3N)_{t=3/2} + \Sigma$ channel is important for 1^+ state. In the cas of ${}^4_\Lambda$He, the calculated energies of 0^+ and 1^+ states with ND are -1.48 MeV and -0.96 MeV, respectively. Those with NF are -0.88MeV and -0.42 MeV, respectively. The results of ${}^4_\Lambda$He are very similar for those of ${}^4_\Lambda$H.

We calculated the probability of the $3N + \Sigma$-channel admixture in ${}^4_\Lambda$H with the use of ND and NF. Tendency of the admixture is different between the two cases. In the case of ND, the probability is ~1 % for the 0^+ state and ~0.5 % for the 1^+ state. But, in the case of NF, it is ~0.7 % for the both states. Precise percentage of the Σ-channel component is listed in Table 1. In the 0^+ state, the probability of the $(3N)_{t=1/2} + \Sigma$ channel is much larger than that of the the $(3N)_{t=3/2} + \Sigma$ channel, but, in the 1^+ state, they are nearly the same. We therefore understand that $(3N)_{t=3/2} + \Sigma$ channel is specially important in the 1^+ state. This tendency holds in the case of NF, too.

Also listed in Table 1 are the S-state and D-state probabilities of the channels. As long as the $3N + \Sigma$ channel in the 0^+ and 1^+ states is concerned, the S-state component is dominant in the case of ND while the D state is dominant in the case of NF. Therefore, we understand that the $\Lambda N - \Sigma N$ coupling is central-force type in ND, but is tensor-force type in NF. This is verified, as seen in Table 2, from the expectation value of the $\Lambda N - \Sigma N$ coupling potential. Namely, the central-force contribution is dominant in the case of ND, while the tensor-force one is dominant in NF. The calculated results of probability of the $3N + \Sigma$-channel admixuture and the expectation value of the $\Lambda N - \Sigma N$

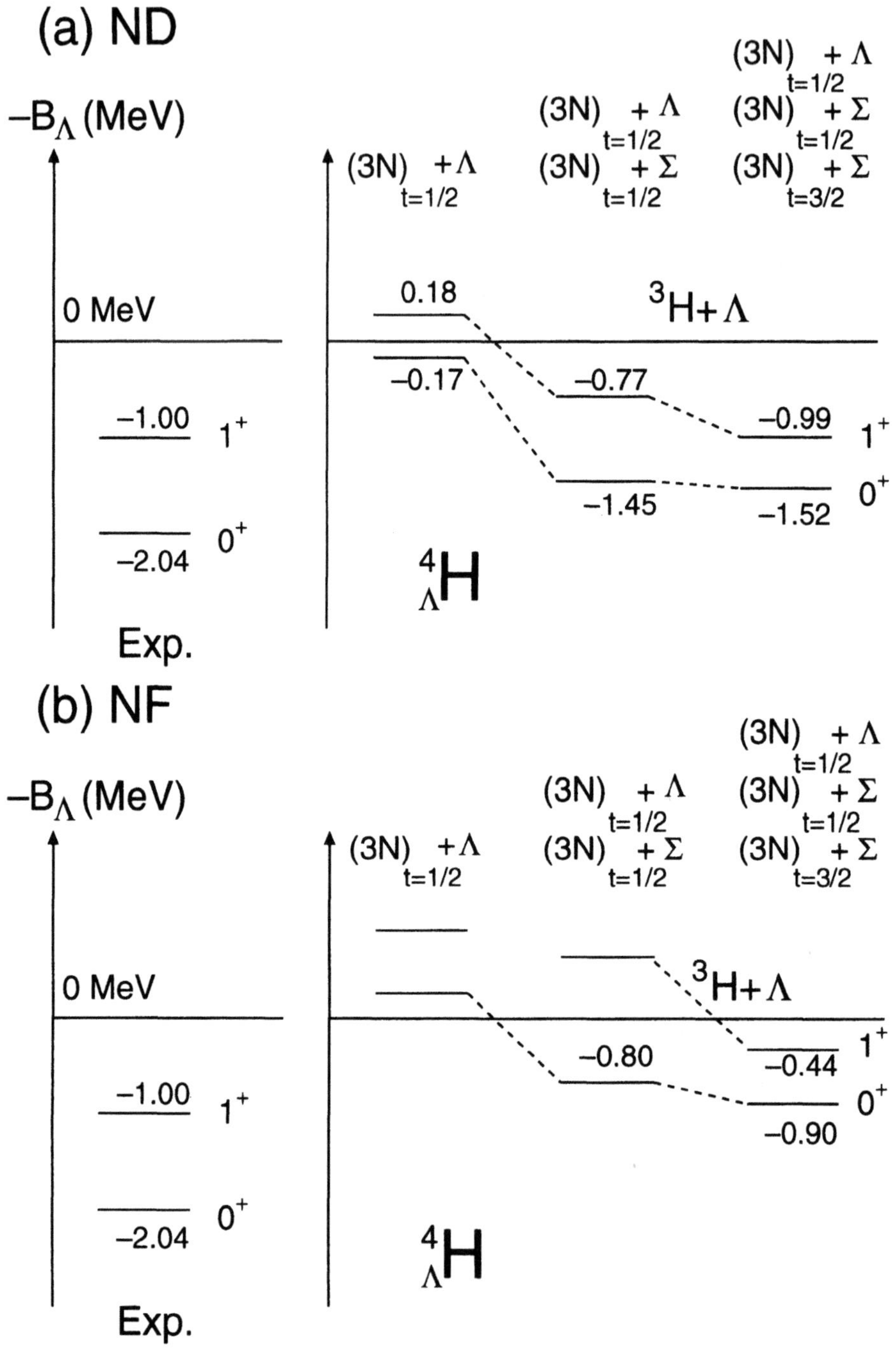

Figure 2. Calculated energy levels of $^4_\Lambda$H with the use of (a) ND and (b) NF for three cases of the Λ and Σ channels.

Table 1. Probabilities of the $(3N)_{t=1/2} + \Lambda$, $(3N)_{t=1/2} + \Sigma$ and $(3N)_{t=3/2} + \Sigma$ channels in the 0^+ and 1^+ states of ${}^4_\Lambda$H with (a) ND and (b) NF. The *S*-state and *D*-state probabilities are also listed.

(a) ND

	$(3N)_{t=1/2} + \Lambda$	$(3N)_{t=1/2} + \Sigma$	$(3N)_{t=3/2} + \Sigma$
0^+ (−1.52 MeV)	98.8 %	1.11 %	0.03 %
S-state	90.9	0.91	0.02
D-state	7.9	0.20	0.01
1^+ (−0.99 MeV)	99.5 %	0.24 %	0.23%
S-state	91.5	0.18	0.17
D-state	8.0	0.06	0.06

(b) NF

	$(3N)_{t=1/2} + \Lambda$	$(3N)_{t=1/2} + \Sigma$	$(3N)_{t=3/2} + \Sigma$
0^+ (−0.99 MeV)	99.3 %	0.69 %	0.03 %
S-state	91.8	0.10	0.02
D-state	7.5	0.59	0.01
1^+ (−0.44 MeV)	99.3 %	0.33 %	0.40%
S-state	91.5	0.01	0.01
D-state	7.8	0.32	0.39

Table 2. Expectation values of the central and tensor terms of the $\Lambda N - \Sigma N$ coupling potential, $2\times\langle V_{\Lambda N-\Sigma N} \rangle$, in ${}^4_\Lambda$H with ND and NF

	ND		NF	
$2\times\langle V_{\Lambda N-\Sigma N} \rangle$ (MeV)	0^+	1^+	0^+	1^+
Central	−3.20	−2.19	−0.33	−0.11
Tensor	−0.62	−0.59	−5.54	−6.74

coupling potential in ${}^4_\Lambda$He are almost similar for those in ${}^4_\Lambda$H.

In summary, we have developed the calculational method of four-body bound-state problems so that it becomes possible to make complicated four-body calculation of ${}^4_\Lambda$H and ${}^4_\Lambda$He taking both the $(3N + \Lambda)$ and $(3N + \Sigma)$ channels explicitly with the use of realistic NN and YN interactions. We are now able to investigate precisely the role of the $\Lambda N - \Sigma N$ coupling in ${}^4_\Lambda$H and ${}^4_\Lambda$He. However, the observed binding energies of the 0^+ and 1^+ states were not well reproduced with the use of ND and NF. Therefore, we are much interested in using a new version of Nijmegen soft-core YN interaction, NSC97 [11] and in examining this interaction by the four-body calculation of ${}^4_\Lambda$H and ${}^4_\Lambda$He including the $\Lambda N - \Sigma N$ coupling explicitly. This calculation is in progress.

Acknowledgement. This work was supported by the Special Postdoctoral Researchers Program at RIKEN (E.H.) and by the Grant-in-Aid for Scientific Research in Priority Areas.

References

1. M. Kamimura: Phys. Rev. **A38**, 621 (1988)
2. H. Kameyama, M. Kamimura and Y. Fukushima: Phys. Rev. **C40**, 974 (1989)
3. M. Kamimura and H. Kameyama: Nucl. Phys. **A508**, 17c (1990)
4. E. Hiyama and M. Kamimura: Nucl. Phys. **A588**, 35c (1995)
5. E. Hiyama et al.: Phys. Rev. **C53**, 2075 (1996)
6. E. Hiyama et al.: Prog. Theor. Phys. **97**, 881 (1997)
7. Y. Kino, M. Kamimura and H. Kudo: Nucl. Phys. **A631**, 649c (1998)
8. E. Hiyama et al.: Nucl. Phys. **A639**, 169c and 173c (1998)
9. E. Hiyama: RCNP Physics Report RCNP-P-132, 35 (1994)
10. Y. Yamamoto: Private Communications (1998)
11. T. A. Rijken, V. G. J. Stoks, Y. Yamamoto: Phys. Rev. **C59**, 21 (1999)

Few-Body Systems Suppl. 12, 353–357 (2000)

Few-Body Systems

Λ-Neutron Scattering Lengths

W.R. Gibbs[1], S.A. Coon[1], H.K. Han[1], B.F. Gibson[2]

[1] Physics Department, New Mexico State University, Las Cruces, NM 88003, USA

[2] Group T-5, Los Alamos National Laboratory, Los Alamos, NM 87545, USA

Abstract. Radiative capture of the K$^-$ on the deuteron is examined as a reaction for the measurement of the Λ-neutron scattering lengths. It is shown that a measurement of the deviation of the vector deuteron polarization asymmetry from -1 provides a direct determination of the difference of the singlet and triplet scattering lengths.

1 Introduction

The interaction of the nucleon with the Λ hypernucleon is of great theoretical interest. Since the one-pion-exchange interaction is absent, the focus falls on the shorter range contributions to the potential. The singlet and triplet scattering lengths are predicted to be equal by SU(6) symmetry. While experiments are very difficult to perform on this system, because of the short lifetime of the Λ, some results do exist[1]. Rijken *et al.*[2] have recently obtained a set of potentials which fit these data for which the range of scattering lengths is rather large.

A determination of the Λn scattering lengths from the photon spectrum of the reaction K$^-$d→nΛγ , where the capture takes place from an atomic state, was suggested by Gibson *et al.*[3]. A study of this process, performed by Workman and Fearing [4], concluded that the different hadronic routes considered by Akhiezer *et al.*[5] had little impact on a possible measurement. They also concluded that the dominant operator for the conversion of a proton to a Λ through the radiative capture of a kaon was of the Kroll-Ruderman form

$$\mathcal{O} = \boldsymbol{\sigma} \cdot \boldsymbol{\epsilon},$$

where ϵ is the polarization of the outgoing photon and $\boldsymbol{\sigma}$ is the spin operator for the proton on which the capture takes place.

Workman and Fearing used asymptotic Λ-neutron wave functions. In the parallel analysis of a_{nn} from $\pi^- d \to nn\gamma$, the lack of knowledge of the short range final-state wave function was found[6] to be the dominant theoretical uncertainty. For the neutron-neutron final state the spin state of the two neutrons is restricted to be singlet by the Pauli principle while the Λ-neutron final state consists of a mixture of the singlet and triplet states. The dependence of the scattering lengths on the spin could provide valuable clues to the structure of the interaction.

2 Expressions for the Rate

To compare with the measurement one needs the theoretical prediction for the shape of the spectrum only; the absolute rate not usually being measured in this type of experiment. The matrix element for the reaction which proceeds from an initial deuteron with spin-projection S_z to a final state of the Λ-neutron system in a total spin state (S', S'_z) is

$$M^{S',S'_z,S_z} = \int d\mathbf{r}_1 d\mathbf{r}_2 < S'S'_z|\psi^-_{S'}(\mathbf{p},\mathbf{r})e^{-i\mathbf{K}\cdot(\mathbf{r}_1+\mathbf{r}_2)}|\mathcal{O}e^{i\mathbf{k}\cdot\mathbf{r}_1}|\Psi^{S_z}_D(\mathbf{r}) >,$$

where $\psi^-_{S'}(\mathbf{p},\mathbf{r})$ is the final state wave function of the neutron-Λ system and the variable S' is 0 or 1 for the singlet or triplet case. $\mathbf{K}$ is the center-of-mass momentum of the baryon pair. The quantity $|\Psi^{S_z}_D(\mathbf{r}) >$ is the deuteron wave function with an initial spin projection S_z. We take the radial deuteron wave function from the solution of the Schrödinger equation for a one-pion-exchange potential.

The capture rate for the various initial and final states (with only the dominant $\ell = 0$ terms) is given, for left circularly polarized photons, by

$$S_z = 1,\ 0;\ S_z = 0,\ \left|S_1 - \sqrt{2}D_1\right|^2;\ S_z = -1,\ \left|S_1 + S_0 + \frac{1}{\sqrt{2}}[D_1 + D_0]\right|^2$$

and, for right circularly polarized photons, by

$$S_z = 1,\ \left|S_1 - S_0 + \frac{1}{\sqrt{2}}[D_1 - D_0]\right|^2;\ S_z = 0,\ \left|S_1 - \sqrt{2}D_0\right|^2;\ S_z = -1,\ 0.$$

The quantities $S_{0,1}$ and $D_{0,1}$ are integrals over the s- and d-states of the deuteron using the singlet or triplet final state Λn wave functions. The D state of the deuteron contributes coherently, with different coefficients depending on whether the initial state is $S_z = 0$ or $S_z = \pm 1$, so that the shape of the spectrum is different for the two cases as illustrated in Fig. 1.

Rather than use an asymptotic final-state Λn wave function, we employ one with a realistic behavior at short distances obtained from potentials parameterized as the sum of four exponentials. These potentials were fit to the phases shifts of both the singlet and triplet states of each of the six cases of Ref. [2]. We have defined continuous variables which reproduce the phase shifts

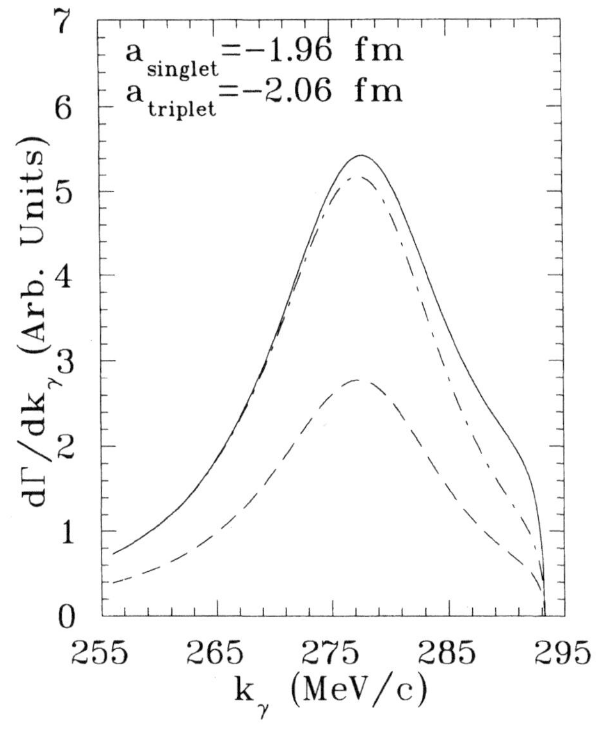

Figure 1. Comparison of singlet (dashed) and triplet (solid) spectra for a full calculation with realistic wave functions. The dash-dot curve shows the singlet spectrum renormalized to the triplet at the lowest energy.

at the integers 1 through 6 (corresponding to cases a through f) and allow an interpolation and extrapolation.

The smallest mass taken in this expansion has an inverse range of 1.41340 fm^{-1} which corresponds very well to two pion masses. We have also considered potentials defined by other rescaling of the ranges which are used to provide estimates of the model dependence of the calculation.

3 Results

If no spin degrees of freedom are measured, as was the case of the data of Gall *et al.*[7], the singlet and triplet states contribute incoherently to the rate and it might seem impossible to separate them. However, because the interference of the s and d states of the deuteron depends on the initial spin projection, the shapes of the singlet and triplet spectra are different (Fig. 1). Because the triplet gives a larger contribution than the singlet, a greater sensitivity to the triplet scattering is expected.

Figure 2 shows an analysis of the data of the feasibility study of Gall *et al.*[7]. We see that the scattering lengths are in the range expected from the

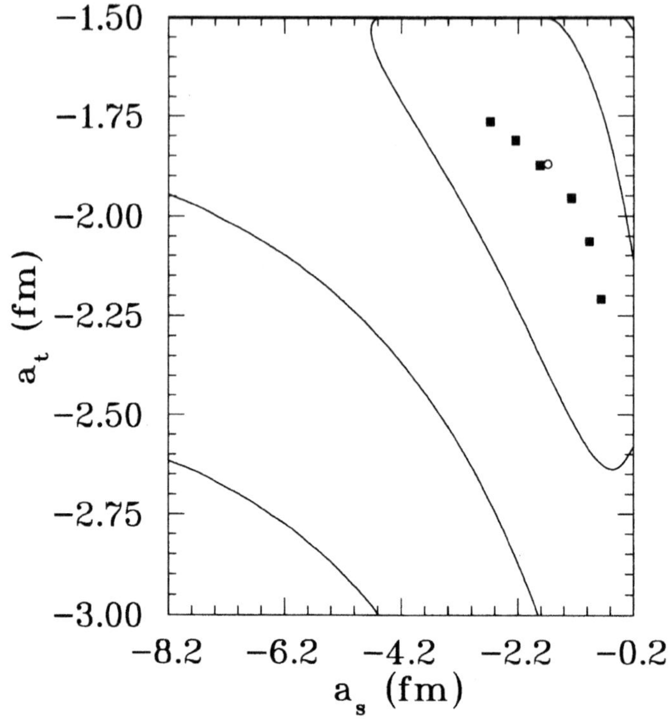

Figure 2. Analysis of the data of Gall *et al.*. The contours are lines of constant χ^2 with the innermost curve corresponding to one standard deviation, the next to two, etc.

values of Ref. [2] (shown by the black squares). The open circle shows the point of minimum χ^2. One-standard-deviations limits can be read from the graph with the triplet length lying between −1.3 and −2.6 fm and the singlet length between −0.2 and −6.3 fm. An analysis of pseudo data allows us to quote model errors of ±0.28 fm for the singlet and ±0.15 fm for the triplet.

The $S_z = 1$ capture rate depends on the difference of the singlet and triplet scattering states. While this cancellation leads to a small rate if the singlet and triplet scatterings are similar, this small quantity is a direct measure of the *difference between the singlet and triplet lengths.*

Figure 3 shows the spectrum expected from a pure vector polarized deuteron target. We see that a measurement of accuracy 10% is needed near the upper part of the spectrum. What is being searched for is a deviation of the asymmetry from −1 or, in other words, a non-null value of the rate from the initial state $S_z = +1$. Such an observation would prove that the two scattering lengths are unequal. This result is essentially model independent.

This work was supported by the U. S. Department of Energy and the National Science Foundation. HKH thanks the Korean Scientific and Engineering Foundation for their support.

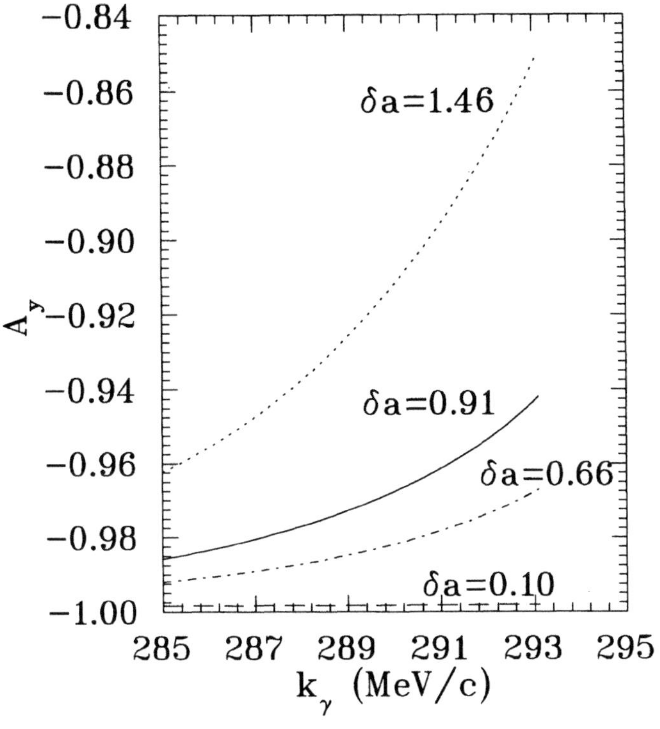

Figure 3. Vector polarization asymmetry for the deuteron initial spins. The labels denote the difference between the singlet and triplet scattering lengths.

References

1. See Ref. [2] for references to the experimental data.
2. Th.A. Rijken, V.G.J. Stoks, Y. Yamamoto: Phys. Rev. **C59**, 21 (1999)
3. B.F. Gibson, G.J. Stephenson, V.R. Brown, M.S. Weiss: BNL Report 18335, 1973
4. R.L. Workman and H.W. Fearing: Phys. Rev. **C41**, 1688 (1990)
5. A.I. Akhiezer, G.I Gakh, A.P. Rekalo, M.P. Rekalo: Yad. Fiz. **27**, 214 (1978) [Sov. J. Nucl. Phys. **27**, 115 (1978)]
6. W.R. Gibbs, B.F. Gibson, G. Stephenson: Phys. Rev. **C11**, 90 (1975)
7. K.P. Gall et al.: Phys. Rev. **C42**, R475 (1990)

Few-Body Systems Suppl. 12, 358–361 (2000)

Few-Body Systems

Study of Doubly Strangeness Nuclear System with Cylindrical Detector System

E906 Collaboration
presented by J.P. Nakano[1,*]

[1] Department of Physics, the University of Tokyo, 7-3-1 Hongo, Bunkyo, Tokyo 113-0033, Japan

Abstract. We have carried out an experiment to study doubly strangeness nuclear system at BNL-AGS(E906). We will describe the principle of the experiment and report the present status of the analysis.

1 Introduction

The S=-2 system, which is composed of "uuddss", gives us a unique opportunity to study various aspects of the strong interaction. First of all, we have much interest in $\Lambda\Lambda$ hypernuclei. Mass of $\Lambda\Lambda$ hypernuclei will give us information about the interaction between two Λ 's, which is essential for the better understanding of baryon-baryon interaction. However we have only three experimental reports so far [1]. On the other hand, the $\Lambda\Lambda$ hypernuclei are related to the existence of the 6-quark bound state, H dibaryon (uuddss), which was predicted by Jaffe. It's originated from the color magnetic interaction of QCD. So far, many attempts to search for the bound H directly have been done, but there is no conclusive evidence. KEK-E224 group recently published a report suggesting an H-dibaryon resonance state, that is, enhanced production of $\Lambda\Lambda$ pairs near 2 Λ mass threshold [2]. They claimed that the data was not reproduced well by any theoretical predictions without $\Lambda - \Lambda$ interaction, but it's hard to get more information due to their limited statistics.

It's certainly necessary to study the "uuddss" system with higher statistics than ever, and therefore we proposed a unique way for the spectroscopy of S=-2 system using the Cylindrical Detector System(BNL-AGS-E906) [3].

* *E-mail address:* joe.nakano@kek.jp

2 Experimental Principle

There may be two mechanisms to produce double Λ hypernuclei as well as other S=-2 systems. One is the stopped Ξ^- process, where quasi-free Ξ^- 's are emitted in the (K^-, K^+) reaction and brought to rest by ionization loss in the target material. After Ξ^- -atom is formed, the Ξ^- is eventually absorbed by the nucleus. Ξ^- in the nucleus will be converted into two Λ 's via strong interaction, $\Xi^- p \rightarrow \Lambda\Lambda$. In the specific cases the double-Λ compound state is formed with a "sticking" probability of more than ~10%. The Ξ^- emission probability is about 70-80% of the total quasi-free events and about 0.3% of the quasi-free events will be converted to the double Λ compound state[4].

On the other hand, the remaining 20~30% of the quasi-free Ξ^- production will be followed by the secondary reaction of a Ξ^- inside a nucleus. A theoretical attempt[5] to estimate the production rate of the $\Lambda\Lambda$ compound nucleus via such a process gives about 0.3% per quasi-free reaction for a ^{12}C target. This assumed a Ξ^- -nucleus potential of 15 MeV depth. In both cases, the $\Lambda\Lambda$ compound nucleus will subsequently decay into hyperfragments containing one or two hyperons per fragment, for example ${}^{6}_{\Lambda\Lambda}\mathrm{He}$ and ${}^{5}_{\Lambda\Lambda}\mathrm{H}$.

The $\Lambda\Lambda$ hypernuclei will be identified with detecting two characteristic decaying pions, in analogy to conventional $\beta - \gamma$ spectroscopy. ${}^{5}_{\Lambda\Lambda}\mathrm{H}$ case serves as an example. ${}^{5}_{\Lambda\Lambda}\mathrm{H}$ will decay in ${}^{5}_{\Lambda\Lambda}\mathrm{H} \rightarrow {}^{5}_{\Lambda}\mathrm{He} + \pi^-$ and ${}^{5}_{\Lambda}\mathrm{He} \rightarrow \mathrm{He} + p + \pi^-$; ${}^{5}_{\Lambda\Lambda}\mathrm{H}$ can be identified by observing two characteristic pions from the successive pion decays. The π^- 's momentum from ${}^{5}_{\Lambda}\mathrm{He}$ is 99.2MeV/c and the width(~1MeV/c) is enough small to be identified. From the momentum spectrum of π^- in coincidence with the π^- with momentum of around 100MeV/c, the mass of ${}^{5}_{\Lambda\Lambda}\mathrm{H}$ can be determined. Note that the branching ratios of pion decay rate $\Gamma_{\pi^-}/\Gamma_{tot}$ are large enough to be observed; ~ 0.21 and ~ 0.39 for ${}^{5}_{\Lambda\Lambda}\mathrm{H}$ and ${}^{5}_{\Lambda}\mathrm{He}$ [6], respectively.

It's also noted that we chose ^{9}Be as a target because the proton number of produced hypernuclei is limited to be less than three. A calculation for hyperfragment formation rate based on a statistical model shows a concentration of ${}^{5}_{\Lambda\Lambda}\mathrm{H}$ (~0.4 per double-Λ compound nucleus)[6].

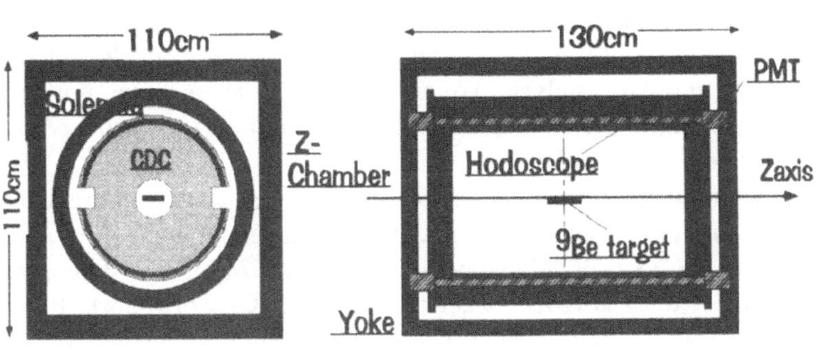

Figure 1. A schematic drawing of the Cylindrical Detector System

Decaying pions will be detected by using the Cylindrical Detector System(CDS). (Figure.1) CDS is composed of solenoid magnet, Cylindrical Drift Chamber(CDC), cathode readout MWPC called Z-Chamber, and hodoscope. CDS has a solid angle over 60% of 4π, and the momentum resolution for 100 MeV/c pion is expected to be around 7% in FWHM.

3 Preliminary Results

The experiment was carried out from Sep. to Nov. in 1998 at the 2GeV/c K^- beam line at the BNL-AGS. In total, $0.9 \times 10^{12} K^-$ was exposed on a ^9Be plate, $6''long \times 2''wide \times 1/2''high$. By analyzing the data from the beam line and the outgoing K^+spectrometers, we have clearly seen the quasi-free Ξ^- production peak in the missing mass spectrum. Eventually, 1.12×10^6 (K^-, K^+) reaction has been identified as Ξ^- production. According to our Monte Calro simulation, our data should contain a hundred of $\Lambda\Lambda$ hypernuclei and a larger number of $\Lambda\Lambda$ pairs than E224 by more than one order of magnitude.

At first, we analyze "two negatively-charged track events", which are considered to contain two-π^- coincidence events from the decay of the $\Lambda\Lambda$ hypernuclei. In Figure.2, we show a spectrum of the higher momentum pions gated by the lower momentum pions between 90-110 MeV/c, where π^- 's from the the most of the relevant single Λ hypernuclei should be included. The other cut are; 1) the chi-square of tracking ≤ 5, 2) the distance of the closest approach between two tracks $\leq$ 3cm and 3) the vertex of the two tracks should be within the target boundary. The energy loss corrections in the target have been applied, and the momentum scale is verified with the decaying π^+ (189MeV/c) from Σ^+ as produced in the (K^-, π^-) reaction on a polyethylene target. The histogram is overlaid by a background estimation which is assumed to come from free Ξ^- decay with a proton out of the CDS. The vertical magnitude of the curve is calibrated by comparing Ξ^- counts reconstructed by CDS with the simulation data without a proton missing. The shape of histogram above 140 MeV/c agree with the background estimation, where the contribution of free Ξ^- decay should be dominant. We tentatively conclude that the excess counts over the background curve are due to the two-π^- decays from S=-2 nuclear systems. However, we cannot specify the $\Lambda\Lambda$ hypernuclei in this histogram because of ambiguities in the energy loss correction inside the target at the present stage of the analysis. We are now trying to improve momentum correction of decaying π^- by studying vertex information between beam line spectrometer system and CDS.

Secondly, we analyze "two negatively- and two positively- charged track events", which should include $\Lambda\Lambda$ pairs. An enhancement near two Λ mass threshold similar to KEK-E224's report[2] has been observed. The present statistics of our $\Lambda\Lambda$ data is twice larger than the KEK-E224 report. However it's smaller than our estimate, because the track finding efficiency is low at present. For more quantitative discussion, we are now trying to improve track finding efficiency and also to correct the CDS acceptance.

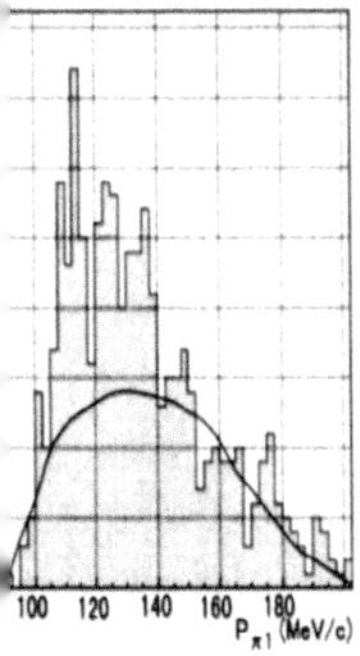

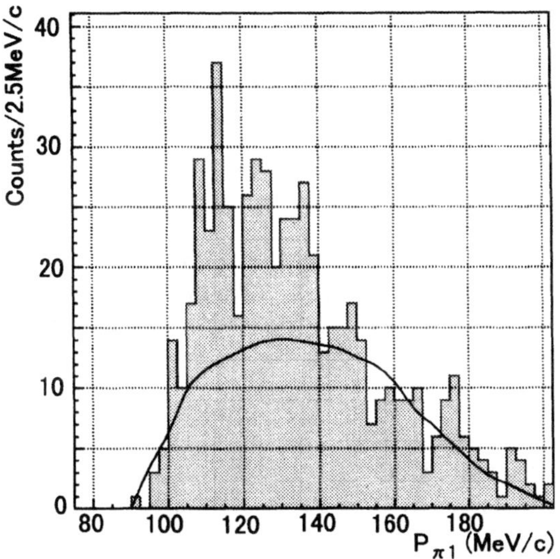

Figure 2. Spectrum of the higl momentum pions gated by t lower momentum pions with ! 110 MeV/c. See the text for t details.

Finally, it should be noted that E906 is the first measurement to dete decay particles from a large number of S=-2 nuclear systems, so that we c expect to get better information on S=-2 nuclear systems by this new techniq

References

1. M. Danysz et al.: Nucl. Phys. **49**, 121 (1963); P.J. Prowse: Phys. Rev. Le **17**, 782 (1996): S. Aoki et al.: Prog. Theor. Phys. **85**, 1287 (1991)

2. J.K. Ahn et al.: Phys. Lett. **B444**, 267 (1998)

3. T. Fukuda and R. E. Chrien: AGS proposal E906 (1994)

4. K. Nakazawa: *Proc. of the 23rd INS Int. Symp. on Nuclear and Parti Physics with Meson Beams in the 1 GeV/c Region*, Tokyo, 1985, p.261

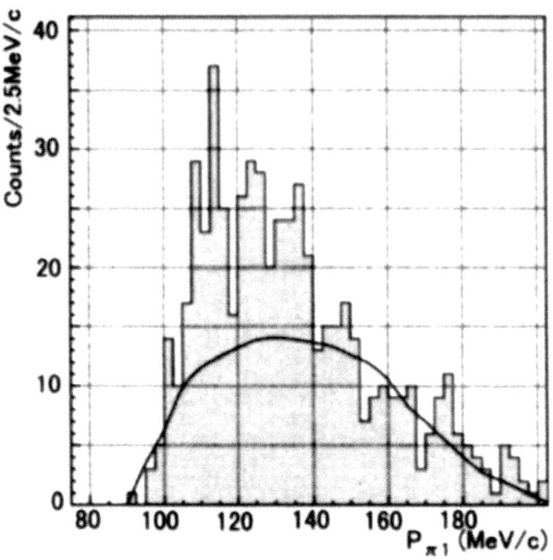

Figure 2. Spectrum of the higher momentum pions gated by the lower momentum pions with 90-110 MeV/c. See the text for the details.

Finally, it should be noted that E906 is the first measurement to detect decay particles from a large number of S=-2 nuclear systems, so that we can expect to get better information on S=-2 nuclear systems by this new technique.

References

1. M. Danysz et al.: Nucl. Phys. **49**, 121 (1963); P.J. Prowse: Phys. Rev. Lett. **17**, 782 (1996): S. Aoki et al.: Prog. Theor. Phys. **85**, 1287 (1991)

2. J.K. Ahn et al.: Phys. Lett. **B444**, 267 (1998)

3. T. Fukuda and R. E. Chrien: AGS proposal E906 (1994)

4. K. Nakazawa: *Proc. of the 23rd INS Int. Symp. on Nuclear and Particle Physics with Meson Beams in the 1 GeV/c Region*, Tokyo, 1985, p.261

5. Y. Yamamoto, M. Wakai, T. Fukuda, M. Sano: Prog. Theor. Phys. **88**, 1163 (1992)

6. Y. Yamamoto, M. Wakai, T. Motoba, T. Fukuda: Nucl. Phys. **A625**, 107 (1997)

Table 1. Calculated binding energies and channel components for the ${}^{6}_{\mathrm{YY}}\mathrm{He}$, ${}^{10}_{\mathrm{YY}}\mathrm{Be}$ and ${}^{13}_{\mathrm{YY}}\mathrm{B}$ g.s. with the Nijmegen model-D and Ehime potentials, where No (Yes) denotes the results with only the $\Lambda\Lambda$ channel ($\Lambda\Lambda$ and ΞN channels).

	$B^{exp}_{\Lambda\Lambda}$	Model-D pot. $B^{cal}_{\Lambda\Lambda}$		$P_{\Lambda\Lambda}$	P_{Ξ}	Ehime pot. $B^{cal}_{\Lambda\Lambda}$		$P_{\Lambda\Lambda}$	P_{Ξ}
	(MeV)	No	Yes	(%)	(%)	No	Yes	(%)	(%)
${}^{6}_{\mathrm{YY}}\mathrm{He}$	10.9 ± 0.5	9.4	9.8	99.7	0.3	9.7	9.8	99.9	0.1
${}^{10}_{\mathrm{YY}}\mathrm{Be}$	17.7 ± 0.4	16.9	17.4	99.7	0.3	17.3	17.5	99.9	0.1
${}^{13}_{\mathrm{YY}}\mathrm{B}$	27.6 ± 0.7	27.7	28.0	99.7	0.3	27.5	27.7	99.9	0.1

and [core+Ξ+N] model (see §2). Then, the Kyoto-Niigata RGM-F and FSS versions are employed to study the hyperon mixing in the ${}^{10}_{\mathrm{YY}}\mathrm{Be}$ ground state with the core+$(3q)$+$(3q)$ model (see §3). The brief summary is given in §4.

2 [Core+Λ+Λ]+[Core+Ξ+N] Model

In the [core+Λ+Λ]+[core+Ξ+N] model, we use the simulated Nijmegen model-D potential [1] and the Ehime potential [2] for the interaction between the two valence baryons, where the former is a Gaussion-type potential equivalent to the T-matrix of the original model-D potential. Concerning the potential between the core and valence baryon, we employ the folding potential, whose strength is adjusted so as to reproduce the experimental binding energy for the ground state of the [core+baryon] system with the potential model. The Pauli principle between the core and valence nucleon in the core+Ξ+N channel is taken into account with the orthogonality condition model (OCM). The coupled-channel three-body problem is solved in terms of the coupled-rearrangement-channel Gaussian-Lobe basis variational method developed by the Kyushu group.

The calculated results are shown in Table 1. In case of the Nijmegen model-D case, the calculated binding energies for the three $S = -2$ nuclei are reproduced well, and the binding energy gain due to the coupling with the Ξ channel and the Ξ component are as small as $0.3 \sim 0.5$ MeV and 0.3 %, respectively. On the other hand, the Ehime potential also reproduces the experimental binding energies, and the binding energy gain due to the coupling with the Ξ channel ($0.1 \sim 0.2$ MeV) and the Ξ component (0.1%) are smaller than those for the model-D case. This indicates that the strength of the $\Lambda\Lambda$-ΞN coupling potential for the Ehime potential is weaker than that for the Nijmegen model-D potential. From the above results, both the model-D and Ehime potentials predict the $\Lambda\Lambda$-hypernuclear state for the light $S = -2$ nuclear ground states.

The reason why the Ξ component is small is understood as follows; 1) the energy difference between the $\Lambda\Lambda$- and Ξ-hypernuclear states ($\sim \Delta M(\Xi N - \Lambda\Lambda)$ + nucleon separation energy) is as large as $30 \sim 40$ MeV, which is much larger than the coupling matrix element ($\sim$a few MeV), 2) in the core+Ξ+N channel, due to the Pauli principle between the core and nucleon, the valence nucleon as well as the Ξ particle must be in the p state, then the coupling matrix element which is originally small in the free space becomes smaller, and consequently, the Ξ component in the ground states is as small as $0.1 \sim 0.3$ %.

3 Core+(3q)+(3q) Model

3.1 Formulation

The core+$(3q)$+$(3q)$ model is an applied version of the $(3q)$+$(3q)$ quark cluster model [3] formulated in the resonating group method (RGM) to nuclear systems. The total wave function of the core+$(3q)$+$(3q)$ system within the frame of the RGM, $\Phi = \Phi_{\Lambda\Lambda} + \Phi_{\Xi N} + \Phi_{\Sigma\Sigma}$, is represented as

$$\Phi = \sum_{\alpha} \int d\boldsymbol{r} d\boldsymbol{R} \left[\mathcal{A}' \left\{\phi_\alpha \delta(\boldsymbol{r}_\alpha - \boldsymbol{r})\right\} \delta(\boldsymbol{R}_\alpha - \boldsymbol{R})\right] \chi_\alpha(\boldsymbol{r}, \boldsymbol{R}), \tag{1}$$

where ϕ_α denotes the $(3q)$+$(3q)$ intrinsic wave function with the orbital-spin-flavor-color part for the α channel ($\alpha = \Lambda\Lambda, \Xi N, \Sigma\Sigma$), $\mathcal{A}'$ is the antisymmetrization operator among quarks belonging to different clusters, and χ is the relative wave function referring to the $(3q) - (3q)$ relative coordinate $\boldsymbol{r}$ and the core$-[(3q) + (3q)]$ relative coordinate $\boldsymbol{R}$. The core part is treated as a point particle, and the antisymmetrization between the valence quark and core is not taken into account except for the core$-N$ part in the core+Ξ+N channel as discussed below. The total Hamiltonian is given as $H = H^{(3q+3q)}_{\alpha\beta} + \delta_{\alpha\beta} T_{\boldsymbol{R}_\alpha} + \sum_{q=1}^{6} v_{q-core} + V_{\rm Pauli} + \delta_{\alpha\beta} \Delta M_{\alpha\beta}$, where $H^{(3q+3q)}$ denotes the Hamiltonian of the $(3q)+(3q)$ system including the quark mass, kinetic energy, confinement potential, Fermi-Briet interaction and meson-exchange potential between quarks, which is equivalent to the Kyoto-Niigata one [3]. $T_{\boldsymbol{R}}$ and v_{q-core} represent, respectively, the kinetic energy referring to the coordinate $\boldsymbol{R}$ and quark-core interaction. $V_{\rm Pauli}$ is the pseudo-potential to remove the Pauli-forbidden state for the core-N part in the core+Ξ+N channel. The mass shift matrix ΔM is introduced to reproduce the experimental mass differences among $\Lambda\Lambda$, ΞN and $\Sigma\Sigma$. The equation of motion is derived from the variational principle, and then it becomes the three-body RGM equation, $\sum_\beta \int d\boldsymbol{r}' d\boldsymbol{R}' \left\{H_{\alpha\beta}(\boldsymbol{r}, \boldsymbol{R}; \boldsymbol{r}', \boldsymbol{R}') - E N_{\alpha\beta}(\boldsymbol{r}, \boldsymbol{R}; \boldsymbol{r}', \boldsymbol{R}')\right\} \chi_\beta(\boldsymbol{r}', \boldsymbol{R}') = 0$, where $H_{\alpha\beta}$ and $N_{\alpha\beta}$ are the Hamiltonian and norm kernels within the RGM framework, respectively. For the RGM kernel of the quark-core interactions, only the direct kernel is taken and is replaced with the phenomenological core+baryon potentials (see §2). The relative wave function χ_β is expanded with the Gaussian basis, and then the RGM equation is solved with the variational method.

3.2 Results and Discussion

First, we use the RGM-F version of the Kyoto-Niigata group [3] as the $(3q)$ + $(3q)$ Hamiltonian. In the original RGM-F Hamiltonian, a bound H-dibaryon state appears with $B_{\Lambda\Lambda} \sim 19$ MeV. However, since the mass differences among $\Lambda\Lambda$, ΞN and $\Sigma\Sigma$ are not reproduced well, the mass-shift matrix ΔM (diagonal

Table 2. Calculated binding energies and channel components for the free H-dibaryon state and ${}^{10}_{YY}$Be g.s. in case of the RGM-F version.

	H-dibaryon state				${}^{10}_{YY}$Be g.s.				
channel	$B_{\Lambda\Lambda}$	$P_{\Lambda\Lambda}$	$P_{\Xi N}$	$P_{\Sigma\Sigma}$	$B_{\Lambda\Lambda}$	$P_{\Lambda\Lambda}$	$P_{\Xi N}$	$P_{\Sigma\Sigma}$	P_H
only $\Lambda\Lambda$	–	100%	0%	0%	16.0	100%	0%	0%	25%
full	12.2	31	44	25	24.0	70	14	16	70

and constant matrix) is added to the original RGM-F Hamiltonian, and then we call it as the modified RGM-F version. The calculated results are shown in Table 2. We found a bound state with $B_{\Lambda\Lambda} \sim 12$ MeV, whose channel components are similar to those of the original RGM-F version, and represent the characteristic of the flavor SU(3) singlet state. Hereafter, the state is called as the H-dibaryon state. The channel probability densities with respect to the $(3q) - (3q)$ relative coordinate are shown in Fig. 1(a), where they are defined as the squared renormalized radial wave function times r^2. From Fig. 1(a), we can see the strong hyperon mixing in the short-range region.

The calculated results for applying the RGM-F version to the ${}^{10}_{YY}$Be g.s. are also shown in Table 2. In case of only the $\Lambda\Lambda$ channel, the hypernuclear state is bound with $B_{\Lambda\Lambda} = 16$ MeV, due to the strong attractive potential between the core and Λ particle. (Note that B_Λ for ${}^{9}_{\Lambda}$Be is 6.7 MeV.) This situation is different from the case of the $(3q) + (3q)$ system, in which there is no bound state for only the $\Lambda\Lambda$ channel. When we switch on the full channel coupling, the hypernuclear state is overbound by about 6 MeV in comparison with the experimental data ($B^{exp}_{\Lambda\Lambda} = 17.7 \pm 0.4$ MeV), reflecting the strong coupled-channel effect. Since the ΞN and $\Sigma\Sigma$ components are as large as about 15 %, the hypernuclear state has a strongly hyperon-mixed configuration. This result is in contrast to those in case of the OBE-model potentials; the Ξ component

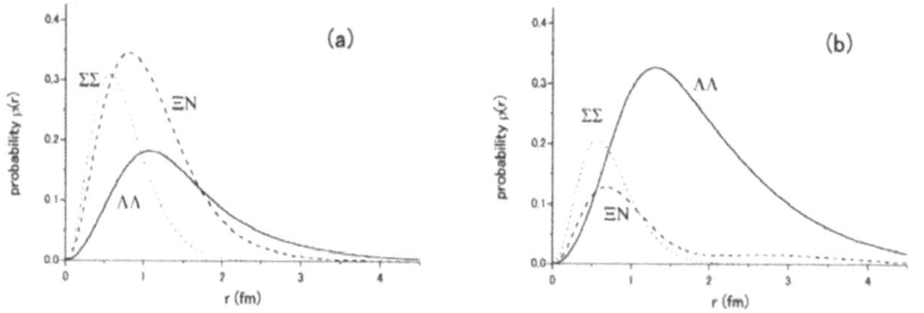

Figure 1. Channel probability densities referring to the $(3q)-(3q)$ relative coordinate in case of the (a) free H-dibaryon state and (b) ${}^{10}_{YY}$Be ground state.

is as small as 0.1 $\sim$ 0.3 % in the light $S = -2$ nuclei (see Table 1).

The calculated component of the free H-dibaryon state in the $^{10}_{\rm YY}$Be ground state is about 70 % (see Table 2). This indicates that the ground state is a free H-dibaryon state bound with the core nucleus, namely, H-nuclear state. However, a remarkable difference between them can be seen, when we compare the channel components of the free H state and hypernuclear state in Table 2; the ΞN channel is main in the free H state, while the $\Lambda\Lambda$ channel is main in the hypernuclear state. The difference can also be seen in the channel probability densities shown in Figs. 1(a) and 1(b), where we find out that the structure of the H-dibaryon state changes significantly in a nucleus in comparison with that of the free H one. The origins of the difference are given as follows: 1) When only the $\Lambda\Lambda$ channel is switched on, the $^{10}_{\rm YY}$Be ground state is bound by about 16 MeV, while the free H state is not bound (see Table 2), and therefore, the energy difference between the core+Λ+Λ and core+Ξ+N (core+Σ+Σ) channels becomes large. 2) In the core+Ξ+N channel, the valence nucleon must be occupied in p orbit as well as the Ξ particle because of the Pauli principle between the core and valence nucleon, and then, the coupling matrix element between the $\Lambda\Lambda$ and ΞN channels (as well as the ΞN and $\Sigma\Sigma$ channels) becomes weak in comparison with the case of the free H state. The above two facts lead to the significantly different structures of the $(3q)+(3q)$ part between the $S = -2$ nuclear ground state and the free H state.

Finally, we apply the FSS version [3] to the $^{10}_{\rm YY}$Be ground state. The FSS is known to have no H-dibaryon state and weak $\Lambda\Lambda$-ΞN potential. We found that the calculated binding energy of $^{10}_{\rm YY}$Be is a little underbound by about 2 MeV in comparison with the experimental data. The channel components are $P_{\Lambda\Lambda} = 98\%$, $P_{\Xi N} = P_{\Sigma\Sigma} = 1\%$, which results indicate that the $S = -2$ nucleus is a double-Λ hypernuclear state with the similar structure to the OBEP case.

4 Summary

The OBE-model potentials and quark-cluster-model (QCM) interactions are applied to study the structure of light $S = -2$ nuclei. We found that the OBE-model potentials predict a double-Λ hypernuclear state for the ground states of light $S = -2$ nuclei, while the QCM interaction with a bound H-dibaryon predicts a bound H-nuclear state, although the structure of the H-dibaryon state in a nucleus is significantly different from the free one.

References

1. M.M. Nagel, T.A. Rijken, J.J. de Swart: Phys. Rev. **D15**, 2547 (1977)
2. T. Tominaga, T. Ueda, M. Yamaguchi, N. Kijima, D. Okamoto, K. Miyagawa, T. Yamada: Nucl. Phys. **A642**, 483 (1998)
3. C. Nakamoto,Y. Suzuki and Y. Fujiwara: Prog. Theor. Phys. **97**, 761 (1997); Y. Fujiwara, C. Nakamoto, Y. Suzuki: Phys. Rev. **C54**, 2180 (1996); C. Nakamoto: private communication, 1999

Few-Body Systems Suppl. 12, 367–371 (2000)

Few-Body Systems

Λ-Λ Interferometry in (K^-, K^+) Reactions at $P(K^-) = 1.8$ GeV/c

A. Ohnishi[1], Y. Hirata[1], Y. Nara[2], S. Shinmura[3], Y. Akaishi[4]

[1] Division of Physics, Graduate School of Science, Hokkaido University, Sapporo 060-0810, Japan

[2] Advanced Science Research Center, Japan Atomic Energy Research Institute, Tokai, Ibaraki 319-11, Japan

[3] Faculty of Engineering, Gifu University, Yanagido 1-1, Gifu 501-11, Japan

[4] Institute of Particle and Nuclear Study, KEK (Tanashi Branch), Midori-machi 3-2-1, Tanashi 188-8501, Japan

Abstract. We study (K^-, K^+) reactions on nuclear targets (^{12}C and ^{9}Be) at incident K^- momentum of $P(K^-)$ =1.65 and 1.8 GeV/c by using the intranuclear cascade model combined with the correlation function technique. We compare the calculated results with two different sets of elementary meson-baryon cross sections, but we cannot find significant differences in K^+ spectra and $\Lambda\Lambda$ invariant mass spectra. The calculated $\Lambda\Lambda$ invariant mass spectrum in ^{9}Be$(K^-, K^+\Lambda\Lambda)$ reaction at $P(K^-) = 1.8$ GeV/c shows a comparable yield to that in ^{12}C target reaction.

1 Introduction

The recent measurement of $\Lambda\Lambda$ invariant mass spectrum in KEK-E224 experiment [1] has made a great impact on strangeness nuclear physics. It not only provides us of the oppotunity to search for the H dibaryon state above the $\Lambda\Lambda$ threshold, but also can be regarded as the first "$\Lambda\Lambda$ scattering" data. The measured data show two striking features concering the above two points. The first one is the enhancement just below the ΞN threshold, which might come from the ΞN bound state partially having features of the flavor singlet H dibaryon resonance. The second one is the strong enhancement at low energies, $E_{\Lambda\Lambda} = M_{\Lambda\Lambda} - 2M_\Lambda \leq 10$ MeV. At these energies, two emitted Λ particles are considered to scatter through the pairwise $\Lambda\Lambda$ final state interaction (FSI).

Recently, we have analyzed the $\Lambda\Lambda$ invariant mass spectrum data of KEK-E224 by using a combined framework of IntraNuclear Cascade (INC) model and the correlation function technique, and we have shown that the observed enhancement at low-invariant masses can be well reproduced with appropriate attractive $\Lambda\Lambda$ interactions, as shown in Fig. 1 [2]. Therefore, it would be possible to extract precise information of the $\Lambda\Lambda$ interaction, *if* precise $\Lambda\Lambda$ correlation data are available, *and* theoretical ambiguities other than the $\Lambda\Lambda$ interaction is small. In the BNL-E906 experiment, the estimated number of detected $\Lambda\Lambda$ pairs exceeds that in KEK-E224 by one to two order of magnitude [4, 5], then high quality correlation data will be available soon. Theoretically, it is required to explain various data other than $\Lambda\Lambda$ correlation based on the well-established knowledges of elementary hadron-hadron cross sections and the mean field potentials. Especially, since various two-step processes ($K^-N \to MY, MN \to K^+Y$) significantly contribute to the concidence yield of ($K^-, K^+\Lambda\Lambda$) [6], theoretical estimates of experimentally unknown cross sections such as $\eta N \to KY$ may cause a large ambiguity. Therefore we focus our attention to the elementary meson-baryon cross sections in this work.

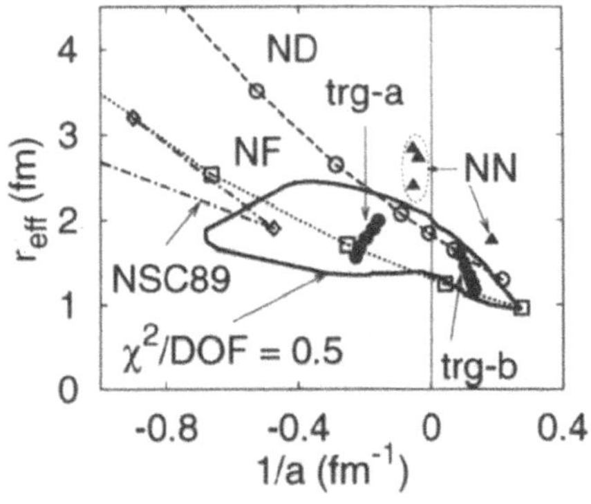

Figure 1. Extracted Λ-Λ scattering parameters [2]. Fitted two-range gaussian potentials (solid circles) are compared with Nijmegen models (open marks) [3]. Inside the thick solid line, potential with χ^2/DOF ≤ 0.5 exists with longer gaussian range parameters in $0.6 \leq \mu_i \leq 1.0$ fm.

2 Meson-Baryon Cross Sections

Meson-baryon reactions have been extensively studied mainly in 60's and 70's, and various baryon resonances have been found. In ref. [6], experimentally unknown cross sections were estimated by using the so-called Generalized Breit-Wigner (GBW) formula [7], which assumes that the s-channel formation of these baryon resonances is dominant and that the interferences between resonances can be ingored. Phenomenologically, this prescription works reasonably well, but if we compare the cross sections in more detail, it has some defects. For example, the GBW formula gives the same cross section in $K^-p \to \pi^-\Sigma^+$ and $K^-p \to \pi^+\Sigma^-$ reactions, while the latter cross section falls much faster than the former experimentally, as shown in Fig. 2.

The reason of this difference can be easily understood by considering t- and u-channel diagram. In the case of $K^-p \to \pi^-\Sigma^+$, K^* exchange is possible, while the baryon exchange is necessary for $K^-p \to \pi^+\Sigma^-$. The baryon exchnage amplitude decrease much faster than that of meson exchange, as explained in the Regge theory [8]. The Regge theory tells us that at high energy limit,

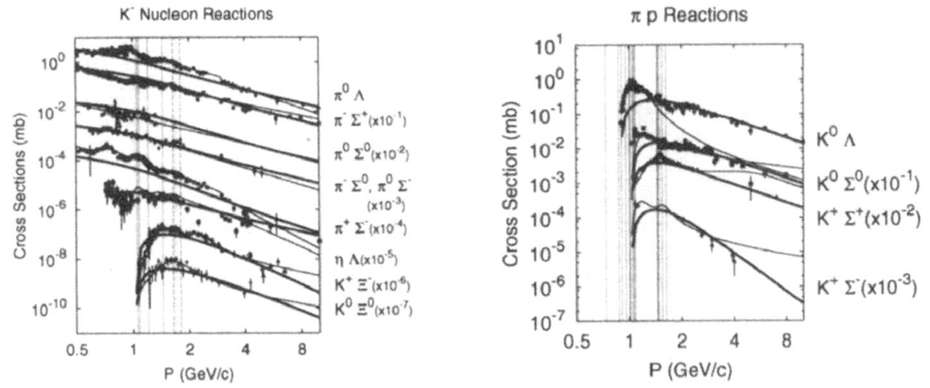

Figure 2. K^-N (left panel) and $\pi^\pm p$ cross sections. Thin lines show the cross sections which we used in ref. [6], while thick lines show the Reggeon exchange cross sections.

the t-channel meson exchange amplitude is approximately given by $\mathcal{M}(s,t) \simeq F(t)\,(s/s_0)^{\alpha_R(t)}$ (thick lines in Fig. 2), where $J = \alpha_R(M^2) = \alpha_R(0) + \alpha'_R(0)\,M^2$ represents the angular momentum of hadron resonances in the range $M^2 > 0$.

In this work, we have adopted the meson-baryon cross section, which is dominated by the s-channel reaction at resonance region, and gragually converges to the Reggeon exchange cross section at higher energies. This treatment reproduces the data very well (energy-dependent weighted average of thin and thick lines in Fig. 2), and guarantees the high energy behavior. It should be noted that we here ignore coherence and duality [9]. However, since we are more interested in the many-body or few-body problems rather than in the hadron reaction itself, we do not go into the detailed underlying problems of hadron-hadron reacion, and proceed to the (K^-, K^+) reactions on nuclei in the next section.

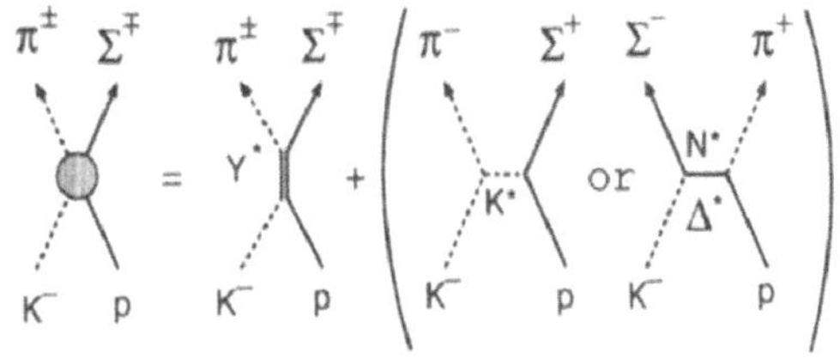

Figure 3. Diagrams which contribute to $K^-N \to \pi^\pm\Sigma^\mp$ reactions.

3 Results

We have made the INC simulation calculation for ^{12}C(K^-, K^+) ($P(K^-) = 1.65$ GeV/c, corresponding to KEK-E176 and E224 experiments) and ^{9}Be(K^-, K^+) ($P(K^-) = 1.8$ GeV/c, corresponding to BNL-E906 experiment). The number of generated events is around 10^8 for each reaction, and the mean field effects for baryons are included as the external field.

The left panel of Fig. 4 shows the calculated momentum distribution of K^+ particle, in comparison with the KEK-E176 data (^{12}C(K^-, K^+)) [10] and the BNL-E906 preliminary data (^{9}Be(K^-, K^+)) [5]. We also compare the calcu-

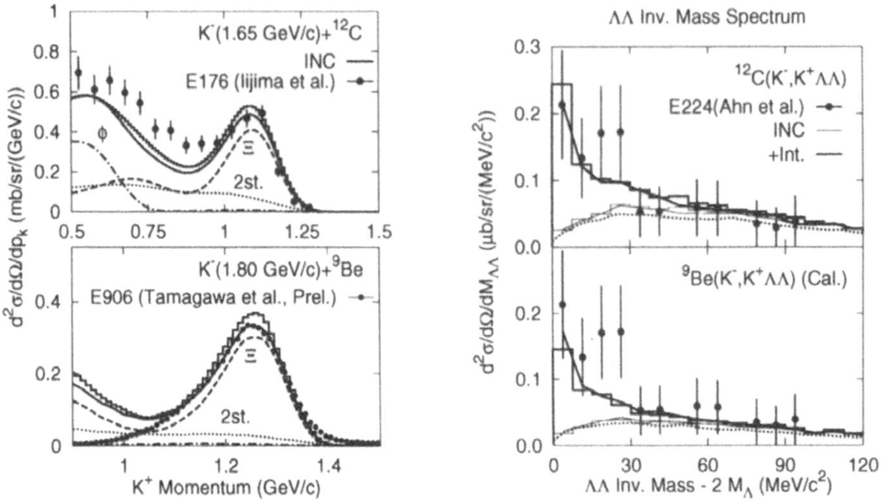

Figure 4. K^+ spectra (left panel) and $\Lambda\Lambda$ invariant mass spectra (right panel) in (K^-, K^+) reactions. Solid histograms and curves show the calculated results with elementary meson-baryon cross sections which are used in ref. [6] and those with Regge behavior (see text). Experimental data are taken from refs. [10, 5, 1]. The detector efficiency correction has not been done for E906 data yet. Thin and thick lines in the right panel show the calculated results without and with correlation effects generated by FSI between $\Lambda\Lambda$. Dotted curves show two-step contributions without FSI effects.

lated results with the elementary cross sections used in ref. [6] (solid histograms) and those explained in the previous section including the Reggeon exchange behavior (solid curves). In both of the reactions, the difference between the two calculated results are small.

In the right panel of Fig. 4, we show the calculated $\Lambda\Lambda$ invariant mass spectrum without (thin) and with (thick) the FSI correlation effects [11], in comparison with the KEK-E224 data (^{12}C(K^-, K^+) reaction). The difference caused by the elementary cross section choice (histograms and curves as in the left panel) is not large.

4 Summary

In this work, we have studied (K^-, K^+) reactions at $P(K^-) = 1.65$ and 1.8 GeV/c, and $\Lambda\Lambda$ correlation in these reactions. One of the problems to extract $\Lambda\Lambda$ interaction from the correlation is the elementary cross section ambiguity, but we have shown that, there will be no serious differences in K^+ spectra and $\Lambda\Lambda$ correlation caused by this ambiguity, if we are interested in the high momentum range of K^+ particle. This is mainly because the dominant processes to produce K^+ with high momentum are quasi-free Ξ production, $K^- N \to K\Xi$, and two-step processes with pions in the intermediate stage,

$K^-N \to \pi Y$ followed by $\pi N \to KY$, whose elementary cross sections are experimentally well known.

In ^{9}Be(K^-,K^+) reaction at $P(K^-) = 1.8$ GeV/c, the quasi-free Ξ production is dominant in high momentum region of K^+, but the two-step contributions are much more important in the coincidence yield of $(K^-, K^+\Lambda\Lambda)$ (dotted curves in Fig. 4). By using the $\Lambda\Lambda$ interaction which well reproduces the $\Lambda\Lambda$ correlation data of KEK-E224 experiment, the present model predicts the invariant mass spectrum of $\Lambda\Lambda$ for the BNL-E906 experiment as shown in the right panel of Fig. 4, which is comparable in magnitude to that in KEK-E224 data.

Acknowledgement. The authors would like to thank Mr. T. Tamagawa, J. K. Ahn and Prof. T. Fukuda for supplying us of the experimental data and also for fruitful discussions. This work was supported in part by the Grant-in-Aid for Scientific Research (Nos. 07640365, 08239104 and 09640329) from the Ministry of Education, Science and Culture, Japan.

References

1. J.K. Ahn et al. (KEK-E224 collaboration): Phys. Lett. **B444**, 267 (1998)
2. A. Ohnishi, Y. Hirata, Y. Nara, S. Shinmura, Y. Akaishi: *Proc. of KEK-Tanashi Int. Symp. on Physics of Hadrons and Nuclei*, Tokyo, Japan, December 14-17, 1998, Nucl. Phys. **A**, in press, nucl-th/9903021
3. M.M. Nagels et al.: Phys. Rev. **D15**, 2547 (1977); Phys. Rev. **D20**, 1633 (1979); P.M. Maessen et al.: Phys. Rev. **C40**, 2226 (1989)
4. J. Nakano et al. (BNL-E906 collaboration): in this proceedings
5. T. Tamagawa et al. (BNL-E906 collaboration): Private Communication
6. Y. Nara et al.: Nucl. Phys. **A614**, 433 (1997)
7. G.E. Brown et al.: Phys. Rev. **C43**, 1881 (1991); H. Sorge et al.: Phys. Lett. **B271**, 37 (1991); Z. Phys. **C59**, 85 (1993)
8. V.D. Barger and D.B. Cline: *Phenomenological Theories of High Energy Scattering*, New York: Benjamin 1969; H. Sorge: Phys. Rev. **C52**, 3291 (1995)
9. G. Veneziano: Nuov. Cim. **57A**, 190 (1968)
10. T. Iijima et al. (KEK-E176 collaboration): Nucl. Phys. **A546**, 588 (1992)
11. For example, W. Bauer et al.: Annu. Rev. Nucl. Part. Sci. **42**, 77 (1992), and references therein; I. Slaus, Y. Akaishi, H. Tanaka: Phys. Rep. **173**, 257 (1989)

Few-Body Systems Suppl. 12, 372–377 (2000)

Few-Body Systems

The $\Lambda N, \Lambda\Lambda$ and ΞN Attractions in the Intermediate Force Range —Ehime Approach to YN and YY Forces—

T. Ueda, K. Tominaga*, M. Yamaguchi

Faculty of Science, Ehime University, Matsuyama, Ehime 790-8577, Japan

Abstract. Ehime $\Lambda N, \Lambda\Lambda, \Sigma N$ and ΞN OBEP's are reviewd.

1 Ehime Approach

Recently theoretical YN and YY interactions have been provided variously[1]: Among all are noted the Nijmegen models with soft or hard cores, the Jülich models with the two-meson-exchange contribution due to the isobar degree of freedom and the Niigata-Kyoto models with pseudoscalar and scalar meson exchanges in hybrid with the quark exchange for short distance . These aim a microscopic description of the YN and YY interactions. However difficulties are in uncertainties in mesonic form factors and meson-meson correlations. As a result, none of these attempts is fully consistent with the experimental data for the ΛN scattering and the single and double Λ hypernuclei.

Very recently the NN OBEP due to Ueda, Riewe and Green[2] has been extended to the OBEP for ΛN, $\Lambda\Lambda$ and ΞN systems, where an SU(3) nonet meson scheme is considered for pseudoscalar, vector and scalar mesons: (π, η, K, η'), $(\rho, \omega, K^*, \phi)$ and $(\delta, S^*, K_0^*(1429)\ f_0(1581))$, respectively. Additionally, the isosinglet scalar meson σ is taken into account which represents the $I = 0, J^P = 0^+$ part of the correlated and uncorrelated 2π exchanges[3].

The aim of this approach is to build an effective model for the YN and YY interactions so as to be consistent with related hypernuclei. We allow that the scalar and vector meson exchanges involve effectively correlated and uncorrelated two octet-meson exchanges. The uncertainties arising from this treatment are removed in combined analyses with hypernuclei[3].

In Section 2 single particle channel approach is reviewed , where the $\Lambda N - \Sigma N$ channel coupling is effectively involved in the σ exchange. In Section 3 the

*Present address: RCNP, Osaka University, Osaka 567-0047, Japan

coupling is explicitly introduced. In Section 4 the vertex form factors of $NN\pi$, $\Sigma\Sigma\pi$ and $\Xi\Xi\pi$ are discussed.

2 Single Particle Channel Approach

The SU(3) parameters are the following four for each set of the nonet mesons: The coupling constants of the octet and singlet mesons, g_8 and g_1, respectively, the parameter α for the F and D couplings, $\alpha = F/(F+D)$, and the mixing parameter θ between the neutral octet and singlet mesons.

To determine the coupling constants for the nonet mesons, the model II by Ueda, Riewe and Green is employed[2]. For pseudoscalar mesons we use the coupling constants determined in the NN analysis: $g_{NN\pi}, g_{NN\eta}$ and $g_{NN\eta'}$. The mixing parameter θ_P for the pseudoscalar mesons is determined by the mass formula. Then the SU(3) relation determines the g_8, g_1 and α.

For the vector mesons the coupling constants are similarly treated as the pseudoscalar mesons except the mixing parameter. This is made as adjustable, since the effect of the correlated and uncorrelated exchanges of the two octet mesons which have totally the quantum number of the vector meson, $J^P = 1^-$, produces the additional SU(3) components. The effect makes a shift of the mixing parameter θ_V from the value determined by the mass formula. For the scalar mesons too, we make the mixing parameter θ_S as adjustable from the same reason with the vector meson case. Behind this treatment we interpret the mesons as involving the ingredients of not only $(q\bar{q})$, but also $(q\bar{q}q\bar{q})$. The latter ingredient makes obviously the shift of the mixing parameter.

The $g_{\Lambda\Lambda\sigma}$ is assumed as being independent of the $g_{NN\sigma}$. This is determined by the NN analysis. Then we have three adjustable parameters left: θ_V, θ_S and $g_{\Lambda\Lambda\sigma}$. These parameters are determined in fitting the Λp total cross sections at $P_L = 150 - 400$ MeV/c and the binding energy of the hypertriton. In this case the separation energies of the double hypernuclei, ${}^{6}_{\Lambda\Lambda}He$, ${}^{10}_{\Lambda\Lambda}Be$ and ${}^{13}_{\Lambda\Lambda}B$, are also respected. Thus we fix all the parameters for ΛN and $\Lambda\Lambda$ OBEP's.

As a result, our ΛN and $\Lambda\Lambda$ OBEP are consistent with the data on ΛN scattering, the hypertriton and the double Λ hypernuclei. No $\Lambda\Lambda$ bound state is found. (See Figure 1, Tables 1 and 2.)

Furthermore, we have a ΞN OBEP where only $g_{\Xi\Xi\sigma}$ is left as unknown. We make a potential for $\Xi^- - {}^{12}C$ atomic states by folding the G matrix due to the ΞN OBEP over ${}^{12}C$ nuclei. Adjusting the $g_{\Xi\Xi\sigma}$ so as to fit the Yokohama event for $\Xi^- - {}^{12}C$ atomic bound state, we find that it reproduces reasonablly the data[4].

(1) It is found that the attraction of $\Lambda\Lambda$ is approximately equal with or a little stronger than that of ΛN in the intermediate force range in the ${}^1S_0$ state.

(2) The coupling constants of σ which are most responsible to the $\Lambda N, \Lambda\Lambda$ and ΞN attractions in the intermediate force range take similar values one another as: $g_{NN\sigma} = 1.78$, $g_{\Lambda\Lambda\sigma} = 1.88$ and $g_{\Xi\Xi\sigma} = 1.82$.

The structures of the ΛN, $\Lambda\Lambda$ and NN ${}^1S_0$ potentials at rest and their meson components are shown for mutual comparisons in Figures 2, 3 and 4.

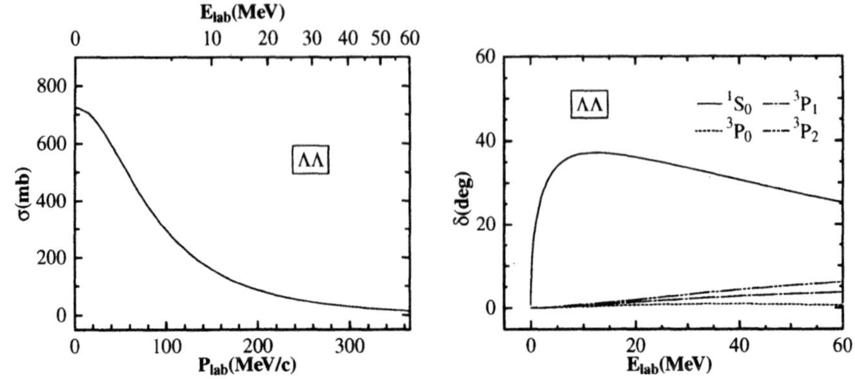

Figure 1. The cross sections (left) and the $^1S_0-$ and $^3P-$wave phase shifts (right) for the $\Lambda\Lambda$ scattering

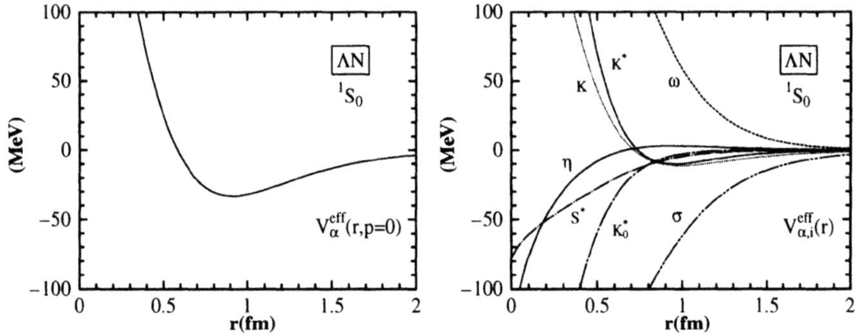

Figure 2. The effective potential $V^{eff}(r)$ at rest and its meson components for the ΛN $^1S_0$ state

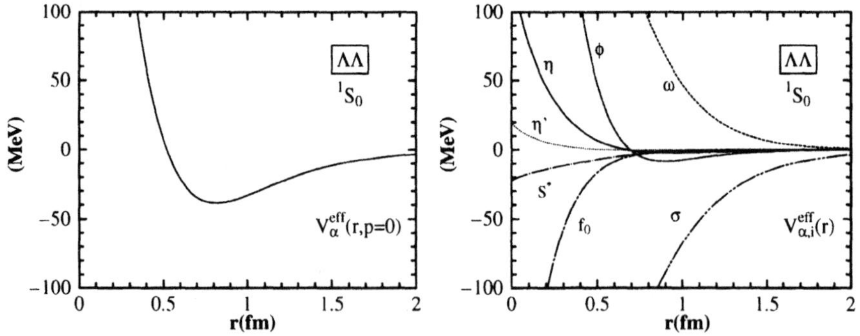

Figure 3. The effective potential $V^{eff}(r)$ at rest and its meson components for the $\Lambda\Lambda$ $^1S_0$ state

Table 1. The scattering lengths a and the effective ranges r in the ${}^1S_0$ and ${}^3S_1$ states for ΛN, $\Lambda\Lambda$ and ΞN scatterings

	ΛN	$\Lambda\Lambda$	$\Xi N(I=0)$	$\Xi N(I=1)$	NN(experiment)
$a({}^1S_0)$ (fm)	-2.71	-3.40	-0.66	-0.27	-23.675 ± 0.095
$r({}^1S_0)$ (fm)	3.21	2.79	7.3	2.0×10	2.69 ± 0.18
$a({}^3S_1)$ (fm)	-1.95		-0.43	-0.56	5.399 ± 0.011
$r({}^3S_1)$ (fm)	3.56		13.5	9.0	1.82 ± 0.05

Table 2. The hypertriton binding energy, the binding energy difference $\Delta B_{\Lambda\Lambda}$ of the double Λ hypernuclei ${}^{6}_{\Lambda\Lambda}He$, ${}^{10}_{\Lambda\Lambda}Be$ and ${}^{13}_{\Lambda\Lambda}B$, and the binding energy of the atomic $\Xi^- - {}^{12}C\ (1P)$ state

	${}^3_{\Lambda}H$	${}^{6}_{\Lambda\Lambda}He$	${}^{10}_{\Lambda\Lambda}Be$	${}^{13}_{\Lambda\Lambda}B$	$\Xi^- - {}^{12}C\ (1P)$
Theory (MeV)	2.35	3.60	4.14	4.94	0.57
Experiment (MeV)	2.355 ± 0.05	4.7	4.3	4.9	0.57 ± 0.19

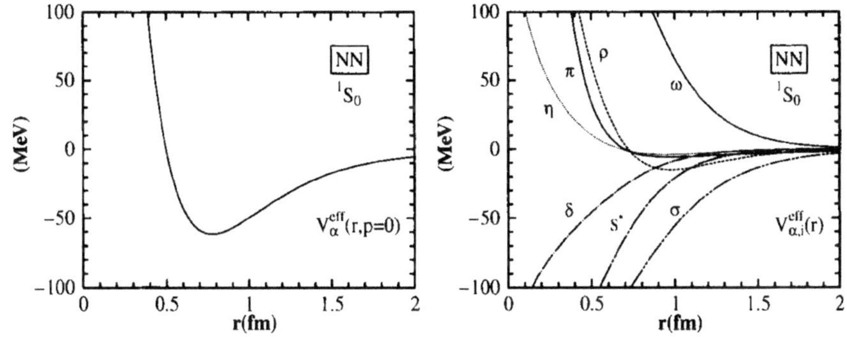

Figure 4. The effective potential $V^{eff}(r)$ at rest and its meson components for the $NN\ {}^1S_0$ state

The potentials have a fundamental structure in common: The large cancellation between the σ attraction and the ω repulsion and the balance results in attraction at medium range ($r = 0.7 - 1.5$ fm) and repulsion at short distance ($r \lesssim 0.6$ fm). The strengths of the attraction are in the relations: $|V(NN)| >> |V(\Lambda\Lambda)| \gtrsim |V(\Lambda N)|$. The relation, $|V(\Lambda\Lambda)| \gtrsim |V(\Lambda N)|$, comes primarily from the relations:$g_{NN\omega}(= 3.17) > g_{\Lambda\Lambda\omega}(= 2.56)$ and $g_{NN\sigma}(= 1.78) < g_{\Lambda\Lambda\sigma} = (1.88)$. Furthermore the relation, $|V(NN)| >> |V(\Lambda\Lambda)|$, comes from the two sources:(1) The different S^* attractions due to $g_{NNS^*}(= 2.6) > g_{\Lambda\Lambda S^*}(= 0.56)$ and (2) the π and ρ contribute to the attraction at $r = 0.8 - 1.0$ fm in NN, while do not in ΛN or $\Lambda\Lambda$.

3 $\Lambda N - \Sigma N$ Coupled Channel Approach

In this approach[5] the four adjustable parameters are used:$g_{\Lambda\Lambda\sigma}$, $g_{\Sigma\Sigma\sigma}$, θ_S and θ_V for ΛN and ΣN OBEP's. They are determined in fitting the total cross sections for $\Lambda p - \Lambda p$, $\Sigma^+ p - \Sigma^+ p$, $\Sigma^- p - \Sigma^- p$,$\Sigma^- p - \Sigma^o n$, $\Sigma^- p - \Lambda n$ and the two ΛN S wave phase shifts due to the single particle channel approach[3]. These observables and each differential cross sections are reasonably reproduced by the coupling constants listed in Table 3 except the total cross section of $\Lambda p - \Lambda p$. For this observable the model gives a Λp resonance at $p_{lab} \sim 450$ MeV/c against the experimental data. This is due to the strong attraction which is generated by the iterated one-pion exchange in the $I = \frac{1}{2}$ $\Sigma N - \Sigma N$ intermediate process. How to remove the unwanted resonance is discussed in the next section.

In this coupled channel treatment the $\Lambda\Lambda\sigma$ coupling constant is found as $g^c_{\Lambda\Lambda\sigma} = 1.25$. This is compared with $g_{\Lambda\Lambda\sigma} = 1.88$ in the single particle channel approach. The difference between the $g_{\Lambda\Lambda\sigma}$ and the $g^c_{\Lambda\Lambda\sigma}$ is interpreted in terms of the contribution from the $\Lambda N - \Sigma N$ coupling.

Table 3. Coupling constants

M		NNM	ΛNM	$\Lambda\Lambda M$	ΣNM	$\Sigma\Sigma M$	$\Lambda\Sigma M$
π	g	3.7736				3.9328	2.0867
η	g	2.4021		-1.9741		2.1308	
η'	g	0		0.8036		0.0498	
K	g		-4.4493		-0.1593		
ρ	g	0.8426				1.8806	-0.1128
	f	4.1625				2.0813	3.6048
ω	g	3.1718		2.3365		2.2167	
	f	0		-1.9151		1.9151	
ϕ	g	0		-1.3320		-1.5231	
	f	0		-3.0541		3.0541	
K^*	g		-1.5722		-1.0379		
	f		-3.6048		2.0813		
δ	g	1.6248				2.3221	0.5355
S^*	g	2.6038		1.0782		1.7953	
f_0	g	0		-1.6927		-0.8971	
K_0^*	g		-2.2788		-0.6973		
σ	g	1.7776		1.2496		0.8788	

4 Pionic Form Factors of $NN\pi$, $\Sigma\Sigma\pi$ and $\Xi\Xi\pi$ vertexes

In early NN OBEP the pionic form factor in πNN vertex was a hard type: The parameter Λ in the monopole parametrization takes about 2-3 GeV much larger than that of the electro-magnetic form factor about 0.7 GeV[6]. This puzzle was solved by introducing the Π which represents effectively the $I = 1$ and $J^P = 0^-$ part of the correlated three-pion state with mass 730 MeV[7]. It was proved that the one-pion excange potential with the hard pionic form factor is approximately equivalent with the sum of the one-pion exchange potential with a soft pionic form factor such as the electro-magnetic one and the one-Π exchange potential. The soft pionic form factor is interpreted in terms of the contribution of the Π which intermediates between the source nucleon and the emitted (or absorbed) pion.

The $\Lambda N - \Sigma N$ coupled channel calculation shows that an unwanted ΛN resonance appears. This calculation employs a hard pionic form factor of about 2 GeV for the $\Sigma\Sigma\pi$ vertex. However the resonance is removed with a smaller value for the form factor parameter such as about 1.5 GeV. Thus the strength of $\Sigma\Sigma\pi$ vertex must be reduced, compared with that of the hard form factor.

On the other hand in the application of the ΞN OBEP to $\Xi^- - {}^{12}C$ atomic states it is found that, to reproduce the experimental data for the spectra of $\Xi^- - {}^{12}C$ atomic bound states, $g_{\Xi\Xi\sigma}$ is approximetely equal to $g_{NN\sigma}$. Since the σ represents effectively the $I = 0$ and $J^P = 0^+$ part of the two-π exchange, the $\Xi\Xi\pi$ coupling is required to be similarly strong as the $NN\pi$ coupling. However the SU(3) invariance requires that $g_{\Xi\Xi\pi} << g_{NN\pi}$[3]. Therefore the strength of the $\Xi\Xi\pi$ vertex form factor must be enhanced.

These two problems encountered in the $\Sigma\Sigma\pi$ and $\Xi\Xi\pi$ vertexes are simultaneously solved by introducing the contribution from the Π which is able to make approximately vanishing contribution to the $\Sigma\Sigma\pi$ vertex, while a sizable contribution to the $\Xi\Xi\pi$ vertex. For this purpose plausible parameter values for Π are found as:$g_\Pi = 4.82, \alpha_\Pi = 0.0$.

References

1. T.A. Rijken: this conference; Y. Fujiwara: this conference; S. Shinmura: this conference; A. Reuber, et. al.: Czech. J. Phys. **42**, 1115 (1992)
2. T. Ueda, F.E. Riewe, A.E.S. Green: Phys. Rev. **C17**, 1763 (1978)
3. K. Tominaga, T. Ueda, M. Yamaguchi, N. Kijima, D. Okamoto, K. Miyagawa, T. Yamada: Nucl. Phys. **A642**, 483 (1998)
4. M. Yamaguchi, T. Ueda, Y. Yamamoto: this conference
5. K. Tominaga: PhD thesis of Ehime University, 1999, unpublished
6. T. Ueda, A.E.S. Green: Phys. Rev. **174**, 1304 (1968)
7. T. Ueda: Phys. Rev. Lett. **68**, 142 (1992)

Few-Body Systems Suppl. 12, 378–382 (2000)

Few-Body Systems

^{4}He(K^-,π^-) Spectra and the $^4_\Sigma$He Bound State

T. Harada*

Department of Social Information, Sapporo Gakuin University
11-Bunkyodai, Ebetsu, Hokkaido 069-8555, Japan

Abstract. We examine phenomenologically the ^{4}He(K^-,π^-) spectra in flight and at rest within the distorted-wave impulse approximation by using a coupled ($3N$-Λ)+($3N$-Σ) model with a spreading potential. The resultant spectra can explain consistently all the available data. We show that there exists a pole of the $^4_\Sigma$He unstable bound state on the second Riemann sheet of the complex energy plane.

1 Introduction

The strangeness-exchange reaction by the K^- capture in a helium target nucleus is important to study the structure of the $^4_\Sigma$He *unstable* bound state [1]. The final states with π^- are classified as

$$K^- + {}^4\text{He} \rightarrow \begin{array}{l} (a)\, \pi^- + {}^3\text{He} + \Lambda, \\ (b)\, \pi^- + {}^3\text{H} + \Sigma^+, \\ (c)\, \pi^- + {}^3\text{He} + \Sigma^0, \\ (d)\, \pi^- + X^{++} + \Lambda, \end{array} \tag{1}$$

where X^{++}=d+p or p+p+n. The reaction cross section (d) is connected with the secondary $3N$ breakup process, $^4_\Sigma\text{He} \rightarrow X^{++} + \Lambda$, because of the $\Sigma N \rightarrow \Lambda N$ conversion interaction.

It has been debated for about 10 years whether there exists the $^4_\Sigma$He unstable bound state or not [2]. Recently, Nagae et al.[3] confirmed the existence of the $^4_\Sigma$He bound state in the in-flight ^{4}He(K^-,π^-) reaction at BNL. In this note, we calculate hypernuclear production distributions from the K^- capture in flight and at rest on the helium target, with the distorted wave impulse approximation (DWIA) by using a coupled ($3N$-Λ)+($3N$-Σ) model with a spreading potential. Our purpose here is to give a unified understanding for $^4_\Sigma$He, in comparison with the available experimental data at BNL [3], ANL [2] and KEK [4, 5].

* *E-mail address:* harada@earth.sgu.ac.jp

2 Calculations

The inclusive spectra for the ^{4}He(K^-,π^-) reaction in DWIA are estimated by using the Green's function method [6]. We employ the Fermi-averaged amplitudes $\overline{f}_{\pi Y}$ for $K^- N \to \pi Y$ in nuclear medium, and meson distorted-waves for the outgoing π^- and the initial K^-. Our calculation has assumed that the $3N$ core-nucleus is rigid, where $3N$ denotes the (^{3}He, ^{3}H)-doublet. Thus Green's function for the $3N$-Y systems in Eq.(1) is obtained by solving a coupled equation with ^{3}He+Λ, ^{3}H+Σ^+ and ^{3}He+Σ^0 channels, and has natural-parities, 0^+, 1^-, 2^+,$\cdots$, because the spin-flip processes are known to be negligible in (K^-,$\pi^\mp$) reactions. The $3N$-Y potential matrix with J^π=0^+ on the isospin bases is given by

$$\hat{U} = \begin{pmatrix} U_{\Lambda,1/2} & U_{X,1/2} & 0 \\ U_{X,1/2} & U_{\Sigma,1/2} & 0 \\ 0 & 0 & U_{\Sigma,3/2} \end{pmatrix}, \tag{2}$$

where the Λ-Σ coupling potential $U_{X,1/2}$ is assumed to be real. The diagonal potentials for Y=Σ, Λ and T=1/2, 3/2 are given as $U_{Y,T} = V_{Y,T} + iW_{Y,T}$, where the imaginary part $W_{Y,T}$ is a *spreading* potential, which describes effectively all the $3N$ breakup processes.

We have determined parameters of the potential strength $U_{\Lambda,1/2}$, $U_{X,1/2}$, $U_{\Sigma,1/2}$ and $U_{\Sigma,3/2}$ by making a fit to the in-flight BNL data [7]. Here we made $U_{X,1/2}$ satisfy a plausible Σ admixture of about 2% in the $^4_\Lambda$He ground state [8], and its Λ-binding energy of B_Λ= 2.39 MeV. In addition, since the Fermi-averaged amplitudes $\overline{f}_{\pi Y}$ remain in ambiguities, the relative phase of φ_Λ (φ_Σ) for $\pi\Lambda$ ($\pi\Sigma$ I=0) to $\pi\Sigma$ I=1 channels should be determined from the present fit. As a result, we have obtained the $3N$-Y potential for the best fit, and the suitable relative phases for the in-flight data [7].

Moreover, we also calculate the at-rest spectra with the $3N$-Y potential we determined, in order to explain the data of KEK [5] and ANL [2]. Since atomic orbits where the K^- is absorbed on the K^--^{4}He atom are still unknown experimentally, we assume the K^- capture rate of the 30% s- and 70% p-orbits, as suggested by Koike and Akaishi [9].

3 Results and Discussion

Figure 1 shows the spectra from the K^- capture in ^{4}He, together with the data of KEK, ANL and BNL. The resultant spectra are consistent with all the data from Λ to Σ regions; the peak for $^4_\Lambda$He is in good agreement with the data. We confirm the Λ-binding energy B_Λ=2.4 MeV and the Σ^+ (Σ^0) admixture of 1.30% (0.63%). For the Σ regions, the peak in the BNL (bottom) data can be reproduced, which corresponds to the $^4_\Sigma$He unstable bound state with J^π=0^+, $T \simeq \frac{1}{2}$(99%). We recognize it observed as a bump just below the Σ^+^{3}H threshold in the KEK (top) data. Moreover, we can reproduce the HeBC events of the ANL (middle) data, including the π^-+X^{++}+Λ spectrum. It implies that the

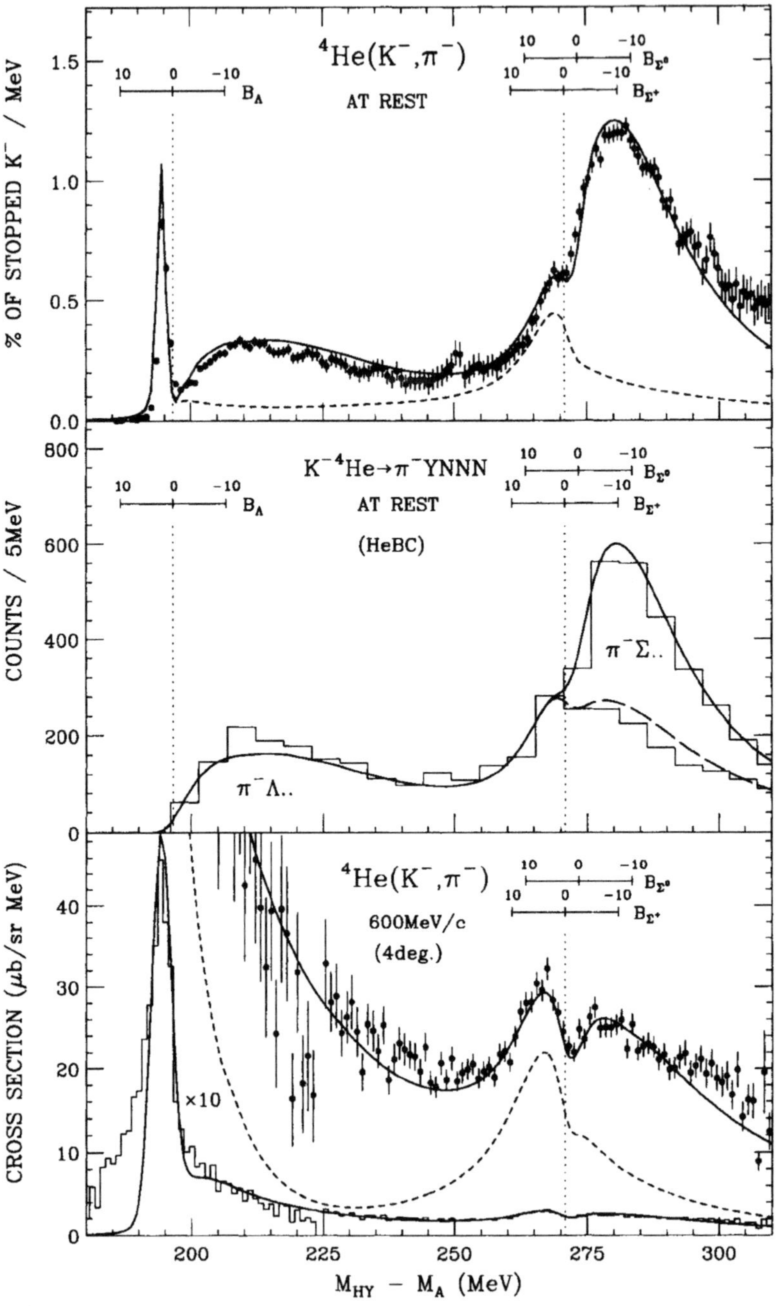

Figure 1. The calculated π^- spectra from the K^- capture in liquid helium, together with the data plots at KEK(top) [5], ANL (middle) [2] and BNL (bottom) [3], as a function of the hypernuclear energy $\Delta M_{\rm HY}= M_{\rm HY}-M_{\rm A}$. A peak at 194 MeV corresponds to the ${}^4_\Lambda$He ground state. The dashed curves denote the $J^\pi=0^+$ component in the π^- spectrum. The long-dashed curve denotes the π^- distribution with the Λ emission.

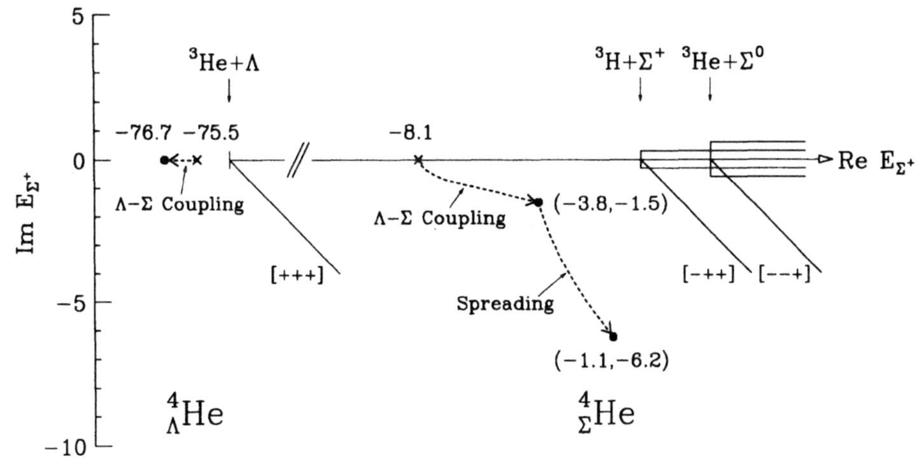

Figure 2. Pole positions for the $3N$-Y system in the complex energy plane of E_{Σ^+} measured from the Σ^{+3}H threshold. The pole for $^4_\Lambda$He ($^4_\Sigma$He) resides on the [+++] ([−++]) sheet. The black circle and cross denote pole positions with and without the Λ-Σ coupling potential of $U_{X,1/2}$, respectively. The spreading effects come from the imaginary part of the $3N$-Y potential.

spreading potential $W_{\Sigma,1/2}$ describing the $^4_\Sigma$He decay processes is reasonable. Then we obtain the complex *eigenvalue* for $^4_\Sigma$He measured from the Σ^{+3}H threshold as

$$E^{(\text{pole})}_{\Sigma^+} = (k^{(\text{pole})}_{\Sigma^+})^2/2\mu_{\Sigma^+} = -1.1 - i6.2 \text{ MeV} \tag{3}$$

on the *second* Riemann sheet [−++][1] of the complex energy plane.

Figure 2 illustrates the resultant pole positions of the $3N$-Y system near the Λ and Σ thresholds. Quantum mechanically, a peak structure in the energy spectrum is associated with a pole in the scattering amplitude. The pole position corresponds to the eigenvalue for the $3N$-Y system on the complex energy plane. To investigate how the poles move on the complex energy plane depending on the $3N$-Y potential, we vary the strength of the Λ-Σ coupling $U_{X,1/2}$. The pole for $^4_\Lambda$He resides at point (−76.7, 0.0) on the *first* Riemann sheet [+++]. When the coupling is turned off, the pole moves to point (−75.5, 0.0); $U_{X,1/2}$ makes the Λ-binding energy shifted by about 1.2 MeV. On the other hand, the pole for $^4_\Sigma$He is at point (−1.1, −6.2) on the *second* Riemann sheet [−++], as mentioned above. If both $U_{X,1/2}$ and $W_{Y,T}$ are turned off, this pole moves to point (−8.1, 0.0). After they are turned on, the pole can finally reach to point (−1.1, −6.2) via point (−3.8, −1.5). The spreading potential plays significant roles in describing the shape and width of the peak for $^4_\Sigma$He even if the pole resides on the sheet [−++].

[1]The sheet is identified by three signs of [Imk_Λ, Imk_{Σ^+}, Imk_{Σ^0}], where k_Y denotes the c.m. momentum of the $3N$-Y system in the complex momentum plane.

Table 1. The observed peak position and width for the ${}^{4}_{\Sigma}$He bound state

	Experiment		Calculation	
Data	B_{Σ^+} (MeV)	Γ_Σ (MeV)	B_{Σ^+} (MeV)	Γ_Σ (MeV)
BNL [3]	4.4±0.3±1	$7\pm0.7^{+1.2}_{-0.0}$	3.7	~10
KEK [5]	2.8±0.7	12.1±1.2	2.5	~10

We realize that the real part of $E^{(\mathrm{pole})}_{\Sigma^+}$ does not necessarily correspond to the position of the peak in the π^- spectrum, as seen in Table 1, because this pole resides near the Σ^{+3}H threshold and far from the physical axis [7]. Hence a position of the observed peak is shifted within about 1 MeV and its shape is asymmetric, because of the Σ branch cut associated with opening the Σ channel, and of the strong interference between Λ and Σ.

4 Summary

We have examined phenomenologically the ^{4}He(K^-,π^-) spectra in flight and at rest, by using the coupled-channel framework among ^{3}He+Λ, ^{3}H+Σ^+ and ^{3}He+Σ^0 channels with a spreading potential. The analysis including continuum spectra gives us valuable information about the $3N$-Y potential. The calculated spectra can explain consistently all the available data of BNL, ANL and KEK. The pole of the ${}^{4}_{\Sigma}$He unstable bound state with $J^\pi=0^+$, $T\simeq1/2$ (99%) is located on the second Riemann sheet [$-$++].

Acknowledgement. The author thanks Professor Y. Akaishi, Professor T. Nagae and Mr. H. Outa for many valuable discussion and comments. This work was supported by the Grants-in-Aid for Scientific Research of the Japanese Ministry of Education, No. 08239104.

References

1. T. Harada et al.: Nucl. Phys. **A507**, 715 (1990)
2. R. H. Dalitz and A. Deloff: Nucl. Phys. **A585**, 303c (1995)
3. T. Nagae et al.: Phys. Rev. Lett. **80**, 1605 (1998)
4. R. S. Hayano et al.: Phys. Lett. **B231**, 355 (1989)
5. H. Outa et al.: Prog. Theor. Phys. Suppl. **117**, 171 (1994)
6. O. Morimatsu and K. Yazaki: Nucl. Phys. **A483**, 1493 (1988)
7. T. Harada: Phys. Rev. Lett. **81**, 5287 (1998)
8. Y. Akaishi, T. Harada, S. Shinmura: (to be submitted)
9. T. Koike and Y. Akaishi: Nucl. Phys. **A639**, 521c (1998)

Few-Body Systems Suppl. 12, 383–386 (2000)

Few-Body Systems

Neutron Rich Lambda Hypernuclei

Khin Swe Myint[1], T. Harada[2], S. Shinmura[3], Y. Akaishi[4]

[1] Physics Department, Mandalay University, Mandalay, Myanmar
[2] Sapporo Gakuin University, Ebetsu, Hokkaido 069-8555, Japan
[3] Gifu University, Gifu 501-1112, Japan
[4] Institute of Particle and Nuclear Studies, KEK, Tanashi 188-8501, Japan

Abstract. It is found that the suppression due to ΛN-ΣN coupling solves the overbinding problem in ${}^5_\Lambda$He but it, in turn, causes a severe underbinding in the four-body systems. The shortage of this binding is overcome by introducing the Λ-Σ coupling explicitly which is equivalent to the ΛNN three-body force. This force becomes strong in the 0^+ states of ${}^4_\Lambda$H and ${}^4_\Lambda$He according to the coherently added enhancement. No significant effect of the Λ-Σ coupling is found in He isotope Λ hypernuclei while a strong Λ-Σ coupling effect is found in ${}^6_\Lambda$H.

1 Introduction

There is a long standing problem of fitting the experimental Λ-separation energies of ${}^3_\Lambda$H, ${}^4_\Lambda$H, ${}^4_\Lambda$He and ${}^5_\Lambda$He consistently. Dalitz et al.[1] determined a reasonable ΛN central force that will produce the correct Λ-separation energies in ${}^3_\Lambda$H and ${}^4_\Lambda$H. Then this potential is found to be so strong for ${}^5_\Lambda$He that it gives the binding energy value of 5.46 MeV which is much larger than the experimental value of 3.12 ± 0.02 MeV. The significant effect of coupling between Σ and Λ was proposed in theoretical treatments of the hypernuclear problem. Recently Akaishi et al.[2] has succeeded in fitting consistently all the experimental Λ binding energies of the s-shell Λ hypernuclei based on realistic YN interactions. We extend the calculation to neutron rich He isotope Λ hypernuclei.

2 Realistic Hyperon-Nucleon Interactions

In the work of ref.(2) various types of hyperon-nucleon interactions derived by Shinmura are used. All these potentials reproduce the phase shifts equivalent to the Nijmegen D and the recent soft core potentials. The potential which is designated as D0 has only central part of ΛN interaction and D1 has both central and tensor parts of ΛN channel interaction. The potential D2 has central parts of both ΛN and ΣN channels.

The Λ separation energies are calculated by the Brueckner-Hartree-Fock method on Gaussian basis: Reaction matrices from YN interaction are expanded on 20-range Gaussian functions and are used to obtain a hyperon-nucleus Hartree-Fock potential by folding procedure with nucleon wave functions of Gaussian combination. This hyperon-nucleus potential is again expanded on Gaussian bases, and is used to solve the relative motion between the hyperon and the core nucleus. The center-of-mass motion of the system is removed by using the hyperon-nucleus reduced mass.

The D0 potential causes the overbinding of 6.66 MeV in ${}^5_\Lambda$He as it was calculated by Dalitz et al. By employing the D2 potential which includes the ΛN - ΣN coupling of central type, the binding energy in the five-body system is reduced to 3.01 MeV due to the large suppression of the ΛN - ΣN coupling. While this coupling remedies the overbinding problem in ${}^5_\Lambda$He, the suppression effect in turn causes a serious underbinding problem in the four-body systems. Then the three body force ΛNN due to the ΛN-ΣN coupling is introduced which was first proposed by Gibson et al.[3]. This three body force effect is taken into account by including explicitly Σ-space components into the wave function of ${}^4_\Lambda$He as

$$|{}^4_\Lambda\mathrm{He}>= \Phi_\Lambda(\vec{r})|{}^3\mathrm{He}> +\sqrt{\frac{2}{3}}\Phi_{\Sigma^+}(\vec{r})|{}^3\mathrm{H}> -\sqrt{\frac{1}{3}}\Phi_{\Sigma^0}(\vec{r})|{}^3\mathrm{He}> . \quad (1)$$

The coupling strength is found stronger in the 0^+ state than in the 1^+ state. This is due to the coherently added enhancement where contributions from each pair of hyperon nucleon coupling are added constructively. On the other hand, contributions from the two pairs are canceled out in the 1^+ state and the resulting coupling weight is three times smaller than in the 0^+ state. That is why the three body force coupling strength is about one order stronger in the 0^+ state. Thus the potential D2 cooperating with the three body force Λ-Σ coupling reproduces the existing experimental Λ binding energies of all s-shell Λ hypernuclei. It is found that the energy splitting between the spin doublet states is due to Λ-Σ coupling for the D2 potential while for the Nijmegen soft core SC89 and SC97f potentials, it is caused by the spin-spin interaction plus the Λ-Σ coupling. Large effect of the Λ-Σ coupling is expected to be found in neutron rich nuclei and we extend our study to various He isotope hypernuclei with the Nijmegen model D, the soft core SC89 and SC97f potentials[4].

3 He Isotope Λ Hypernuclei

It is found that the ΛN-ΣN coupling strength becomes stronger for larger number of neutrons in He isotope Λ hypernuclei. But the coupling effect is small due to the spread out of the p-shell neutron distributions.

The result of Λ binding energies for H and He isotope hypernuclei for the different interactions are shown in Fig.1.

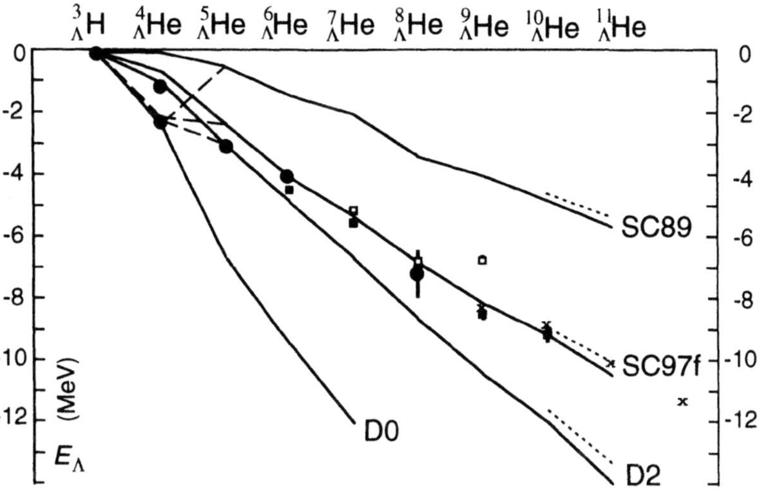

Figure 1. The Λ binding energies for H and He isotope Λ hypernuclei with various YN interactions. Li(■), Be(□), B(×) data are also shown.

The central type ΛN interaction alone (D0) cannot solve the large overbinding problem. In the case of D2, though it reproduces very well the s-shell Λ hypernuclear experimental data it has also overbinding problem for larger A hypernuclei. The recent soft core SC97f potential has both central and tensor type ΛN-ΣN coupling and it fits well all the experimental binding energies. Another soft core potential SC89 also has both central and tensor type ΛN-ΣN coupling but it produces binding energies lower than the experimental values. To understand the nature of the different potentials we investigate the relative strength of coupling terms by measuring the volume integral of each term in $^{4}_{\Lambda}$He. The soft core potentials SC89, SC97f and the D2 potential have nearly an equal strength of ΛN-ΣN central coupling which gives rise an attractive three body ΛNN force in the four body systems. The large suppression due to ΛN-ΣN tensor coupling and ΛN tensor force are found in the soft core potentials. Without these suppression, the D2 potential has overbinding problem when A becomes larger than 5. The proper ratios of these three couplings, namely ΛN-ΣN central coupling, ΛN-ΣN tensor coupling and ΛN tensor force would be significant in YN interactions. It is found that the Nijmegen SC97f meets this requirement.

It is also found from our calculations that there are particle stable bound states in $^{8}_{\Lambda}$He, $^{9}_{\Lambda}$He and $^{11}_{\Lambda}$He except $^{10}_{\Lambda}$He whereas ^{7}He, ^{9}He and ^{10}He have only resonance states. Nuclei near neutron drip line become more stable by adding a Λ hyperon and these Λ hypernuclei would give an opportunity to study neutron rich nuclei in more favorable way than the resonances of the ordinary nuclei.

4 $^{6}_{\Lambda}$H System

The large effect of three body type Λ-Σ coupling is found in the 0^+ state of $^{4}_{\Lambda}$H. If more neutrons are added to that state the Λ-Σ coupling would be enhanced by the contribution from the added neutrons. Since there is a resonance in ^{5}H at 2.1 MeV [5] above the threshold of ^{3}H+2n it is interesting to study the $^{6}_{\Lambda}$H hypernucleus. The YN interaction for this system is the sum of the contributions from the 0^+ state of $^{4}_{\Lambda}$H which we consider as a seed and that from the two p-shell neutrons. It is found that there is a particle stable bound state with binding energy of 5.8 MeV from the threshold of ^{5}H+Λ. The three body force Λ-Σ coupling is found to be 1.4 MeV with 1.6% of Σ component. Without this coupling force the state would come very close to the threshold of $^{4}_{\Lambda}$H+2n and the existence of this state would become ambiguous. The possible reaction to produce the $^{6}_{\Lambda}$H hypernucleus is through ^{6}He(K^-,π^0) or ^{6}Li(K^-,π^+) reaction and the observation of this bound state would indicate the significance of the three body force Λ-Σ coupling.

5 Summary

The suppression due to two body ΛN-ΣN coupling solves the overbinding problem in $^{5}_{\Lambda}$He but it, in turn, causes a severe underbinding in the four body systems. The shortage of this binding is overcome by introducing the three body force ΛNN due to the Λ-Σ coupling. This three body force is strong in the 0^+ states of $^{4}_{\Lambda}$H and $^{4}_{\Lambda}$He according to the coherently added enhancement. The spin doublet splittings in $^{4}_{\Lambda}$H and $^{4}_{\Lambda}$He are found partly due to the ΛN spin-spin interaction and partly due to the Λ-Σ coupling in the soft core SC97f model. Our calculation based on this realistic potential SC97f has justified Gibson's statement that this energy splitting would not be a measure of the ΛN spin-spin interaction [6]. He isotope Λ hypernuclear study shows that ΛN-ΣN central coupling, ΛN-ΣN tensor coupling and ΛN tensor force are significant in YN interaction. The Nijmegen soft core SC97f has a proper combination of these three couplings and fits well the experimental data. The $^{6}_{\Lambda}$H has a particle stable bound state with large Λ-Σ coupling effect whose structure is based on the 0^+ state of $^{4}_{\Lambda}$H.

References

1. R.H. Dalitz, R.C. Herndon, Y.C. Tang: Nucl. Phys. **B47**, 109 (1972)
2. Y. Akaishi, T. Harada, S. Shinmura: Proc. SNP99 in Seoul, (1999)
3. B.F. Gibson, A. Goldberg, M.S. Weiss: Phys. Rev. **C6**, 741 (1972)
4. Th.A. Rijken, V.G.J. Stoks, Y. Yamamoto: Phys. Rev. **C59**, 21 (1999)
5. E.G. Adelberger et al.: Phys. Lett. **25B** 595 (1967)
6. B.F. Gibson: Proc. U.S.-Japan Seminar, Maui, 55 (1993)

Few-Body Systems Suppl. 12, 387–390 (2000)

Few-Body Systems

Final State Interaction in Kaon Photoproduction from the Deuteron

H. Yamamura[1], K. Miyagawa[1], T. Mart[2], C. Bennhold[3], W. Glöckle[4]

[1] Department of Applied Physics, Okayama University of Science, Ridai-cho Okayama 700, Japan

[2] Jurusan, Fisika, FMIPA, Universitas Indonesia, Depok 16424, Indonesia

[3] Center for Nuclear Studies, Department of Physics, The George Washington University, Washington, D.C. 20052, USA

[4] Institute für theoretishe Phsik II, Ruhr-Universität Bochum, D-44780 Bochum, Germany

Abstract. We calculate inclusive $d(\gamma, K^+)$, exclusive $d(\gamma, K^+Y)$ cross sections and hyperon polarization for the reaction $\gamma + d \to K^+ + \Lambda(\Sigma) + N$. Modern hyperon-nucleon (YN) forces and a recently updated production operator for the $\gamma + N \to K^+ + \Lambda(\Sigma)$ process are used. Sizable effects of YN final state interaction (FSI) have been found around the $K^+\Lambda N$ and $K^+\Sigma N$ threshold. Around the $K^+\Sigma N$ threshold FSI effects are especially interesting and the two YN forces show quite different effects.

1 Introduction

Recent analyses of few-baryon systems with strangeness have fairly clarified the property of the hyperon-nucleon (YN) interaction[1]. One prominent feature of the YN interaction is the existence of the $\Lambda N - \Sigma N$ coupling which yields an enhancement around the ΣN threshold in the ΛN elastic total cross sections. Our previous work[2] demonstrated that the enhancement is not a simple threshold effect, but caused by a t-matrix pole near the ΣN threshold. The position of the pole depends on YN force models. However, the available data are not enough to determine its actual location.

We then analyze the reaction $\gamma + d \to K^+ + \Lambda(\Sigma) + N$ which is experimentally accessible at the TJLAB and SPring-8. The Nijmegen soft-core YN interactions NSC97f[4] and NSC89[5] and a recently updated production operator for the $\gamma + N \to K^+ + \Lambda(\Sigma)$ process are used[3].

2 Formulation

The T-matrix for the reaction $\gamma + d \rightarrow K^+ + \Lambda(\Sigma) + N$ are defined by

$$T_i|\Psi_d> = \sum_j U_{ij} \cdot t_{\gamma K}^{(j)}|\Psi_d> \qquad i,j = \Lambda N, \Sigma N. \tag{1}$$

The operator $t_{\gamma K}^{(i)}$ describes the elementary process $\gamma + N \rightarrow K^+ + \Lambda(\Sigma)$. It includes not only tree level diagrams but also some effects of nucleon resonances. The state $|\Psi_d>$ represents the deuteron which is generated by Nijmegen 93 NN interaction [6]. U_{ij} corresponds to the YN final state interaction processes. It is represented by

$$\begin{aligned} U_{ij} &= \delta_{ij} + V_{ij}G_0^{(j)} + \sum_{j'} V_{ij'}G_0^{(j')}V_{j'j}G_0^{(j)} + \cdots \\ &= \delta_{ij} + \sum_{j'} V_{ij'}G_0^{(j')}U_{j'j}. \end{aligned} \tag{2}$$

Where V_{ij} is the YN interaction including $\Lambda N - \Sigma N$ conversion. By using Eq.(1) and (2), the components T_i obey the coupled set of integral equations

$$T_i|\Psi_d> = \left[t_{\gamma K}^{(i)} + \sum_{j'} V_{ij'}G_0^{(j')}T_{j'} \right] |\Psi_d> . \tag{3}$$

We solve this set, Eq.(3), partial-wave decomposed in momentum space. The three elementary process $\gamma + p \rightarrow K^+ + \Lambda(\Sigma^0)$ and $\gamma + n \rightarrow K^+ + \Sigma^-$ are property incorporated in the driving term in Eq.(3). These T-matrices are described on isospin bases $(\Lambda N, \Sigma N)$. They are transformed into particle bases $(\Lambda n, \Sigma^0 n, \Sigma^- p)$. Thus, the inclusive $d(\gamma, K^+)$, exclusive $d(\gamma, K^+ Y)$ cross sections and hyperon polarization are calculated. For the details, we refer the readers to ref.[7].

3 Result

We assume the following condition in the present calculations. First, the photon direction is along z axis, and K^+ meson laboratory scattering angle is 0 degree $(\theta_{K^+} = 0°)$. Second, we ignore the K^+ meson interaction with the nucleon and hyperon at final states.

In Fig.1 the prediction by the two YN forces, NSC89 and NSC97f are compared to the plane wave inpulse approximation (PWIA). The inclusive cross section is contributed additively by the $K^+\Lambda n$, $K^+\Sigma^0 n$ and $K^+\Sigma^- p$ processes. The two pronounced peak around $p_K = 945$ and 809 MeV/c can be understood in PWIA. They are due to quasi-free processes, where one of the nucleons in the deuteron is a spectator and has zero momentum in the laboratory system. The FSI results deviate from the PWIA around the two threshold. The FSI effects

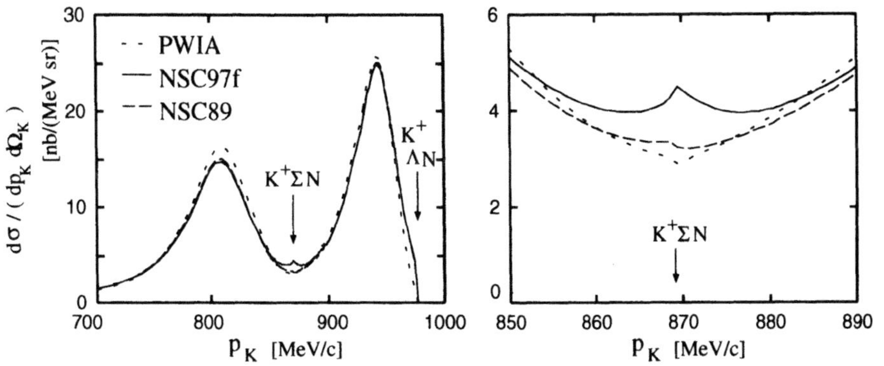

Figure 1. The inclusive $d(\gamma, K^+)$ cross section as a function of lab momentum p_K for $\theta_{\mathrm{K,lab}} = 0°$ and photon lab energy $E_\gamma = 1.3$ GeV. The two thesholds $K^+\Lambda N$ and $K^+\Sigma N$ are indicated by the arrows.

near the $K^+\Lambda N$ threshold enhance the cross section by up to 86%. Around the $K^+\Sigma N$ threshold, there are large differences between PWIA and FSI results and also between the predictions of YN force models.

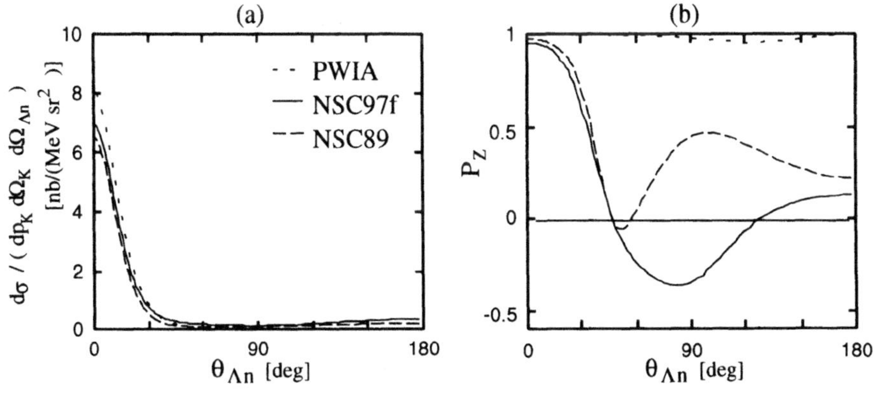

Figure 2. The exclusive $d(\gamma, K^+\Lambda)$ cross section (a) and Λ recoil polarization with incoming polarized photon (b) as a function of the c.m. angle in the Λn system. Photon lab energy is $E_\gamma = 1.3$ GeV. Outgoing kaon lab momentum and angle are $p_{K^+} = 870$ MeV/c and $\theta_{K^+} = 0°$ respectively.

We also calculate other observables. Fig.2 illustrates the result for exclusive $d(\gamma, K^+\Lambda)$ and Λ recoil polarization with incoming polarized photons just below the $K^+\Sigma N$ threshold. The result for $d(\gamma, K^+\Sigma^-)$ just above the $K^+\Sigma N$ threshold is in Fig.3. Significant FSI effects are seen and the prediction by NSC97f is quite different from that of NSC89. Precise data would help to clarify the strength and property of the $\Lambda N - \Sigma N$ interaction.

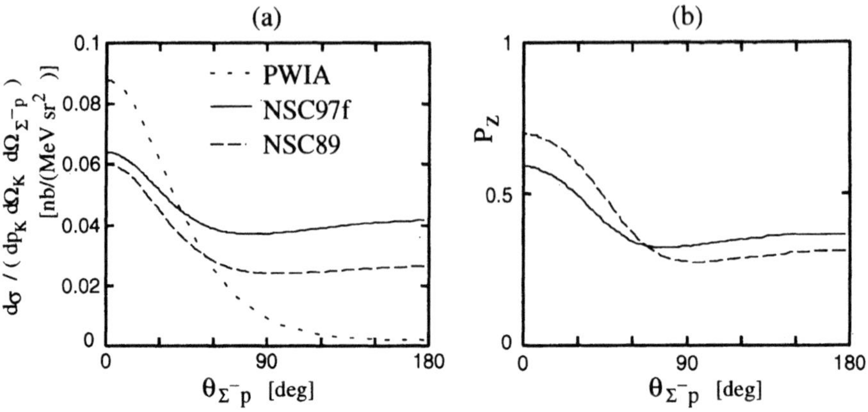

Figure 3. The exclusive $d(\gamma, K^+\Sigma^-)$ cross section (a) and Σ^- recoil polarization with incoming polarized photon (b) as a function of the c.m. angle in the $\Sigma^- p$ system. Photon lab energy is $E_\gamma = 1.3$ GeV. Outgoing kaon lab momentum and angle are $p_{K^+} = 865$ MeV/c and $\theta_{K^+} = 0°$ respectively.

References

1. K. Miyagawa, H. Kamada, W. Glöckle, V. Stoks: Phys. Rev. **C51**, 2905 (1995)

2. K. Miyagawa, H. Yamamura: Phys. Rev. **C60**, 024003 (1999); nucl-th/9904002

3. C. Bennhold, T. Mart, A. Waluyo, H. Haberzettl, G. Penner, T. Feuster, U. Mosel: in *Proceedings of the Workshop on Electron-Nucleus Scattering*, Elba, Italy, 1998, edited by O. Benhar, A. Fabrocini, R. Schiavilla, p.149. Pisa: Edizioni ETS 1999; nucl-th/9901066

4. Th.A. Rijken, V.G.J. Stoks, Y. Yamamoto: Phys. Rev. C **59**, 21 (1999)

5. P.M.M. Maessen, Th.A. Rijken, J.J. de Swart: Phys. Rev. **C40**, 2226 (1989)

6. V.G.J. Stoks, R.A.M. Klomp, C.P.F. Terheggen, J.J. de Swart: Phys. Rev. **C49**, 2950 (1994)

7. H. Yamamura, K. Miyagawa, T. Mart, C. Bennhold, W. Glöckle: Phys. Rev. C, in press; nucl-th/9907029

Few-Body Systems Suppl. 12, 391–394 (2000)

Few-Body Systems

Quark Pauli Principle in Λ-Hypernuclear Systems

H. Nemura[1,*], Y. Suzuki[2]

[1] Graduate School of Science and Technology, Niigata University, Niigata 950-2181, Japan

[2] Department of Physics, Niigata University, Niigata 950-2181, Japan

Abstract. We explore a question of whether quark Pauli principle plays an important role or not in the binding mechanism of Λ hypernuclei. We introduce a special five-baryon configuration which is forbidden by the Pauli principle at the quark level. By excluding the forbidden state, we show that the Λ separation energy of ${}^{5}_{\Lambda}$He or ${}^{6}_{\Lambda\Lambda}$He is significantly reduced for a baryon size of 0.86 fm, while it is unchanged for a smaller size of 0.6 fm.

The Pauli principle plays a vital role in the binding mechanism for systems of identical particles. Its importance is well known in atoms and atomic nuclei. However, in Λ hypernuclei where the Λ-particle is embedded in nuclei the role of the Pauli principle has not yet been explored from the point of view of the binding mechanism. Though Λ can be distinguished from the nucleon N at the baryon level, they are subjected to the Pauli principle at the quark level. We estimate how much the Λ separation energy of $A = 5, 6$ Λ hypernuclei changes by the quark Pauli effect.

In a naive baryon picture of ${}^{5}_{\Lambda}$He, for example, Λ is distinguishable from the nucleon so that ${}^{5}_{\Lambda}$He is considered a spatially compact system as ^{4}He is. In such a compact system, quark effects might appear due to the strong overlapping of each baryon. If we assume that a baryon is described as a three quark wave function of $(0s)^3$ and all five baryons completely overlap, such a state must be forbidden due to quark Pauli principle. (This system contains seven u and d quarks.) This situation derived from the quark picture must be reflected in ${}^{5}_{\Lambda}$He based on the baryon model. Namely, there should exist some forbidden states in a model space describing ${}^{5}_{\Lambda}$He in a baryon picture. Even in case of the system consisting of less than five baryons, there exists the effect of the quark Pauli principle, as was discussed by Takeuchi and Shimizu[3]. Such a light system, however, does not produce a quark Pauli forbidden state mentioned above. It is important to note, however, that quark effects might be included into an

* *E-mail address:* nemura@nt.sc.niigata-u.ac.jp

appropriate effective baryon-baryon potential. The quark Pauli effect including five baryons cannot be represented by two- or three-body effective force.

Taking into account of the quark Pauli effect into a framework of baryon few body calculation amounts to constraining the model space of the variational calculation. In order to perform such a calculation, we must define a five-baryon configuration which is forbidden by the Pauli principle at the quark level. As explained in Ref.[1, 2] in detail, we give briefly our idea below. We assume that a baryon is a $3q$ system, and the orbital motion of each quark is described by a $(0s)$ harmonic-oscillator function, $\exp\{-\rho^2/2b^2\}$, with a size parameter b. The relative motion function between the baryons of a 15-quark, five-baryon wave function becomes Pauli-forbidden when the function takes the form

$$\psi_{\rm F} = \left(\frac{\pi b^2}{3}\right)^{-\frac{3}{4}} \exp\left\{-\frac{3}{2b^2}\sum_{i=1}^{5}(\boldsymbol{r}_i - \boldsymbol{R})^2\right\}, \tag{1}$$

where $\boldsymbol{R} = \frac{1}{5}\sum_{i=1}^{5}\boldsymbol{r}_i = \frac{1}{15}\sum_{i=1}^{15}\boldsymbol{\rho}_i$ is the center-of-mass coordinate of the five baryons. The ground state wave function Ψ must be subjected to remain in the subspace which is orthogonal to the forbidden state, that is, $\langle\psi_F|\Psi\rangle = 0$. The constraint applies to the hypernuclei with $A \geq 5$.

In order to evaluate quantitatively the binding energy reduction due to the quark Pauli effect, the variational trial function has to be chosen flexibly enough. We assumed that the trial function with the total orbital angular momentum $L = 0$ is given by

$$\Psi_{JMIM_I} = \sum_{k=1}^{K} c_k\Big(\varphi_k - \sum_a \Gamma_a\langle\Gamma_a|\varphi_k\rangle\Big),$$

$$\varphi_k = \mathcal{A}_{\rm B}\{G(\boldsymbol{x}, A_k)\chi_{kJM}\eta_{kIM_I}\}, \quad G(\boldsymbol{x}, A_k) \equiv \exp\left\{-\frac{1}{2}\boldsymbol{x}A_k\boldsymbol{x}\right\}, \tag{2}$$

where Γ_a is an orthonormal set of functions constructed from the Pauli forbidden space. $\mathcal{A}_{\rm B}$ is the operator antisymmetrizing identical baryons, $\boldsymbol{x} = (\boldsymbol{x}_1, ..., \boldsymbol{x}_{A-1})$ stands for a set of relative (e.g., Jacobi) coordinates, A_k is positive definite matrix characterizing the k-th basis, and χ_{kSM_S} (η_{kIM_I}) is the spin (isospin) function of the system.

The basis element is increased one by one in the stochastic variational method[4, 5] until the solution reaches a practical convergence. Here the basis set φ_k is chosen so as to minimize the energy, whereas the set Γ_a is chosen to maximize the energy. The accuracy of solutions is vital in the present calculation because we have to take account of very small components produced by the quark Pauli principle.

To consider the effect of the quark structure of baryons on the binding energy of Λ hypernuclei, it is desirable to use a ΛN potential that reproduces the Λ separation energies of $A = 3, 4$ Λ hypernuclei. The NN and ΛN potentials were the same as employed in Ref. [1]. The NN potential reproduces the binding energies and sizes of ^{3}H, ^{3}He, and ^{4}He very well[5], and likewise the ΛN

potential is set to well reproduce the energies of $A = 3, 4$ hypernuclei. The $\Lambda\Lambda$ potential was taken from Ref. [7].

Table 1 compares the Λ separation energies calculated for $A = 5$ systems ($^5_\Lambda$He, $^5_{\Lambda\Lambda}$H, $^5_{\Lambda\Lambda}$He). Spin, parity and isospin (J^π, I) are $(\frac{1}{2}^+, 0)$, $(\frac{1}{2}^+, \frac{1}{2})$, and $(\frac{1}{2}^+, \frac{1}{2})$, respectively. Table 1 also shows the result of $A = 6$ system ($^6_{\Lambda\Lambda}$He), which will be discussed later. Experimental search for $^5_{\Lambda\Lambda}$H is in progress.[6]

As shown in Table 1, the quark Pauli effect leads to a significant reduction in the binding energy $^5_\Lambda$He at $b = 0.86$ fm. Although the value of b is chosen referring to the experimental proton's charge radius, a reasonable value of b could be slightly smaller than 0.86 fm because we know that a baryon is made from not only three quarks but also other contributions, e.g. meson clouds. In the case of $b = 0.6$ fm, the energy reduction becomes small. The quark Pauli effect becomes much less in $^5_{\Lambda\Lambda}$H and $^5_{\Lambda\Lambda}$He, which reflects the fact that they are spatially extended more than $^5_\Lambda$He. In fact, the root mean square (rms) radius of $^5_{\Lambda\Lambda}$H (and also $^5_{\Lambda\Lambda}$He) turns out to be 2.0 fm, while that of $^5_\Lambda$He is 1.6 fm.

These results can be qualitatively understood as follows. Let Ψ_0 and E_0 denote the wave function and the energy of the ground state which is obtained without the quark Pauli effect. When the Pauli effect is taken into account, the wave function may be written according to the perturbation theory as

$$\Psi = \frac{1}{\sqrt{1-\alpha^2}}(\Psi_0 - \alpha\psi_F) \quad \text{with} \quad \alpha = \langle\psi_F|\Psi_0\rangle. \tag{3}$$

Assuming that α is small, the energy difference is given by

$$E - E_0 = \langle\Psi|H|\Psi\rangle - E_0 \approx \alpha^2\langle\psi_F|H|\psi_F\rangle. \tag{4}$$

The energy difference is determined by the product of α^2, the probability of finding the forbidden component in Ψ_0, and the energy expectation value of the Pauli forbidden state. Let us use the above equation to estimate how much the energy difference is produced by the quark Pauli principle. In the case of $^5_\Lambda$He, $\alpha^2 \approx 0.004$, $\langle\psi_F|H|\psi_F\rangle \approx 500$ MeV for $b = 0.86$ fm, and thus $E - E_0 \approx 2$ MeV, while in the case of $^5_{\Lambda\Lambda}$H $\alpha^2 \approx 0.001$, $\langle\psi_F|H|\psi_F\rangle \approx 800$ MeV, and $E - E_0 \approx 1$

Table 1. The Λ separation energies, in MeV, of $A = 5$ and 6 systems as a function of the baryon size b (in fm). The case of $b = 0$ indicates that no quark Pauli effect is imposed in the calculation. Set A of ΛN potential parameters are used. (See Ref.[1].)

	Theory				Expt.
	$b = 0$	0.6	0.7	0.86	
$B_\Lambda(^5_\Lambda\text{He})$	4.98	4.90	4.70	3.63	3.12±0.02
$B_{\Lambda\Lambda}(^5_{\Lambda\Lambda}\text{H})$	5.63	5.59	5.52	5.12	
$B_{\Lambda\Lambda}(^5_{\Lambda\Lambda}\text{He})$	5.53	5.52	5.45	5.05	
$B_{\Lambda\Lambda}(^6_{\Lambda\Lambda}\text{He})$	14.3	13.9	12.9	9.4	10.9 ± 0.6

MeV. This estimation is in qualitatively good agreement with the result of Table 1.

We have calculated the binding energy of ${}^{6}_{\Lambda\Lambda}$He in a complete six-body treatment. This system has three types of forbidden states (1), containing $ppnn\Lambda$, $ppn\Lambda\Lambda$, or $pnn\Lambda\Lambda$ five baryons. The energy obtained without the quark Pauli effect ($b = 0$) is apparently overbound. The elimination of the forbidden states leads to a surprisingly large reduction in the binding energy. The rms radius of ${}^{6}_{\Lambda\Lambda}$He is calculated to be 1.6 fm, considerably smaller than that (2.0 fm) of ${}^{5}_{\Lambda\Lambda}$H, which leads to a larger quark Pauli effect in ${}^{6}_{\Lambda\Lambda}$He.

We studied the consequence of the quark structure of the baryons on the binding energies of Λ- and $\Lambda\Lambda$ hypernuclei. We showed that, though the Λ particle can be distinguished from the nucleon, the quark model predicts a special forbidden state involving five baryons. By excluding the forbidden states we found that the binding energies started to decrease at the baryon size of $b = 0.6$ fm and reduced significantly at the size of free baryons, $b = 0.86$ fm. Since the forbidden states considered here are many-body configurations involving five baryons, the Pauli effect cannot be revealed through a study on the baryon-baryon interaction alone. In other words, a full understanding of the binding mechanism of the hypernuclei may not be attainable only through the knowledge of the two- and three-body interactions of the baryons.

Acknowledgments

We are thankful to Dr. Fujiwara and Dr. Nakamoto for useful discussion. This work was in part supported by Grant-in-Aid for Scientific Research (No. 10640255) of the Ministry of Education, Science and Culture (Japan).

References

1. H. Nemura, Y. Suzuki, Y. Fujiwara, C. Nakamoto: Prog. Theor. Phys. **101**, 981 (1999)
2. Y. Suzuki, H. Nemura: Prog. Theor. Phys. **102**, 203 (1999)
3. S. Takeuchi, K. Shimizu: Phys. Lett. **B179**, 197 (1986)
4. V. I. Kukulin, V. M. Krasnopol'sky: J. Phys. **G3**, 795 (1977)
5. K. Varga, Y. Suzuki: Phys. Rev. **C52**, 2885 (1995); Y. Suzuki, K. Varga: *Stochastic Variational Approach to Quantum-Mechanical Few-Body Problems* (Lecture Notes in Physics, Vol. m54). Berlin: Springer-Verlag 1998
6. J. Nakano, E906 collaboration: This proceeding
7. E. Hiyama, M. Kamimura, T. Motoba, T. Yamada, Y. Yamamoto: Prog. Theor. Phys. **97**, 881 (1997)

Few-Body Systems Suppl. 12, 395–398 (2000)

Few-Body Systems

An SU(3) Symmetric OBEP with Gaussian Core between Octet Baryons

S. Shinmura[1], I. Arisaka[2], K. Nakagawa [3], M. Wada[3]

[1] Department of Information Science, Gifu University, Gifu 501-1193, Japan

[2] Chiba Institute of Technology, Narashino 274-0023, Japan

[3] College of Science and Technology, Nihon University, Funabashi 274-8501, Japan

Abstract. An SU(3) symmetric one-boson-exchange potential(OBEP) between two octet baryons is presented. In the core region, the Gaussian smooth cutoff of OBEP and the Gaussian soft core are introduced.There are 24 parameters(12 of SU(3) coupling constants and 12 of Gaussian core strengths and range), which are determined by the χ^2-fitting to NN phase shifts, deuteron and YN data. The fitting is quite satisfactory on the whole. Some discussions to show characters of the potential are presented.

1 Introduction

The recent developement in hypernuclear physics is remarkable and the realistic hyperon-nucleon(YN) and hyperon-hyperon(YY) interactions are neccesary for us to understand and to predict the exparimental data. In recent years, various types of combined models of nucleon-nucleon(NN) and hyperon-nucleon(YN) interactions have been proposed. Nijmegen group [1] has proposed various types of NN and YN potentials, those with hard cores and with soft cores, on the basis of the SU(3) symmetric one-boson-exchange potential(OBEP). Bon-Jülich group [2] has proposed the OBEP including two pion exchange contribution. Recently, Kyoto-Niigata group [3] has presented potential models based on the quark model.

At present, various types of OBEP are possible theoretically and phenomenologically. For the short range part, hard cores, soft cores or form factors are possible. For exchanged scalar mesons, nonet mesons, only one singlet meson or a fictitious σ meson are used in OBEP. In addition, potential forms and velocity-dependences are not unique. As a result, the coupling constants and other physical parameters adjusted so as to reporduce the NN and YN data are different more or less for model by model.

The purpose of the present work is to extend the Funabashi NN potential [4] to the baryon-baryon(BB) potential. The potential describes inetractions between all pairs of octet baryons in a unified way.

2 Framework of Model

The framework of our potential can be summarized as follows;

(1) We consider the exchange of nonets of scalar, pseudoscalar and vector mesons with physical masses and widths. The mass and width of the scalar singlet meson f_0 are much controversial and we assume m = 760MeV and Γ = 640MeV, which are used in the Nijmegen models. To evaluate the effects of meson width, which are relevant for f_0, K^{0*} and ρ mesons, we use the original integral formula [5] instead of the two-pole approximation.

(2) We assume exactly SU(3)-symmetric coupling constants. In our model, the following four parameters are fixed,

$$\theta_{ps} = -23.92^\circ, \quad \theta_v^e = \theta_v^m = 36.44^\circ, \quad \alpha_v^e = 1.0,$$

based on the GMO mass formula and the ideal mixing. The other 12 parameters are determined through the χ^2 fitting.

(3)The velocity-dependence and the retardation effects are taken into account up to the first order of $\boldsymbol{k}^2/M$, $\boldsymbol{q}^2/M$ and $(\boldsymbol{kq})^2/M^2$. The details will be discussed elsewhere.

(4)SU(3) symmetric Gaussian soft cores are introduced in the short range region. Gaussian soft cores have a common range r_G = 0.51fm for all cases. Strenghts(Hights) of the cores are assumed to have SU(3) symmetry on the basis of SU(3) representations of two octet baryons. As the state-dependence, we consider the spin dependence(singlet(1) and triplet(3)) and the spatial parity dependence(even(E) and odd(O)). Then, we have 12 parameters as shown in Table 2. The strengths in 1E, 3O states of $\{1\}$ are not determined in this work. For 3S1 of NN, we introduce an additional strength so as to reproduce the accurate deuteron properties.

(5)The smooth cutoff of OBEP with the Gaussian shape in the short range region is introduced. The functional form is given by

$$F(r) = \{1 - \exp(-(r/r_{cut})^2)\}^n,$$

where,

$$r_{cut} = 0.4\text{fm}, \qquad n = 4.$$

The range r_{cut} = 0.4 fm is determined to satisfy the conditions that the cutoff has no effect for $r > 1$ fm and that it cuts a half of OBEP around $r \sim 0.5$ fm. The index $n = 4$ is selected to avoid the sigularity of OBEP at $r = 0$(highest singularity coming from the quatratic LS force).

In our model, 24 parameters are adjustable. They are determined by the χ^2 fitting. As optimization algorithm, we employed the simplex method, which is essentially the method of the local optimization. We tried the optimizations from many starting points.

Table 1. SU(3) parameters of coupling constants

meson	$g_1/\sqrt{4\pi}$	$g_8/\sqrt{4\pi}$	α	θ
scalar	4.85012	2.34306	1.22897	-1.44
pseudoscalar	0.19234	0.26600	0.49117	-23.92
vector(e)	2.67578	0.92093	1.00000	36.44
vector(m)	1.26422	5.87956	0.76414	36.44

Table 2. Strengths of Gaussian soft cores in GeV

state	{27}	{10*}	{10}	{8a}	{8s}	{1}
1E	3.59900				0.32232	-
1O		1.53284	0.57651	0.35910		
3E		2.60205	0.37693	1.75103		
3S1		1.86127				
3O	0.00000				0.04571	-

3 Results and Discussions

In the χ^2 fitting, we use the single-energy NN phase shift analysis by Arndt et al.(SAID) [6], deuteron properties and YN scattering data. The resultant values of the SU(3) coupling constants and the core parameters are shown in Tables 1 and 2, respectively.

The NN phase shifts calculated with our potential are in good agreement with values of SAID except the $^1F_3$. This partial wave is not included in our fitting, because it has not the single-energy phase shift analysis. For deuteron, we obtain $B = 2.22457\text{MeV}$, $Q = 0.266\text{fm}^2$, $P_D = 5.558\%$ (the experimental values:$B = 2.224575 \pm 0.000009\text{MeV}$, $Q = 0.2860 \pm 0.0015\text{fm}^2$).

Results for YN scattering are also in very good agreement with experimental data. The experimental information for the spin-dependence of the ΛN interaction is very limited. Therefore we use the result of phenomenological study of s-shell hypernuclei as a guideline. For the $\Sigma^- p$ inelastic capture ratio, our result 0.455 reproduces successfully the experimental data, that is, 0.465 ± 0.011, 0.474 ± 0.016 and 0.33 ± 0.05.

To clarify the properties of our new potential, we discuss here only about two subjects. The first is the effect of anti-symmetric LS force(ALS). ALS causes the $^1L_J - ^3L_J$ mixing. The $^1P_1 - ^3P_1$ mixing parameters in $\Sigma^+ p$ and Λp scattering calculated with our potential are considerably large, that is, about $-2°$ and $+3°$ at $p_Y = 400$ MeV/c, respectively. These values are larger than those of the Nijmegen model D by about 10 times. If we use the same coupling constants as used in the model D, we obtain small values of the mixing angles, which are larger than model D only by two times. This suggests that the main part of large ALS comes from difference of coupling constants.

The second subject to see the properties of our new potential is the existence of two-baryon bound states. For $^1S_0$ states of uncoupled baryonic pairs, which all belong the {27} representation, our potential gives strong attractions. As a result, all states with $S < -1$ have a bound state. On the other hand, for $^3S_1$ states of uncoupled baryonic pairs, we find two types of interaction, which correspond to the {10} and {10*} representations, respectively. np, $\Xi^-\Sigma^-$ and $\Xi^0\Sigma^+$ belong the {10*} and have a bound state, and Σ^-n, Σ^+p and $\Xi^-\Xi^0$ belong the {10} and have repulsive interactions. Our potential has surprisingly a simple symmetric property in spite of the SU(3) breaking of meson masses.

4 Summary

We proposed the extended version of the Funabashi potential. This potential describes the interactions between octet baryons in a unified way. The SU(3) breaking only comes from the breaking in masses of mesons and baryons. The other physical parameters including core strengths are exactly SU(3)-symmetric. The ALS force in our potential is considerably large in comparison with the Nijmegen models. Our potential gives strong attractions in $^1S_0$ of {27} and $^3S_1$ of {10*}, and repulsions in $^3S_1$ of {10}.

At present, two parameters remain free, that is, the core strenghts in 1E and 3O states of {1}. These appear only in the $\Lambda\Lambda$ interaction. We wish that the experimental data for the $\Lambda\Lambda$ interaction would be improved.

References

1. M.M. Nagels, Th.A. Rijken, J.J. de Swart: Phys. Rev. **D15**, 2547(1977); **D20**, 1633 (1979); P.M.M. Maessen, Th.A. Rijken, J.J. de Swart: Phys. Rev. **C40**, 1633 (1989); Th.A. Rijken, V.G.J. Stoks, Y. Yamamoto: Phys. Rev. **C59**, 21 (1999)

2. B. Holzenkamp, K. Holinde, J. Speth: Nucl. Phys. **A500**, 485 (1989); A. Reuber, K. Holinde, J. Speth: Nucl. Phys. **A570**, 543 (1994)

3. Y. Fijiwara, C. Nakamoto, Y. Suzuki: Phys. Rev. Lett. **76**, 2242 (1996); Phys. Rev. **C54**, 2180 (1996)

4. T. Obinata, M. Wada: Prog. Theor. Phys. **53**, 732 (1975); **57**, 1984 (1977); **61**, 1697 (1979)

5. J. Schwinger: Phys. Rev. **D3**, 1967 (1971); J. Binstock, R. A. Bryan: Phys. Rev. **D4**, 1341 (1971)

6. R. A. Arndt et al.: SAID program (clsaid.phys.vt.edu)

Few-Body Systems Suppl. 12, 399–402 (2000)

Few-Body Systems

Scattering Observables of the YN Interaction in the SU_6 Quark Model

T. Fujita[1], Y. Fujiwara[1], C. Nakamoto[2], Y. Suzuki[3], M. Kohno[4]

[1] Department of Physics, Kyoto University, Kyoto 606-8502, Japan
[2] Suzuka National College of Technology, Suzuka 510-0294, Japan
[3] Department of Physics, Niigata University, Niigata 950-2181, Japan
[4] Physics Division, Kyushu Dental College, Kita-Kyushu 803-8580, Japan

Abstract. Kyoto-Niigata group has recently developed a new version of the quark model for the simultaneous description of the NN and YN systems. YN scattering observables by this model are investigated and compared with the results by the old versions of our model, FSS and RGM-H. It is shown that the essential features of the ΛN-$\Sigma N(I = 1/2)$ channel coupling are common for all the versions of our model. Our results show that the tensor force from the 1 π exchange and the $LS^{(-)}$ force from the Fermi-Breit interaction are important in this coupling.

1 Introduction

One of the most successful realistic description of the baryon-baryon interaction is based on the resonating-group method (RGM) applied to the constituent quark model. Kyoto-Niigata group developed a simultaneous description of the NN and YN interactions based on this framework, the $(3q)$-$(3q)$ RGM + effective meson exchange potentials (EMEP). Although two versions of the model, FSS [1, 2] and RGM-H [2], give almost satisfactory description for the NN phase shifts and for the YN scattering cross sections, there still exists some room for improvement. Based on the analysis of the interaction properties of FSS and RGM-H, we have recently upgraded our EMEP in several aspects. Regarding the NN phase shifts, quality of the description by the new version of our model is found to be almost comparable to that by OBEP models [3].

In this report, we proceed to investigate the YN interaction by the new version. Various scattering observables are studied, and are compared with the results by the old versions of our model. We especially pay attention to the features of the ΛN-ΣN channel coupling.

Table 1. Introduced meson species in each version of our quark model. (C: central part, SS: spin-spin part, T: tensor part, LS: spin-orbit part, QLS: quadratic spin-orbit part, BS: Bryan-Scott term)

model	int. type		mesons
FSS, RGM-H	S	C	ϵ, S^*, δ, κ
	PS	$SS+T$	η', η, π, K
present	S	$C(BS)+LS$	ϵ, S^*, δ, κ
	PS	$SS+T$	π, K
	V	$C(BS)+LS+QLS$	ω, ϕ, ρ, K^*

2 Formulation

Our Hamiltonian consists of the non-relativistic kinetic-energy term, the quadratic confinement potential, the full FB interaction with explicit quark-mass dependence and EMEP acting between quarks:

$$H=\sum_{i=1}^{6}(m_ic^2+\frac{p_i^2}{2m_i})+\sum_{i<j}^{6}(U_{ij}^{\rm Cf}+U_{ij}^{\rm FB}+\sum_{\beta}U_{ij}^{{\rm S}\beta}+\sum_{\beta}U_{ij}^{{\rm PS}\beta}+\sum_{\beta}U_{ij}^{{\rm V}\beta})\ .$$

The new version is different from the old ones in the EMEP sector, which is summarized in Table 1. While FSS and RGM-H have only the S and PS meson exchange, the V meson exchange is also introduced in the new version. Main benefit of the V meson exchange comes from its QLS component, which remedies the description of the NN $^3D_2$ phase shifts. We note, however, that the introduction of the V meson exchange is not an essential change in the framework of our model, but just an improvement of EMEP. Coupling constants of the V meson exchange are comparatively small ($f_1^e, f_1^m \sim 1$ and $f_8^e=0$). Another new point is to include the Bryan-Scott term (q^2 dependent term) in the central part of the S and V meson exchange potentials. This term brings repulsion to the single-particle potentials in high momentum region.

3 Result

We investigate various scattering observables by our quark model. Fig. 1 (a) displays the total cross sections of the Λp scattering by our quark model. Both FSS and the new version reproduce the experiment over low-energy region. We note that there is a cusp-like energy dependence at the $\Sigma N(I=1/2)$ threshold. The origin of this structure is the strong $^3S_1$-$^3D_1$ coupling by the 1π tensor force, which is common for both versions of our model. Our model gives a sharp rise in the ΛN^3S_1 phase shift at the threshold, while the threshold effect is small in the ΛN^1S_0 channel. Another point to note is the enhancement of the cross sections around the cusp, which can be clearly seen in FSS. This

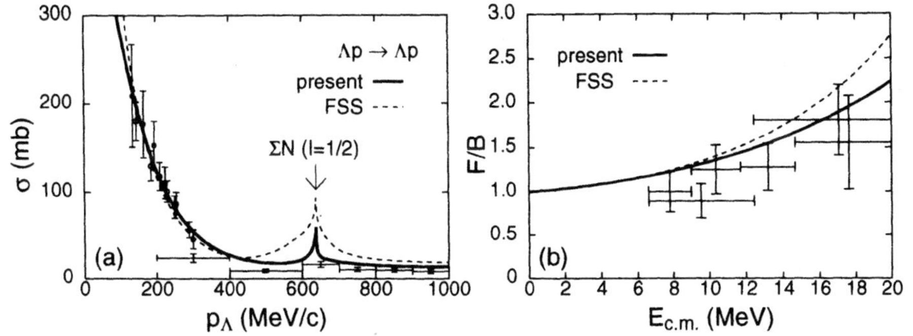

Figure 1. (a) Λp total cross sections and (b) the forward to backward ratio of the Λp differential cross sections by the quark model. Solid curves denote the results by the new version of our model, dashed curves denote those by the old version FSS.

enhancement comes from a rapid increase of the ΛN^1P_1-ΛN^3P_1 transition by $LS^{(-)}$ force around the ΣN threshold. The increase is more significant in FSS, but we can see the same coupling feature also in the new version of our model. The forward to backward ratio of the Λp differential cross section is displayed in Fig. 1 (b). Both versions of our model give $F/B > 1$, which implies attractive P state interaction [4]. This is also due to the threshold effect in the ${}^1P_1$-${}^3P_1$ coupling by the $LS^{(-)}$ force, which enhances the ${}^1P_1$ and ${}^3P_1$ phase shifts even at lower energies.

Total cross sections of the Σ^-p scattering by the new version of our model are displayed in Fig. 2. (The final results are denoted by the solid curves.) The fit to the experiment is reasonably well. Another observable to investigate is the Σ^-p inelastic capture ratio at rest (r_R):

$$r_R = \frac{1}{4}r_{S=0} + \frac{3}{4}r_{S=1}, \quad r_{S=0,1} \equiv \frac{\sigma_B(S=0,1)}{\sigma_B(S=0,1)+\sigma_C(S=0,1)}|_{p_{\Sigma^-}=0} ,$$

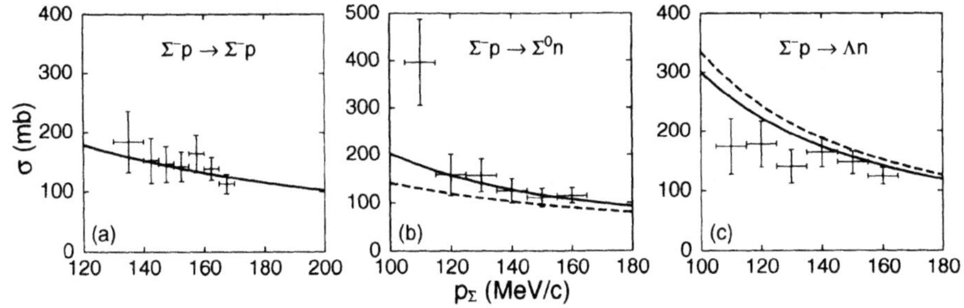

Figure 2. Total cross sections of the Σ^-p scattering by the new version. The dashed curves denote the result without isospin symmetry breaking (The Coulomb effect and baryon mass difference). Final results are denoted by the solid curves.

Table 2. Contribution of each spin state to the $\Sigma^- p$ inelastic capture ratio at rest.

	$r_{S=0}$	$r_{S=1}$	r_R
present	0.962	0.289	0.457 (without Coulomb)
FSS	0.959	0.296	0.462 (without Coulomb)
	0.955	0.296	0.461 (with Coulomb)
exp.			0.33 ± 0.05 [6], 0.474 ± 0.016 [7], 0.465 ± 0.011 [8]

where B denotes the process $\Sigma^- p \to \Sigma^0 n$, C the process $\Sigma^- p \to \Lambda n$ [5]. Contribution to r_R from each spin state is listed in Table 2. We see that $r_{S=0} \sim 1$, while $r_{S=1}$ is significantly smaller than 1 in both versions of our model. This indicates that the ΛN-ΣN coupling mainly occurs in the spin triplet state, which again shows the importance of the 1π tensor force in this channel coupling. Since the Coulomb effect almost disappears for the ratio of the cross sections (See the results by FSS.), present result is consistent with the most recent experimental data.

4 Summary

We have investigated the NN and YN scattering observables by the new version of our quark model. The phase-shift parameters in the NN sector are greatly improved [3]. Quality of the fit to the low-energy YN cross sections is similar to that by the old versions. The essential feature of the ΣN-ΛN coupling is unchanged. The tensor force from the 1π exchange and the $LS^{(-)}$ force from the Fermi-Breit interaction are found to play an important role in this channel coupling.

References

1. Y. Fujiwara, C. Nakamoto, Y. Suzuki: Phys. Rev. Lett. **76**, 2242 (1996)
2. Y. Fujiwara, C. Nakamoto, Y. Suzuki: Phys. Rev. **C54**, 2180 (1996)
3. Y. Fujiwara et al.: Contribution to this conference APFB99
4. R.H. Dalitz, R.C. Herndon, Y.C. Tang: Nucl. Phys. **B47**, 109 (1972)
5. Y. Fujiwara, T. Fujita, C. Nakamoto, Y. Suzuki: Prog. Theor. Phys. **100**, 957 (1998)
6. R.R. Ross: Bull. Ann. Phys. Sic. **3**, 335 (1958)
7. V. Hepp, H. Schleich: Z. Phys. **214**, 71 (1968)
8. D. Stephen: Ph. D. thesis, Univ. of Massachusetts 1970

Few-Body Systems Suppl. 12, 403–408 (2000)

Few-Body Systems

Precise Measurement of *dp* Scattering at 270 MeV

H. Sakai[1,2,*], K. Sekiguchi[1], H. Kato[1], A. Tamii[1], T. Ohnishi[1], K. Yako[1], S. Sakoda[1], Y. Maeda[1], M. Hatano[1], N. Sakamoto[2], T. Uesaka[2], Y. Satou[2], H. Okamura[3], K. Suda[3], T. Wakasa[4]

[1] Department of Physics, University of Tokyo, Hongo 7-3-1, Bunkyo, Tokyo 133-0033, Japan

[2] RIKEN, Wako, Saitama 351-0198, Japan

[3] Department of Physics, Saitama University, 338-8570, Japan

[4] Research Center for Nuclear Physics, Osaka University, Osaka 567-0047, Japan

Abstract. Precise data were obtained for the cross sections as well as the vector and tensor analyzing powers for the deuteron-proton elastic scattering at 270 MeV covering the scattering angles $\theta_{cm} = 10^\circ - 180^\circ$. In addition, the tensor-to-vector polarization transfer coefficients $K_{xx}^{y'} - K_{yy}^{y'}$ have been measured for the first time at $\theta_{cm} = 120^\circ, 130^\circ$ and 150°. These results have been compared to the Faddeev calculations with and without the three nucleon force.

1 Introduction

Three nucleon force (3NF) effects in nucleon-deuteron (Nd) scattering system attracts a lot of interest. This is partly because exact numerical solutions of the Faddeev equations for the *nd* elastic scattering using modern nucleon-nucleon(NN) forces together with 3NF are now available to compare with accurate scattering data.

We have shown in Ref.[1] that the significant discrepancy between the data and the Faddeev calculation without 3NF exists in the region where the cross section becomes minimum in the $\vec{d}+p$ elastic scattering at 270 MeV. Recently it is indeed shown by two groups [2, 3] that this discrepancy can be removed if the 3NF are included in the calculations[2,3]. The Bochum group[2] employs the 2π exchange Tuscon-Melbourne (TM) 3NF[5] and CD-Bonn NN -force while the Hannover group [3] uses the coupled-channel formalism taking into account

* *E-mail address:* sakai@phys.s.u-tokyo.ac.jp

the Δ-isobar degree of freedom explicitly which effectively creates a major part of 3NF.

Recently we have extended the angular range of measurement to cover $\theta_{cm} = 10^\circ - 180^\circ$. In Ref.[1] it is limited only $\theta_{cm} = 57^\circ - 138^\circ$. Since the detailed report on this extension will be given by Kimiko Sekiguchi elsewhere in this proceedings[4], we show only the analyzing power data in spherical coordinate representation in Section 2.

In order to examine the spin dependent part of 3NF, the polarization transfer (PT) measurement for the $\vec{d} + p \rightarrow \vec{p} + d$ elastic scattering at 270 MeV has been performed for the first time. The section 3 deals with this new measurement.

2 Cross Section and Vector- and Tensor-Analyzing Powers for the *dp* Elastic Scattering at $E_d = 270$ MeV

The experiment was performed at the RIKEN Accelerator Research Facility (RARF) using vector and tensor polarized deuteron beams of 270 MeV. The beam polarization was monitored by using the *dp* scattering and it was 60–80% of the ideal value throughout the experiment. The CH_2 target with a thickness of 46.7 mg/cm^2 was bombarded and either the scattered proton or deuteron was momentum analyzed by the magnetic spectrometer SMART. In this system the spectrometer is fixed to the ground so that the beam swinger system is employed to change the incident beam angle with respect to the target for the angular distribution measurement. Here the reaction plane is vertical contrary to an ordinary spectrometer system which is usually horizontal. This feature introduces extra unnecessary complication into the polarization measurement. The polarization axis of deuteron has to be arranged to a proper direction whenever the reaction angle is changed. To meet this requirement the Wien filter is equipped just after the polarized ion source prior to the acceleration to control the beam polarization axis to an arbitrary direction on the target. This is possible due to the single-turn extraction feature of the RARF cyclotron.

2.1 *Results and Discussion*

The measurements for the *dp* elastic scattering at 270 MeV were made for the cross sections (σ) and the complete deuteron analyzing powers (AP) in the angular range of $\theta_{cm} = 10^\circ - 180^\circ$. Since σ and AP's in Cartesian coordinate representation are shown by K. Sekiguchi *et al.*[4] elsewhere in this proceedings, we show here only the AP's in spherical coordinate system (iT_{11}, T_{20}, T_{21} and T_{22}) in Fig. 1.

Three theoretical calculations in terms of Faddeev theory are presented here.

pd The dashed curve[1, 6] is the calculation for the *dp* scattering in which the Argonne v_{14} (AV14) potential is used for the *NN* force. The Coulomb force is taken into account approximately while the 3NF is not included.

Bochum The solid curve[2] is the calculation for the *dn* scattering in which the CD Bonn potential is used for the *NN* force and the TM 3NF is included with no Coulomb force.

Hannover The dot-dashed curve[3] is the calculation for the *dn* scattering in which the Paris A4 potential is used for the *NN* force and the Δ-isobar is included with no Coulomb force.

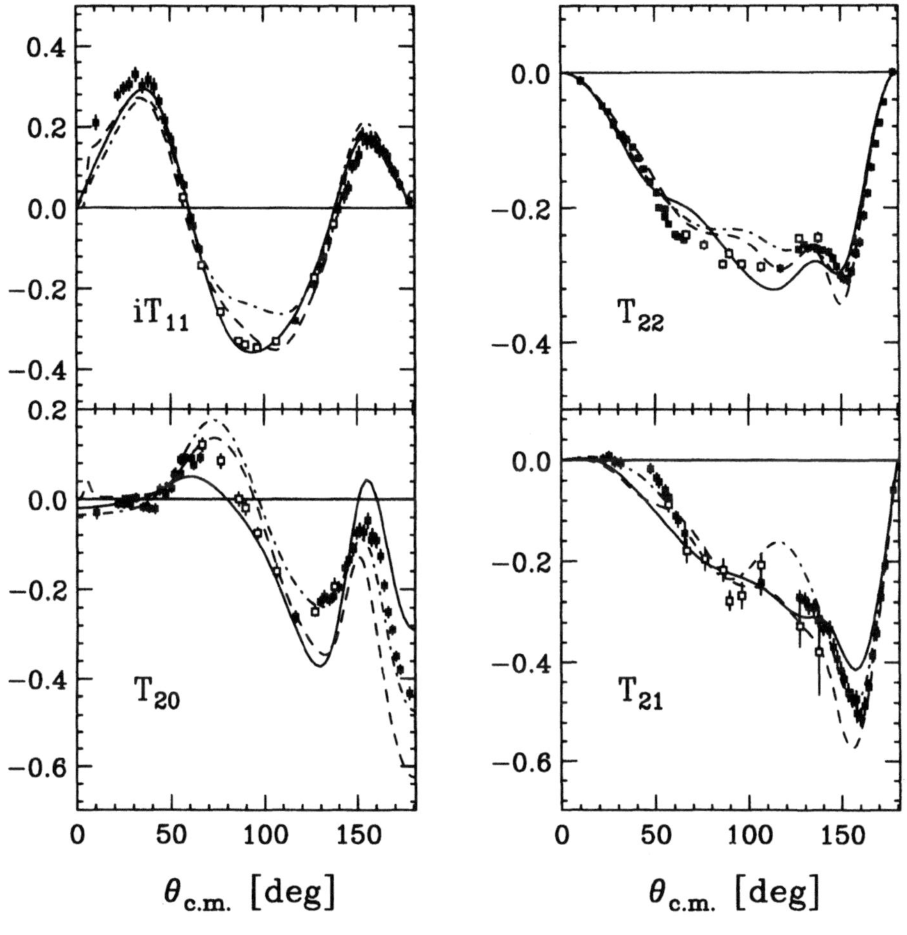

Figure 1. The analyzing powers iT_{11}, T_{20}, T_{21} and T_{22} for the $\vec{d}+p$ elastic scattering at 270 MeV. The dash, solid and dot-dash curves are *dp*, Bochum and Honnover calculations, respectively. See text for detail.

The vector AP iT_{11} is well reproduced by the Bochum group calculation which includes the TM-3NF. However the fit of the tensor AP's T_{ij} in the region at where the 3NF effect appears drastically, namely the cross section minimum region $\theta_{cm} = 60° - 140°$, becomes even worse when the 3NF is introduced. In the backward region $\theta_{cm} = 150° - 180°$ at where the nucleon exchange mechanism is believed to be dominant, the tensor AP's seem to be well reproduced by the Hannover group calculation.

These tensor AP behaviors indicate that the spin dependent part of the 3NF may not be well determined while the magnitude of them seems to be adequate.

3 Polarization Transfer Measurement for the *dp* Elastic Scattering at 270 MeV

In order to examine the spin dependent part of 3NF, the polarization(spin) transfer measurement for the $\vec{d} + p \to \vec{p} + d$ elastic scattering at 270 MeV has been performed.

The polarization of scattered protons was measured by using the focal-plane polarimeter DPOL[7] which is designed and optimized for deuteron polarization measurements with the deuteron energies $\geq$ 230 MeV.

3.1 *Effective Analyzing Power Measurement for DPOL*

In order to calibrate the effective analyzing power of DPOL for proton we need the polarized proton beam whose polarization is known precisely beforehand. Since the polarized proton beam is not available at RIKEN, we used the induced polarization $P^{y'}$ through the $^{12}\mathrm{C}(p, \vec{p})^{12}\mathrm{C}$ elastic scattering at 200 MeV for the calibration purpose. The analyzing powers A_y for the $^{12}\mathrm{C}(\vec{p}, p)^{12}\mathrm{C}$ elastic scattering at 200 MeV has been accurately measured at IUCF[8]. This is the time reversed process so that $A_y = P^{y'}$.

The graphite target with a thickness of 284 mg/cm^2 in the SMART scattering chamber was bombarded by the unpolarized proton beam of 200 MeV. Then the scattered protons to $\theta_{cm} = 16°$ and $\theta_{cm} = 23°$ expected to have the polarization of 0.93 and -0.33, respectively, bombarded the polarization analyzer target of DPOL which consisted of a graphite plate and a Plastic scintillator with thicknesses of 5 cm and 13 mm, respectively. The left-right asymmetry of these double scattered protons was used to extract the effective analyzing power $<A>$ of DPOL by selecting the elastic scattering events. The result is

$$<A> = 0.48 \pm 0.01 \qquad (1)$$

at $T_p = 198$ MeV. The energy dependence of $<A>$ of DPOL has been estimated by the simple Monte Carlo simulation.

3.2 *Results and Discussion*

In Fig. 2, the results of the tensor-to-vector polarization transfer coefficient $K_{xx}^{y'} - K_{yy}^{y'}$ are plotted with solid dots. Experimentally $K_{xx}^{y'}$ and $K_{yy}^{y'}$ were measured separately at angles corresponding to $\theta_{cm} = 120°, 130°$ and $150°$.

The solid and dashed curves are the predictions by Bochum group[2] with and without TM-3NF, respectively.

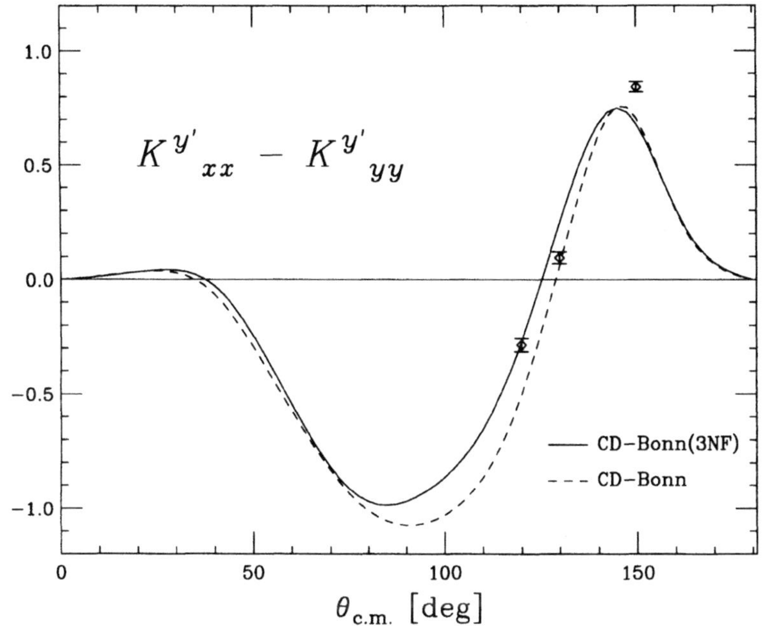

Figure 2. Polarization transfer coefficient $K^{y'}_{xx} - K^{y'}_{yy}$ for the $\vec{d}+p$ elastic scattering at 270 MeV.

At 120° the prediction with 3NF agrees well with the measurement, while at 130° the measured value is much closer to that without 3NF. Furthermore both predictions with and without 3NF are much smaller than the measured value at 150°. Therefore it is very difficult to draw a definitive conclusion from this result.

4 Summary

We have measured the cross sections as well as the vector and tensor analyzing powers for the *dp* elastic scattering at 270 MeV covering the wide scattering angle, $\theta_{cm} = 10° - 180°$. We also measured the tensor-to-vector polarization transfer coefficients $K^{y'}_{xx} - K^{y'}_{yy}$ at $\theta_{cm} = 120°, 130°$ and $150°$ by using the SMART focal-plane polarimeter DPOL.

Although it is not shown here, the discrepancy observed in the cross section minimum region $\theta_{cm} = 80° - 150°$ between data and the calculation without 3NF can be removed if 3NF is taken into account. The vector and tensor analyzing powers are plotted in terms of spherical coordinate representation in Fig.

1. The vector analyzing powers are well reproduced by the Bochum calculation with 3NF. However the fits to the tensor analyzing powers are generally poor and even deteriorated if the 3NF is included.

From the polarization transfer measurement, no definitive conclusion can be drawn, since neither calculation reproduces the experimental values.

Acknowledgement. The authors would like to thank the Bochum group, W. Glöckle, H. Witała and H. Kamada and Hannover group, P.U. Sauer and S. Nemoto for providing them their Faddeev calculations prior to the publication. They are also grateful to K. Itoh for his collaboration in the experiment.

They thank RIKEN Accelerator group for delivering excellent proton and deuteron beams. This work is supported financially in part by the Grant-in-Aid for Scientific Research No. 10304018 of Ministry of Education, Science, Culture and Sports of Japan.

References

1. N. Sakamoto et al.: Phys. Lett. **B**367, 60 (1996)
2. H. Witała et al.: Phys. Rev. Lett. **8**1, 1183 (1998)
3. S. Nemoto et al.: Phys. Rev. **C**58, 2599 (1998)
4. K. Sekiguchi et al.: elsewhere in this proceedings
5. S.A. Coon et al.: Nucl. Phys. **A**317, 242 (1979)
6. Y. Koike, private communication
7. S. Ishida et al.: AIP Conf. Proc., **3**43, 182 (1995)
8. H.O. Meyer et al.: Phys. Rev. **C**23, 616 (1981)

Few-Body Systems Suppl. 12, 409–414 (2000)

Few-Body Systems

Nucleon-Deuteron Scattering with Δ-Isobar Excitation

P.U. Sauer[1], K. Chmielewski[1], S. Nemoto[2], S. Oryu[2]

[1] Institut für Theoretische Physik, Universität Hannover, Appelstraße 2, D-30167 Hannover, Germany

[2] Department of Physics, Faculty of Science and Technology, Science University of Tokyo, Noda, Chiba 278, Japan

Abstract. A coupled-channel formulation with Δ-isobar excitation is used for the description of nucleon-deuteron scattering. The three-particle scattering equations are solved by a separable expansion of the two-baryon transition matrix. The effect of Δ-isobar excitation on spin-averaged and spin-dependent cross sections is studied.

1 Introduction

Virtual Δ-isobar excitation yields an effective three-nucleon force in the nuclear medium. A Δ-isobar gets created when a nucleon (N), through two-nucleon scattering in the medium, is internally excited. Theoretically, the process is described by a coupled-channel two-baryon interaction. Such a coupled-channel interaction is applied in ref. [1] to the three-nucleon bound state. References [2, 3, 4] extend its application to nucleon-deuteron scattering. That extension is the subject of this talk.

The calculational scheme is described in ref. [2]. The Hilbert space of the calculation has a purely nucleonic sector and a sector in which a nucleon is turned into a Δ-isobar. The Δ-isobar is considered a stable baryon with mass 1232 MeV and with spin and isospin $\frac{3}{2}$; that treatment of the Δ-isobar is appropriate below the pion-production threshold; the considered scattering energies stay well below. The coupled-channel two-baryon potential has an almost phase-equivalent purely nucleonic reference potential; comparison of results with both potentials allows to isolate the Δ-isobar effect on observables. The three-particle AGS equations [5] with channel coupling are solved in momentum space without Coulomb interaction; predictions therefore refer to neutron-deuteron scattering, though they will also be compared to the easier accessible proton-deuteron data. The AGS equations are solved by a separable expansion of the two-baryon transition matrix. That separable expansion is judged in ref. [2] to be valid for

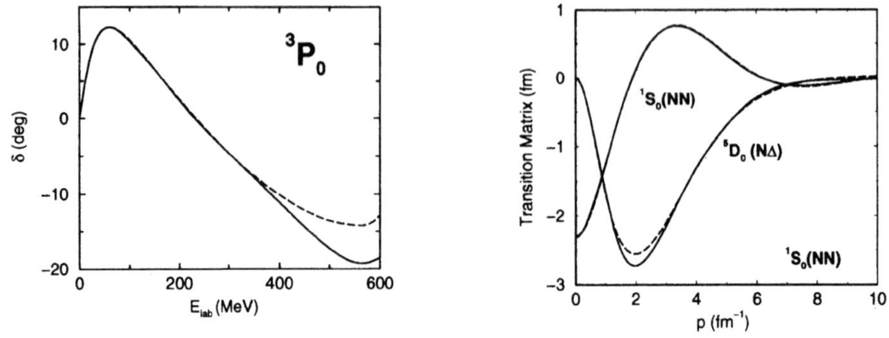

Figure 1. Two-nucleon phase shift $^3P_0$ as function of the nucleon lab energy $E_{\rm lab}$ (left) and off-shell partial-wave-projected coupled-channel transition matrix $^1S_0(NN)$-$^5D_0(N\Delta)$ (right). On the right, matrix elements coupling $^1S_0$ with $^1S_0$ and with $^5D_0$ for the characteristic momentum $p_0 = 0.25$ fm^{-1} and for the characteristic available energy $-p_0^2/m_N$, m_N being the nucleon mass, are compared as function of the second momentum p. The comparisons [2] are carried out for the separably expanded coupled-channel interaction (solid curves) and for its unexpanded form (dashed curves)

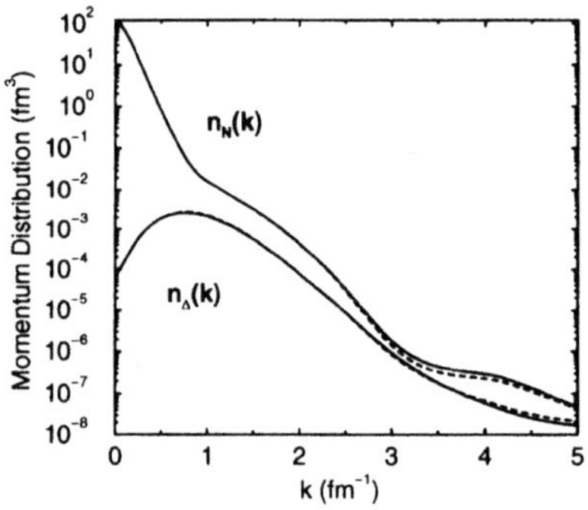

Figure 2. ^{3}H nucleon and Δ-isobar momentum distributions $n_N(k)$ and $n_\Delta(k)$ as function of the magnitude of the single-particle momentum k. Results [2] derived from the separably expanded coupled-channel interaction (solid curves) are compared with results of its unexpanded form (dashed curves)

the considered scattering energies. Figures 1 and 2 give examples. The examples use the coupled-channel two-baryon potential based on the Paris potential [6] nucleonic reference potential. Figure 1 shows that the separably expanded form of the coupled-channel potential has almost the same two-nucleon phase shifts as the unexpanded one in the considered energy domain; the same holds true

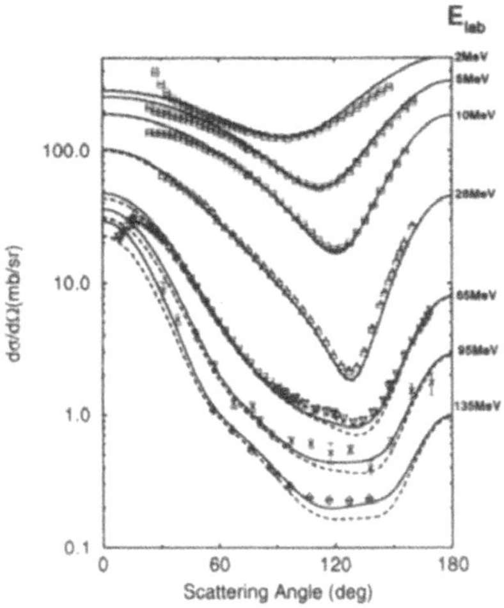

Figure 3. Differential cross sections of elastic nucleon-deuteron scattering as function of the c.m. scattering angle for several nucleon lab energies. The resulting cross sections derived from the coupled-channel potential with Δ-isobar excitation (solid curves) are compared with results [4] based on a purely nucleonic reference potential (dashed curve). For the shown results, the reference potential is the Paris potential [6]

for the off-shell coupled-channel transition matrix at characteristic available energies. Figure 2 shows that even detailed trinucleon bound-state properties of widely varying magnitude are described by the separably expanded form of the coupled-channel interaction with high precision.

Our study of Δ-isobar effects in three-nucleon scattering has by now yielded three results:

1. *Sagara Discrepancy.*
   The data on the spin-averaged differential cross section of elastic proton-deuteron scattering were poorly described in the past: Calculations, based on traditional two-nucleon potentials, fail in the region of the diffraction minimum. This failure is often called Sagara discrepancy. In contrast to purely nucleonic potentials, the Δ-isobar makes a substantial and beneficial contribution to the differential cross section [4] as Fig. 3 proves. Reference [7] sees the same beneficial effect arising from a standard irreducible three-nucleon force.

2. *Spin Observables at Lower Energies.*
   The so-called A_y-puzzle, i.e., the failure of the theoretical description of the neutron analyzing power A_y at 10 MeV neutron lab energy and related problems, was the underlying motivation for our calculation with

the Δ-isobar. In the coupled-channel interaction $N\Delta$ partial waves couple strongly to the two-nucleon P-waves which determine the shape of the analyzing power in a complicated superposition mechanism. We are disappointed to find that the Δ-isobar has almost no effect on spin observables at low energies [3], except for those strongly correlated with the trinucleon binding energy.

3. *Spin Observables at Higher Energies.*
   The Δ-isobar can have sizeable effects on spin observables at higher energies. Figure 4 gives an example. The effects are not generally successful [8] for the known data.

As observed already for the trinucleon binding energy, there is competition between different Δ-isobar effects, i.e., between the two-nucleon dispersive and the three-nucleon force effects. That balancing of contributions is demonstrated in Fig. 5 for the neutron analyzing power A_y. Furthermore, the Δ-isobar is introduced [1, 2] in a very particular form; the Δ-isobar effects are therefore almost independent from the underlying nucleonic reference potential. That result is illustrated by Fig. 6. All shown numerical calculations include the interaction up to the total two-baryon angular momentum $I = 4$, whenever necessary; the coupling of the Δ-isobar is only considered in partial waves up to $I = 2$. Convergence characteristics are being investigated.

The importance of the Δ-isobar for break-up in nucleon-deuteron scattering [12] and for electromagnetic three-nucleon processes are being studied.

References

1. C. Hajduk, P.U. Sauer, W. Strueve: Nucl. Phys. **A405**, 581 (1983)
2. S. Nemoto et al.: Few-Body Systems **24**, 213 (1998)
3. S. Nemoto et al.: Few-Body Systems **24**, 241 (1998)
4. S. Nemoto, K. Chmielewski, S. Oryu, P.U. Sauer: Phys. Rev. **C58**, 2599 (1998)
5. E.O. Alt, P. Grassberger, W. Sandhas: Nucl. Phys. **B2**, 167 (1967)
6. M. Lacombe et al.: Phys. Rev. **C21**, 861 (1980)
7. H. Witała et al.: Phys. Rev. Lett. **81**, 1183 (1998)
8. S. Nemoto, K. Chmielewski, S. Oryu, P.U. Sauer: talk at this conference
9. R. Machleidt: Adv. Nucl. Phys. **19**, 189 (1998)
10. N. Sakamoto, H. Okamura, T. Uesaka, S. Ishida, H. Otsu, T. Wakasa, Y. Satou, T. Niizeki, K. Katoh, T. Yamashita, K. Hatanaka, Y. Koike, H. Sakai: Phys. Lett. **B367**, 60 (1996)
11. H. Sakai: talk at this conference; K. Sekiguchi: talk at this conference
12. K. Chmielewski, A.C. Fonseca, S. Nemoto, P.U. Sauer: talk at this conference

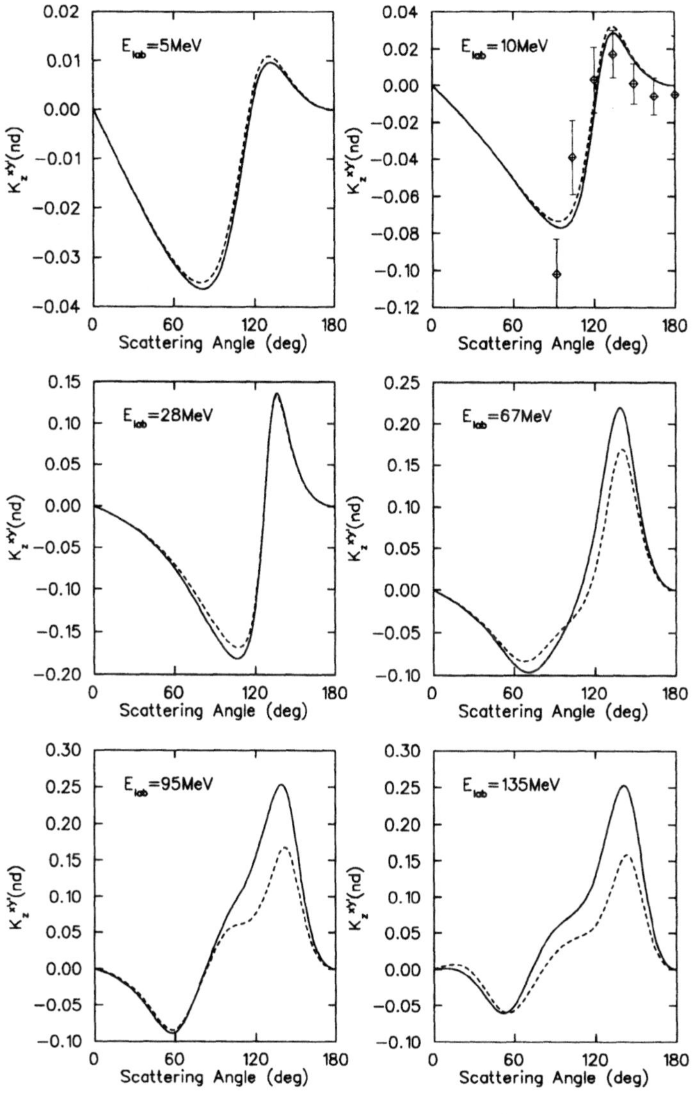

Figure 4. Spin observable $K_z^{xy}(nd)$ of elastic nucleon-deuteron scattering as function of the c.m. scattering angle for several nucleon lab energies. The results derived from the coupled-channel interaction with Δ-isobar excitation (solid curves) are compared with results based on a purely nucleonic reference potential (dashed curves). For the shown results, the reference potential is the Paris potential [6]

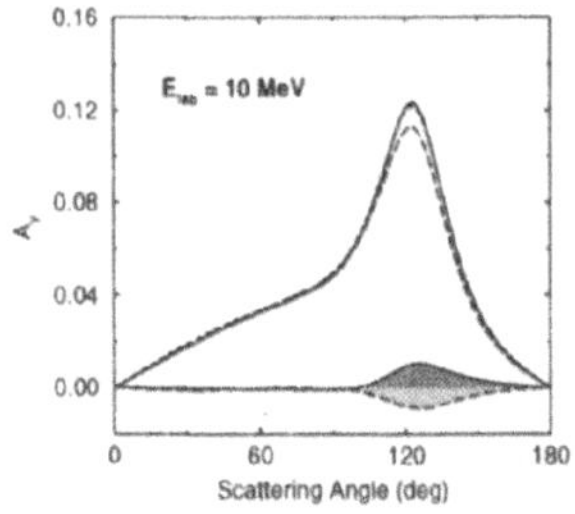

Figure 5. Neutron analyzing power A_y of elastic nucleon-deuteron scattering as a function of the scattering angle at 10 MeV nucleon lab energy. The contribution from the dispersive effect is indicated by the lightly shaded area, its effect on A_y by the long-dashed curve. The contribution from the effective three-nucleon force is indicated by the heavily shaded area. The total result for A_y arising from the coupled-channel potential with Δ-isobar (solid curve) and from a purely nucleonic reference potential (dashed curve) are almost indistinguishable. The Paris potential [6] is the reference potential

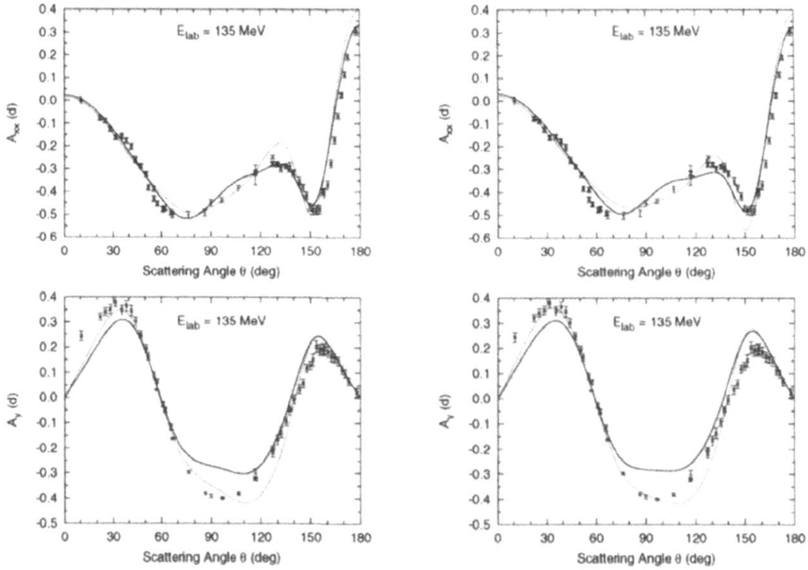

Figure 6. Spin observables $A_{xx}(d)$ and $A_y(d)$ of elastic nucleon-deuteron scattering as function of the c.m. scattering angle for the nucleon lab energy of 135 MeV. The results derived from the coupled-channel potential with Δ-isobar excitation (solid curves) are compared with results based on a purely nucleonic reference potential (dashed curves). On the left side, the shown results belong to the Paris potential [6] as reference potential. On the right side, the shown results belong to the Bonn B potential [9] as reference potential. The theoretical results with the Bonn B potential are still preliminary. The experimental data are from refs. [10, 11]; the data of ref. [11], denoted by $\times$, are also still preliminary

Few-Body Systems Suppl. 12, 415–420 (2000)

Few-Body Systems

Some Highlights of NN and $3N$ Studies at TUNL

W. Tornow

Duke University and Triangle Universities Nuclear Laboratory, Durham, NC 27708-0308, USA

Abstract. We report on a recent determination of the $^3S_1$ - $^3D_1$ mixing parameter ϵ_1 in neutron-proton scattering between $E_n = 5$ and 20 MeV. The results follow the trend of previous determinations at 25 and 50 MeV, thus indicating a stronger NN tensor force than predicted by NN potential models and the Nijmegen NN partial-wave analysis. In the field of $3N$ physics we summarize the ongoing search for an explanation of the $3N$ analyzing power puzzle.

Let me start off with some remarks about the nucleon-nucleon (NN) tensor force. This topic is quite appropriate for a Few-Body Conference. Few-nucleon systems are bound to a large extent by the NN tensor force, which is the result of a delicate balance between an attractive and a repulsive component.

What do we know experimentally about the tensor force? In NN phase-shift analyses the $^3S_1$ - $^3D_1$ mixing parameter ϵ_1 is a measure of the strength of the NN tensor force. Shown in Fig. 1 (left side) by the band is ϵ_1 and its uncertainty given by the Nijmegen [1] multi-energy partial-wave analysis (PWA 93) between 0 and 200 MeV neutron energy. All the recent NN potential models like the new Nijmegen potentials, AV18, and CD Bonn are fitted to this band. Obviously, the uncertainty of ϵ_1 is very small, especially at energies below 40 MeV. However, the two single-energy analyses performed by the Nijmegen group at $E_n = 25$ and 50 MeV (see dots with error bars) are somewhat inconsistent with the global analysis. Although the Nijmegen group has repeatedly pointed out that single-energy phase-shift analyses are not meaningful, this statement is hard to swallow from an experimentalist's point of view. The associated uneasiness was the prime motivation for TUNL's Polarized Target Group to apply a novel scheme for determining ϵ_1 at low energies. From a theoretical point of view the TUNL approach [2] is practically free of uncertainties related to other NN phase-shift parameters. The TUNL results are given in Fig. 1 (right side) by solid dots in comparison to other experimental determinations of ϵ_1, and phase-shift analyses (solid [1] and dot-dashed curves [3]) and the Bonn B NN potential-model prediction (dashed curve). Clearly, the new data give larger values for ϵ_1 than predicted by these curves. However, the trend of the TUNL data supports the results at $E_n = 25$ [4] and 50 MeV [5]. The TUNL

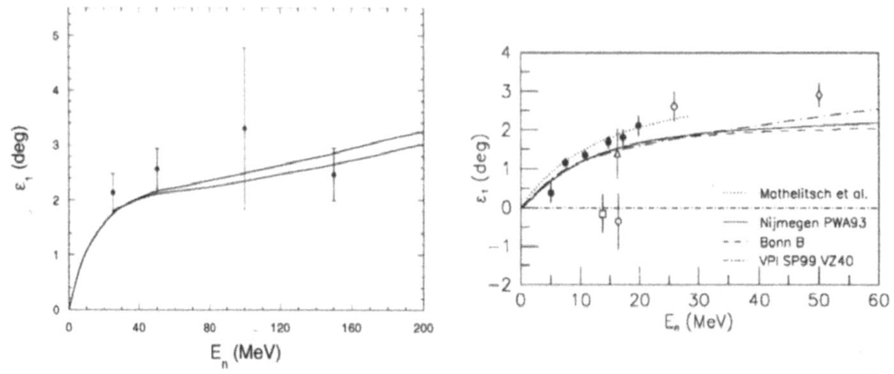

Figure 1. The $^3S_1$-$^3D_1$ mixing parameter ϵ_1. The band in the left panel presents the multi-energy Nijmegen PWA 93 prediction and the data are the results of single-energy analyses of the Nijmegen group. The right panel shows values for ϵ_1 from the present (solid dots) and previous experiments in comparison to theoretical predictions.

data agree well with predictions obtained from the effective-range parameterization of Mathelitsch and VerWest [6] (dotted curve). This parameterization involves one-pion exchange (OPE), but contrary to the more recent NN potential models and phase-shift analyses, here the OPE tensor force is not reduced by incorporating a πNN form factor and the ρ-meson exchange contribution. As has been pointed out by Machleidt [7], our data and the deuteron properties can be described well by an OPE NN potential with a CD Bonn type πNN form factor, but without ρ- and heavier meson-exchange contributions. Finally, our data are inconsistent with the two anomalously low values for ϵ_1 near $E_n =$ 15 MeV [8, 9]. Summarizing TUNL's NN tensor force results we conclude that our studies support a NN tensor interaction that is stronger than predicted by all modern meson-exchange based NN potential models and the Nijmegen PWA 93 [1] and VPI SP 99 [3] phase-shift analyses. Interestingly, the recent effective field theory approach of Kaplan *et al.* [10] predicts even larger values for ϵ_1 than found in the present work. Clearly, additional studies of the tensor interaction are required to obtain further insight into this important NN force component.

Now let me move on to the three-nucleon ($3N$) scattering system. Here, I will focus on the analyzing power puzzle. This puzzle has a long history. It dates back to 1986 when the Bochum-Cracow group presented one of their first rigorous calculations for *n-d* elastic scattering [11]. It should be noted that in 1986 Koike and Haidenbauer [12] also reported a sizeable discrepancy between their *n-d* calculations and experimental data. However, in this case it was not clear how much of the discrepancy was due to a real physical effect and how much was of calculational nature. Contrary to the Bochum-Cracow group, Koike and Haidenbauer used a separable expansion method, and therefore a number of approximations. Shown in Fig. 2 (left side) are data for the

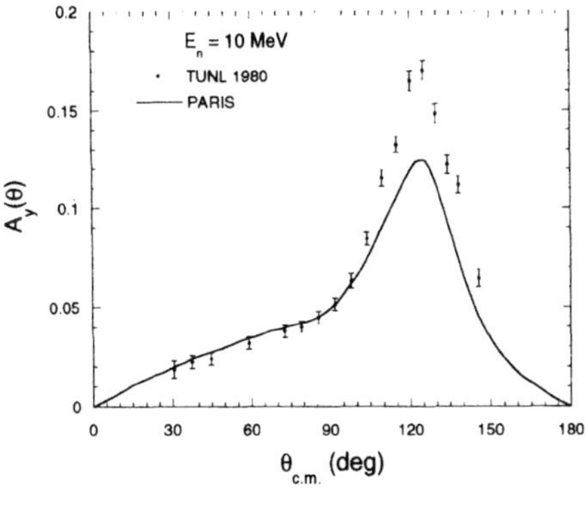

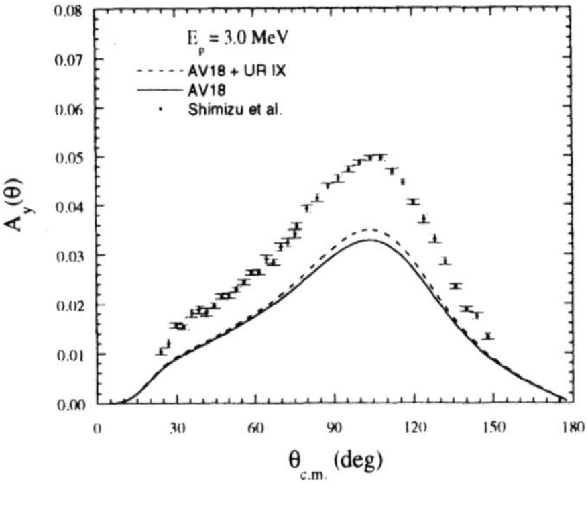

Figure 2. Left panel: Neutron-deuteron $A_y(\theta)$ data at $E_n = 10$ MeV in comparison to a rigorous $3N$ calculation using the Paris NN potential. Right panel: Comparison of theoretical predictions and data for the *p-d* analyzing power at $E_p = 3.0$ MeV. The rigorous theoretical calculations employ the AV18 NN potential.

analyzing power $A_y(\theta)$ at $E_n = 10$ MeV [13] and a rigorous calculation of the Bochum-Cracow group using the Paris NN potential. Clearly, the calculations underestimate the data considerably. Is this discrepancy due to a yet unknown three-nucleon force ($3N$F) or is it simply the manifestation of a problem on the two-nucleon level, i.e., a problem related to the $^3\mathrm{P}_j$ NN interactions? Although additional experimental evidence was published in the following years, it was not until about 10 years later that the analyzing power puzzle has developed into one of the most referred to phenomena in low-energy few-nucleon physics. In 1995 the Pisa group published their first results for *p-d* elastic scattering below the deuteron breakup theshold with exact inclusion of the Coulomb interaction [14]. The solid curve in Fig. 2 (right side) shows their calculation for $A_y(\theta)$ at $E_p = 3.0$ MeV in comparison to the data of Shimizu *et al.* [15]. This breakthrough clearly revealed the magnitude of the analyzing power puzzle (a 20 standard deviation effect), which now includes in addition to the nucleon analyzing power $A_y(\theta)$ also the deuteron vector analyzing power $iT_{11}(\theta)$ in ***d**-p* scattering. Interestingly, the differential cross section and the tensor analyzing power data are reasonably well described by the calculations of the Pisa group.

Let me reformulate the question posed above: Is the physics that is responsible for the differences shown in Fig. 2 the same as the physics observed for the differential cross section and the analyzing power at higher energies, for example at $E_n = 135$ MeV? In Fig. 3 the solid curves are calculations [16] of the Bochum-Cracow group at $E_n = 135$ MeV using the CD Bonn NN potential. Although not shown in Fig. 3, the *d-p* data of Sakamoto *et al.* [17] at $E_d = 270$ MeV are in serious disagreement with the calculations (solid curves). However, both $\sigma(\theta)$ and $A_y(\theta)$ are described very well when the Tucson-Melbourne $3N$F is added to the CD Bonn NN potential (dashed curves). This is in contrast to Fig. 2, where the addition of the Urbana IX or other $3N$Fs like Tucson-

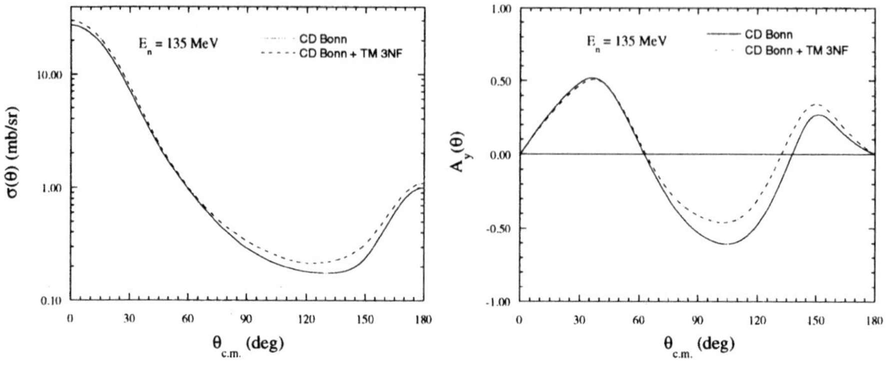

Figure 3. Neutron-deuteron differential cross-section (left side) and analyzing power (right side) calculations of Witała at $E_n = 135$ MeV using CD Bonn (solid curve) and CD Bonn + TM 3*N*F (dashed curve).

Melbourne or Brazil, improves the discrepancy only slightly.

Our question can be phrased in even broader terms. Are the differences observed in Fig. 2 between data and calculations using *NN* interactions only (i.e., ignoring 3*N*Fs) related to the about 30% underprediction of the calculated spin-orbit splitting of nuclear levels found by the Argonne-Los Alamos group [18] in their state-of-the-art bound-state calculations for light nuclei? Clearly, 3*N*F effects must be important when the three nucleons are close together, like in bound states or in high-energy elastic scattering and reactions. However, nucleon-deuteron scattering below $E_N = 10$ MeV does not qualify for this situation. I think there are also other reasons why the low-energy analyzing power puzzle has not much to do with the physics that is responsible for the results given in Fig. 3.

First, one reason is based on the following statement: whenever there is a 3*N*F effect on $\sigma(\theta)$, there ought to be an effect on $A_y(\theta)$, unless the 3*N*F is of purely spin-orbit type. After all, the two observables are related to each other. It is a well established observation in nucleon-nucleus elastic scattering that a sign change in the derivative of the differential cross section coincides with a zero-crossing of $A_y(\theta)$. Therefore, we have the approximate relations: $d\sigma(\theta)/d\theta \sim A_y(\theta)$ or more accurately $d[\log \sigma(\theta)]/d\theta \sim A_y(\theta)$. Although the deuteron is not a standard nucleus, the different zero-crossing angles of $A_y(\theta)$ shown in Fig. 3 (right side) for the solid and dashed curves near $\theta_{c.m.} = 135°$ and their comparison to the $\sigma(\theta)$ calculations given in Fig. 3 (left side) clearly demonstrate this feature. At $E_p = 3.0$ MeV there is only a 5% difference between the calculated and measured $\sigma(\theta)$ near $\Theta_{c.m.} = 90°$ and the associated 3*N*F effect is therefore not related to the large difference observed between data and calculations (solid curve) for $A_y(\theta)$ in this angular range. In fact, the dashed curve in Fig. 2 (right side), which in addition to AV18 includes the Urbana IX 3*N*F, is quite consistent with the 3*N*F influence on $\sigma(\theta)$ at this

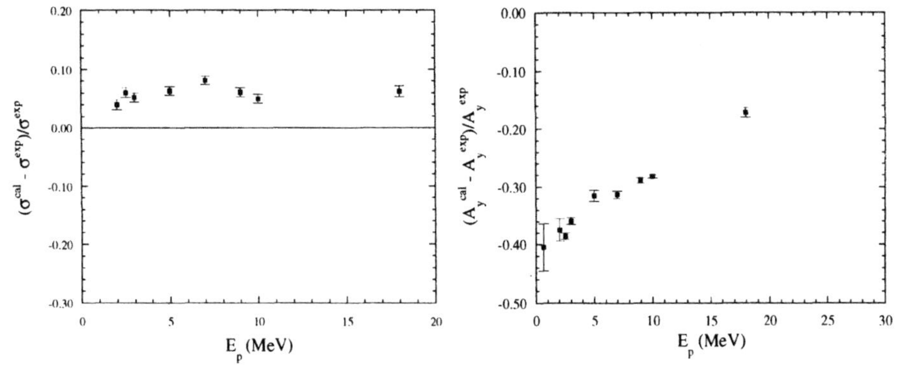

Figure 4. Energy dependence of $\sigma(\theta)$ (left side) and $A_y(\theta)$ (right side) discrepancy in the angular range of the *p-d* cross-section minimum. The calculations are from the Pisa group.

energy.

Second, the about 5% difference between the *p-d* $\sigma(\theta)$ data and the calculations of the Pisa group is independent of energy in the 2 to 18 MeV energy range (see Fig. 4, left side), while the $A_y(\theta)$ difference decreases from 40% at $E_p = 650$ keV to about 18% at 18 MeV (see Fig. 4, right side). Both this lack of tracking between the $\sigma(\theta)$ and $A_y(\theta)$ discrepancies and the counter-intuitive energy dependence of the $A_y(\theta)$ discrepancy clearly supports the notion that the analyzing power puzzle must have a complicated physical origin, either a new spin-orbit type $3N$F or a more basic phenomenon on the two-nucleon level. Either way, rigorous *p-d* calculations in the 20 to 50 MeV energy range are needed to confirm the conjecture concluded from Fig. 4 (right side) that the analyzing power puzzle may disappear near 30 or 40 MeV. This feature has been suggested earlier [19] from the comparison of rigorous *n-d* calculations with *p-d* $A_y(\theta)$ data in this energy range. In addition, the 60 MeV *n-d* $A_y(\theta)$ [20] data and the associated rigorous calculations [19] are in close agreement. Accurate *n-d* $A_y(\theta)$ data in the 20 to 50 MeV energy range are of considerable importance, since the current variational approach of the Pisa group experiences convergence problems for $E_p > 20$ MeV. Rigorous calculations for *n-d* scattering are available for the entire energy range of interest.

The recent chiral pertubation theory guided $3N$F studies [21] revealed already that it is not easy to reproduce the energy dependence of the analyzing power puzzle. The calculations of Kievsky [22], who introduced an ad hoc spin-orbit $3N$F in the S = 1, T = 1 two-nucleon channel should be extended beyond 10 MeV in order to gain more insight into this interesting approach. The work of Kievsky constitutes the only example of a successful description of the low-energy analyzing power puzzle in terms of $3N$Fs without negatively affecting other $3N$ observables.

The analyzing power puzzle together with the problem of insufficient spin-orbit splitting of nuclear levels calculated by the Argonne-Los Alamos group for light nuclei clearly shows that few-nucleon physics continues to be an exciting and important laboratory for testing and refining our understanding of the nuclear force.

Acknowledgement. This work was supported in part by the U.S. Department of Energy, Office of High Energy and Nuclear Physics, under grant No. DE-FG02-97ER41033. We gratefully acknowledge discussions with A. Kievsky and H. Witała.

References

1. V.G.J. Stoks et al.: Phys. Rev. **C48**, 792 (1993)
2. B.M. Raichle et al.: Phys. Rev. Lett. **83**, 2711 (1999)
3. R.A. Arndt: interactive computer code SAID, 1999
4. M. Ockenfels et al.: Nucl. Phys. **A526**, 109 (1991)
5. R. Henneck: Phys. Rev. **C47**, 1859 (1993)
6. L. Mathelitsch, B.J. VerWest: Phys. Rev. **C29**, 739 (1984)
7. R. Machleidt: Private Communications
8. M. Schöberl et al.: Nucl. Phys. **A489**, 284 (1998)
9. P. Clotten et al.: Phys. Rev. **C58**, 1325 (1998)
10. D.B. Kaplan, M.J. Savage, M.B. Wise: Nucl. Phys. **B534**, 329 (1998)
11. H. Witała, W. Glöckle, T. Cornelius: Few-Body Syst. Suppl. **2**, 555 (1987)
12. Y. Koike, J. Haidenbauer: Nucl. Phys. **A463**, 365c (1987)
13. W. Tornow et al.: Phys. Rev. Lett. **49**, 312 (1982)
14. A. Kievsky, M. Viviani, S. Rosati: Phys. Rev. **C52**, R15 (1995)
15. S. Shimizu et al.: Phys. Rev. **C52**, 1193 (1995)
16. H. Witała: Private Communication
17. N. Sakamoto et al.: Phys. Lett. **B367**, 60 (1996)
18. R.B. Wiringa: Private Communication
19. W. Glöckle et al.: Phys. Rep. **274**, 107 (1996)
20. H. Rühl et al.: Nucl. Phys. **A524**, 377 (1991)
21. D. Hüber, J.L. Friar: Private Communication
22. A. Kievsky: Phys. Rev. **C60**, 034001 (1999)

Few-Body Systems Suppl. 12, 421–426 (2000)

Expectations in Systematic 3N Experiments

K. Sagara*

Department of Physics, Kyushu University, Hakozaki, Higashi-ku, Fukuoka 812-8581, Japan

Abstract. Experimental efforts to find out the $\pi - \pi$ exchange 3N force effects and non-3N effects in Nd scattering, Nd breakup reactions, and Nd capture reactions are briefly described and foreseen. Systematic precise experiments are essential to discover and to confirm the effects.

1 Introduction

Basic purpose in the study of 3N systems is to clarify the difference between the NN force in 3N systems and the NN force in 2N systems, and to determine the strength and character of the 3N force. Many accurate 3N experiments and many exact 3N calculations have been made for the purpose. Now we are aware of the existence of several kinds of discrepancy between the experiments and the calculations.

A long-term problem is the shallow binding of 3N systems calculated using realistic NN potentials. Addition of a $\pi - \pi$ exchange 3N force with an appropriate parameter is one way to reproduce the 3N binding energy [1]. This fact suggests the existence of the $\pi - \pi$ exchange 3N force, though existence of other type forces can not be excluded.

Recently large discrepancy of about 20-30% between the experiment and the calculation at the *pd* differential cross section minimum has been found first below 18 MeV [2] and also at higher energies [3]. The discrepancy is considered to be solved by the same 3N force that can reproduce the 3N binding energy [4]. New experiments on *pd* and *nd* scattering at the higher energies are being made and systematic trend will be precisely investigated.

At the energy around 10 MeV, large discrepancy between the experiment and calculation for the differential cross section of the *pd* and *nd* breakup reactions at especially the space star configuration has been found [5, 6, 7]. The $\pi - \pi$ exchange 3N force has small effect at the low energy and can not explain the discrepancy. Other scalar type force is expected to exist.

* *E-mail address:* sagara@kutl.kyushu-u.ac.jp

The well-known A_y puzzle is the 20-30% difference between the experiment and the calculation for the vector analyzing power A_y and iT_{11} of the N-d scattering below about $E_N = 30$ MeV. The $\pi - \pi$ exchange 3N force has little effect on the A_y puzzle that is a vector-type low-energy phenomena. Theoretical attempts have been made to solve the A_y puzzle by modifying NN P-state interactions [8, 9], however, the A_y puzzle has been an open problem for about 15 years since it was first pointed out.

One more problem is in A_{xx} and A_{yy} of *pd* breakup reactions. So far a large discrepancy in A_{xx} and A_{yy} has been found at one angle pair at $E_d = 52$ MeV [10]. The discrepancy has possibility to call for some improvement in NN D-state interactions. Confirmation of the discrepancy by a new measurement is very interesting.

Al the S-, P- and D- state NN interactions have possibility to be found out to bear non-negligible modifications in 3N system. To investigate the amplitudes of the modifications, we need accurate and systematic experiments together with challenging theoretical analyses. Also we need one more experimental fact to prove the existence of the $\pi - \pi$ exchange 3N force.

Some of the planned experiments are described in the following sections.

2 Nd Scattering

2.1 Differential Cross Section of Nd Scattering

Large discrepancy of about 20-30% between the experiment and the calculation at the *pd* differential cross section minimum at E_p = 65-200 MeV has been nicely explained by introducing the $\pi - \pi$ exchange 3N force in the nd calculations [4]. The existence of the 3N force is strongly suggested.

At low energy around 10 MeV where the discrepancy in the cross section minimum was first found, the nd calculation predicts no sizable contribution of the $\pi - \pi$ exchange 3N force to the cross section [4], and the nd calculation curiously reproduces the pd experimental cross section.

On the other hand, pd calculations with Coulomb force correctly treated have been made above the deuteron breakup threshold. The calculated pd cross section minima are about 5% greater than the experimental ones around 10 MeV [11, 12]. This possibly suggests that there remains a new type of discrepancy in the cross section minimum which can not be explained by the $\pi - \pi$ exchange 3N force.

The existence of the remaining discrepancy is not astonishing, because we know that the A_y puzzle can not be explained by the $\pi - \pi$ exchange 3N force and that the A_y puzzle suggests the necessity for the modification of P-wave NN interactions. The remaining discrepancy, if confirmed, possibly suggests that S-wave NN interactions is modified in 3N system.

To confirm the existence of the remaining discrepancy, pd calculations with various NN potentials have to be made to make systematic comparison with the existing pd data. Precise experiments on the nd scattering cross section are also necessary to compare with the existing nd calculations.

The existing data for the nd cross section have uncertainties of 3-5 %, which are a little large to investigate the remaining discrepancy of about 5 %. Precise nd data of about 1% uncertainty are necessary to investigate the remaining discrepancy.

For precise measurements of the nd cross section, we need to precisely know the n-beam intensity as well as the n-detection efficiency. Measurement of $D(d,{}^3He)n$ cross section at 0° is planned at Kyushu University tandem laboratory (KUTL) to precisely know the intensity of the n-beam at 5 - 20 MeV. If the n-beam of precisely known intensity is available, the efficiency of the n-detector can be precisely determined.

2.2 Vector Analyzing Powers of Nd Scattering

Systematic and precise experimental data for A_y of pd scattering are available in the energy range of E_p below 18 MeV [2, 13]. Precise experimental data for nd A_y are available in the energy range of E_n below 16 MeV. The experimental data to investigate the A_y puzzle is sufficiently prepared. If a new idea to explain the A_y puzzle is proposed, we can examine in detail the energy and angular dependence of the explanation.

The investigation of the origin of the A_y puzzle is very interesting. First, the A_y puzzle can not be explained by the $\pi - \pi$ exchange 3N force. The fact can be understood from the scalar property of the $\pi - \pi$ exchange 3N force.

The new force necessary to solve the A_y puzzle should have the vector property, that is, the force should be a P-wave interaction. So far, modification of P-state NN interaction [9] and NN LS force [8] have been proposed. Recently phenomenological P-wave 3N force are introduced to explain the A_y puzzle. The discrepancy of the A_y puzzle can be reduced in any of these ways, however, the theoretical origin of the discrepancy is still unclear.

It is very interesting whether the A_y puzzle is solved by the introduction of a new type P-wave 3N force or by the modification of the P-state NN interactions. In other word, whether the A_y puzzle is the 2-body problem or the 3-body problem is a fundamental problem in the study of 3N systems. The answer determines the direction of the future study of the 3N systems.

It is also the basic problem whether the A_y puzzle in P-state interaction is a exclusively characteristic problem in P-state NN interactions alone, or a universal problem and the similar problems appear in also S-, and D-state NN interactions. It seems natural to expect that the similar problem exists in all the interactions.

Therefore, to see whether the similar problems is found also in S-state and D-state NN interactions or not is a very important subject. We already pointed out in the preceding subsection that the Nd cross section minimum around 10 MeV seems to have some discrepancy that can not be explained by the $\pi - \pi$ exchange 3N force. The fact suggests that the S-state NN interaction has some problem.

On the other hand, if P-state NN interactions are modified in 3N system, the effects may appear not only in A_y but also in other observables sensitive to

P-state NN interactions. We must be careful to catch the signals of the effects.

2.3 Tensor Analyzing Powers of pd Scattering

If D-state NN interactions are modified in 3N systems, the tensor analyzing powers of Nd scattering may bear the influence. The tensor analyzing powers of pd scattering have been measured in a wide energy range. On the other hand, no available data exist for the tensor analyzing powers of nd scattering because of the difficulty both in the measurement of the polarization of the d-recoil and in the preparation of the tensor-polarized d-target.

When the pd experiment is compared with the pd calculation in which Coulomb force is approximately treated in Doreshall's method [14], the ratio (calculation-experiment) / experiment for T_{22} minimum varies from +7 % to -7 % in the energy range from $E_d = 5$ MeV to 18 MeV. Although the Coulomb effect is expected to be small in T_{22} which is proportional to $A_{xx} - A_{yy}$, quantitative study can not be made before the pd calculations with correct Coulomb force are extensively made above the deuteron breakup threshold.

When such pd calculations are made and compared with the existing systematic T_{20}, T_{21} and T_{22} data, we can get some insight on the modification of D-state NN interactions in 3N systems.

3 Nd Breakup Reaction

3.1 Cross Section of Nd Breakup Reaction

Kinematically complete measurement of the differential cross section of the pd breakup reaction have been made in special kinematical conditions, such as the star, the space star, the collinear, the quasi-free scattering, the final state interaction conditions, because the cross section is small and the phase space is very large.

Remarkable discrepancy has been reported at the space star configuration. The nd breakup cross section at the space star configuration at $E_n = 13$ MeV [6, 7] is about 20 % greater than the nd calculation, and the pd breakup cross section at $E_p = 13$ MeV [5] is about 15 % smaller than the nd calculation. The $\pi - \pi$ exchange 3N force has no sizable effects on the cross section. One of the possible origin of the discrepancy may be modification of S-state NN interaction in 3N systems. It is interesting to see whether the discrepancies in the breakup cross section and in the Nd scattering cross section both around 10 MeV have the same origin.

Systematic measurement of the pd breakup cross section around the space star condition may be very useful to investigate the origin. Measurement of D(p,pp)n cross section at 13 MeV are planned at KUTL at nine angle pairs of

$\theta_1 = 47.5°$, $50.5°$ and $53.5°$,
$\theta_2 = 47.5°$, $50.5°$ and $53.5°$,
with the azimuthal angle difference ϕ_1-$\phi_2 = 120°$.
The space star configuration is satisfied at $\theta_1 = \theta_2 = 50.5°$.

Syatematic measurement of pd breakup cross section at all the configuration space ("4π experiment") is also planned at KUTL. The long-term experiment will be made using the KUTL home machine. After the experiment, we get the discrepancy map for all the configuration space which is very useful for the study of NN forces and 3N force in 3N systems.

3.2 Analyzing Powers of pd Breakup Reaction

It was reported that tensor analyzing powers A_{xx} and A_{yy} of the H(d,pp)n reaction at E_d = 52 MeV at $\theta_1 = \theta_2 = 40.38°$ and ϕ_1-ϕ_2 = 180° show the energy spectra definitely different from the theoretical prediction, while A_{xx} and A_{yy} at $\theta_1 = \theta_2 = 38.72°$ are fairly well reproduced by the calculation [10].

Systematic measurements of A_{xx} and A_{yy}, for example,
$\theta_1 = 38.4°$, 40.4° and 42.4°,
$\theta_2 = 38.4°$, 40.4° and 42.4°, with ϕ_1-ϕ_2 = 180°,
are useful to see the systematic behavior of the difference. If the large difference is confirmed, modification of D-state NN interactions may be considered as a possible origin.

So far no large difference in A_y or iT_{11} of the pd breakup reaction between the experiment and calculation has been reported, partly due to large statistical errors in the experiments. It is natural to consider that the counterpart of the A_y puzzle in Nd scattering may exists in Nd breakup reaction. However, "4π experiment" for pd breakup A_y or iT_{11} is not feasible due to the small cross section.

4 Nd Capture Reaction

The Nd capture reaction is the transition between the Nd scattering state and the 3N bound state. The 3N binding energy and the Nd scattering cross section above E_N = 50 MeV are largely affected by the $\pi - \pi$ exchange 3N force, therefore the Nd capture, especially of energy above 50 MeV, is expected to be suitable to find the third indication for the $\pi - \pi$ exchange 3N force. The Nd capture is dominated by E1 transition at low energy and E2 and M1 transition become large with the energy. Therefore the vector and tensor analyzing powers of the Nd capture may be suitable to study the 3N force effects.

Precise measurement of iT_{11}, T_{20} and T_{22} of pd capture reaction has been made at E_d = 17.5 MeV at KUTL. The comparison with the calculation indicates that the iT_{11} is sensitive to the $\pi - \pi$ exchange 3N force [15].

To obtain one more concrete indication for the $\pi - \pi$ exchange 3N force, measurement of iT_{11}, T_{20} and T_{22} at E_d = 200 MeV is planned.

5 Conclusion

Efforts to find out the third indication for the $\pi - \pi$ exchange 3N force are in progress to establish the existence of the 3N force. Systematic experiments to investigate the discrepancies that can not be explained by the $\pi - \pi$ exchange

3N force have been made and are being planned. Extensive pd calculations with the Coulomb force correctly treated are highly required to extract meaningful information from the extensive pd data.

Acknowledgement.

The author thanks members of the 3N experimental group at KUTL for collaboration in long-term, elaborate many experiments. He also thanks Koike, Kamada, Ishikawa, Glöckle, Witala, Alt, Kievsky and Takemiya for their informative suggestions and calculations.

References

1. T. Sasakawa, S. Ishikawa, Y. Wu, T-Y. Saito: Phys. Rev. Lett. **68**, 3503 (1992)
2. K. Sagara et al.: Phys. Rev. **C50**, 576 (1994)
3. Y. Koike and S. Ishikawa: Nucl. Phys. **A631**, 683 (1998)
4. H. Witała et al.: Phys. Rev. Lett. **81**, 1183 (1998)
5. G. Rauprich et al.: Nucl. Phys. **A535**, 313 (1991)
6. J. Strate et al.: Nucl. Phys. **A501**, 51 (1989)
7. C.R. Howell et al.: Nucl. Phys. **A631**, 692 (1998)
8. T. Takemiya: Prog. Theor. Phys. **86**, 975 (1991)
9. H. Witała and W. Glöckle: Nucl. Phys. **A528**, 48 (1991)
10. L.M. Quin et al.: Nucl. Phys. **A587**, 252 (1995)
11. E.O. Alt, A.M. Mukhamedzhanov, A.I. Sattarov: Phys. Rev. Lett. **81**, 4820 (1998)
12. A. Kievsky: Private Communication
13. S. Shimizu et al.: Phys. Rev. **C52**, 1193 (1995)
14. P. Doleschall et al.: Nucl. Phys. **A380**, 72 (1982)
15. H. Kamada: Private Communication

Few-Body Systems Suppl. 12, 427–432 (2000)

Few-Body Systems

Three-Nucleon Scattering: New Physical Insights and Applications

H. Witała[1], W. Glöckle[2], J. Golak[1], H. Kamada[2], A. Nogga[2]

[1] Institute of Physics, Jagellonian University, PL- 30059 Cracow, Poland

[2] Institut für theoretische Physik II, Ruhr-Universität Bochum, D-44780 Bochum, Germany

Abstract. Recent results on three-nucleon (3N) scattering below the pion production threshold are reviewed. New physical insights on the role played by a three nucleon force (3NF) and present experimental evidence for it will be discussed by comparing new, high precision 3N data and theory based on modern nucleon-nucleon (NN) forces and including the 2π-exchange Tucson Melbourne (TM) 3NF. As an application of 3N bound and scattering states new results for the pd capture reaction will be presented. They are based on an improved nuclear current which includes π and ρ meson exchange contributions. First results for the effects of 3NF's in the pd capture process will be shown.

1 Introduction

New experimental data for the pd system at higher energy taken with polarized protons or deuterons as well as precise measurements of the pd elastic cross section brought new insights on the importance of the 3NF in the 3N system. In section 2 we give an overview on the present state of 3N continuum calculations and discuss these new results. In section 3 we present the application of rigorous 3N bound and scattering states to the pd capture process with and without initially polarized protons or deuterons. Different forms of the hadronic current operator are used and effects of meson exchange currents (MEC) are presented. We end with a brief outlook in section 4.

2 Successes and Failures in the 3N Continuum

The 3N system is the first nontrivial case to test the nucleonic Hamiltonian. Traditionally it is taken in a nonrelativistic form with realistic NN forces, which are adjusted to the NN data. We use the most modern NN forces AV18 [1], CD Bonn [2], and Nijm I and Nijm II [3], which are very well tuned to the

NN data base up to about 350 MeV. In general such a Hamiltonian gives a quite good description for 3N scattering observables and the predictions show stability against exchanges of modern NN forces [4]. If we add the TM 3NF [5] we adjust the form factor parameter Λ in that force together with each of the NN forces to the triton binding energy [6]. These Hamiltonians are then used to provide estimates for 3NF effects in the 3N continuum.

A rather complete overview of our former results and their comparison to many data is shown in [4]. In this contribution we would like to point out the new results based on recent experimental data. For the total nd cross section up to about 100 MeV there is a nice agreement of the pure 2N force predictions and the data [7, 8]. A discrepancy developes at higher energies and calculations underestimate the data by about 11 % at 300 MeV. The effect of the 3NF enhances the total nd cross section at the higher energies only by about 4 % [8]. A simple estimate of a relativistic kinematical effect gives an increase of the cross section of about 3 % at 100 MeV and about 7 % at 250 MeV [8]. This probably indicates the onset of relativity.

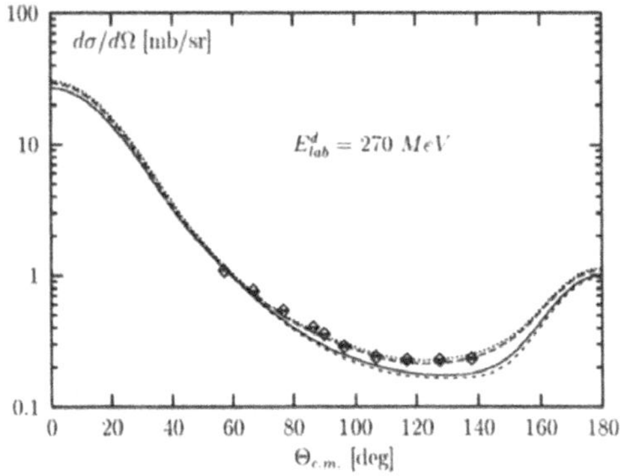

Figure 1. The Nd elastic scattering angular distribution at E_d=270 MeV. The solid line is the CD Bonn potential prediction. The long dashed line results when TM 3NF is included. The corresponding results for AV18 potential are given by short dashed and dotted lines.

The elastic Nd scattering cross section is quite well described with NN forces only at lower energies but there developes a strong discrepancy, starting at about 30 MeV, in the minimum. Very probably it is caused by 3NF effects, which fill this minimum [9]. This expectation is supported by recent precise data [10] at 135 MeV shown in Fig.1. The nucleon analyzing power A_y in low energy elastic Nd scattering poses a still unsolved puzzle [11]. In [12] arguments are given that changes in the NN forces, excluding the well established property of the one-pion exchange, are not capable of reproducing A_y. There are, however, still doubts whether the ^3P_J NN force components, on which A_y is extremely sensitive,. have been constrained sufficiently well by the NN data basis [13]. Present day 3N forces have insignificant effects at those low energies. If a 3NF would be responsible for the low energy A_y puzzle it must be of

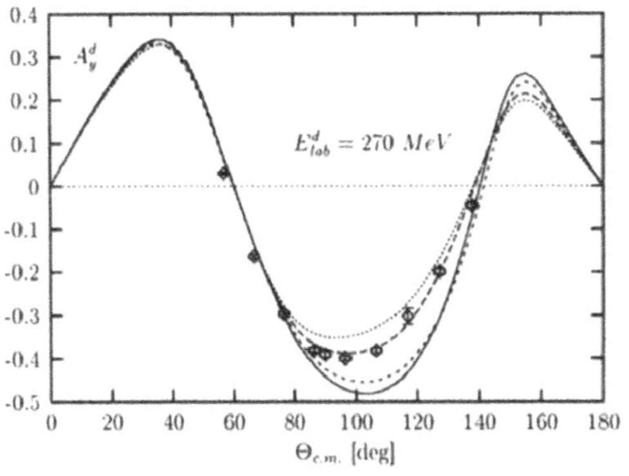

Figure 2. The Nd elastic scattering analyzing power $A_y^{(d)}$ at E_d=270 MeV. For the description of lines see Fig.1.

different structure than the TM 3NF. Some overview for possible 3N forces is given in [14]. Recently another interesting possibility for ρ meson modifications in the nuclear medium (Brown-Rho scaling) and its impact on the A_y puzzle have been investigated in [15].

New precise pd polarization data taken at higher energies opened recently a new region to test the 3N Hamiltonian and the properties of a 3NF. At those energies large 3NF effects, as given by the 2π-exchange TM model, are predicted for some observables. For some of them a discrepancies between pure NN force predictions and data have been found [16, 17]. Fig.2 shows the case of the vector analyzing power $A_y^{(d)}$ taken at E_d = 270 MeV [10]. Adding the 3NF leads to a good description of the data. A similar behavior show the $A_y^{(d)}$ and C_{yy} pd data at E_p = 200 MeV [17]. However, for the proton analyzing power $A_y^{(p)}$ the effects predicted by the TM 3NF are too large [16] indicating an failure of the TM 3NF.

Also for polarization observables in the breakup process at higher energy large 3NF effects are predicted in some kinematical configurations and require a precise data basis [4]. At low energy the five-fold differential cross section for the space-star geometry remains a problem. At 13 MeV nd [18] and pd [19] data deviate strongly from each other and exhibit an unexpectedly large Coulomb force effect. Our theoretical prediction lies in between. Including our TM 3NF does not change the theory significantly. It is therefore clear that the exact inclusion of the pp Coulomb force together with realistic NN forces above the deuteron breakup threshold is an urgent theoretical problem to be solved.

3 The p-d Capture and the Photodisintegration Processes

The nuclear matrix element from which all observables can be calculated

$$N_\tau = \; < \Psi_{bound} M_b | j_\tau(\vec{Q}) | \Psi_{pd}^{(+)} \vec{q}_0 M_d M_p >, \tag{1}$$

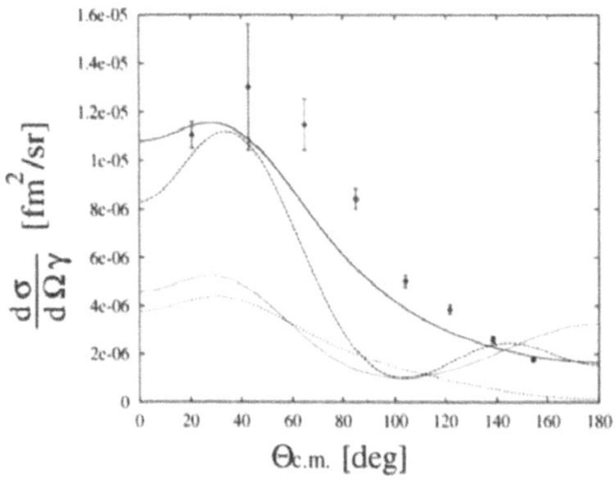

Figure 3. The pd capture cross section at $E_p = 200$ MeV. The data are from ref.[27]. The full calculation with Siegert terms is shown by the solid line and the plane wave impulse approximation symmetrized (PWIAS) [21] by the long dashed line. Corresponding results without Siegert terms are shown by dotted and short dashed lines.

contains in addition to the initial interacting state (IIS) $|\Psi_{pd}^{(+)}\vec{q}_0 M_d M_p>$, which accounts for the proton-deuteron motion with asymptotic relative momentum $\vec{q}_0$ and spin projections M_d, M_p, the ^3He bound state $|\Psi_{bound} M_b>$ with spin projection M_b, and the spherical components $j_\tau(\vec{Q})$ $(\tau = \pm 1)$ of the nuclear current operator defined in relation to the photon momentum $\vec{Q}||z$. Using Faddeev decomposition of the IIS the nuclear matrix element N_τ can be calculated by directly applying the numerical methods used in the 3N continuum [21].

The most simple assumption for the current operator is the single nucleon approximation. To see effects of two-body currents we use Siegert's idea [20], which, however, covers not all two-body currents. Our way of using the Siegert approximation is presented in ref.[21].

Another way is to include two-body current contributions explicitly. As a first step we consider π and ρ MEC's [22, 23, 24].

At lower energies the angular distribution for pd capture or photodisintegration is rather well described [21, 25, 26]. Thereby the effects of initial state interactions (ISI) are very strong as well as the inclusion of the subset of MEC's via Siegert. At high energy (E_p = 200 MeV, Fig.3) there are significant discrepancies. Also for proton vector analyzing power we find discrepancies at low and higher energies [21].

The most probable candidates to account for the existing discrepancies are 3NF's and explicit MEC's. As a first step in this direction we show in Fig.4 results for A_{yy} at E_d = 17.5 MeV. Only small 3NF effects are seen. Similar effects are seen for A_{zz}. For A_{xx} and $A_y^{(d)}$ the inclusion of our 3NF has essentially no effect at all. An explicit inclusion of π and ρ MEC's is not sufficient and works worse than the Siegert approximation in interpreting the data.

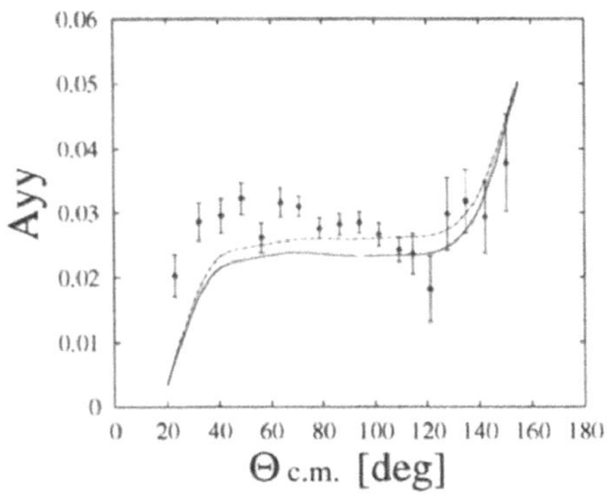

Figure 4. 3NF effects for the pd capture tensor analyzing power A_{yy} at $E_d = 17.5$ MeV. The full calculation with Siegert terms is shown by the dashed-dotted line. Including the TM 3NF results in solid line. The data are from ref.[28].

4 Summary and Outlook

Using modern computers 3N scattering can be solved exactly in a numerical sense up to high energies just below the pion production threshold. It allows to test 3N Hamiltonians containing modern realistic NN interactions and 3NF's by comparing the theoretical predictions to precise 3N data. Especially the higher energy region seems to be a promising ground in hunting for 3NF effects. They are predicted to be rather large at these energies when the 2π-exchange TM 3NF model, properly adjusted together with particular NN interaction to the experimental triton binding energy is used in the 3N Hamiltonian. Some recent elastic Nd scattering cross sections and polarization data support these predictions, some, however, also indicate an deficiency of the TM 3NF model. A good knowledge and a well founded understanding of the 3N Hamiltonian is a prerequisite to theoretically analyse electroweak processes with three participating nucleons. Their initial or final state interactions can in most cases not be neglected and have to be exactly accounted for by using exact 3N bound and scattering states. In such processes the restriction to a single nucleon current operator will be in general insufficient and for a convincing picture mesonic exchange currents have to be included and current conservation has to be fulfilled. The Siegert approximation can be a first step and accounts for a subset of two body currents. At higher energies there are indications for the onset of relativity, whose incorporation will be a challenging task for the theory.

Acknowledgement. This work was supported in part by the Polish Committee for Scientific Research under Grant No. 2P03B03914, by the Science and Technology Cooperation Germany-Poland, and by the Deutsche Forschungsgemeinschaft. The numerical calculations have been performed on the CRAY T90 and T3E of the John von Neumann Institute for Computing, Jülich, Germany.

References

1. R.B. Wiringa et al.: Phys. Rev. **C51**, 38 (1995)
2. R. Machleidt et al.: Phys. Rev. **C53**, R1483 (1996)
3. V.G.J. Stoks et al.: Phys. Rev. **C49**, 2950 (1994)
4. W. Glöckle et al.: Phys. Rep. **274**, 107 (1996)
5. S.A. Coon and W. Glöckle: Phys. Rev. **23**, 1790 (1981)
6. A. Nogga et al.: Phys. Lett. **B409**, 19 (1997)
7. W.P. Abfalterer et al.: Phys. Rev. Lett. **81**, 57 (1998)
8. H. Witała et al.: Phys. Rev. **C59**, 3035 (1999)
9. H. Witała et al.: Phys. Rev. Lett. **88**, 1183 (1998)
10. N. Sakamoto et al.: Phys. Lett. **B367**, 60 (1996)
11. H. Witała, D. Hüber, W. Glöckle: Phys. Rev. **C49**, R14 (1994)
12. D. Hüber and J.L. Friar: Phys. Rev. **C58**, 674 (1998)
13. T. Tornow and W. Tornow: Few-Body Systems **26**, 1 (1999)
14. J.L. Friar et al.: Phys. Rev. **C59**, 53 (1999)
15. F. Sammarruca, H. Witała: to be published
16. R. Bieber et al.: to be published
17. R.V. Cadman et al.: to be published
18. H.R. Setze et al.: Phys. Lett. **B388**, 229 (1996)
19. G. Rauprich et al.: Nucl. Phys. **A535**, 313 (1997)
20. A.J.F. Siegert: Phys. Rev. **52**, 787 (1937)
21. W. Glöckle et al.: In: *Workshop on Electron Nucleus Scattering*, EIPC 1998, ed. O. Benhar et al., p.261. Edizioni ETS 1998
22. D.O. Riska: Phys. Scr. **31**, 471 (1985)
23. R. Schiavilla et al.: Phys. Rev. **C40**, 2294 (1989) and references therein
24. V.V. Kotlyar, H. Kamada, W. Glöckle, J. Golak: to be published
25. W. Sandhas et al.: Nucl. Phys. **A631**, 210c (1998)
26. H. Anklin et al.: Nucl. Phys. **A636**, 189 (1998)
27. M.A. Pickar et al.: Phys. Rev. **C35**, 37 (1987)
28. K. Sagara: Private communication

Few-Body Systems Suppl. 12, 433–438 (2000)

Few-Body Systems

A Practical Method for Relativistic 3N-Scattering Calculations with Realistic Potentials

H. Kamada*,†

Institut für theoretische Physik II, Ruhr-Universität Bochum, D44780 Bochum, Germany

Abstract. Using a scale transformation in momentum space a phase equivalent relativistic potential is generated from the nonrelativistic potential. By that transformation a practical method for the relativistic 3N scattering with realistic nucleon-nucleon potentials is introduced. The formalism can be applied to any realistic nonrelativistic potential. We also discuss the locations of the moving logarithmic singularities of the free relativistic three-body Green's function. It enters in the relativistic 3N Faddeev equations, which have been formulated long time ago for a 3-boson bound state and which we propose to also use for 3N scattering. Finally we compare relativistic deuteron wave functions to nonrelativistic ones.

1 Introduction

Recently the total cross section for nd scattering has been measured for incident neutron energies from 50 MeV to 600MeV [1]. The data have been analyzed [2] up to 300 MeV by rigorous 3N Faddeev calculations [3] based on the CD-Bonn potential [4] and the Tucson-Melbourne three-nucleon force (3NF) [5]. Two nucleon forces alone are not sufficient to describe the data above about 100 MeV. The discrepancy is very likely filled by 3NF effects and relativistic corrections. In [2] we simply estimated one relativistic kinematical effect related to the incident flux. This leads to an increase of the total cross section by about 3% at 100MeV and about 7% at 250MeV. The rest of the discrepancy of about the same amount turned out to be understandable as a 3NF contribution. 3NF effects can be seen already at lower energies around 65 MeV nucleon lab energy in the minimum of the differential cross section in elastic pd scattering. This is more pronounced at higher energies as recent measurements at RIKEN

* *Alternative address:* Institut für Strahlen und Kernphysik der Universität Bonn, D53115 Bonn, Germany

† *E-mail address:* kamada@hadron.tp2.ruhr-uni-bochum.de

[6, 7] and IUCF [8] show. Those discrepancies to NN force predictions alone are known as Sagara discrepancy [9, 10, 11]. This very simplistic estimate of a relativistic correction just mentioned is not a substitute for a consistent treatment. The dynamical equation should be relativistic, not only the kinematics.

Relativistic effects in a three-boson bound state have been discussed in [12]. Following those concepts several questions arise:
(1) How do we formulate the relativistic three-body equations for scattering processes?
(2) To what extent does relativity require modifications of the conventional nuclear forces?

With respect to (1) the answer is given in [12] for three bosons by appropriately supplementing the relativistic Faddeev equation presented in [12] for scattering. An important ingredient in these equations is the boost transformation of the two-body t-matrices. What is not treated in [12] are the Wigner rotations of the nucleon spin. This has to be added. This formalism in [12] is of the instant form of relativistic dynamics and has been proposed by F. Coester in [14].

In order to answer the second question one has to introduce new forces which inserted into the relativistic two-body Schrödinger equation reproduce the two-body observables as precisely as the nonrelativistic potentials.

Recently Glöckle and the author found an analytical momentum-transformation [15] which provides a mathematical relation between the relativistic and nonrelativistic two-body Schrödinger equation and thus between the potentials in the two equations. This transformation is such that the NN phase shifts and the binding energy of the deuteron do not change. In [16] the parameters of a given NN potential (AV18) have been readjusted when used together with the relativistic form of the kinetic energy in order to guarantee the same NN phase shifts to some degree of accuracy. Our transformation given in [15] guarantees exactly the same phase shifts and no refitting is required. These tranformed NN forces will be the dynamical input for the properly supplemented relativistic Faddeev equation in our treatment of relativistic 3N scattering.

In section 2 we will briefly review the formalism proposed in [12] for a relativistic 3N Faddeev equation and extend it from bound state to scattering calculations. Thereby the treatment of the logarithmic moving singularities arising from the free 3N propagator requires special considerations. In section 3 we address the transformation which generates from the nonrelativistic NN potential a new one adequate to the relativistic form of the Schrödinger equation and compare the resulting relativistic deuteron wave function with the one established in a recent study of the Urbana group [16]. The summary is given in section 4.

2 Glöckle - Lee - Coester Relativistic Faddeev Equation

In the article [12] a relativistic Faddeev equation is introduced for a bound state of three bosons. In the Bakamjian-Thomas scheme [13] the two-body potential v_{ij} which is defined in the two-body c.m. system (2CM), is transformed into V_{ij} which belongs to the three-body c.m. system (3CM), as

$$V_{ij} \equiv \sqrt{(\omega_{ij} + v_{ij})^2 + \boldsymbol{p}_{ij}^2} - \sqrt{\omega_{ij}^2 + \boldsymbol{p}_{ij}^2} \tag{1}$$

where $\omega_{ij}= 2\sqrt{m^2 + \boldsymbol{k}^2}$ and $\boldsymbol{p}_{ij}$ are the energy and total momentum of the noninteractiong pair of particles ij. The usual nonrelativistic relative momentum $\frac{1}{2}(\boldsymbol{k}_i - \boldsymbol{k}_j)$, where $\boldsymbol{k}_{i,j}$ are the individual momenta, is replaced now by the Lorentz tranformed individual momenta $\boldsymbol{k}$ and $-\boldsymbol{k}$ in the 2CM :

$$\boldsymbol{k} = \frac{1}{2}(\boldsymbol{k}_i - \boldsymbol{k}_j) - \frac{\boldsymbol{p}_{ij}}{2} \frac{E_i - E_j}{E_i + E_j + \sqrt{(E_i + E_j)^2 - \boldsymbol{p}_{ij}^2}} \tag{2}$$

The individual energies are $E_i = \sqrt{m^2 + k_i^2}$. The "boosted" potential V_{ij} depends on the total momentum of the two-body subsystem as is obvious from Eq. (1). Using the complete set of two-body states, $< \boldsymbol{k}|\phi_b >$ and $< \boldsymbol{k}|\phi(k_0) > \equiv < \boldsymbol{k}|\boldsymbol{k}_0 >^{(+)}$, V_{ij} is explicitly given as ,

$$\begin{aligned} &< \boldsymbol{p}_{ij}\boldsymbol{k}|V_{ij}|\boldsymbol{p}_{ij}'\boldsymbol{k}' >= \\ &\delta(\boldsymbol{p}_{ij} - \boldsymbol{p}_{ij}') \int d\boldsymbol{k}_0 < \boldsymbol{k}|\boldsymbol{k}_0 >^{(+)} \sqrt{\omega^2(k_0) + \boldsymbol{p}_{ij}^2} \; ^{(+)} < \boldsymbol{k}_0|\boldsymbol{k}' > \\ &-\delta(\boldsymbol{p}_{ij} - \boldsymbol{p}_{ij}')\delta(\boldsymbol{k} - \boldsymbol{k}')\sqrt{\omega^2(k) + \boldsymbol{p}_{ij}^2} + \ldots \end{aligned} \tag{3}$$

where

$$< \boldsymbol{k}|(\omega(k) + v_{ij})|\phi(\boldsymbol{k}_0) >= \omega(k_0) < \boldsymbol{k}|\phi(k_0) > . \tag{4}$$

The dots denote the two-body bound state contribution. Instead of using (3) one can also express directly the two-body boosted transition operator T_{ij} in the 3CM by the t-matrix in the 2CM, as shown in [12].

The free three-body Greens function is singular above the three-body breakup threshold. It can be written as

$$\begin{aligned} G_0^{-1} &= E - \sqrt{\omega^2(k) + \boldsymbol{p}_{ij}^2} - \sqrt{\boldsymbol{p}_{ij}^2 + m^2} + i\epsilon \\ &= E - \sqrt{\boldsymbol{p}_{ij}^2 + m^2} - \sqrt{\boldsymbol{p}_{jk}^2 + m^2} - \sqrt{\boldsymbol{p}_{ij}^2 + \boldsymbol{p}_{jk}^2 + 2p_{ij}p_{jk}x + m^2} \end{aligned} \tag{5}$$

which exhibits an angular dependence ($x = \hat{p}_{ij} \cdot \hat{p}_{kj}$) After integration over x logarithmic singularities appear in the Faddeev equation. By substituting $y = G_0^{-1}(x)$ the integration can be carried through with the result $\int dx/y = [y + (\sqrt{\boldsymbol{p}_{ij}^2 + m^2} + \sqrt{\boldsymbol{p}_{jk}^2 + m^2} - E\ )ln(y)]/p_{ij}p_{jk}$. The singularities occur under the conditions $x = \pm 1$ and $G_0^{-1} = 0$. Their locations are displayed in Fig. 1 for the example of $E_{lab} = 250$ MeV. We see a shift of the relativistic singularity lines in comparison to the nonrelativistic ones.

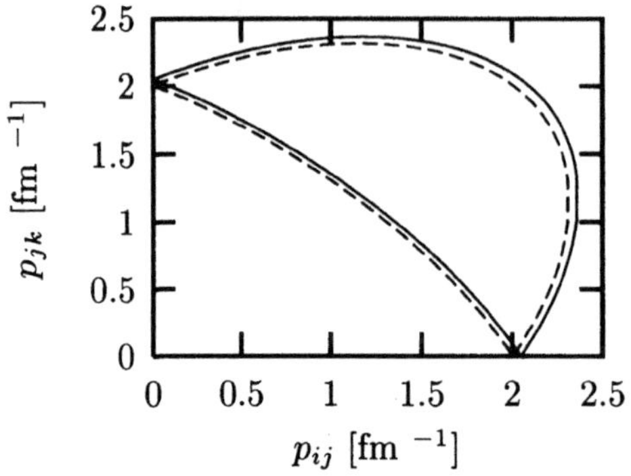

Figure 1. The loci of the Green's function singularity at E_{lab} =250 MeV. The solid (dashed) line is for relativistic (nonrelativistic) kinematics.

3 Momentum Scale Transformation

In [15] a momentum scale transformation has been introduced by which the nonrelativistic potential v^{NR} can be rewritten into a "relativistic" one , v, as used in Eq. (1) without changing the on-the-energy-shell properties in the two-body system. The nonrelativistic potential v^{NR} enters the nonrelativistic Schrödinger equation,

$$< \boldsymbol{k}^{NR}|(\frac{\boldsymbol{k}^{NR^2}}{m} + 2m + v_{ij}^{NR})|\psi(k_0^{NR}) >= (\frac{k_0^{NR^2}}{m} + 2m) < \boldsymbol{k}^{NR}|\psi(k_0^{NR}) > \quad (6)$$

The corresponding relativistic equation is given in Eq. (4). The momentum transformation connecting relativistic and nonrelativistic momenta.

$$\omega(k) = 2\sqrt{k^2 + m^2} \equiv \frac{k^{NR^2}}{m} + 2m \quad (7)$$

leads to the relation between the relativistic and nonrelativistic wave functions:

$$\phi(k) = \frac{\sqrt{2}m}{\sqrt{\sqrt{m^2+k^2}\sqrt{2m(m+\sqrt{m^2+k^2})}}}\psi^{NR}(\sqrt{2m(\sqrt{m^2+k^2}-m)}). \quad (8)$$

Fig. 2. shows the S- and D-wave components of the deuteron wave function for the AV18 [17] model potential in configuration space.

At short distances the relativistic wave function becomes larger, a tendency which agrees with the result of a recent study [16]. In [16] an approximation to the ratio between the relativistic and nonrelativistic D-wave functions (see Eq. (2.27)) has been considered:

$$R \equiv \frac{\phi_D(k)}{\psi_D^{NR}(k)} = \frac{k^2}{2(\sqrt{m^2+k^2}-m)\sqrt{m^2+k^2}} \quad (9)$$

We compare that ratio to the one given in Eq.(8) in Fig. 3 and thereby neglect the momentum transformation of the argument in ψ^{NR}. Both results agree well, especially if one takes into account, that Eq. (9) is only an approximation and that our result is even closer to the real calculated ratio in [16].

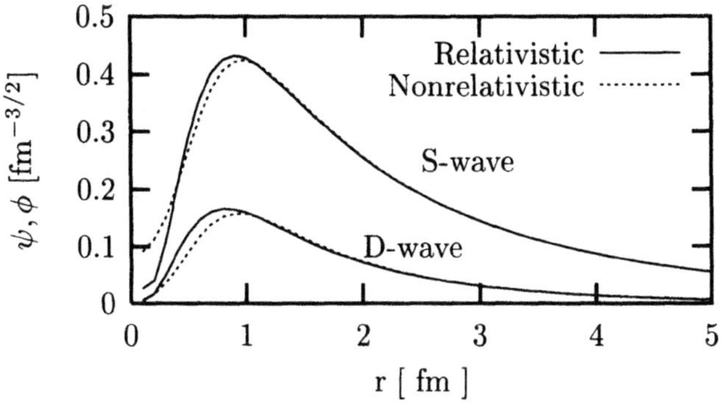

Figure 2. The deuteron wave functions. The solid (dotted) lines are the relativistic (nonrelativistic) wave functions.

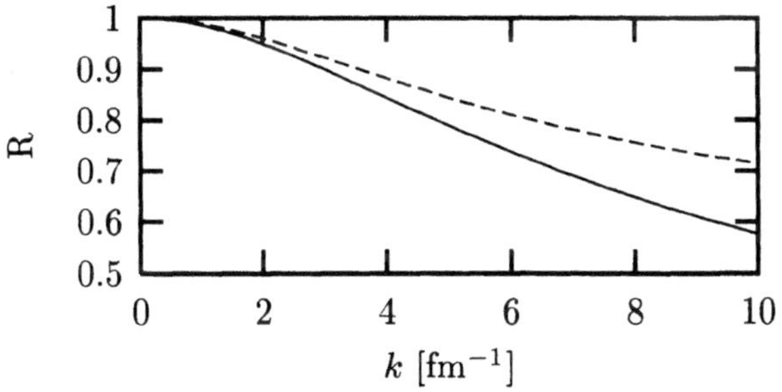

Figure 3. The ratio between the relativistic D-wave deuteron wave function and the nonrelativistic one. The solid and dashed curves are due to Eq. (8) and Eq. (9), respectively.

4 Summary

We reviewed briefly the formalism in [12], which was devoted to a relativistic treatment of a three-boson bound state in the instant form of relativistic dynamics. The relativistic Faddeev equation given there is extended to three-nucleon scattering and the necessary changes in the free propagator singularities are pointed out. This scheme requires NN forces which are tuned to NN phase-shifts together with the relativistic form of the kinetic energy. We proposed to use the momentum scale transformation from [15], which provides an analytical connection between the potentials in the nonrelativistic and relativistic two-body Schrödinger equations. No refitting like in [16] is required. We did not yet address the Wigner rotation of the spin states, but we expect this to be achieved along the line given in [18]. We also plan to proceed without partial wave decomposition and then the first steps formulated by [19] will be useful.

Acknowledgement. I would like to dedicate this paper to Prof. Walter Glöckle on the occasion of his 60th birthday. I would like to thank Prof. Hideyuki Sakai and Prof. Kichiji Hatanaka for fruitful discussions of the 3N scattering measurements at RIKEN and RCNP. It was a pleasure for me to join Prof. Shinsho Oryu, Prof. Masayasu Kamimura, Prof. Souichi Ishikawa and Prof. Yasuro Koike in the organization of the first Asia-Pacific Few-Body conference. This work is financially supported by the Deutsche Forschungsgemeinschaft.

References

1. W.P. Abfalterer et al.: Phys. Rev. Lett. **81**, 57 (1998)
2. H. Witała et al.: Phys. Rev. **C59**, 3035 (1999)
3. W. Glöckle et al.: Phys. Rep. **274**, 107 (1996)
4. R. Machleidt, F. Sammarruca, Y. Song: Phys. Rev. **C53**, R1483 (1996)
5. S.A. Coon, W. Glöckle: Phys. Rev. **C23**, 1790 (1981)
6. N. Sakamoto et al.: Phys. Lett. **B367**, 60 (1996)
7. H. Sakai, K. Sekiguchi: Private communications
8. H. Rohdjeß et al.: Phys. Rev. **C57**, 2111 (1998)
9. K. Sagara et al.: Phys. Rev. **C50**, 576 (1994)
10. S. Nemoto et al.: Phys. Rev. **C58**, 2599 (1998)
11. H. Witała et al.: Phys. Rev. Lett. **81**, 1183 (1998)
12. W. Glöckle, T.-S. H. Lee, F. Coester: Phys. Rev. **C33**, 709 (1986)
13. B. Bakamjian and L.H. Thomas: Phys. Rev. **92**, 1300 (1953)
14. F. Coester: Helv. Phys. Acta. **38**, 7 (1965)
15. H. Kamada, W. Glöckle: Phys. Rev. Lett. **80**, 2547 (1998)
16. J.L. Forest et al.: Phys. Rev. **C60**, 014002 (1999)
17. R. B. Wiringa et al.: Phys. Rev. **C51**, 38 (1995); **C54**, 646 (1996)
18. B.D. Keister and W.N. Polyzou: Adv. Nucl. Phys. **20**, 225 (1990)
19. I. Fachruddin : see contribution to this conference

Few-Body Systems Suppl. 12, 439–444 (2000)

Few-Body Systems

Solutions of Faddeev-Yakubovsky Equations in Configuration Space for the 4N Scattering States

J. Carbonell

Institut des Sciences Nucléaires, 53 Av. des Martyrs, 38026 Grenoble, France

Abstract. The Faddeev-Yakubovsky equations in configuration space have been solved for the four nucleon system with special interest in their scattering states. We present results concerning the structure of the first ^{4}He, scattering states in different (T,S) channels including the first inelatic threshold N+3N→NN+NN and the n-t cross section using realistic potentials.

1 Introduction

The theoretical description of the A=4 continuum constitutes, in addition to its technical difficulty, a serious challenge for the NN interaction models. The remarkable success encountered in the description of the 3N system and in the 4N bound state [1] is largely based on the three nucleon interaction (TNI). The two body dynamics alone looses ≈ 10% in the 3N binding energy and ≈ 15% in the 4N. TNI compensate this defect but, a part from an energy scaling, are found to have small influence in the low energy scattering states [2]. Contrary to the A=3 case, these states present several nearthreshold structures, which can be hardly sensible to the TNI and should be reproduced with the two-body forces only. A first example is provided by the n+t cross section which manifests a much more vivid behaviour than the flat n-d one, with a resonant structure at $T_{lab} \approx 3$ MeV (figure 1). A second interesting accident is the existence of quasidegenerate n-^{3}He and p-^{3}H thresholds with, in their middle, the first ^{4}He excitation and all that in an energy gap of 0.76 MeV. The position of such a state, which dominates the inter-threshold dynamics, has to be ensured with an accuracy better than 100 KeV in order to reproduce the experimental cross sections, what is a strong requirement for an excited state.

2 Formalism

In the case of four identical particles interacting via a two-body potential V, the Faddeev-Yakubovsky (FY) equations result into two integrodifferential equa-

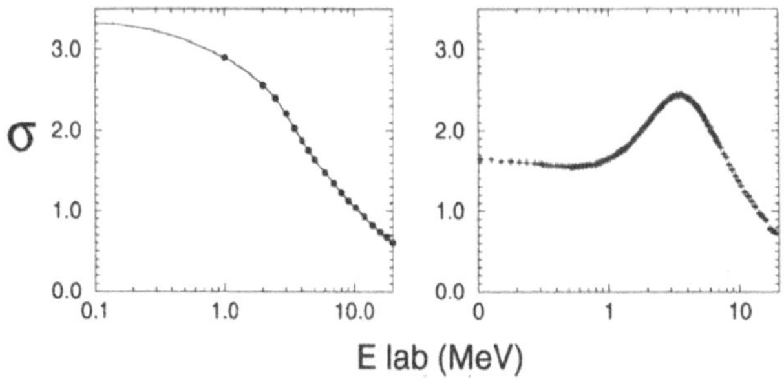

Figure 1. Comparison between the nd and nt elastic cross sections (in barns)

tions, coupling two amplitudes denoted K and H:

$$\begin{aligned} (E-H_0-V)K &= V\left[(P_{23}+P_{13})\,(\varepsilon+P_{34})\,K+\varepsilon(P_{23}+P_{13})\,H\right] \\ (E-H_0-V)H &= V\left[(P_{13}P_{24}+P_{14}P_{23})\,K+P_{13}P_{24}\,H\right] \end{aligned} \tag{1}$$

in which P_{ij} are the permutation operators and $\varepsilon=\pm1$ depending on we deal with bosons or fermions. The total wavefunction is reconstructed according to:

$$\Psi = \Psi_{1+3}+\Psi_{2+2}$$

$$\Psi_{1+3} = [1+\varepsilon(P_{13}+P_{23})]\ [1+\varepsilon(P_{14}+P_{24}+P_{34})]\,K \tag{2}$$

$$\Psi_{2+2} = [1+\varepsilon(P_{13}+P_{23}+P_{14}+P_{24})+P_{13}P_{24}]\ H \tag{3}$$

Each amplitude F=K,H considered as a function of its own set of Jacobi coordinates, is expanded in the corresponding angular variables according to

$$\langle \boldsymbol{xyz}|F\rangle = \sum_{\alpha} \frac{F_\alpha(xyz)}{xyz}\, Y_\alpha(\hat{x},\hat{y},\hat{z}). \tag{4}$$

F_α are the unknowns and Y_α are tripolar harmonics containing spin, isospin and angular momentum variables. Label α holds for the set of quantum numbers defined in a given coupling scheme and includes the type of amplitude K or H.

Boundary conditions for scattering states are of Dirichlet-type. In the 1+3 elastic case e.g. they are implemented by imposing at large enough value of z

$$\begin{aligned} K(x,y,z) &= t(x,y) \\ H(x,y,z) &= 0 \end{aligned}$$

$t(x,y)$ being the triton Faddeev amplitudes previously determined. FY equations ensure a solution which, for n+t S-waves, behaves asymptotically like

$$K(x,y,z) \sim t(x,y)\sin(qz+\delta)$$

where δ is the phaseshift and q is related to the center of mass kinetic energy $T_{\rm cm}$ and to the physical momentum k by $T_{\rm cm}=\frac{\hbar^2}{m}q^2=\frac{2}{3}\frac{\hbar^2}{m}k^2$. A more detailed explanation of the formalism and numerical methods used can be found in [3].

3 Results

We have been interested in the first 0^+ excitation of ^{4}He, experimentally manifested as a Γ=0.5 MeV resonant state, 0.4 MeV above the pt threshold [4]. Using MT I-III potential [5] we got [3] the binding energies (B) and r.m.s. radius (R) displayed in table 1, what means a bound state with $B_{^4He^*}$=0.26

Table 1. Binding energy (MeV) and r.m.s. radius (fm)

	^{4}He	^{4}He*	^{3}H
B	30.30	8.79	8.53
R	1.44	4.95	1.72

Table 2. Low energy N+3N parameters (fm)

S	T	a	r_0	v_0	k_0
0	0	14.75	6.75	0.462	-
1	0	3.25	1.82	0.231	-
0	1	4.13	2.01	0.308	0.505
1	1	3.73	1.87	$\simeq 0$	-

MeV below the N+NNN threshold. Using AV14 [6] and Nim II [7] potentials we found close results and similar conclusions were also reached using separable interactions [8]. To investigate the structure of this state, we looked at the two-body correlation functions. The results (figure 2) show a superposition of two structures with different length scales, the short distance part being similar to the triton one. This suggests – rather than a breathing mode as it is widely considered in the literature – a 3+1 structure formed by an almost unperturbed triton plus a fourth nucleon orbiting around. This state, with a r.m.s. $R \approx 5$ fm, constitutes actually the first "halo-nucleus". It seems difficult for a strong interaction to generate a first excitation in the continuum and the open question is whether or not Coulomb forces, when added to the existing models, would be able to put it in the right place with the required accuracy.

Within the same model we have calculated (table 2) the parameters of the effective range expansion

$$g(k) \equiv k\,cot(\delta) = \left[1 - \left(\frac{k}{k_0}\right)^2\right]^{-1} \left[-\frac{1}{a} + \frac{1}{2} r_0 k^2 + v_0 k^4 + o(k^6)\right] \tag{5}$$

for N+NNN S-states. Expansion (5) provides an accurate representation of the below breakup scattering although the existence of the ^{4}He* pole makes necessary the explicit inclusion of a singular term for the T=S=0 channel. It is worth noticing the coherence between the ^{4}He* binding energy and the scattering parameters from table 2. If expansion (5) is inserted in the S-wave scattering amplitude $f_0^{-1}(k) = g(k) - ik$, 2 imaginary poles follow and the nearest to threshold $k_- = 0.095$ fm^{-1} corresponds to $E_- = -0.25$ MeV, in agreement with the direct calculations of the ^{4}He* binding.

The 4N scattering states with two open channels N+NNN→NN+NN has been calculated for the S=T=0 final state [3]. Of particular interest is the extraction of the imaginary part of the strong NN+NN scattering length which controls the fusion rate in the process d+d → n+^{3}He. We found $a_R = +4.91 \pm$

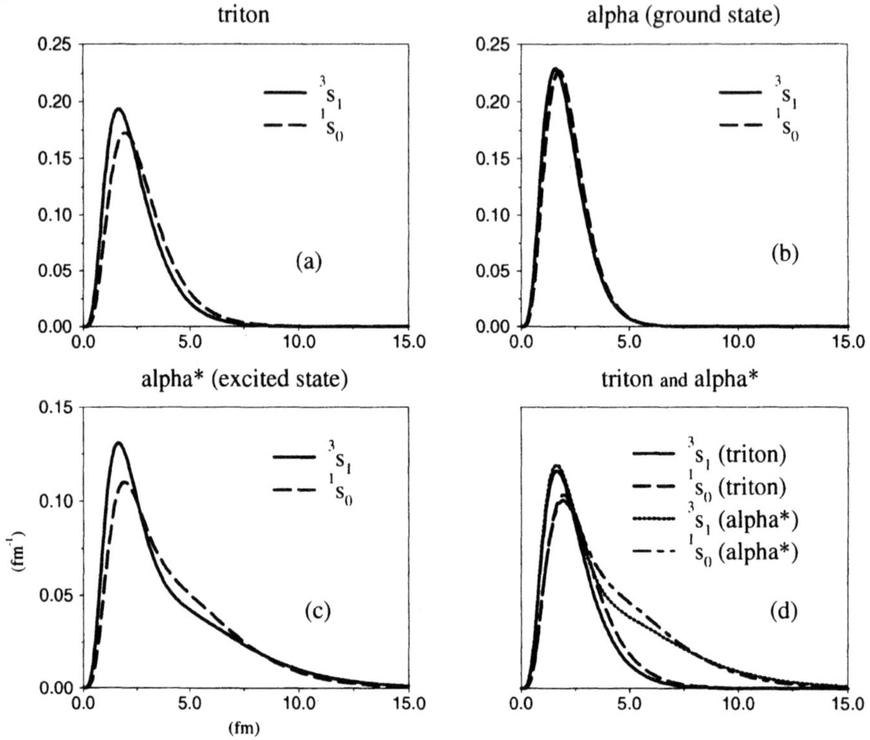

Figure 2. Two-body correlations for (a) triton, (b) ^{4}He ground and (c) first excited states ^{4}He*. Solid (dashed) line denotes the triplet (singlet) contributions. In (d), the results of ^{4}He* are compared to the triton correlation function suitably scaled.

0.02 fm and a very small value of $a_I = -0.0115 \pm 0.0001$ fm which should be only slightly modified once the Coulomb interaction is switched on. This value is due to the small overlapping between the K and H configurations which respectively govern the N+NNN and NN+NN asymptotic states. Other calculations of the dd cross section at very low energy exist [9] but no values of a_I were given.

Special attention was paid to the n+t scattering. We found [3] that the description provided by the simple MT I-III model fits remarkably well the data, as it can be seen in figure 3 where the differential cross sections at different energies are displayed. Using realistic potentials, we have performed [10] the same calculations for the $J^\pi = 0^+, 1^+, 0^-, 1^-, 2^-$ states with $^1S_0, ^3S_1, ^3D_1$ waves in V_{NN} and expansion (4) limited to $l_x, l_y, l_z = 0, 1, 2$. The singlet (a_0), triplet (a_1), coherent (a_c) scattering length and zero energy cross section (σ_0) are in table 3, compared with experimental values taken from [11]. They are in very good agreement with those obtained by Pisa group [12] $a_0 = 4.32$ and $a_1 = 3.80$ but the infered a_c and $\sigma(0)$ values are well far from experiment.

Calculations have been pursued beyond zero energy. One can see from figure 4 (dashed curve), that two-body realistic potentials alone fail in describing the

Table 3. n+t scattering length with realistic interactions

	a_0	a_1	$a_c = \frac{1}{4}a_0 + \frac{3}{4}a_1$	$\sigma(0) = \pi(a_0^2 + 3a_1^2)$
AV14	4.31	3.79	3.92	194
Nijm II	4.31	3.76	3.90	192
MT I-III	4.10	3.63	3.75	177
exp	–	–	3.59±0.02	170± 3

low energy cross section as they failed in giving the binding energies B_3 and B_4. We would like to emphasize that this failure is in fact already present in the n+d case, although somehow hidden because the sizeable disagreement concerns only the doublet scattering length which turns to be very small compared to the quartet one and it is furthermore suppressed by statistical factors.

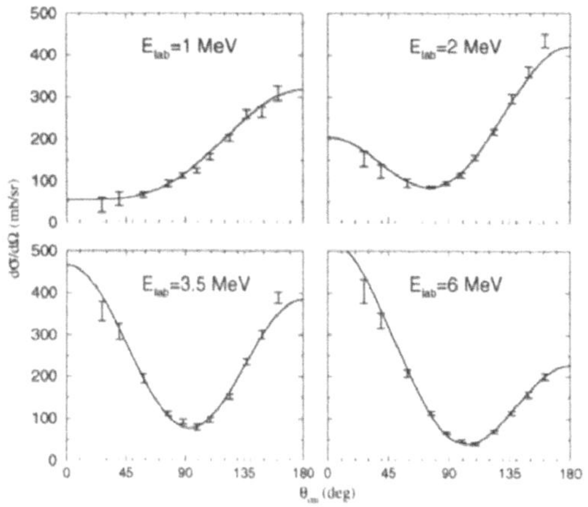

Figure 3. n-t differential cross section with MT I-III at different n laboratory energies

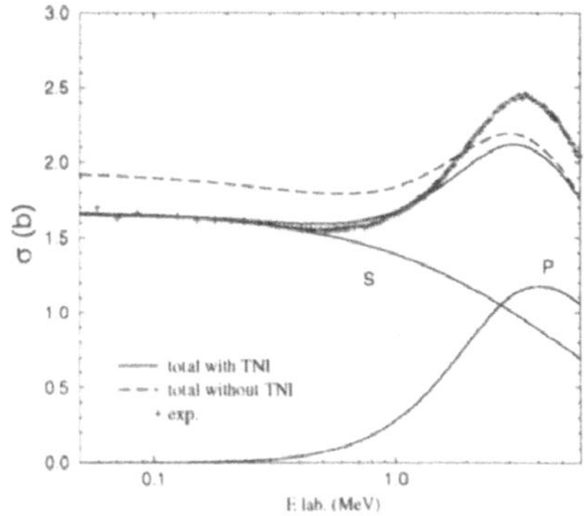

Figure 4. n-t total cross section with AV14 compared to experiment

To improve this situation we introduced the phenomenological TNI:

$$W(\rho) = W_r \frac{e^{-2\mu\rho}}{\rho} - W_a \frac{e^{-\mu\rho}}{\rho} \qquad \rho = \sqrt{x^2 + y^2}$$

Using the parameter set $W_r = 500 \quad W_a = 174 \quad \mu = 2.0$, we got an overall low energy agreement with the values B_3=8.48 MeV, B_4=29.0 MeV, a_0=4.0 fm, $a_1 = 3.53$ fm, $a_c = 3.65$ fm and $\sigma(0)$=168 fm^2 and a satisfactory n+t S-wave cross section (solid line). We point out that some dispersion exists in the experimental results which makes difficult the comparison with theoretical values. If one assumes the two last measurements [11] of a_c to be valid, our calculations are not far from the proposed value $a_c = 3.60$. In this case, however, our scattering lengths are not compatible with those infered from experiment, which are in their turn incompatible with each other. A more precise measurement of

$\sigma(0)$ would be helpful. For that purpose the possibilities offered by the neutron beam facility, recently approved at CERN [13], could deserve some attention from the nuclear Few-Body community.

If the inclusion of TNI is determinant in the n+t S-wave region, it doesn't improve the resonance peak, on the contrary. In a first tentative to explain this discrepancy we have included P-waves in the NN interaction [10]. We found that although separated P-waves had significant influence in the cross section, their global effect was very small and they could not explain the disagreement. A recent work done by A. Fonseca [14] using AGS equations and 1-rank approximation, was able to push expansion (4) further than we could do and found a satisfactory description. It would be interesting to have independent confirmation of this result and go beyond the first obstacle of the A=4 continuum.

Acknowledgement. The numerical calculations have been performed with the Cray-T3E of the CGCV (CEA) and IDRIS (CNRS).

References

1. W. Glöckle, H. Witała, H. Kamada, D. Huber, J. Golak: AIP Conf. Proc. 334, Ed. F. Gross, Williamsburg, 45 (1994) ; Phys. Rep. **274**, 107 (1996)
2. H. Witała, W. Glöckle, D. Huber, J. Golak, H. Kamada: Phys. Rev. Lett **81**, 1183 (1998)
3. F. Ciesielski, J. Carbonell: Phys. Rev. **C58**, 58 (1998)
4. D.R. Tilley, H.R. Weller, G.M. Hale: Nucl. Phys. **A541**, 1 (1992)
5. R.A. Malfliet, J. Tjon: Nucl. Phys. **A127**, 161 (1969)
6. R.B. Wiringa, R.A. Smith, T.L. Ainsworth: Phys. Rev. **C29**, 1207 (1984)
7. V.G.J. Stoks, R.A.M. Klomp, C.P.F. Terheggen, J.J. de Swart: Phys. Rev. **C49**, 2950 (1994)
8. A.C. Fonseca: Phys. Rev. **C40**, 1390 (1989)
9. E. Uzu, S. Oryu, M. Tanifuji: Few-Body Syst. Suppl. **8**, 97 (1995)
10. F. Ciesielski, J. Carbonell, C.Gignoux: Phys. Lett. **B447**, 199 (1999)
11. T.W. Phillips, B.L. Berman, J.D. Seagrave: Phys. Rev. **C22**, 384 (1980); H. Rauch, D. Tuppinger, H. Wolwitsch, T. Wroblewski: Phys. Lett. **B165**, 39 (1985); G.M. Hale, D.C. Dodder, J.D. Seagrave, B.L. Berman, T.W. Phillips: Phys. Rev. **C42**, 438 (1990)
12. M. Viviani, S. Rosati, A. Kievsky: Phys. Rev. Lett. **81**, 1580 (1998)
13. http://www.cern.ch/CERN/Divisions/SL/EET/tof/
14. A.C. Fonseca: Phys. Rev. Lett. (in press); contribution to this conference

Few-Body Systems Suppl. 12, 445–450 (2000)

Few-Body Systems

Use of Realistic N-N Interactions to Calculate 4N Scattering Observables

A.C. Fonseca

Centro de Física Nuclear da Universidade de Lisboa, Av. Prof. Gama Pinto, Nº 2, 1649-003 Lisboa, Portugal

Abstract. The four-body equations of Alt, Grassberger and Sandhas are solved for a system of four nucleons, using the Paris, Argonne V14, Bonn-B and Bonn-A potentials in channels $j \leq 2$. The results of the calculation are compared with data for the reactions $dd \to dd$, $dd \to p^3H$ and $n^3H \to n^3H$. The calculations indicate that the NN P-waves have a strong effect on 4N observables but remaining disagreement with data shows the need for a 3N force.

1 Introduction

The present work involves the solution of Alt, Grassberger, and Sandhas (AGS) equations [1] for the transition operators involving all (2N)+(2N) and N+(3N) channels. For local NN potentials such equations are three-vector variable integral equations which after partial wave decomposition reduce to a set of coupled equations in three continuous scalar variables. Since scattering calculations require a great number of channels for convergence, we follow an approach based on the separable representation of subsystem amplitudes in order to reduce the equations to two or one continuous variable. The integral equations we use are the same as in ref. [2] and result from the modified AGS equations [3] after one has: (a) represented the original NN t-matrix by an operator of rank one; (b) represented the resulting 3N t-matrix by a finite rank operator and taken as many terms as needed for convergence. Since in the modified AGS equations the 2N+2N subamplitudes are expressed in terms of a convolution integral involving two non-interacting pair-propagators, as first proposed in ref. [4], the sole approximation in this approach involves a rank one representation of the 2N t-matrix which may be obtained from the well-known method of Ernst, Shakin, and Thaler (EST) [5]. The multi-term representation of the 3N t-matrix is done using the Energy Dependent Pole Expansion (EDPE) method developed in ref. [6]. This latter representation for the 3N t-matrix is well under control since one may check the convergence rate of 4N observables for increasing rank in the 3N t-matrix.

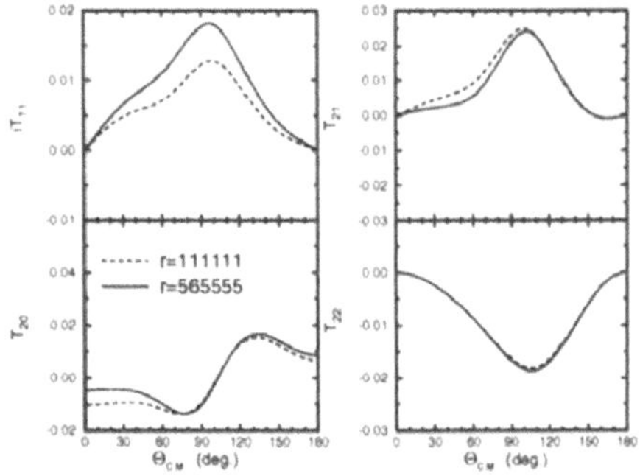

Figure 1. TAP's for ${}^1H(\vec{d},d)^1H$ at $E_d = 3$ MeV for Bonn-B potential in partial waves $j \leq 1 +{}^3P_2$. The solid lines correspond to the use of a high rank ($r \geq 5$) EST representation for the 2N t-matrix in each partial wave, while the dashed lines to $r = 1$.

This method was first used in ref. [2] to calculate the binding energy of ${}^4He$ and later confirmed to be accurate by the exact work of Kamada and Glöckle [7]. More recently [8] the results of our calculations for n-t elastic scattering were shown to agree with the exact results of the Grenoble group [9], for both MT and AV14 potentials taken in 2N partial waves with $j \leq 1^+$ (${}^1S_0, {}^3S_1 - {}^3D_1$). This benchmark was performed in the energy range $0 \leq E_n < 7$ MeV and shows that both works agree within 3% or better. In addition, as shown in Fig. 1 at $E_d = 3$ MeV, $d - n$ elastic observables calculated with the rank one EST representation of the 2N t-matrix are in very good agreement with the high rank results which are known [10] to coincide with the exact calculations; the differential cross sections agree within 1%, and the tensor observables T_{20}, T_{21}, and T_{22} within 10% or better over a broad angular range. Only iT_{11} is reduced by 30% at the peak. Therefore, in the energy region below four-body breakup threshold, we expect the rank one approximation to account for the dominant physics.

2 Results

The four-nucleon calculations we present here make use of different realistic NN interactions in channels ${}^1S_0, {}^3S_1 - {}^3D_1, {}^1P_1, {}^3P_0, {}^3P_1, {}^3P_2 - {}^3F_2, {}^1D_2$, and ${}^3D_2$. The first two channels correspond to including all 2N partial waves with $j \leq 1^+$, the first five channels to $j \leq 1$ and all of them to $j \leq 2$. The Coulomb repulsion between protons is neglected. Independently of the number of 2N partial waves that are included for a given NN interaction, one needs to set upper limits for given subsystem quantum numbers in order to reach a converged 4N result. In particular one has to decide the largest 3N total angular momentum J to

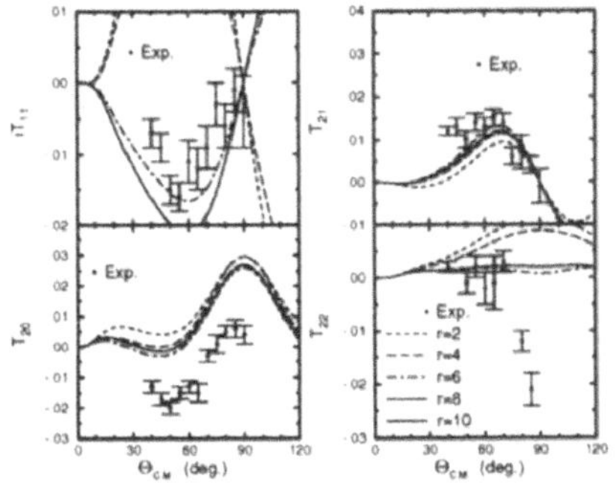

Figure 2. TAP'S for $dd \rightarrow dd$ at $E_d = 6.1$ MeV, using Bonn-B potential in partial waves $j \leq 1^+$. The rank r of the EDPE expansion of each 3N subamplitude $J \leq 7/2^+$ runs from $r = 2$ to $r = 10$. The data is from Ref. [11].

be included and the rank of the corresponding EDPE expansion. In Fig. 2 we show how all four tensor analysing powers (TAP's) for $dd \rightarrow dd$ at $E_d = 6.1$ MeV converge with the rank "r" in the EDPE expansion of the 3N t-matrix, which indicates that $r = 8$ is sufficient for convergence. A similar conclusion may be reached by studying the TAP's for $dd \rightarrow pt$. In general we take all 3N subamplitudes up to $7/2^+(1/2^\pm, 3/2^\pm, 5/2^\pm, 7/2^+)$ for 3N total isospin I = 1/2 and up to $5/2^+$ for isospin I = 3/2. For a given (J,I) the resulting 3N subamplitude is calculated using all underlying 3N channels with N-(2N) orbital angular momentum L≤3 and rank "r" equal eight for I = 1/2 amplitudes and six for I = 3/2. As expected I = 3/2 3N subamplitudes only contribute to isospin $\mathcal{I} = 1$ 4N reactions such as $^3H(n,n)^3H$. Finally in all $\mathcal{I} = 1$ ($\mathcal{I} = 0$) calculations we solve AGS equations for each 4N total angular momentum $\mathcal{J}$ up to three (six) in all corresponding N+(3N) and (2N)+(2N) channels with relative orbital angular momentum $\mathcal{L} \leq 3(\mathcal{L} \leq 5)$. The above limits for 3N and 4N channels are less restrictive than those used in ref. [9] and for this reason we expect this work to be fully converged on all relevant angular momentum channels and subchannels. To illustrate the size of our calculation we mention here that for $\mathcal{I} = 0$ and $\mathcal{J} = 3^-$ we take 29 N+(3N) channels (in the framework of ref. [9] this corresponds to 684), and 158 (2N)+(2N) channels (249 in ref. [9]) for NN partial waves up to $j \leq 2$.

In order to study the sensitivity of 4N observables to different off-shell behaviour of the NN interaction, we show in Fig. 3 the TAP's for $dd \rightarrow dd$ and $dd \rightarrow pt$ for different 2N potentials, taken in partial waves $j \leq 1^+$. Although the results are sensitive to the choice of NN potential, all the curves are very similar in shape. Next in Fig. 4 we show the same tensor observables for Bonn-B potential, but with higher partial waves up to $j \leq 2$ included. To compare with data all nuclear amplitudes were multiplied by the appropriate Coulomb phases and in the case of $dd \rightarrow dd$ the Coulomb amplitude was also added.

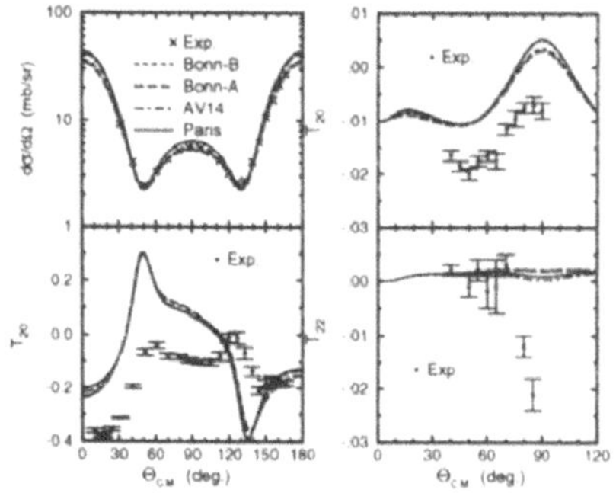

Figure 3. $d\sigma/d\Omega$ and T_{20} for $dd \to pt$ (left), and T_{20} and T_{22} for $dd \to dd$ (right) at $E_d = 6.1$ MeV for different 2N interactions in partial waves $j \leq 1^+$. The experimental points are from Ref. [11].

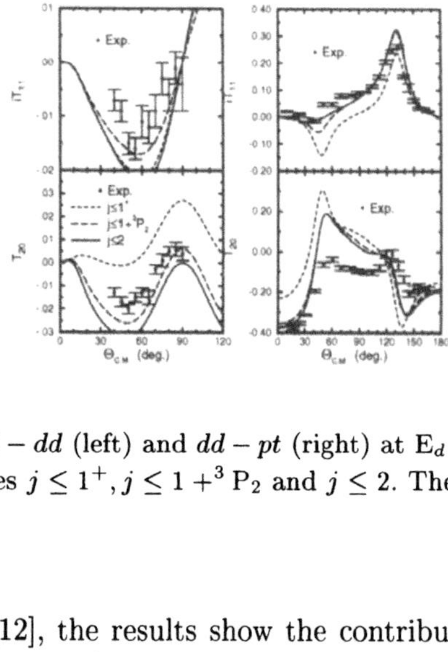

Figure 4. TAP's for $dd - dd$ (left) and $dd - pt$ (right) at $E_d = 6.1$ MeV for Bonn-B potential in partial waves $j \leq 1^+, j \leq 1 +^3 P_2$ and $j \leq 2$. The data is from Ref. [11].

As discussed in Ref. [12], the results show the contribution from higher partial waves which, as expected, are very important to improve the agreement with data. Nevertheless, unlike in $d - n$ scattering, where elastic and breakup observables, to the exception of elastic A_y at low energy, are extremely well described by realistic 2N forces, we find large disagreement with data. If the present results are confirmed by others, this may be a clear signature for a missing 3N force (see Ref. [12] for further discussion).

Finally in Fig. 5 we show $nt \to nt$ elastic observables at different neutron (proton) laboratory energy for Bonn-B potential and in Fig. 6 the elastic $n - t$ phase shifts as a function of energy. The NN P-waves have a strong effect on the differential cross sections and an even bigger contribution to A_y. The resonant behaviour of the ^3P_J $n - t$ partial waves may be directly linked to the J^- 3N

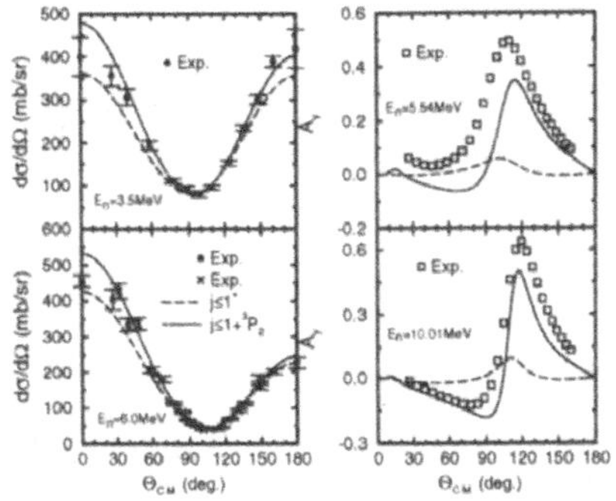

Figure 5. Differential cross sections for $^3H(n,n)^3H$ (left) and analysing power A_y for $^3He(\vec{p},p)^3He$ (right) versus center of mass angle. The experimental points are from Ref. [13] (full circles and squares), Ref. [14] (crosses), and Ref. [15] (open squares).

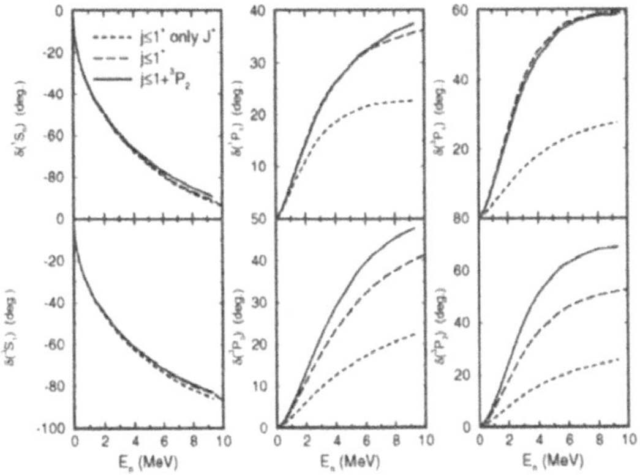

Figure 6. Elastic phase shifts for $n-t$ scattering versus neutron laboratory energy for different S-, P- and D-waves

subamplitudes (compare the dashed lines with the long-dashed lines in Fig. 6) which are enhanced even further by the presence of NN P-waves (solid lines).

3 Conclusions

We find that NN P-waves are extremely important to describe 4N scattering observables, particularly the $n-t$ differential cross section and A_y in the resonance region. As discussed in Ref. [12] the NN P-waves contribute 10% to the total $n-t$ cross section at $E_n = 3.5$ MeV. Nevertheless remaining discrepancies indicate the need for a 3N force.

References

1. P. Grassberger and W. Sandhas: Nucl. Phys. **B2**, 181 (1967); E.O. Alt, P. Grassberger, W. Sandhas: Phys. Rev. **C1**, 85 (1970); ibidem JINR Report No. E4-6688 (1972)

2. A.C. Fonseca: Phys. Rev. **C40**, 1390 (1989)

3. H. Haberzettl and W. Sandhas: Phys. Rev. **C24**, 359 (1981)

4. A.C. Fonseca and P.E. Shanley: Phys. Rev. **D13** 2255 (1976); ibidem, Phys. Rev. **C14**, 1343 (1976)

5. D.J. Ernst, C.M. Shakin and R.M. Thaler: Phys. Rev. **C8**, 46 (1973)

6. S. Sofianos, N.J. McGurk and H. Fiedeldey: Nucl. Phys. **A318**, 295 (1979)

7. H. Kamada and W. Glöckle: Nucl. Phys. **A548**, 205 (1992)

8. F. Ciesielsky et al.: Few-Body System Supplement, **10**, 367 (1999)

9. F. Ciesielski, J. Carbonell and C. Gignoux: Phys. Lett. **B447**, 199 (1999)

10. T. Cornelius et al.: Phys. Rev. **C41**, 2538 (1990)

11. W. Gruebler et al.: Nucl. Phys. **A193**, 129 (1972); ibidem **A193**, 149 (1972); ibidem **A369**, 381 (1981)

12. A.C. Fonseca: *Contribution of Nucleon-Nucleon P-waves to* $nt-nt, dd-pt$ *and* $dd-dd$ *Scattering Observables*, Phys. Rev. Lett. (accepted for publication)

13. J.D. Seagrave, L. Crauberg and J.E. Simmons: Phys. Rev. **119**, 1981 (1960)

14. T.W. Phillips, B.L. Berman and J.D. Seagrave: Phys. Rev. **C22**, 384 (1980)

15. M.T. Alley and L.D. Knutson: Phys. Rev. **C48**, 1890 (1993)

Few-Body Systems Suppl. 12, 451–456 (2000)

Few-Body Systems

Recent High-Accuracy Measurements of the $^1S_0$ Neutron-Neutron Scattering Length

C.R. Howell[1,*], Q. Chen[1,*], D.E. González Trotter[1,*], F. Salinas[1,*], T.S. Carman[2], A.S. Crowell[1,*], A. Hussein[3], W.R. Gibbs[4], B.F. Gibson[5], G. Mertens[6,†], C.F. Moore[7,**], C. Morris[5], A. Obst[5], E. Pasyuk[8], C.D. Roper[1,*], I. Šlaus[1,9,††], S. Sterbenz[5], H. Tang[10], W. Tornow[1,*], R.L. Walter[1,*], C.R. Whiteley[7,**], M. Whitton[5], Z. Zhou[10], W. Glöckle[11], H. Witała[12,***]

[1] Duke University and Triangle Universities Nuclear Lab., Durham, NC
[2] Lawrence Livermore National Laboratory, Livermore, CA
[3] University of Northern British Columbia, Canada
[4] New Mexico State University, Las Cruces, NM
[5] Los Alamos National Laboratory, Los Alamos, NM
[6] University of Tübingen, Tübingen, Germany
[7] University of Texas, Austin, TX
[8] Joint Institute of Nuclear Research, Dubna, Russia
[9] Rudjer Boskovic Institute, Zagreb, Croatia
[10] Chinese Institute for Atomic Energy, Beijing, Peoples Republic of China
[11] Institut für Theoretische Physik, Ruhr-Universität Bochum, Germany
[12] Institute of Physics, Jagellonian University, PL-30059 Cracow, Poland

Abstract. This paper reports two recent high-accuracy determinations of the $^1S_0$ neutron-neutron scattering length, a_{nn}. One was done at the Los Alamos National Laboratory using the $\pi^- d$ capture reaction to produce two neutrons with low relative momentum. The neutron-deuteron (nd) breakup reaction

*This work was supported in part by the U.S. Department of Energy, Office of High Energy and Nuclear Physics, under grant No. DE-FG05-91ER40619.
†Supported in part by the Deutsche Forschungsgemeinschaft Contract No. Me544/8-1.
**Supported in part by the Robert A. Welch Foundation.
††Supported in part by the USA-Croatia NSF Grant JF129 and by the European Community Contract No. CI1*-CT-91-0894.
***Supported in part by the Maria Sklodowska-Curie II Fund under grant No. MEN/NSF-94-161.

was used in other measurement, which was conducted at the Triangle Universities Nuclear Laboratory. The results from the two determinations were consistent with each other and with previous values obtained using the $\pi^- d$ capture reaction. The value obtained from the *nd* breakup measurements is $a_{nn} = -18.7 \pm 0.1(statistical) \pm 0.6(systematic)$ fm, and the value from the $\pi^- d$ capture experiment is $a_{nn} = -18.50 \pm 0.05 \pm 0.53$ fm. The recommended value is $a_{nn} = -18.5 \pm 0.3$ fm.

1 Introduction

The up-down quark mass difference (QMD) leads to a difference in the masses of the neutron and proton and in many hadronic isospin partners, and to a difference between the neutron-neutron (*nn*) and proton-proton (*pp*) $^1S_0$ scattering lengths, $\Delta a = a_{nn}$-a_{pp} [1]. Therefore, an experimental determination of Δa, gives a direct measure of charge-symmetry breaking and can be related to the QMD. The a_{pp} has been measured directly using two-nucleon scattering. The main uncertainty in a_{pp} comes from the large theoretical factors needed to correct for the influences of the Coulomb force. The situation is quite different for a_{nn}. The recommended value [2] for a_{nn} is based on studies of reactions with two strongly interacting neutrons in the exit channel. The two reactions that give a_{nn} with the smallest theoretical uncertainty are pion-deuteron capture ($\pi^- + d \rightarrow n + n + \gamma$) and neutron-deuteron breakup ($n + d \rightarrow n + n + p$). Curiously, measurements using these two reactions give significantly different values of a_{nn}: from $\pi^- d$ capture measurements the average value for a_{nn} is -18.4 ± 0.4 fm and from kinematically-complete (KC) *nd* breakup experiments the average value is -16.9 ± 0.3 fm. It was suggested that the difference in these values has its origin in the three-nucleon force (3NF), which would act in the *nd* breakup reaction [2] but not in the other. Another interesting feature is that the average value of a_{nn} obtained from kinematically-incomplete (KI) *nd* breakup experiments differ significantly from that determined from KC *nd* breakup measurements. This discrepancy is also suspected to be caused by 3NF effects [2]. New measurements of a_{nn} using all three reactions are needed to resolve these issues. In this paper, we report the results of two recent determinations of a_{nn} using the $\pi^- d$ capture [3] and KC *nd* breakup [4] reactions.

2 Measurement of a_{nn} by $\pi^- d$ Capture

The most recent reported measurement of a_{nn} using the $^2H(\pi^-, n\gamma)n$ reaction was made by Howell *et al.* [3] at Los Alamos National Laboratory. In their measurements a beam of pions was stopped in a liquid deuterium target, and the neutron and the associated gamma ray from the $^2H(\pi^-, n\gamma)n$ reaction were detected in coincidence. The neutrons were detected in an array of 24 12-cm diameter $\times$ 5-cm thick liquid organic scintillators, and the γ rays were detected in a position-sensitive γ-ray calorimeter. The neutron and γ-ray detectors were positioned on opposite sides of the pion beam axis about 180° apart.

The value of a_{nn} was determined from the shape of the measured neutron

Table 1. Values of a_{nn} (fm) obtained with increasely more realistic MC simulations. The symbols PG and FG mean point geometry and finite geometry, respectively. The $\theta_{n\gamma}$ angle specifies the center of the 50-mr wide bin. The results in this table are from the the analysis of the NTOF spectrum for the entire neutron-detector array. The uncertainties are statistical only. The last row gives the weighted average of the first five rows.

$\theta_{n\gamma}$	PG	FG	FG + Scatt.	Full MC
25	-17.40 ± 0.15	-17.70 ± 0.15	-18.60 ± 0.15	-18.43 ± 0.12
75	-17.40 ± 0.10	-17.80 ± 0.15	-18.70 ± 0.15	-18.34 ± 0.09
125	-17.30 ± 0.15	-17.90 ± 0.15	-18.80 ± 0.15	-18.47 ± 0.09
175	-17.10 ± 0.15	-17.70 ± 0.15	-18.60 ± 0.20	-18.92 ± 0.13
225	-17.30 ± 0.20	-17.50 ± 0.25	-18.30 ± 0.20	-18.60 ± 0.22
275	-17.60 ± 0.35	-17.90 ± 0.45	-18.30 ± 0.25	-18.48 ± 0.24
wt. avg.	-17.32 ± 0.06	-17.75 ± 0.07	-18.61 ± 0.07	-18.50 ± 0.05

time-of-flight (NTOF) spectrum of the $n-\gamma$ coincidence data. The NTOF was determined from the time difference between the detection of the γ ray and the detection of a neutron in the scintillator array. A total of 1.9×10^6 $n-\gamma$ coincidence events were accumulated.

A number of controls were used to obtain realistic estimates of systematic errors in their determination of a_{nn}. To evaluate their error due to uncertainties in the neutron detection efficiency, four different designs of neutron detectors were used. The detector types differed in thickness and scintillator liquid (three different fluids and three different thicknesses). The efficiency functions of two neutron detectors in each type were measured in a separate experiment at the Triangle Universities Nuclear Laboratory. The relative efficiencies were measured to about $\pm 1\%$ accuracy. A value of a_{nn} was determined separately for each detector type. The results for the four types were in agreement within statistical uncertainities.

Another control employed was that the data were analyzered in six bins of the opening angle between the detected neutron and γ ray ($\theta_{n\gamma}$). Because the shape of the predicted NTOF spectrum is strongly dependent on $\theta_{n\gamma}$, and because the measured value of a_{nn} should be independent of $\theta_{n\gamma}$, statistically significant differences in the measured values for the six angle bins give information about the systematic uncertainty in the angle determination. Each $\theta_{n\gamma}$ bin was 50 mr wide with the first bin starting at $\theta_{n\gamma} = 0$. A value of a_{nn} was obtained for each $\theta_{n\gamma}$ bin using a χ^2 analysis in which the measured NTOF spectrum was compared to a Monte-Carlo (MC) simulation. The simulation used the model in Ref. [5] to generate the probability distributions as functions of $\theta_{n\gamma}$ and the energy of one neutron. It included the finite geometry of the experimental setup, the energy and time resolution of the detection system, the effects due to neutron scattering in the target materials and air, and scattering in the neutron detector array. The results are shown in Table 1. A summary of the systematic uncertainties is given by Howell *et al.* [3]. The value obtained in that work is $a_{nn} = -18.50 \pm 0.05(statistical) \pm 0.44(systematic) \pm 0.30(theoretical)$ fm.

3 Measurement of a_{nn} by *nd* Breakup

The most recent measurement of a_{nn} using the *nd* breakup measurement was reported by González Trotter *et al.* [4]. Two features that distinguish this work from early determinations are: (1) the data were analyzed using rigorous *nd* theory, and (2) the $^1S_0$ neutron-proton scattering length (a_{np}), which has been determined to high accuracy from two-nucleon measurements, was measured in the same experiment to gauge the systematic uncertainties in their methods. Values for a_{nn} and a_{np} were determined from their cross-section measurements of the corresponding nucleon-nucleon final-state interaction (FSI). They measured the cross section for the neutron-neutron (*nn*) FSI at four angles of the emitted *nn* pair and that for the neutron-proton (*np*) FSI at three of the same angles. The dispersion in the values of a_{nn} obtained at the four angles gave information about systematic (experimental and theoretical) uncertainties in their techniques. The deviation of their value of a_{np} from the value obtained in two-nucleon measurements was used to infer the size of the shift in the scattering length due to 3NF effects in *nd* breakup.

All measurements were made at the Triangle Universities Nuclear Laboratory with an incident neutron beam energy of 13.0 MeV. Backgrounds were reduced by over determining the reaction kinematics. The momentum of the two emitted neutrons and the energy of the recoil proton were measured for each event. The neutrons were detected in an array of liquid organic scintillators. The energy of the recoil proton was measured in a deuterated liquid scintillator C_6D_{12} (NE-232), which served as the deuteron scatterer and will be referred to as the central detector (CD). The energies of the outgoing neutrons were determined by measuring their flight times from the CD to the neutron detectors. The neutron detector efficiencies were determined in a dedicated series of measurements using neutrons from the $^2H(d,n)^3He$ reaction and from a ^{252}Cf source. The energy dependence of the relative detection efficiency of each detector was determined to an accuracy of $\pm 1.0\%$, and the absolute efficiency to $\pm 2.5\%$. The breakup cross sections were normalized to the yields for *nd* elastic scattering, which were collected concurrently with the breakup yields.

The measured cross sections were compared to Monte-Carlo (MC) simulations that included the energy resolution and the finite geometry of the experimental setup. The basis of these simulations was theoretical point-geometry cross-section libraries generated for a range of a_{nn} (a_{np}) values using rigorous solutions to the *nd* breakup problem [6]. A value of a_{NN} and its statistical uncertainty were determined for each detector-pair configuration using a single-parameter (either a_{np} or a_{nn}) minimum-χ^2 fit to the absolute cross-section data. The results are given in Table 2. The uncertainties listed in Table 2 are statistical only. The systematic uncertainties in their determinations were

Table 2. The a_{np} and a_{nn} values obtained by González Trotter *et al.* [4]. The χ^2 per datum for their best fit is given at each angle. The weighted mean of the data for all angles is given in the bottom. All uncertainties are statistical only.

θ_{NN}	$a_{np} \pm \Delta a_{np}$ (fm)	χ^2/pt	$a_{nn} \pm \Delta a_{nn}$ (fm)	χ^2/pt
43.0°	-23.6 ± 0.3	2.1	-18.8 ± 0.4	1.5
35.5°	-23.2 ± 0.3	2.7	-17.7 ± 0.4	0.6
28.0°	-23.7 ± 0.3	3.0	-18.8 ± 0.2	0.1
20.5°			-18.9 ± 0.2	0.8
Mean	-23.5 ± 0.2	2.6	-18.7 ± 0.1	0.6

±0.8 fm for a_{np} and ±0.6 fm for a_{nn}. Uncertainties in the neutron detector efficiencies and the integrated target-beam luminosity accounted for about 80% of the systematic uncertainty.

No signficant angle dependence in a_{np} or a_{nn} was observed. Combining the statistical and systematic uncertainties in quadrature, they obtained a_{np} = -23.5 ± 0.8 fm and a_{nn} = -18.7 ± 0.6 fm. Their result for a_{np} is in agreement with the value of a_{np} (-23.748 ± 0.009 fm [7]) obtained from free *np* scattering measurements. This result was used to set an upper limit on the influence of 3NF on the value of *NN* scattering lengths determined using *nd* breakup, i.e., $\Delta a_{np}^{3NF} \leq a_{np}^{nd} - a_{np}^{free} = 0.2 \pm 0.8$ fm, where *free* and *nd* refer to values obtained from data for free *np* scattering and from *nd* breakup, respectively. This result is consistent with zero.

4 Conclusions

The most recent determinations of a_{nn} from $\pi^- d$ capture and from kinematically-complete *nd* breakup cross-section measurements give results that are consisent with each other and in agreement with previous $\pi^- d$ capture measurements. While these results provide little information about the causes of the long-standing discrepancies between the values for a_{nn} obtained from these two reactions, they do strongly indicate that the problem is with the previous *nd* results. For this reason, we recommend a value for a_{nn} that is the weighted average of the values obtained in four measurements, the two previous $\pi^- d$ capture experiments [8, 9], the new $\pi^- d$ results [3], and the new *nd* value [4]. Because of the correlation in the theoretical uncertainties in the results of $\pi^- d$-capture experiments, the average was computed in two steps. The results are summarized in Table 3. First the weighted average was computed for the $\pi^- d$ experiments. Only the statistical and systematic uncertainities were used in the weighting. The statistical and systematic uncertainties were added in quadrature and are presented in Table 3 as the experimental uncertainties. The theoretical uncertainty of ± 0.30 fm was added in quadrature with the final experimental uncertainty and is presented in Table 3 as the total uncertainty

Table 3. Recommend value for a_{nn}. The uncertainties in the values from the three $\pi^- d$ capture experiments are experimental (combined statistical and systematic) only. The theoretical uncertainty of ± 0.30 fm is added in quadrature to the experimental uncertainty in the average of the $\pi^- d$ values. The recommended value is the weighted average of the of the $\pi^- d$ and nd results.

Reaction	Ref.	a_{nn} (fm)
$\pi^- d \rightarrow nn\gamma$	[8]	-18.60 ± 0.43
	[9]	-18.70 ± 0.57
	[3]	-18.50 ± 0.44
	wt. avg.	-18.58 ± 0.40
$nd \rightarrow nnp$	[4]	-18.7 ± 0.6
	recommended	-18.6 ± 0.3

in the average of the $\pi^- d$ results. Next the weighted average of the mean value from the $\pi^- d$ capture measurements and the value obtained in nd breakup was computed. The total uncertainty (experimental and theoretical) in the $\pi^- d$ results was used in the weighting and in the computation of the uncertainty in the recommended value. The values in Table 3 have not been corrected for nn magnetic interactions.

References

1. G.A. Miller, B.M.K. Nefkens, I. Slaus: Phys. Reports **194**, 1 (1990)
2. I. Šlaus, Y. Akaishi, H. Tanaka: Phys. Reports **173**, 257 (1989)
3. C.R. Howell et al.: Phys. Lett. **B444**, 252 (1998)
4. D.E. Gonzalez Trotter et al.: Phys. Rev. Lett. **83**, 3788 (1999)
5. W.R. Gibbs, B.F. Gibson, G.J. Stephenson, Jr.: Phys. Rev. **C11**, 90 (1975); Phys. Rev. **C12**, 2130 (1975); Phys. Rev. **C16**, 327 (1977); Phys. Rev. **C17**, 856 (1978)
6. H. Witała et al.: Phys. Rev. **C52**, 1254 (1995)
7. L. Koester and W. Nistler: Z. Physik, **272**, 189 (1975)
8. B. Gabioud et al.: Phys. Rev. Lett. **42**, 1508 (1979); Phys. Lett. **103B**, 9 (1981); Nuc. Phys. **A420**, 496 (1984)
9. O. Schori et al.: Phys. Rev. **C35**, 2252 (1987)

Few-Body Systems Suppl. 12, 457–461 (2000)

Few-Body Systems

Phase-Shift Analysis of Elastic *pp* Scattering at High Energies and Catastrophic Energy-Dependence of Spin-Orbit Interaction

J. Nagata[1], K. Harada[2], H. Yoshino[3], M. Matsuda[4]

[1] Hiroshima University, Higashi-Hiroshima 739-8527, Japan
[2] Kurashiki-Sakuyo University, Kurashiki 710-0292, Japan
[3] Hiroshima International University, Kamo-gun 724-0695, Japan
[4] Hiroshima University, Higashi-Hiroshima 739-8521, Japan

Abstract. A possible weak first-order phase transition of sub-nuclear medeium at T_c=100~150 MeV is indicated by means of the phase-shift analyses of *pp* scattering at P_L=1~12 GeV/c. A phenomenological model on the analogy of BCS-theory is proposed for a scenario of the phase transition.

1 Introduction

We have performed the phase-shift analyses (PSA) of elastic *pp* scattering at P_L=1~12 GeV/c[1]. It has been found that the spin-orbit component phase-shift of *p*-wave has such a strong energy dependence that it decreases rapidly at P_L=3~6 GeV/c and tends to zero at higer energies. As regards of such a strong energy-dependence, we suggested a " shrinkage " of spin-orbit interaction and give its threshold radius[1].

Here our PSAs of *pp* scattering at P_L=1~12 GeV/c are reviewed and the implications of the indicated shrinkage of spin-orbit force in the weak first-order phase transition of sub-nuclear medium at T_c= 100~150 MeV are discussed. In order to write a scenario of the weak first-order transition from Quark-Gluon phase (QGP) to Hadron phase(HP), we propose a phenomenological model on the analogy of the BCS-theory of the Fermi liquid.

2 Phase-shift Analyses of *pp* Scattering at 1~12 GeV/c and Spin-orbit Interaction

In order to do the partial wave analyses of *N*-*N* scattering data in this energy region, we have proposed a study of nuclear interaction in which the scattering amplitudes are determined on the basis of the following correspondence principle. This method is a relativistic extension of Taketani's way[2] of studying nuclear force.

Principle I The peripheral part of the amplitude of nucleon-nucleon scattering in the outer region of the distance $r \gtrsim 2.5$ fm is provided with the one-pion-exchange(OPE) contribution.

Principle II In the region $2.5 \gtrsim r \gtrsim 1.0$ fm, the scattering amplitude is evaluated by the modified one-boson-exchange(OBE) contributions[3]. The modification of the OBE-amplitude of nucleon-nucleon scattering is

$$\frac{g^2}{m^2 - t} \longrightarrow \frac{g^2}{m^2}[\frac{\Lambda^2}{\Lambda^2 - t/n}]^n, \tag{1}$$

where t is the squared momentum transfer of nucleons and m the observed mass of the exchanged boson, and g, Λ and n are parameters peculiar to boson. The full amplitude of nucleon-nucleon scattering is represented by

$$M = \sum_{J<J_0} [f_J(\delta_J, \eta_J)] + \sum_{J_0 \leq J \leq J_1} [f_J(\delta_J(\mathrm{OBE}), \eta_J)] + M_{\mathrm{OBE}}(J > J_1). \tag{2}$$

Here $[\ f_J(\delta_J, \eta_J)]$ is the contribution from the phase shift δ_J and the reflection parameter η_J for the angular momentum J. $\delta_J(\mathrm{OBE})$ are the phase shifts obtained from the modified OBE-amplitudes M_{OBE} by the K-matrix method. The boundary angular momentum J_0 and J_1 are chosen by their corresponding impact parameters b, which are equal to about 1.0 and 2.5 fm, respectively.

Principle III The partial wave amplitudes $[f_J(\delta_J, \eta_J)]$ with the angular momentum $J < J_0$ reflect dynamics in the inner region $r \leq 1$ fm. They are determined together with the amplitudes $[f_J(\delta_J(\mathrm{OBE}), \eta_J)]$ and M_{OBE}, by means of the PSA.

The modified OBE-amplitudes are calculated with the well known bosons π, σ, ρ and ω, the observed masses of which are taken as 135, 400, 770 and 770 MeV, respectively. σ-meson is still not observed, but is known to represent the effect of two-pion-exchange between nucleons.

We performed the PSAs of *pp* scattering at P_L=1~12 GeV/c by means of this method. In the assumption of negligibly small quadratic spin-orbit and q-square terms, the triplet-odd (real) phase shift $\delta_{\ell,J}$ is decomposed into the three components,

$$\delta_{\ell,J} = \delta_C^\ell + (S_{12})_{\ell,J}\delta_T^\ell + (L \cdot S)_{\ell,J}\delta_{LS}^\ell, \quad (J = \ell - 1, \ell, \ell + 1). \tag{3}$$

Here δ_C^ℓ, δ_T^ℓ and δ_{LS}^ℓ are the central, tensor and spin-orbit phase-shifts, respectively. The recently obtained solutions of δ_{LS}^1 at 3, 6 and 12 GeV/c are shown in Fig.1 together with the solutions given by the other groups. The energy-dependence of spin-orbit phase-shifts indicates that there exists a repulsive spin-orbit interaction which is so strong as to be superior to the attractive one due to the one vector-boson exchange, and the repulsion of such an interaction becomes prominent at the range $r \lesssim 0.5$ fm.

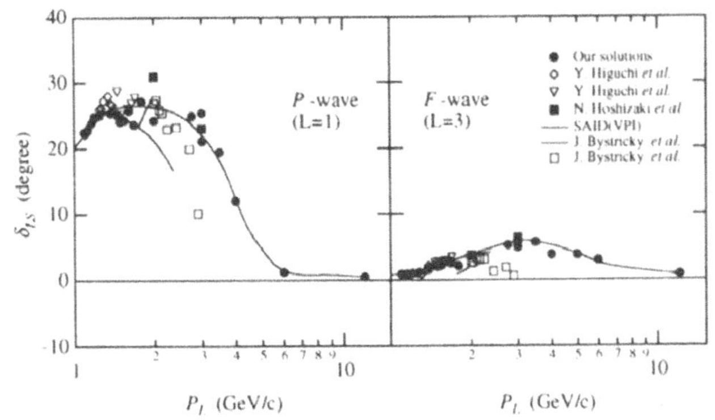

Figure 1. The spin-orbit phase-shifts of P and F-waves. The dots show our obtained solutions and the solid lines the interpolated ones by Spline function.

The scattering matrix M in the two-proton spin space is determined by a PSA of p-p scattering at each energy. In the eikonal model, the scattering matrix has an impact parameter representation in terms of the eikonal $\chi(\boldsymbol{b})$ as follows:

$$M(\boldsymbol{q}) = i(p/2) \int [1 - \exp(i\chi(\boldsymbol{b}))] \exp(-i\boldsymbol{q} \cdot \boldsymbol{b}) d^2\boldsymbol{b}, \tag{4}$$

where $\boldsymbol{b}$ is the impact parameter, $\boldsymbol{q}$ is the momentum transfer $\boldsymbol{q} = \boldsymbol{p}' - \boldsymbol{p}$, and $\boldsymbol{p}$ and $\boldsymbol{p}'$ are the incident and scattered momenta in the c.m. system, respectively.

The eikonal is numerically evaluated by Fourier-Bessel transformation of the scattering matrix $M(\boldsymbol{q})$ of the PSA solutions[4]. The obtained spin-orbit components of eikonal χ_{LS} at 3, 6 and 12 GeV/c are shown in Fig.2, respectively. Because $\mathrm{Re}\chi_{LS}(b)$ is proportional to $-\int_{-\infty}^{+\infty} V_{LS}(r)dz$, where $V_{LS}(r)$ is the spin–orbit part of a local optical potential and $r = (\boldsymbol{b}^2 + z^2)^{1/2}$, Re χ_{LS} presents the information on the b-dependence of the spin–orbit interaction. The solutions of $\mathrm{Re}\chi_{LS}$ at 3, 6 and 12 GeV/c agree among them in the outer region $b \gtrsim 1.0$ fm, which represent the force due to peripheral mesonic contribution. In the region of 0.5~1.0 fm, $\mathrm{Re}\chi_{LS}$ exhibits the nonstatic effect due to the modified one-boson-exchange interaction. In the inner region of $b \lesssim 0.5$ fm, $\mathrm{Re}\chi_{LS}$ gives new indication on an existence of a short-range repulsive force with a strong energy dependence.

It is, however, to be noted that the spin-orbit component phase shifts have a tendency to be zero progressively with energy. These found results can be never understood by the usual hadron physics. Our only possible explanation about them is that a phase transition takes place in the incident energy region of proton P_L= 3.5~12 GeV/c[1, 5]. The threshold radius (R_{th}) for the "shrinkage" of the spin-orbit force is about 0.5 fm, which almost agrees with the bag-radius predicted by the MIT bag model[6]. The critical temperature for the transition from QGP to HP was estimated as 150~200 MeV by using the Gibbs condition,

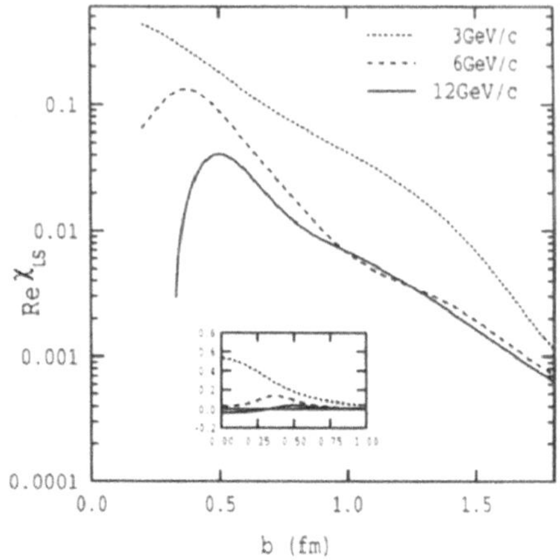

Figure 2. The real part of the spin-orbit eikonal Re $\chi_{LS}(b)$ obtained from the PSAs at 3, 6 and 12 GeV/c.

provided that the transition is approximately same to that from the massless quark-gluon gas to the massless pion gas[7]. The energy density of fire ball at this critical point is ϵ_c= 0.9 - 2.5 GeV/fm^3.

The momentum region P_L=3.5~12 GeV/c, where the catastrophic transition of spin-orbit force is found in our analysis, correspond to the incident-proton energy in c.m. system E_{cm}=1.5~2.5 GeV[1], which is equivalent to T_c=100~150 MeV.

3 A Phenomenological Model for a Scenario of the Weak 1st-order Phase Transition

We propose a phenomenological model[9] in which the quark Cooper-pairs play an essential role in the hadron phase, and an extent of quark pair is supposed to be twice of the threshold radius of the shrinkage, 1fm. By using the model,we estimate a bare quark mass m_q and an order parameter Δ. Here we give the physical meaning of Δ the excitation energy of unpaired quarks from the paired quarks.

In analogy with Weizsäcker mass formula[10] for nucleus, the nucleon mass in our qq-pairing model will be given by $M = n_1 \times m_q + n_2 \times \Delta$, where n_1=(No. of constituent quarks) + (No. of constituent anti-quarks), n_2=(No. of constituent quarks) - (No. of constituent anti-quarks).

The nucleon mass with $n_1 = n_2 = 3, \Delta = 250$ MeV and $m_q = 60$ MeV is evaluated as $M_N = 3 \times 60$ MeV $+3 \times 250$ MeV= 930 MeV.

If this mass formula is valid for the pion mass in $q\bar{q}$ -pairing, the pion mass with $n_1 = 2$, $n_2 = 0$ and $m_q = 60$ MeV is evaluated as $M_\pi = 2 \times 60$ MeV $= 120$ MeV. The constituent-quark mass (u or d) is predicted by $m^{\#}$= (1/3)(3 × 60

MeV $+3 \times 250$ MeV) = 310 MeV.

4 Concluding Remarks

By means of the PSAs of *pp* scattering at P_L=1∼12 GeV/c, we found the catastrophic energy-dependence of spin-orbit interaction in the c.m. energy region of proton $E_{cm} = 1.5 \sim 2.5$ GeV, and suggested the possible phase-transition at $T_c = 100 \sim 150$ MeV. Our estimated T_c-range is somewhat lower than the ones suggested by Shuryak[7] and Witten[8].

We considered for the analogy of the dynamical parameters of the present model to those of the BCS-theory to be more important than the relativistic scheme of the field theory. Such a model should be permitted in a phase transition of the " strong " interaction by the reason mentioned in the above section. In the case of the first-order phase transition, it is meaningless to reinterpret the hadron dynamics by the QCD.

By the present model of the phase transition from hadron phase to quark gluon phase, the shrinkage of spin orbit force is interpreted as follows: (1) The threshold radius of shrinkage (∼0.5 fm) is a half of Pippard length of the *qq*-pair wave function (2) When a distance between two nucleons is shortened and amounts to almost the same as the Pippard length of *qq*-pair, the *qq*-pair composing nucleon starts to disintegrate (3) The critical temperature of the disintegration is about 100∼150 MeV.

References

1. M. Matsuda et al.: Prog. Theor. Phys. **93**, 1059 (1995)
2. J. Iwadare et al.: Prog. Theor. Phys. Suppl. No. 3, 32 (1956) ; M. Taketani, S. Nakamura, M. Sasaki: *ibid.* 169
3. M. Kawasaki, Y. Susuki, M. Yonezawa: Prog. Theor. Phys. **47**, 589 (1972)
4. M. Matsuda, H. Suemitsu, M. Yonezawa: Prog. Theor. Phys. **77**, 497 (1987)
5. M. Matsuda: *Proceedings of XIII International Conference on Few-Body Problems in Physics, Flinders University*, FLAS-R-216, 118 (1992)
6. C. E. Carlson, T. H. Hansson, C. Peterson: Phys. Rev. **D27**, 1556 (1983)
7. E.V. Shuryak: Phys. Rev. **61**, 71 (1980)
8. E. Witten: Phys. Rev. **D30**, 272 (1984)
9. M. Matsuda: Nucl. Phys. **A631**, 436c (1998) ; K. Harada and M. Matsuda: Soryushiron-Kenkyu (mimeographed circular in Japanese) **Vol.96**, No.4, 139 (1998) ; K. Harada, M. Matsuda, Z. YuMing: Proceedings of CCAST Workshop, CIAE(Beijing), **Vol.80**, 187 (1998)
10. A. Bohr, B.R. Mottelson, D. Pines: Phys. Rev. **110**, 936 (1958)

Few-Body Systems Suppl. 12, 462–465 (2000)

Few-Body Systems

Helicity Formalism for NN Scattering without Partial Wave Decomposition

I. Fachruddin*, W. Glöckle

Institut für Theoretische Physik II, Ruhr-Universität Bochum, D-44780 Bochum, Germany

Abstract. At intermediate energies it appears advantageous to work without partial wave decomposition and use directly the momentum vectors. We derived a finite set of coupled Lippmann-Schwinger equations using helicities. They can be used for any type of realistic NN forces, which are given in operator form. As an example we have chosen the Bonn OBEPR NN potential.

1 Introduction

At low energies up to few tenth of MeV very few partial waves are sufficient to describe NN scattering. However, at intermediate (a few hundred MeV) and higher energies many partial waves are needed. Besides in this energy region the scattering amplitude for a high partial wave oscillates strongly in angle. This suggests T-matrix calculations using directly the momentum vectors.

One-boson-exchange NN potentials are derived in momentum space. Instead of using spin operators with a fixed quantisation axis often these potentials are formulated in helicity representation. Moreover, in high energy region taking relativity into account is unavoidable. Thereby, using spin components along a fixed direction is more complicated than using helicities [1].

There are some works [2, 3, 4] preceding ours, which address formulations without partial waves and one of them [3] even using helicities. We propose a modified form of equations in comparison to [3]. We present here the formalism for the two nucleon system avoiding partial waves and using helicities.

In Sect. 2 we describe the Lippmann-Schwinger equations as well as the basis states and the physical T-matrix. In Sect. 3 we show the operators to express NN potentials. In Sect. 4 we show the relation between the partial wave T-matrix and the helicity T-matrix. We conclude in Sect. 5.

* *Permanent address:* Jurusan Fisika, FMIPA, Universitas Indonesia, Depok 16424, Indonesia

2 Basis States, Lippmann-Schwinger Equation and Physical T-Matrix

To calculate observables we need T-matrix elements. Firstly we define basis states based on which T-matrix elements are calculated by solving the Lippmann-Schwinger equation. The basis states are

$$|\boldsymbol{q};\hat{q}S\Lambda;t\rangle^{\pi a} \equiv \frac{1}{\sqrt{2}}(1-\eta_\pi(-)^{S+t})|t\rangle|\boldsymbol{q};\hat{q}S\Lambda\rangle_\pi \tag{1}$$

which are antisymmetric choosing $|\boldsymbol{q};\hat{q}S\Lambda\rangle_\pi \equiv \frac{1}{\sqrt{2}}(1+\eta_\pi P)|\boldsymbol{q};\hat{q}S\Lambda\rangle$ and $|\boldsymbol{q};\hat{q}S\Lambda\rangle \equiv |\boldsymbol{q}\rangle|\hat{q}S\Lambda\rangle$. Here $\boldsymbol{q}$, S, Λ, P and η_π are relative momenta, total spin, total helicity, parity operator and parity eigenvalue, respectively.

Based on these basis states the T-matrix element is defined as

$$T^{\pi St}_{\Lambda'\Lambda}(\boldsymbol{q}',\boldsymbol{q}) \equiv {}^{\pi a}\langle\boldsymbol{q}';\hat{q}'S\Lambda';t|T|\boldsymbol{q};\hat{q}S\Lambda;t\rangle^{\pi a} \tag{2}$$

For $\hat{q} = \hat{z}$ one can show $T^{\pi St}_{\Lambda'\Lambda}(\boldsymbol{q}',\boldsymbol{q}) = e^{i\Lambda(\phi'-\phi)}T^{\pi St}_{\Lambda'\Lambda}(q',q,\theta')$ where $T^{\pi St}_{\Lambda'\Lambda}(q',q,\theta')$ obeys a set of coupled Lippmann-Schwinger equations:

$$T^{\pi St}_{\Lambda'\Lambda}(q',q,\theta') = \frac{1}{2\pi}v^{\pi St,\Lambda}_{\Lambda'\Lambda}(q',q,\theta',0) + \frac{1}{4}\sum_{\Lambda''=0}^{S} 2^{\Lambda''}$$
$$\times \int_0^\infty dq''q''^2 \int_{-1}^{1} d(\cos\theta'')v^{\pi St,\Lambda}_{\Lambda'\Lambda''}(q',q'',\theta',\theta'')G_0(q'')T^{\pi St}_{\Lambda''\Lambda}(q'',q,\theta'') \tag{3}$$

with

$$v^{\pi St,\Lambda}_{\Lambda'\Lambda''}(q',q'',\theta',\theta'') \equiv \int_0^{2\pi} d\phi'' e^{-i\Lambda(\phi'-\phi'')}V^{\pi St}_{\Lambda'\Lambda''}(\boldsymbol{q}',\boldsymbol{q}'') \tag{4}$$

For $S = 0$ it is a single equation but for $S = 1$ it is a set of two coupled equations. We need to solve it only for helicities 0 and 1. Figure 1 shows a three dimensional plot of the real part of the half-shell T-matrix for S = 1, parity = even, $\Lambda' = -1$, $\Lambda = 1$ at $E_{lab} = 100$ MeV.

The physical T-matrix for spin-magnetic and isospin quantum numbers m_1m_2, $\nu_1\nu_2$ is calculated using the following relation:

$$_a\langle\nu_1\nu_2m_1'm_2'\boldsymbol{q}'|T|\nu_1\nu_2m_1m_2\boldsymbol{q}\rangle_a$$
$$= \frac{1}{4}e^{-i(\Lambda_0'\phi'-\Lambda_0\phi)}\sum_{S\pi t}(1-\eta_\pi(-)^{S+t})C(\frac{1}{2}\frac{1}{2}t;\nu_1\nu_2)^2C(\frac{1}{2}\frac{1}{2}S;m_1'm_2'\Lambda_0')$$
$$\times C(\frac{1}{2}\frac{1}{2}S;m_1m_2\Lambda_0)\sum_{\Lambda'\Lambda} d^S_{\Lambda_0'\Lambda'}(\theta')d^S_{\Lambda_0\Lambda}(\theta)T^{\pi St}_{\Lambda'\Lambda}(\boldsymbol{q}',\boldsymbol{q}) \tag{5}$$

3 Operators for the Potential

We need potentials given in operator form. For the example of OBEPR there are five types of operators, called O operators in [3], which cannot be applied

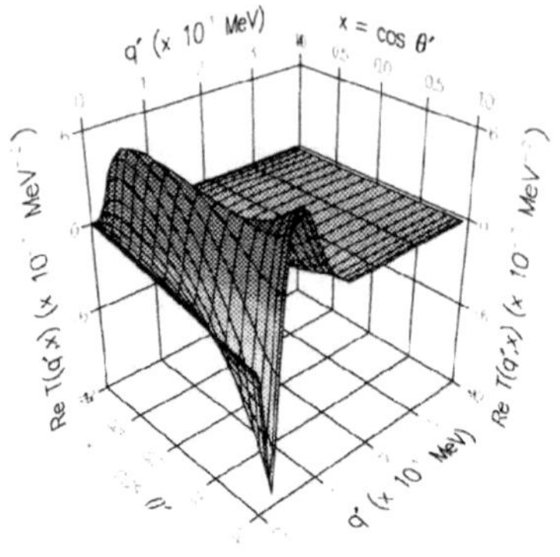

Figure 1. The real part of the half-shell T-matrix for S = 1, parity = even, $\Lambda' = -1$, $\Lambda = 1$ at $E_{lab} = 100$ MeV.

directly to our basis states. We take the six Ω operators from [3] for which this application is possible, but we redefined Ω_5 so that $\Omega_5 = (\boldsymbol{S} \cdot \hat{q}')^2 (\boldsymbol{S} \cdot \hat{q})^2$. The relation between the O and the Ω operators is then

$$
\begin{aligned}
O_1 &= \Omega_1 \qquad\qquad\qquad (\gamma = \hat{q}' \cdot \hat{q}) \\
O_2 &= -3\Omega_1 + 2\Omega_2 \\
O_3 &= 2q'^2(\Omega_3 - \frac{1}{2}\Omega_1) + q^2(\Omega_6 - \frac{1}{2}\Omega_1) - q'q\Omega_4 \\
&\quad + q'q\gamma(2\Omega_1 - \Omega_2) + q'q\frac{1}{\gamma}(\Omega_2 - 2\Omega_3 - 2\Omega_6 + 2\Omega_5) \\
O_4 &= \frac{1}{2}q'q(2\Omega_4 - \gamma\Omega_2 + \frac{1}{\gamma}(\Omega_2 - 2\Omega_3 - 2\Omega_6 + 2\Omega_5)) \\
O_5 &= 2q'^2(\Omega_3 - \frac{1}{2}\Omega_1) + q^2(\Omega_6 - \frac{1}{2}\Omega_1) + q'q\Omega_4 \\
&\quad - q'q\gamma(2\Omega_1 - \Omega_2) - q'q\frac{1}{\gamma}(\Omega_2 - 2\Omega_3 - 2\Omega_6 + 2\Omega_5) \qquad (6)
\end{aligned}
$$

4 The Partial Wave T-Matrix

From the helicity T-matrix we can calculate the partial wave T-matrix.

$$
\begin{aligned}
T_{l'l}^{Sjt}(q) &= \frac{4\pi}{(1 - \eta_\pi(-)^{S+t})(1 + \eta_\pi(-)^{l'})(1 + \eta_\pi(-)^{l})} \frac{\sqrt{2l'+1}\sqrt{2l+1}}{2j+1} \\
&\times \sum_{\Lambda'\Lambda} C(l'Sj; 0\Lambda') C(lSj; 0\Lambda) \int_{-1}^{1} d(\cos\theta') d^j_{\Lambda\Lambda'}(\theta') T^{\pi St}_{\Lambda'\Lambda}(q, q, \theta') \qquad (7)
\end{aligned}
$$

where $T^{Sjt}_{l'l}(q) \equiv \langle q(l'S)jmt|T|q(lS)jmt\rangle$. We calculated phase shifts from the T-matrix (7) using the Bonn OBEPR potential [5]. They are shown in Table 1 together with those calculated using partial wave decomposition.

Table 1. Phase shifts at $E_{lab} = 100$ MeV (first line) and $E_{lab} = 325$ MeV (second line). δ = phase shift from the T-matrix (7) and δ_{pw} = phase shift from partial wave calculation.

States	δ	δ_{pw}	States	δ	δ_{pw}
$^1S_0$	27.38	27.37	$^3P_0$	13.10	13.10
	-7.25	-7.26		-9.43	-9.44
$^1P_1$	-12.87	-12.87	$^3P_1$	-13.72	-13.72
	-20.09	-20.10		-28.01	-28.04
$^1D_2$	3.28	3.28	$^3D_2$	21.56	21.55
	7.13	7.13		36.25	36.20
$^1F_3$	-2.48	-2.48	$^3F_3$	-1.77	-1.77
	-5.24	-5.24		-4.93	-4.93

5 Summary

We have formulated a method to carry out scattering process calculations of the two nucleon system without partial wave decomposition in helicity representation. With this method we can reduce the algebraic and numerical work drastically. We need to solve only one set of Lippmann-Schwinger equations instead of solving many for all partial waves. The results of phase shift calculation shows that this method works well.

Acknowledgement. We would like to thank Charlotte Elster from Ohio University, Athens, Ohio for fruitfull discussions and assistance. The first author also would like to thank dem Deutschen Akademischen Austauschdienst for its financial support.

References

1. M. Jacob, G.C. Wick: Annals of Physics **7**, 404 (1959)

2. J. Holz, W. Glöckle: Phys. Rev. **C37**, 1386 (1988); J. Comput. Phys. **76**, 131 (1988)

3. R.A. Rice, Y. Kim: Few-Body Systems **14**, 127 (1993)

4. Ch. Elster, J.H. Thomas, W. Glöckle: Few-Body Systems **24**, 55 (1998)

5. R. Machleidt, K. Holinde, C. Elster: Phys. Rep. **149**, 1 (1987)

Few-Body Systems Suppl. 12, 466–469 (2000)

Few-Body Systems

The $np \to d(\pi\pi)^0$ Reaction with the σ-Model and $\Delta\Delta$ Model

N. Hiroshige[1], M. Matsuda[2], J. Nagata[3], T. Ueda[4], H. Yoshino[5]

[1] Faculty of Economics, Hannan University, Matsubara, 580-8502, Japan
[2] Faculty of IAS, Hiroshima University, Higashi-Hiroshima, 739-8521, Japan
[3] VBL, Hiroshima University, Higashi-Hiroshima, 739-8527, Japan
[4] Faculty of Science, Ehime University, Matsuyama 790-8577, Japan
[5] Faculty of Health Sciences, Hiroshima International University, 724-0695, Japan

Abstract. A theoretical study of the $np \to d(\pi\pi)^0$ reaction at three energy points (P_L = 1.46, 1.70 and 1.9 GeV/c) with the σ-model and $\Delta\Delta$-model is discussed. Important features of the two models are presented.

1 Introduction

It is interesting to study the existence of the dibaryon resonances with the strangeness $S = 0$ and the isospin $I = 0$. For this purpose, we put an important basis on the phase shift analysis(PSA) of pp[1] and np[2] scattering. The np scattering amplitudes are given by a linear combination of the $I = 1$ and $I = 0$ parts with equal weight. In the PSA of np scattering, the $I = 1$ amplitudes are fixed on the values determined by the PSA of pp scattering and the $I = 0$ ones are determined so as to reproduce the experimental data.

Theoretical study of the np inelastic processes gives important information for determination of the absorption effects of the $I = 0$ amplitudes. Here we present a theoretical study of the $np \to d(\pi\pi)^0$ reaction at three energy points (P_L = 1.46, 1.70 and 1.9 GeV/c) with the σ-model and $\Delta\Delta$-model.

2 Description of the Two Models

The Feynman graph of the σ-model is shown in Fig. 1(a). We assume that the full amplitudes of this model are given by a product of the Born amplitudes of $np \to \sigma d$ and the decay amplitudes of $\sigma \to (\pi\pi)^0$, which are estimated by using

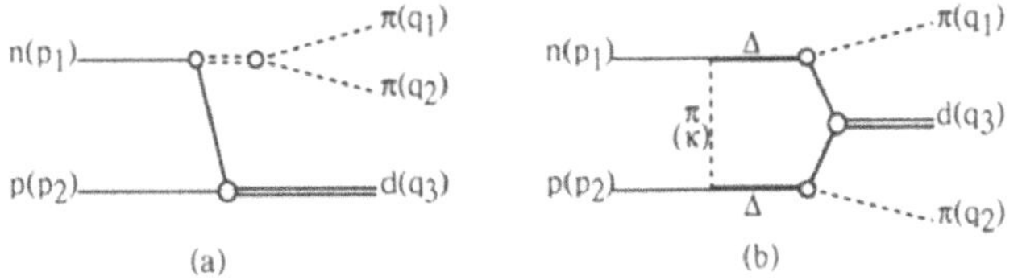

Figure 1. Feynman graphs of (a)The σ-model and (b)The $\Delta\Delta$-model for $np \to d(\pi\pi)^0$ reaction.

the S-wave $\pi\pi$ phase shifts with $I = 0$[3]. For the deuteron wave functions, we take the relativistic invariant 3-pole expansion[4].

The amplitudes of the $\Delta\Delta$-model are calculated by the Feynman graph shown in Fig. 1(b). In the calculation of the scattering amplitudes, we make two approximations according to Bar-Nir et al.[5]: i) the two nucleons at the pnd vertex are on the mass-shell. ii) the pion-nucleon matrix elements are evaluated at $n = 0$ where n is the relative four-momentum of the two nucleons. Moreover, the deuteron wave functions are restricted to the S-state only. As for more detailed discussion, see Refs.[3] and [5]

3 Comparison with the Experimental Data

In Fig. 2, we show the comparison of our theoretical calculations with the double differential cross section of the $np \to d(\pi\pi)^0$ reaction at $P_L = 1.46$

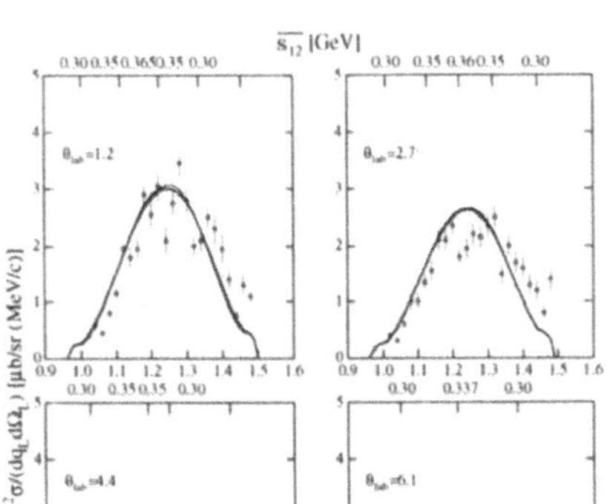

Figure 2. Comparison of the calculated values with the experimental data at $P_L = 1.46$ GeV/c.

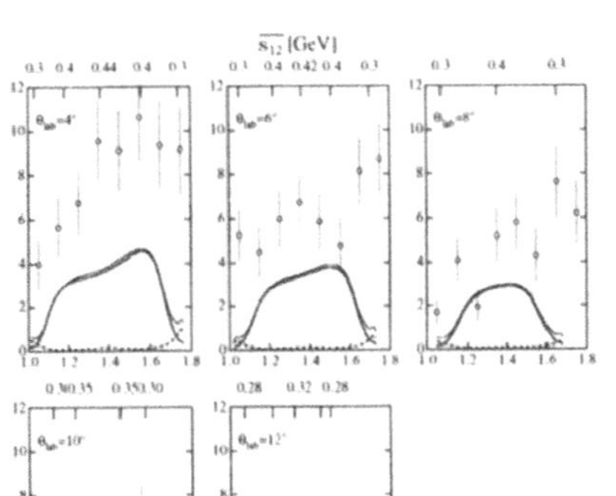

Figure 3. Comparison of the calculated values with the experimental data at $P_L = 1.70$ GeV/c.

GeV/c. The dotted curves denote the σ-model predictions and the solid curves are the sum of the σ-model and the $\Delta\Delta$-model. The contributions from the $\Delta\Delta$-model are too small to see in this figure. As can be seen from this figure, the deuteron momentum spectra decrease with the deuteron lab angle and the experimental data[6] are well reproduced by the σ-model in both the shape and magnitude.

Next, we show our results at 1.70 GeV/c in Fig. 3. The dashed curves show the $\Delta\Delta$-model predictions. The $\Delta\Delta$-model contributions are very small, too. The interference terms between the two models are negligibly small and are not shown in this figure. As can be seen from this figure, the experimental data[7] hardly depend on the deuteron lab angle unlike the data at 1.46 GeV/c, while the theoretical values rapidly decrease and deviate from the data with the lab angle.

Finally, we show the results at 1.9 GeV/c in Fig. 4. At this energy point, the differential cross sections on the $np \to dX$ reaction were measured, where X includes $(1\pi)^0$, $(2\pi)^0$, $(3\pi)^0$ and η. The experimental data clearly show the so-called ABC peaks at the both ends of the deuteron momentum. At $\theta_{\text{lab}} = 1.5°$, the experimental data on the pure 2π production region are well reproduced by the two models. A large deviation of the theoretical results from the experimental data in the central momentum region is due to $(3\pi)^0$ and η productions whose effects are not considered in the present calculation. However, the agreements of the models with the experimental data have the similar feature as that at 1.7 GeV/c.

In order to get the information on the absorption effects of the partial wave in the $I = 0$ amplitudes, we calculated the partial cross sections in the σ-model. It is concluded that the $^3S_1$ initial state is dominant in the σ-model[3].

We show the separate contributions from $I = 0$ and $I = 1$ channels of the $\Delta\Delta$model to the cross section at $P_L = 1.9$ GeV/c and $\theta_{\text{lab}} = 1.5°$ in Fig. 5. It can be seen from this figure that the $I = 0$ channel is dominant.

The $^3S_1$ initial state is dominant in the σ-model and the $I = 0$ is dominant in the $\Delta\Delta$-model. Moreover, as stated before the interference terms between the two models are very small. From these results it is suggested that the $\Delta\Delta$-model is $^1P_1$ dominance in the initial state.

4 Conclusions

Double differential cross sections of the $np \to d(\pi\pi)^0$ reaction are studied with the σ-model and $\Delta\Delta$-model at the three incident lab momenta.

We found that the experimental data at $P_L = 1.46$ GeV/c are well reproduced by the σ-model. The $\Delta\Delta$-model contributions are very small at $P_L = 1.46$ and 1.70 GeV/c.

The so-called ABC peak at $P_L = 1.9$ GeV/c and $\theta_{\text{lab}} = 1.5°$ is reproduced by the two models. The experimental data at $P_L = 1.70$ and 1.9 GeV/c hardly depend on the deuteron lab angle, while the theoretical predictions decrease and deviate from the data with the lab angle.

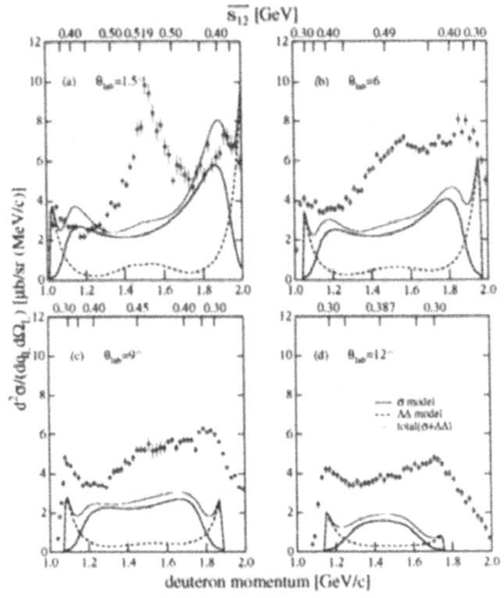

Figure 4. Comparison of the calculated values with the experimental data at $P_L = 1.90$ GeV/c.

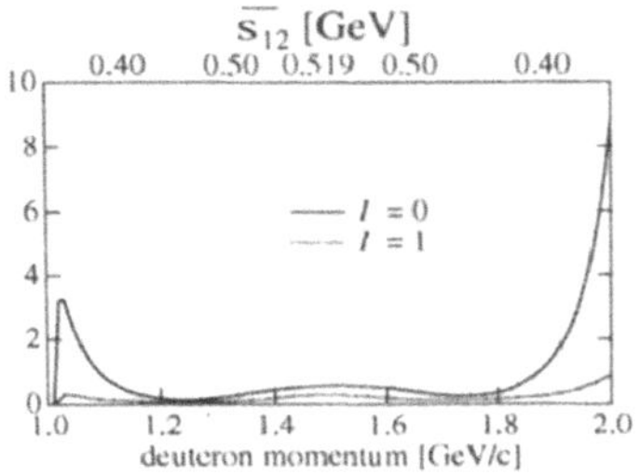

Figure 5. Contributions from $I = 0$ and $I = 1$ channels at $P_L = 1.9$ GeV/c and $\theta_{\text{lab}} = 1.5°$ in the $\Delta\Delta$ model.

The important features of the two models are obtained: i) the σ-model is dominant in the ${}^3S_1$ initial state; ii) the $\Delta\Delta$-model is dominant in the $I = 0$ and the ${}^1P_1$ initial state. These results will provide useful information for determination of the absorption effects of the $I = 0$ amplitudes.

The scattering amplitudes of the $\Delta\Delta$-model are evaluated in the two approximations as discussed in sectiton 2. It is interesting to investigate the effects when these conditions are released. It is also interesting to study the effects of the other intermediate states than the two Δ particles.

References

1. J. Nagata et al.: Prog. Theor. Phys. **93**, 559 (1995); M. Matsuda: Nucl. Phys. **A631**, 436 (1998)
2. H. Yoshino et al.: Prog. Theor. Phys. **95**, 577 (1996)
3. H. Yoshino et al.: Prog. Theor. Phys. **95**, 353 (1996)
4. M.P. Locher and A. Svarc: Z. Phys. **A338**, 89 (1991)
5. I. Bar-Nir et al.: Nucl. Phys. **B87**, 109 (1975)
6. C.L. Hollas et al.: Phys. Rev. **C25**, 2614 (1982)
7. S. Sawada et al.: Nucl. Phys. **A615**, 277 (1997)
8. F. Plouin et al.: Nucl. Phys. **A302**, 413 (1978); F. Plouin et al.: Phys. Rev. Lett. **65**, 690 (1990)

Few-Body Systems Suppl. 12, 470–473 (2000)

Few-Body Systems

Pion Production Mechanism in Nucleon-Nucleon Collisions

K. Tamura[1], Y. Maeda[2], N. Matsuoka[2]

[1] Physics Division, Fukui Medical University, Fukui 910-1193, Japan
[2] Research Center for Nuclear Physics, Osaka University, Osaka 567-0047, Japan

Abstract. The theoretical analysis for the threshold pion production has been performed by using the models of heavy-meson exchange and pion-rescattering including the effect of nucleon resonances. The experimental data of the π^- production support the relative sign between the s-wave and p-wave pion-production amplitudes which is given by these models. The d-wave pion production amplitude is discussed about the angular distribution of π^0 production. The spin correlation observables give information about the partial wave amplitudes.

1 Introduction

The behavior of the pion production near threshold provides us with considerable information on the nature of low-energy strong-interaction physics. Recently several measurements of the reaction $NN \to NN\pi$ near the pion-production threshold have been performed. The cross section of $pp \to pp\pi^0$ was measured very precisely [1, 2]. It has been pointed out that large contribution of the s-wave pion-production amplitude is necessary to reproduce the total cross section of the reaction $pp \to pp\pi^0$. Lee and Riska found that the short range effect for the two-body pion-production operator gave enhancement of the cross section by a factor 3∼5 (HM) [3, 4]. Hernández and Oset took into account the off-shell properties of the πN amplitude in the pion-rescattering process (RES) [5]. They found constructive interference between the amplitudes of the direct pion-production process and the pion-rescattering process. Hanhart *et al.* examined the off-shell πN amplitude with a realistic meson-theoretical model [6]. The works with chiral perturbation method [7] predict the s-wave pion-production amplitude with the opposite sign to the results of the heavy-meson exchange model and the pion-rescattering model.

2 Theoretical Model

The parameters of the heavy-meson-exchange model are taken from Bonn potential. We adopted Hamilton model for the pion-rescattering model [5]. The effects of the Δ and $N(1440)$ resonance and the meson dissociation mechanism ($\rho\pi\pi$, $\omega\rho\pi$) are also taken into account. The partial waves for the initial and final states of the NN system are summed up to $\ell = 6$ incorporating the Coulomb interaction. In Fig.1 are shown the calculated results of the total cross section. The hatched regions show results with the models of heavy-meson exchange and pion rescattering using the NN wave function given by the various nuclear forces (Paris, Bonn, Argonne V14 and Reid soft-core). The calculated results give good fit to the experimental data. Our results also show good explanation for $d\sigma/d\Omega$ and A_y of the reactions $pp \to pn\pi^+$ and $pp \to d\pi^+$.

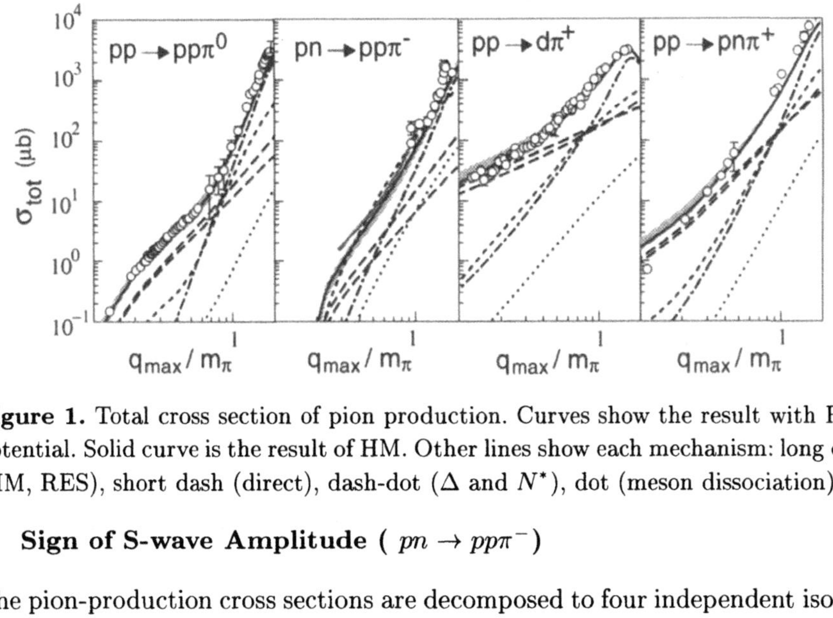

Figure 1. Total cross section of pion production. Curves show the result with Paris potential. Solid curve is the result of HM. Other lines show each mechanism: long dash (HM, RES), short dash (direct), dash-dot (Δ and N^*), dot (meson dissociation).

3 Sign of S-wave Amplitude ($pn \to pp\pi^-$)

The pion-production cross sections are decomposed to four independent isospin amplitudes $T_{I_i I_f}$, each represents the transition from an initial two nucleon state with isospin I_i to a final state with isospin I_f. In the case of the amplitude T_{01}, the main process of the Δ excitation is forbidden from the isospin selection rule. Thus the p-wave (T_{01}) contribution is suppressed and might be a comparable magnitude as the s-wave amplitude (T_{11}) at the threshold energy region. In this meaning the reaction $pn \to pp\pi^-$ is suitable to study the interference between s-wave and p-wave amplitudes. In Fig.2(a) shows the angular distribution of cross section for π^- production by the heavy-meson exchange model and the pion-rescattering model as solid line and hatched region. The p-p relative energy of the final state was limited within 3 MeV. The experimental data point at 180° was extracted from the inclusive measurement of $d(p,pp)p\pi^-$ at RCNP. The data of TRIUMF [8] also show the asymmetric nature caused by the interference between s-wave and p-wave amplitude. The experimental data can be reproduced well by the predictions of both the heavy-meson exchange model and the pion-rescattering model. If we change the relative sign between s-wave and p-wave amplitudes, the angular distributions become the reversed shapes (dashed line). The calculated results with the opposite sign relation of the s-wave and p-wave amplitudes can not explain the experimental data.

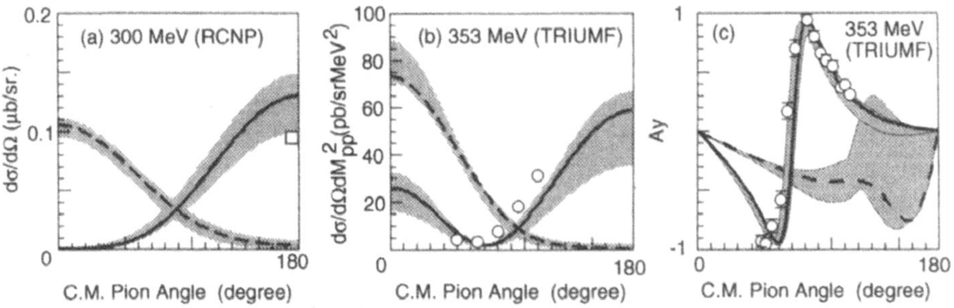

Figure 2. Angular distribution of $pn \to pp\pi^-$. Solid curve shows HM. Dashed curve is the calculation with reversed sign of s-wave amplitude.

4 D-wave Pion Production Mechanism

The angular distribution of π^0 production was measured at RCNP and TSL. The experimental data show convex structure for the angular distribution. This structure mainly comes from the d-wave amplitude ${}^3P_2$-${}^3F_2 \to {}^1S_0$ + d-wave. The state ${}^3P_2$-${}^3F_2$ has strong coupling to the ΔN state. By this reason the angular dependence of the π^0 production is sensitive to the short range structure of Δ excitation mechanism. The pion production operator of Δ excitation contains the factor $\vec{q}^{\,2}/(\vec{q}^{\,2}+\mu^2)$. Here $\vec{q}$ and μ are the momentum and mass of the exchanged meson. If one write this factor as $\vec{q}^{\,2}/(\vec{q}^{\,2}+\mu^2) = 1-\mu^2/(\vec{q}^{\,2}+\mu^2)$, then the term 1 gives a contact interaction. Because we introduced the vertex form factor ($\Lambda_{\pi NN}$=700 MeV), the contact interaction is smeared out and gives sizable effect at short range region ($r \leq 0.5$ fm). In Fig.3, the calculations including the contact interaction (dashed line) failed to explain the convex structure of experimental data. The results without the contact interaction are shown as solid lines which give good fit to the experimental data.

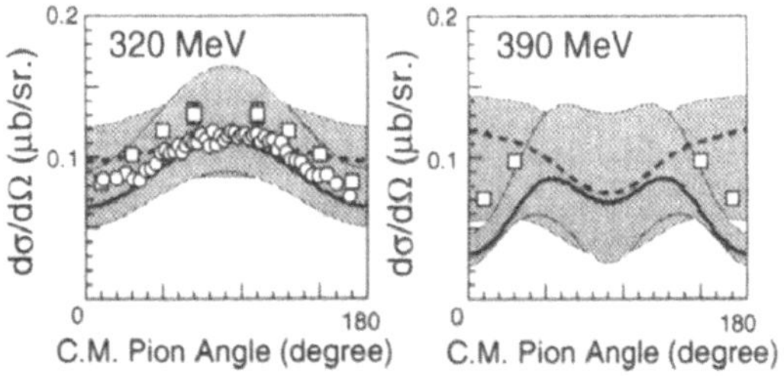

Figure 3. Angular distribution of π^0 production. Solid (Dashed) line is the calculation without (with) the contact interaction. Data are from RCNP(box) and TSL(circle)[9].

5 Spin Correlation Observables ($\vec{p}\vec{p} \to pp\pi^0$)

Recently Meyer *et al.* [10] measured the spin correlation observables with the beam and target polarization. By using the partial wave amplitude Pp, Ps and Ss, we can see the global structure of the observables as

$$A_{xx} + A_{yy} = 2Ss^2/\sigma \ , \qquad A_{xx} - A_{yy} = 4Pp^2/15\sigma \ ,$$
$$A_{zz} = -Ss^2 - Ps^2 + Pp^2/3 \ , \quad A_y = \pi\sqrt{5}Pp \times Ps \sin\delta/10\sigma$$

Here σ means the total cross section and δ is given by the phase shift of NN states. Fig.4 show the experimental data and calculated results. The combination $A_{xx} + A_{yy}$ can be explained by our model satisfactorily. Therefore the model gives reasonable magnitude of the Ss amplitude. But the calculated result of analyzing power is very small. Analyzing power A_y is proportional to the product of p-wave amplitude Pp and s-wave amplitude Ps in which nucleons of the final state couple to P-wave. In the case of Ss amplitude, the main contribution comes from the transition ${}^3P_0 \to {}^1S_0 + s$. Because the ${}^1S_0$ state has large strength at short range region ($r \leq 0.5$ fm), the wave function has good overlap with the Yukawa function of the heavy meson exchange. Contrary to this, in the Ps-amplitude case, the overlap between the wave function and the Yukawa function becomes small because the P-wave NN state has no strength at short range region. As the result the Ps amplitude becomes very small. In other word, we do not have s-wave pion production mechanism which might be effective around 1 fm region in our model.

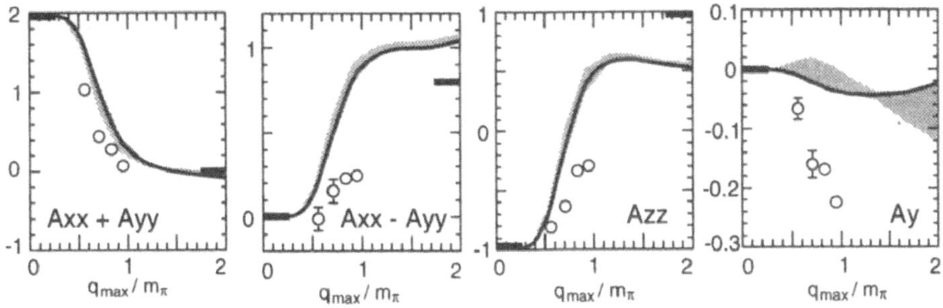

Figure 4. Spin correlation observables of $pp \to pp\pi^0$. Solid line is the calculation of HM with Paris potential. The data are from Ref.[10].

References

1. H.O. Meyer et al.: Phys. Rev. Lett. **65**, 2846 (1990)
2. A. Bondar el al.: Phys. Lett. **B356**, 8 (1995)
3. T.-S.H. Lee and D. Riska: Phys. Rev. Lett. **70**, 2237 (1993)
4. C.J. Horowitz et al.: Phys. Rev. **C49**, 1337 (1994)
5. E. Hernández and E. Oset: Phys. Lett. **B350**, 158 (1995)
6. C. Hanhart et al.: Phys. Lett. **B358**, 21 (1995)
7. T. Sato et al.: Phys. Rev. **C56**, 1246 (1997), and references therein
8. F. Duncan et al.: Phys. Rev. Lett. **80**, 4390 (1998)
9. J. Zlomanczuk et al.: Phys. Lett. **B436**, 251 (1998)
10. H.O. Meyer et al.: Phys. Rev. Lett. **81**, 3096 (1998); nucl-ex/9907017

Few-Body Systems Suppl. 12, 474–478 (2000)

Few-Body Systems

Measurement of Cross Sections and Vector and Tensor Analyzing Powers for d - p Elastic Scattering at $E_d = 270$ MeV

K. Sekiguchi[1], H. Sakai[1], H. Okamura[2], N. Sakamoto[3], A. Tamii[1], T. Uesaka[3], T. Wakasa[4], Y. Satou[3], T. Ohnishi[1], K. Yako[1], S. Fukusaka[1], S. Sakoda[1], K. Suda[2], H. Kato[1], Y. Maeda[1], K. Itoh[3], T. Niizeki[5]

[1] Department of Physics, University of Tokyo, Tokyo 113-0033, Japan,

[2] Department of Physics, Saitama University, Saitama 338-8570, Japan,

[3] The Institute of Physical and Chemical Research, Saitama 351-0198, Japan

[4] Research Center for Nuclear Physics, Osaka University, Osaka 567-0047, Japan

[5] Department of Environmental and Information Studies, Faculty of Home Economics, Tokyo Kasei University, Tokyo 173-8602, Japan

Abstract. Cross sections and vector and tensor analyzing powers A_y, A_{yy}, A_{xx}, and A_{xz} for d–p elastic scattering have been measured at a deuteron energy of 270 MeV in order to confirm and extend the measurement of Sakamoto et al. [Phys. Lett. B **367**, 60 (1996)]. The extended angular range is $\theta_{\rm c.m.} = 10^\circ - 180^\circ$. Highly accurate data have been obtained. The measured values are compared with Faddeev calculations with/without the three nucleon force.

1 Introduction

The measurement of cross sections and vector and tensor analyzing powers A_y, A_{yy}, A_{xx} and A_{xz} for d–p elastic scattering at 270 MeV was reported by Sakamoto et al. [1], and their results were compared with Faddeev calculations without including the three nucleon force (3NF). Although the agreement of the analyzing powers is generally good, the calculated cross sections are smaller by 30% than the observed values at the angles where the cross sections take the minimum. Recently it was pointed out that this discrepancy might be due to the effect of the 3NF [2, 3].

The previous measurement [1] was the first precise one both on the cross sections and on all components of deuteron analyzing powers for d–p elastic scattering at intermediate energies. Unfortunately this measurement covers only a limited angular range $\theta_{c.m.} = 57.0° - 137.8°$ because of the detection set up. Therefore we extended the measurement to a wide angular range $\theta_{c.m.} = 10° - 180°$ by using the RIKEN magnetic spectrometer SMART. In this article, we present the newly obtained data. They will be compared with Faddeev calculations with/without the 3NF. It is interesting to see how well the cross sections as well as analyzing powers are reproduced by the theoretical predictions all over the angles.

2 Experiment

The experiment was performed at the RIKEN Accelerator Research Facility (RARF). Vector and tensor polarized deuteron beams were provided by a polarized ion source. They were accelerated up to 270 MeV by the AVF and RING cyclotrons. The direction of the polarization axis was controlled by a Wien filter before the acceleration [4]. At the RARF, single-turn extraction was available both for the AVF and RING cyclotrons, so that the polarization amplitude was maintained. Beam polarization was monitored by using a beam line polarimeter based on *d*–*p* elastic scattering [1]. Measured magnitudes of the beam polarization were 60–80% of the theoretical maximum values during the experiment. The polyethylene target (46.7 mg/cm^2 thick) was bombarded by polarized deuteron beams in the scattering chamber. Scattered particles were momentum analyzed by a magnetic spectrometer SMART and detected at the focal plane.

3 Results

The experimental results are plotted with solid circles in Fig. 1. The errors are statistical ones only. The open circles in Fig. 1 are the previous measurement. The two independent measurements provide almost equivalent values with each other in the overlapping region.

The statistical errors of the cross sections are better than $\pm 1.3\%$ for all measured angles. The uncertainties of the target thickness and the charge collection of the beam were estimated by comparing the measured cross sections for the $^1H(p,p)$ scattering with the calculated values which were obtained by using the program code SAID [5]. The ratio of the observed cross sections to the calculated ones is 1.010 ± 0.013. Thus the systematic uncertainty is estimated to be 2% at most.

4 Discussion

Three theoretical predictions of the cross sections in terms of Faddeev theory are shown in Fig. 1 together with the observed values.

The dotted curves (PD) are the calculations for *d*–*p* elastic scattering in which the Argonne v_{14} potential (AV14) [7] is used as *NN* interactions [1]. The total angular momentum of *NN* interactions j is four or less ($j \leq 4$). A Coulomb

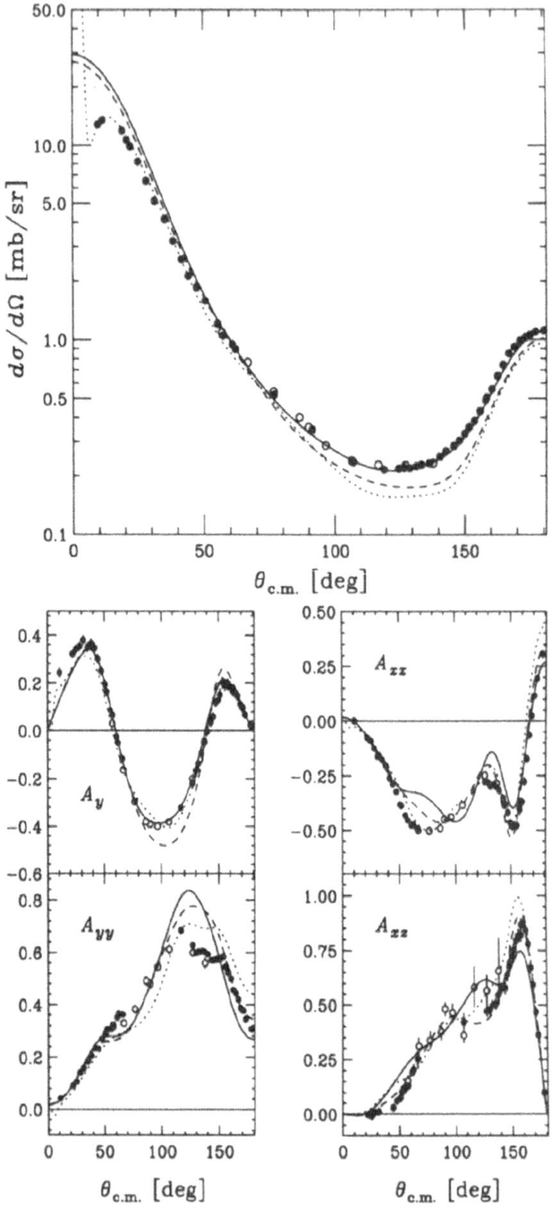

Figure 1. Cross sections and analyzing powers A_y,A_{yy},A_{xx} and A_{xz} for the d–p elastic scattering at $E_d = 270$ MeV. The solid circles are the results of this experiment and the open circles are those of Ref. [1]. The dotted curves are the Faddeev calculations in which the Argonne v_{14} potential is used as *NN* interactions (PD). The dashed curves are the Faddeev calculations in which the CD BONN potential is used as *NN* interactions (ND). The solid curves are the Faddeev calculations including the Tucson-Melbourne 3NF (ND-3NF).

force effect is included, but the 3NF is not included. The dashed curves (ND) are the calculations for *d–n* elastic scattering. *j* is 5 or less ($j \leq 5$) [2, 6]. The CD BONN potential [8] is used as *NN* interactions. Note that this calculations do not include the Coulomb force and the 3NF. The solid curves (ND–3NF) are the calculations in which the Tucson-Melbourne 3NF is added to the ND calculations.

For the cross sections, the ND calculation underestimates the data in the angular range of $\theta_{\mathrm{c.m.}} = 80°$–$180°$. In particular, the discrepancy is conspicuously seen in the angular range where the cross sections take minimum. The ND–3NF calculation has an excellent fit so that this discrepancy seems to be accounted for by including 3NF. However the effect of the difference of *NN* interactions cannot be ignored. Fig. 1 shows that there is a difference between PD and ND calculations around the cross section minimum. The disagreement between these two calculations at very forward angles is due to Coulomb force effect.

The vector analyzing power A_y is reproduced almost perfectly by the ND–3NF calculation. However none of the calculations shows its superiority in describing the tensor analyzing powers A_{yy}, A_{xx} and A_{xz}. It should be noted that the agreement is deteriorated when 3NF is contained. It seems that some key element which improves the fit of the analyzing powers is missing in these theories.

5 Summary

Cross sections and vector and tensor analyzing powers A_y, A_{yy}, A_{xx} and A_{xz} for *d–p* elastic scattering have been measured at a deuteron energy of 270 MeV to confirm and extend the measurement of Ref. [1]. The extended angular range is $\theta_{\mathrm{c.m.}} = 10° - 180°$. Highly accurate data have been obtained. The experimental results have been compared with Faddeev calculations [1, 2, 6] and Faddeev calculations including the Tucson-Melbourne 3NF [2, 9]. The discrepancy between the observed values and the calculations without 3NF in the cross sections might be accounted for by 3NF. However the agreement in the tensor analyzing powers cannot be improved by including 3NF, although the vector analyzing power A_y is well reproduced. These results indicate that there seems to be ambiguity of spin dependent part of 3NF as well as *NN* interactions.

References

1. N. Sakamoto et al.: Phys. Lett. B **367**, 60 (1996)
2. H. Witała et al.: Phys. Rev. Lett. **81**, 1183 (1998)
3. S. Nemoto et al.: Phys. Rev. **C58**, 2599 (1998)
4. H. Okamura et al.: AIP Conf. Proc. **293**, 84 (1993),
   H. Okamura et al.: AIP Conf. Proc. **343**, 123 (1995)

5. R.A. Arndt and L.D. Roper, *Scattering Analysis Interactive Dial–In program* (SAID), Virginia Polytechnic Institute and State University (unpublished)

6. W. Glöckle et al.: Phys. Rep. **274**, 107 (1996)

7. R.B. Wiringa R.A. Smith, and T.L. Ainsworth: Phys. Rev. **C29**, 1207 (1984)

8. R. Machleidt, F. Sammarruca, and Y. Song: Phys. Rev. **C53**, R1483 (1996)

9. H. Kamada: Private Communication

Few-Body Systems Suppl. 12, 479–482 (2000)

Few-Body Systems

Nucleon-Deuteron Break-Up Scattering with Δ-Isobar Excitation

K. Chmielewski[1], S. Nemoto[2], A.C. Fonseca[3], P.U. Sauer[1]

[1] Institut für Theoretische Physik, Universität Hannover, Appelstraße 2, D-30167 Hannover, Germany

[2] Department of Physics, Faculty of Science and Technology, Science University of Tokyo

[3] Centró Física Nuclear da U. L., Av. Prof. Gama Pinto, N°2, 1699 Lisboa, Portugal

Abstract. A two-baryon interaction with explicit Δ-isobar degrees of freedom is used for the description of inelastic nucleon-deuteron scattering. The Δ-isobar yields an effective three-nucleon force in three-nucleon systems. The three-particle scattering equations are solved by employing a separable expansion of the two-baryon transition matrix. The singularities in the scattering equations are treated by the product integration method. Results for nucleon-deuteron break-up with Δ-isobar excitation are presented.

1 Baryon-Baryon Interaction

The description is based on the coupled-channel two-baryon interaction depicted diagrammatically in Fig. 1. It allows for the virtual excitation of a nucleon into a Δ-isobar. In the three-nucleon medium three-nucleon channels are coupled to those in which one nucleon is transformed into a single Δ-isobar. The virtual excitation of the Δ-isobar yields an effective three-nucleon force. Sample processes which contribute to the three-nucleon force are shown in Fig. 2. The coupled-channel interaction was developed in refs. [1, 2] and used there to calculate properties of the three-nucleon bound state. We extent those calculations to elastic and inelastic nucleon-deuteron scattering.

2 Numerical Techniques

The three-particle scattering equations are solved by means of a separable expansion of the two-baryon transition matrix developed in ref. [3]. With the

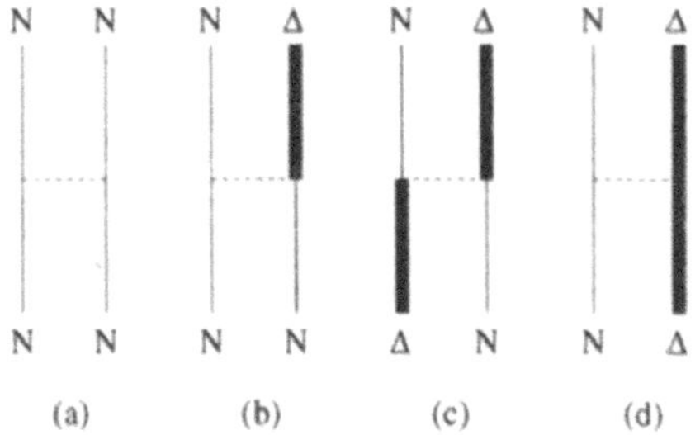

Figure 1. The baryon-baryon interaction employed. A thin solid lines indicates a nucleon (N), a thick solid line indicates a Δ-isobar (Δ). The instantaneous interaction is drawn as a dashed line. Processes (b)–(d) refer to isospin triplet partial waves only.

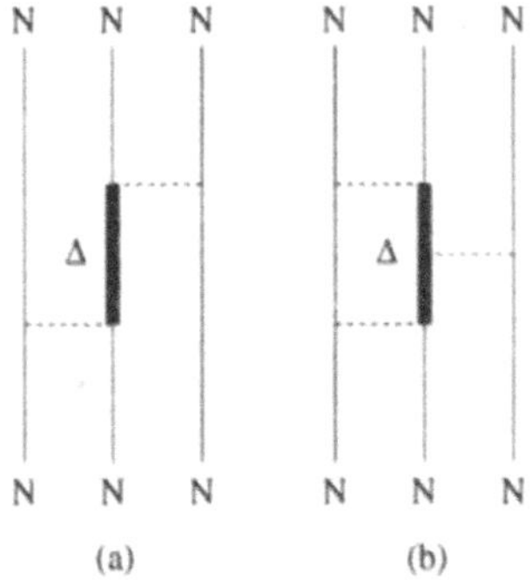

Figure 2. Example processes in the three-nucleon medium which give rise to effective three-nucleon forces. The legend of lines is the same as in Fig. 1.

numerical techniques of *contour deformation* and *real-axis integration* we employ two different and independent methods to solve the three-particle scattering equations for elastic nucleon-deuteron scattering. The equivalence of both methods is demonstrated in Fig. 3.

The real-axis technique treats the singularities of kinematical origin which appear in the scattering equations above break-up threshold by the product integration method [4, 5, 6]. An integral of the form

$$I(a,b) = \int_a^b dx\, w(x) f(x), \tag{1}$$

where $f(x)$ denotes a continuous function and $w(x)$ is absolute integrable in $[a,b]$, is numerically approximated by the sum

$$I(a,b) \approx \sum_{k=1}^{N} w_k f(x_k) \tag{2}$$

for some N and a chosen set $\{x_k\}$ of mesh points. Note, that the function $w(x)$ of Eq. (1) enters into the integration rule (2) only implicitly through the weights w_k. These weights are uniquely determined by the requirement that the integration rule (2) is an exact one for a particular class of functions $h_i(x)$ with $i = 1, \dots N$, e. g. polynomials. Our numerical investigations have shown that the product integration method yields precise results and facilitates the solution of the three-particle scattering equations. Further technical details are given in refs. [7, 8].

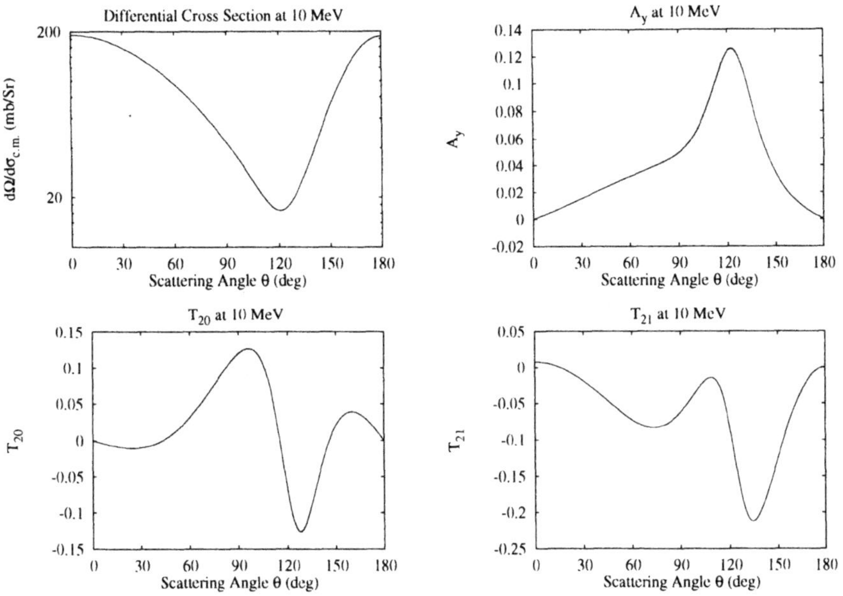

Figure 3. Comparison of numerical techniques for elastic nucleon-deuteron scattering at 10 MeV neutron lab energy. Shown are the results for the differential cross section, the neutron vector analyzing power A_y and the deuteron tensor analyzing power T_{20} and T_{21} as a function of the scattering angle

3 Nucleon-Deuteron Break-Up Scattering

Calculations for elastic nucleon-deuteron scattering based on the real-axis integration technique can be easily extended to nucleon-deuteron break-up scattering. Preliminary results are shown in Figs. 4–5 where calculations for the purely nucleonic Paris potential are compared to calculations with virtual Δ-isobar excitation. Only two-baryon partial waves up to a total two-baryon angular momentum $I = 2$ are taken into account. The few low-energy observables, considered so far, do not exhibit a sizable Δ-isobar effect. At present further calculations are under way which will explore the role of the Δ-isobar in break-up configurations.

4 Acknowledgments

The authors thank C. Chandler for pointing out early references on the product integration method. Numerical calculations were performed at the Regionales Rechenzentrum für Niedersachsen.

References

1. C. Hajduk and P.U. Sauer: Nucl. Phys. **A322**, 329 (1979)

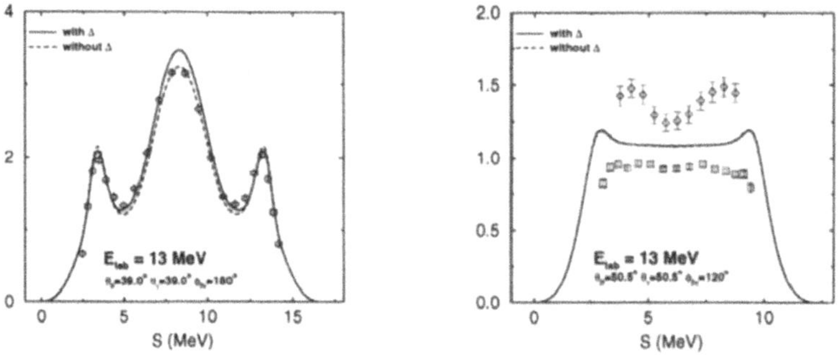

Figure 4. The nucleon-deuteron break-up cross section at 13 MeV nucleon lab energy as a function of the arc length S for two different kinematical configurations. Results for the calculation with Δ-isobar (solid line) and the Paris potential (dashed line) are shown.

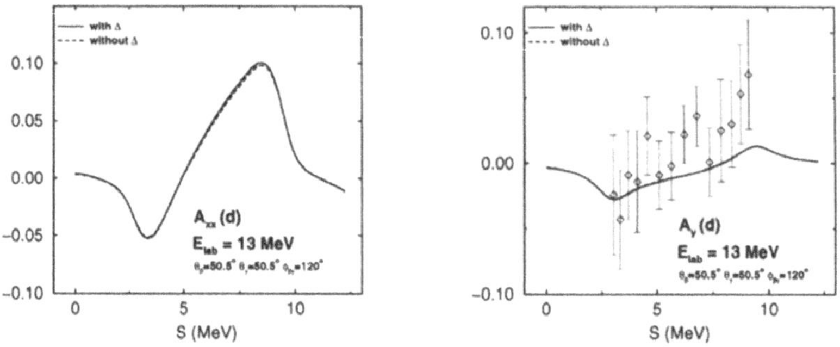

Figure 5. Deuteron analyzing power A_{xx} and vector analyzing power A_y for nucleon-deuteron break-up scattering at 13 MeV nucleon lab energy as a function of the arc length S in the space star configuration. Results for the calculation with Δ-isobar (solid line) and the Paris potential (dashed line) are shown.

2. C. Hajduk, P.U. Sauer, W. Strueve: Nucl. Phys. **A405**, 581 (1983)
3. S. Nemoto et al.: Few-Body Systems **24**, 213 (1998)
4. I.H. Sloan and W.E. Smith: Numer. Math. **30**, 415 (1978)
5. I.H. Sloan: J. Math. Phys. **21**, 1032 (1979)
6. P. Rabinowitz and I.H. Sloan: SIAM J. Numer. Anal. **21**, 149 (1984)
7. K. Chmielewski: Ph.D. thesis, Universität Hannover, 1999
8. K. Chmielewski, A.C. Fonseca, S. Nemoto, P.U. Sauer: (1999), in preparation

Few-Body Systems Suppl. 12, 483–486 (2000)

Few-Body Systems

Effect of Δ-Isobar Excitation on Spin-Dependent Observables of Elastic Nucleon-Deuteron Scattering

S. Nemoto[1], K. Chmielewski[2], S. Oryu[1], P.U. Sauer[2]

[1] Department of Physics, Faculty of Science and Technology, Frontier Research Center for Computational Science, Science University of Tokyo, Noda, Chiba 278-8510, Japan

[2] Institut für Theoretische Physik, Universität Hannover, Appelstraße 2, D-30167 Hannover, Germany

Abstract. Δ-isobar excitation in the nuclear medium yields an effective three-nucleon force. A coupled-channel formulation with Δ-isobar excitation developed previously is used. The three-particle scattering equations are solved by a separable expansion of the two-baryon transition matrix for elastic nucleon-deuteron scattering. The effect of Δ-isobar excitation on the spin-dependent observables is studied at energies above 50 MeV nucleon lab energy.

A coupled-channel description of the three-nucleon system allowing for virtual Δ-isobar excitation is used for elastic nucleon-deuteron scattering. The calculational scheme is taken over from refs. [1] and [2]. The Hilbert space of the calculation has a purely nucleonic sector and a sector in which a nucleon is turned into a Δ-isobar. The Δ-isobar is considered a stable baryon with mass 1232 MeV and with spin and isospin $\frac{3}{2}$. That treatment of the Δ-isobar is appropriate below the pion-production threshold, and the considered scattering energies stay well below. The employed coupled-channel two-baryon potentials have almost phase-equivalent purely nucleonic reference potentials. Potentials based on the Paris potential [3] and the Bonn B potential [4] are used as reference potentials. The calculations are done without Coulomb interaction. The results in all figures derived from the coupled-channel potential with Δ-isobar excitation (solid curves) are compared with results based on the purely nucleonic reference potential (dashed curves) in order to identify Δ-isobar effects.

The calculations except at 135 MeV nucleon lab energy include the interaction up to a total two-baryon angular momentum of $I = 3$ and the calculation at 135 MeV nucleon lab energy even include it up to $I = 4$. The coupling of the Δ-isobar is only considered in partial waves up to $I = 2$. The three-baryon channels are taken into account up to three-baryon total angular momentum $\frac{31}{2}$.

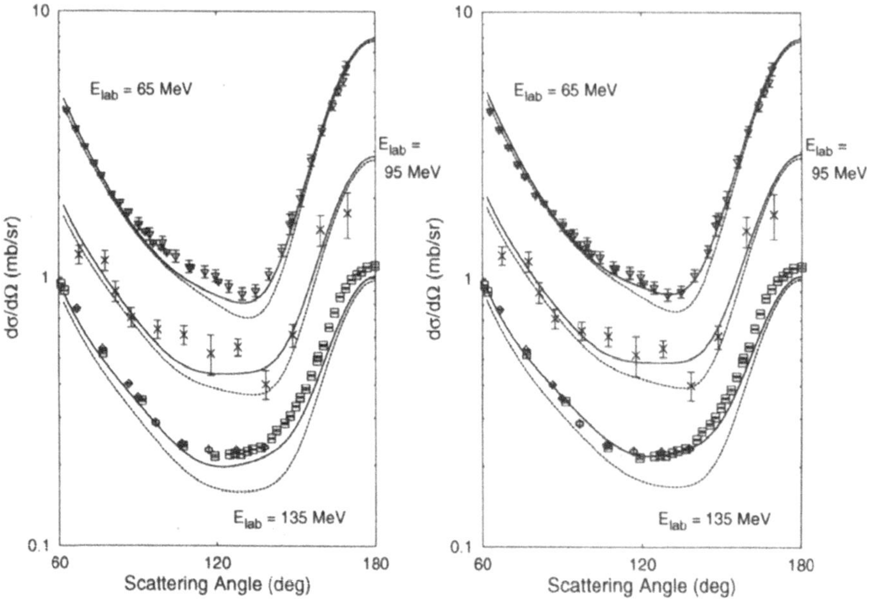

Figure 1. Differential cross sections of elastic nucleon-deuteron scattering as function of the c.m. scattering angle for 65, 95 and 135 MeV nucleon lab energy. The left-hand figure belong to the Paris potential and the right-hand figure belong to the Bonn B potential as reference potential. The experimental data refer to proton-deuteron scattering; they are taken at 65 MeV from ref. [6], at 95 MeV from ref. [7] and at 135 MeV from ref. [8] (◇) and from ref. [9] (□). The experimental data [9] are still preliminary.

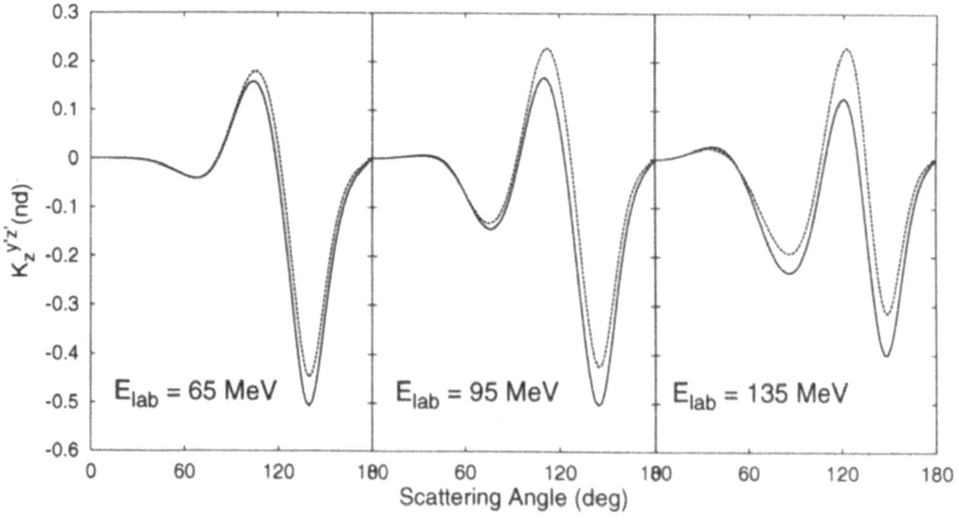

Figure 2. Spin transfer coefficients $K_z^{y'z'}(nd)$ of the incoming nucleon to the outgoing deuteron as function of the scattering angle in the c.m. system for elastic nucleon-deuteron scattering.

The diffraction minimum of the spin-averaged differential cross section at energies above 50 MeV nucleon lab energy are calculated [5]. Results derived from the Paris potential and from the Bonn B potential are shown in Fig. 1. The Δ-isobar effect is in both cases beneficial and in both cases very similar. Therefore we expect the same for spin observables and that is the reason why in the following we will show only results for spin observables calculated with the coupled-channel interaction based on the Paris potential as reference potential.

Figure 2 shows $K_z^{y'z'}(nd)$, one of the spin transfer coefficients of the incoming deuteron to the outgoing nucleon, at 65, 95, and 135 MeV nucleon lab energy. The results demonstrate the increasing importance of Δ-isobar effects with energy.

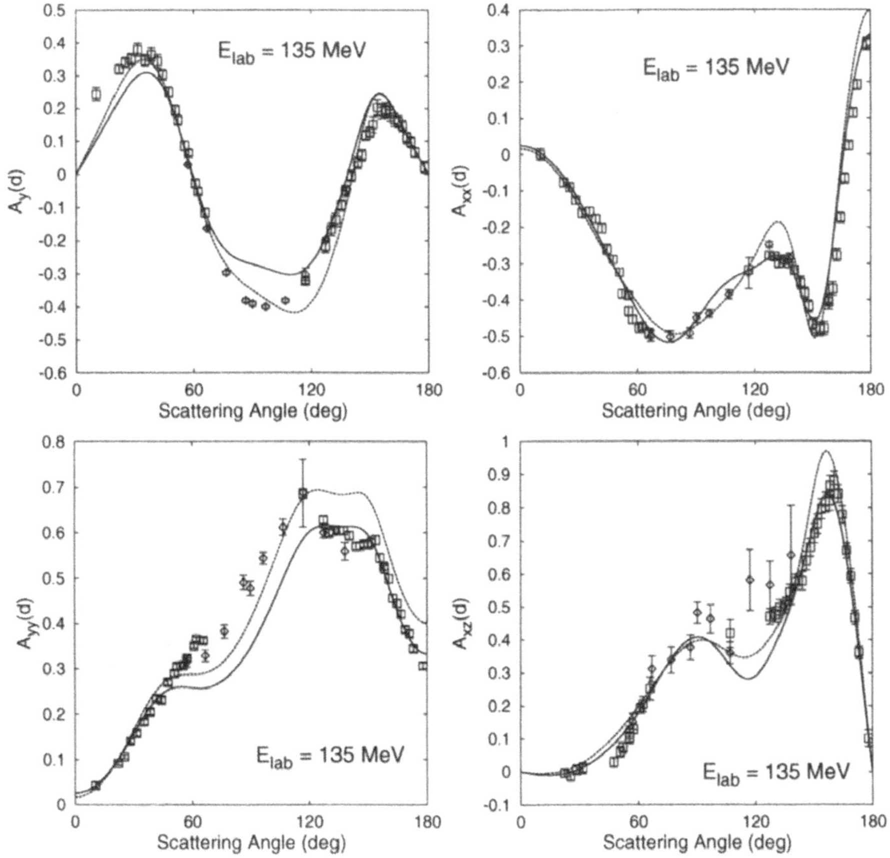

Figure 3. Deuteron vector and tensor analyzing powers of elastic nucleon-deuteron scattering as function of the c.m. scattering angle at 135 MeV nucleon lab energy. The results belong to the Paris potential as reference potential. The experimental data shown by the symbol "◇" are taken from ref. [8] and those shown by the symbol "□" from ref. [9]. The experimental data [9] are still preliminary.

Figure 3 shows the vector and tensor analyzing powers at 135 MeV nucleon lab energy. The Δ-isobar effects are not always beneficial; in part they also disagree [10] with those arising from a standard irreducible three-nucleon force.

References

1. S. Nemoto, K. Chmielewski, U. Meyer, J. Haidenbauer, S. Oryu, P.U. Sauer: Few-Body Systems, **24**, 213 (1998)
2. S. Nemoto, K. Chmielewski, N.W. Schellingerhout, J. Haidenbauer, S. Oryu, P.U. Sauer: Few-Body Systems, **24**, 241 (1998)
3. M. Lacombe, B. Loiseau, J.M. Richard, R. Vinh Mau, J. Coté, P. Pirès, R. de Tourreil: Phys. Rev. **C21**, 861 (1980)
4. R. Machleidt: Adv. Nucl. Phys. **19**, 189 (1998)
5. S. Nemoto, K. Chmielewski, S. Oryu, P.U. Sauer: Phys. Rev. **C58**, 2599 (1998)
6. H. Shimizu, K. Imai, N. Tamura, K. Nisimura, K. Hatanaka, T. Saito, Y. Koike, Y. Taniguchi: Nucl. Phys. **A382**, 242 (1982)
7. O. Chamberlain and M.O. Stern: Phys. Rev. **94**, 666 (1954)
8. N. Sakamoto, H. Okamura, T. Uesaka, S. Ishida, H. Otsu, T. Wakasa, Y. Satou, T. Niizeki, K. Katoh, T. Yamashita, K. Hatanaka, Y. Koike, H. Sakai: Phys. Lett. **B367**, 60 (1996)
9. K. Sekiguchi: Private communication
10. H. Sakai: Talk at this conference

Few-Body Systems Suppl. 12, 487–490 (2000)

Few-Body Systems

Phase-Shift Analyses of p-^{3}He Scattering at Low Energies

Y. Yoshino[1], V. Limkaisang[1], J. Nagata[2], H. Yoshino[3], M. Matsuda[1]

[1] Faculty of IAS, Hiroshima University, Higashi-Hiroshima, 739-8521, Japan
[2] VBL, Hiroshima University, Higashi-Hiroshima, 739-8527, Japan
[3] Fuculty of Health Sciences, Hiroshima International University, 724-0695, Japan

Abstract. The phase-shift analyses of p-^{3}He scattering are performed at the incident-proton energy T_L = 4.0, 5.5, 6.8, 8.5, 9.5 and 19.48 MeV by means of a new developed program.

1 Introduction

Since early 1960s, the experiments of p-^{3}He scattering and their phase-shift analyses (PSAs) have been performed with great interest in the $A = 4$ nuclei [1]. Recently, the experiment of ^{3}He$(d,p)^4$He reaction at the incident-deuteron energy in the laboratory system 270 MeV was carried out at RIKEN [2]. The derivation of nucleon-^{3}He potentials plays an important role in order to analyze those experimental data [3]. In order to do it, the model-independent solutions of p-^{3}He and n-^{3}He phase shifts are required in the wide-energy region (the incident-nucleon energy in the laboratory system $T_L \lesssim 300$ MeV).

Recently the experimental data on the various kinds of spin-correlation coefficients of p-^{3}He scattering have been accumulated in the low-energy region ($T_L \lesssim 20$ MeV), and give us the expectation to perform the single-energy PSA in this energy region more completely. Moreover in the intermediate energy region ($T_L = 20 \sim 300$ MeV), there are some energy points where the spin-correlation coefficients were measured. So it seems that it will be possible to perform the single-energy PSAs in this energy region. In the energy region of $T_L = 19 \sim 48$ MeV, Murdoch et al. [4] roughly performed the energy-dependent PSA, where the inelastic effect is evaluated.

We develop the program of single-energy PSA of p-^{3}He scattering by using the S matrix in the Matsuda-Watari (MW) representation [5] in order to evaluate the inelastic effect, which will be applied for the analyses not only in the low-energy region but also in the intermediate one. Here we perform the single-energy PSAs of p-^{3}He scattering at $T_L = 4.0, 5.5, 6.8, 8.5, 9.5$ and 19.48 MeV and give their results.

2 Representation of S matrix

We perform the single-energy PSA of p-^{3}He scattering by use of the S matrix in the MW representation. In this representation, the S matrix with the total angular momentum J for the coupled waves is expressed as follows,

$$S_J = \begin{pmatrix} \sqrt{1-|\rho_J|^2}\;\eta_+ e^{2i\delta_+} & i\rho_J\,\sqrt{\eta_+\eta_-}\;e^{i(\delta_++\delta_-)} \\ i\rho_J\,\sqrt{\eta_+\eta_-}\;e^{i(\delta_++\delta_-)} & \sqrt{1-|\rho_J|^2}\;\eta_- e^{2i\delta_-} \end{pmatrix}.$$

In this equation, the $\delta_\pm$, ρ_J and $\eta_\pm$ are the phase shifts, the mixing parameters and the reflection parameters, respectively. In the elastic energy region, the reflection parameters should be equal to 1 : $\eta_{\ell,J} = 1$, where the phase-shift parameters in the MW representation become to be equivalent to the nuclear-bar phase shifts introduced by Stapp et al. [6].

3 Results of Phase-Shift Analyses

Our database includes the experimental data on the differential cross section and the spin observables for elastic p-^{3}He scattering at $T_L \le 20$ MeV. Table 1 shows the kinds and the numbers of experimental data used in our PSAs.

We perform the single-energy PSAs at $T_L = 4.0, 5.5, 6.8, 8.5, 9.5$ and 19.48 MeV, where a plentiful data on many kinds of spin observables are accumulated. Sourkes et al. [7] reported the results of their experiments for the total reaction

Table 1. The kinds and the numbers of experimental data used in our phase-shift analyses at each energy point.

T_L(MeV)	4.0	5.5	6.8	8.5	9.5	19.48
$d\sigma/d\Omega$	19	53	34	53	61	25
A_{y0}	95	55	56	42	37	13
A_{0y}	14	13	14	14	14	12
A_{yy}	8	10	10	10	10	11
A_{xx}		5				8
A_{xz}		5		2		
A_{zx}		5				
A_{zz}		5				
σ_r						1
total	136	151	114	121	122	70

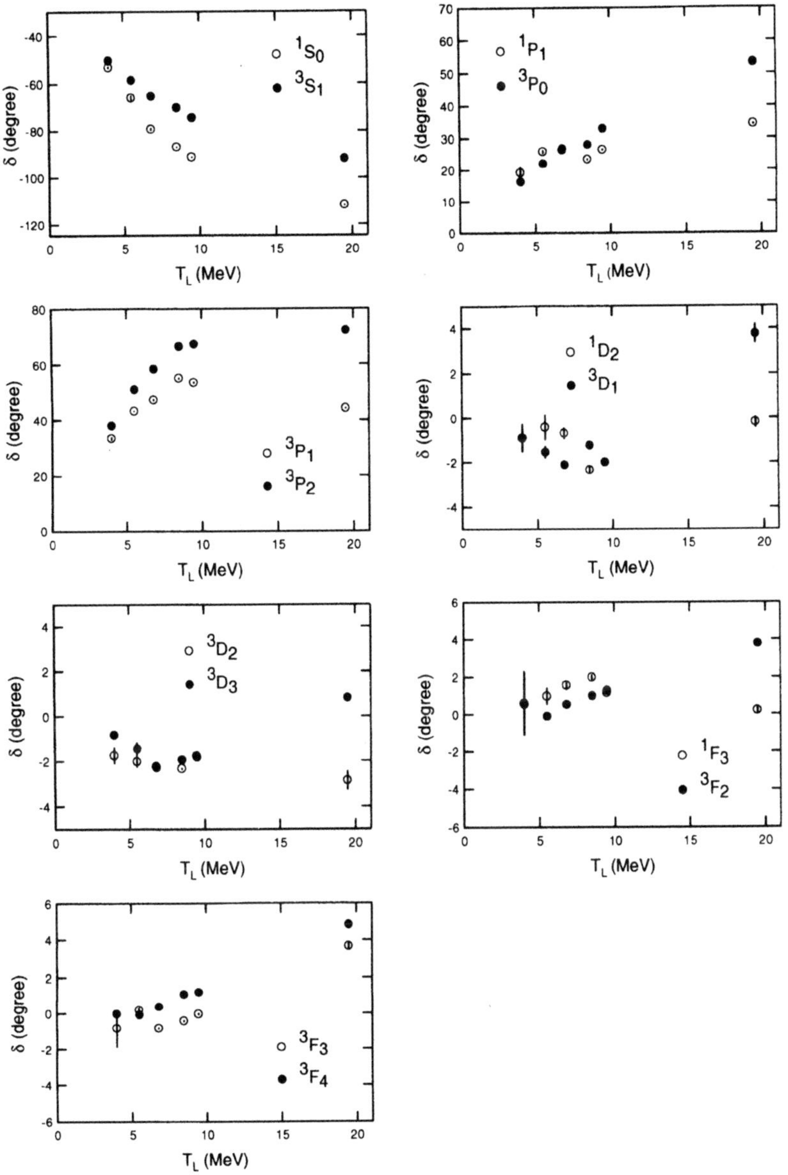

Figure 1. Obtained MW phase-shift parameters.

cross section σ_r at $T_L = 18 \sim 48$ MeV. Therefore we include their σ_r-datum at $T_L = 19.55$ MeV in our database in order to evaluate the inelastic effect.

By the PSAs at 4.0 and 5.5, we obtained the phase-shift solutions, respectively, together with the mixing parameters $\rho_J^P = \rho_1^{\pm}, \rho_2^{\pm}, \rho_3^{-}$ where J is the total angular momentum and P the parity. At 6.8, 8.5 and 9.5 MeV, the present situation of database is not enough to do the complete single-energy PSAs, so that the present PSAs were performed with the approximation of $\rho_2^+ = \rho_2^- = \rho_3^- = 0$, respectively. At 19.48 MeV, we performed the complete PSA with no approximations, and obtained the phase shifts, the reflection parameters and the mixing parameters $\rho_1^\pm, \rho_2^\pm, \rho_3^-$ in spite of the relatively small total-number of data. This result seems to be owing to having the precise data on $d\sigma/d\Omega$ and containing the data on three kinds of spin observables A_{0y}, A_{yy} and A_{xx}.

The obtained solutions at these energy points are plotted in Fig.1.

Our phase-shift solutions at and below 9.5 MeV are very similar to those of energy-dependent PSA given by Alley et al. [8], where the Blatt-Biedenharn representation [9] was used. The phase-shift solution at 19.48 MeV has a good continuation to the solutions at the lower energies $T_L \leq 9.5$ MeV.

4 Concluding Remarks

We performed the single-energy PSAs of elastic p-^{3}He scattering at $T_L = 4.0$, 5.5, 6.8, 8.5, 9.5 and 19.48 MeV and obtained the solutions of the phase shifts, the mixing parameters and the reflection parameters, respectively.

We are going to extend our analysis to the intermediate energy region $T_L = 20 \sim 300$ MeV.

We have a plan to perform the PSAs of n-^{3}He scattering. It is expected that the experiments for n-^{3}He scattering would be carried out near future.

If we determine the phase-shifts of nucleon-^{3}He scattering in the wide energy region $T_L \lesssim 300$ MeV, we could derive the nucleon-^{3}He potentials precisely. We are trying to derive the phenomenological p-^{3}He potential from the present solutions of PSAs. It would be important to take into account the anti-symmetric spin-orbit force correctly to determine the potential, because the obtained mixing parameters for $^1P_1$-$^3P_1$ coupling is larger than the ones between the triplet states, for instance $^3S_1$-$^3D_1$ coupling.

References

1. T. A. Tombrello: Phys. Rev. **138**, B40 (1965)
2. T. Uesaka et al.: Nucl. Instr. and Meth. **A402**, 212 (1998)
3. S. Oryu et al.: Nucl. Instr. and Meth. **A402**, 402 (1998)
4. B. T. Murdoch et al.: Phys. Rev. **C29**, 2001 (1984)
5. M. Matsuda and W. Watari: Lett. Nuovo Cimento **6**, 23 (1973)
6. H. P. Stapp and T. J. Ypsilantis: Phys. Rev. **105**, 302 (1957)
7. A. M. Sourkes et al.: Phys. Rev. **C13**, 451 (1976)
8. M. T. Alley and L. D. Knutson: Phys. Rev. **C48**, 1901 (1993)
9. J. M. Blatt and L. C. Biedenharn: Rev. Mod. Phys. **24**, 258 (1952)

Few-Body Systems Suppl. 12, 491–496 (2000)

Few-Body Systems

Studies of ^{2}H$(d,p)^3$H Reaction at $E_d = 30$ keV by Four-Body Faddeev-Yakubovsky Equation

E. Uzu[1], S. Oryu[1], M. Tanifuji[2]

[1] Department of Physics, Faculty of Science and Technology, and FRCCS, Science University of Tokyo, Yamazaki 2641, Noda, Chiba 278-8510, Japan

[2] Department of Physics, Hosei University, Fujimi 2-17-1, Chiyoda-ku, Tokyo 102-8160, Japan

Abstract. Cross sections and analyzing powers for ^{2}H$(d,p)^3$H reactions are calculated at $E_d = 30$ keV by solving the four-body Faddeev-Yakubovsky equation. The Yamaguchi-type potential is adopted for inter-nucleon forces. The calculated quantities consistently agree with the experimental data. By using the solution, the suppression ratio σ_{pol}/σ_0 for the Yamaguchi potential is calculated.

In this decade we have studied ^{2}H$(d,p)^3$H and ^{2}H$(d,n)^3$He reactions by solving the 4-body Faddeev-Yakubovsky (FY) equations [1]. The calculated differential cross sections for the unpolarized reaction and the tensor analyzing powers for a polarized incident beam at the several 10 keV energy region agreed well with experimental data [2, 3, 4], except for the vector analyzing power [5]. Unfortunately, the previous calculations were carried out by restricting to a limited number of state channels in the [3+1] and the [2+2] subamplitudes. One may expect that the higher spin states of the [3+1] subsystem have large effects on the reaction amplitudes. In order to check the convergence for the higher spin states, the ^{2}H$(d,p)^3$H reaction at a deuteron incident energy of E_d=30 keV is investigated, where the cross sections for an unpolarized beam on an unpolarized target, and the analyzing powers for polarized beam are calculated.

From a technological point of view, $^2\overrightarrow{\mathrm{H}}(\overrightarrow{d},p)^3$H and $^2\overrightarrow{\mathrm{H}}(\overrightarrow{d},n)^3$He reactions are attracting much attention in relation to a neutron-free fusion reactor, where the incident and the target deuterons are polarized simultaneously in the same direction. The suppression ratio is defined as σ_{pol}/σ_0, where σ_0 is the total cross section for the unpolarized reaction, and σ_{pol} is that for the simultaneous polarization of the beam and the target.

We start with the four-body Lippmann-Schwinger (LS) equation which describes a system consisting of four nucleons, which contain six inter-nucleon

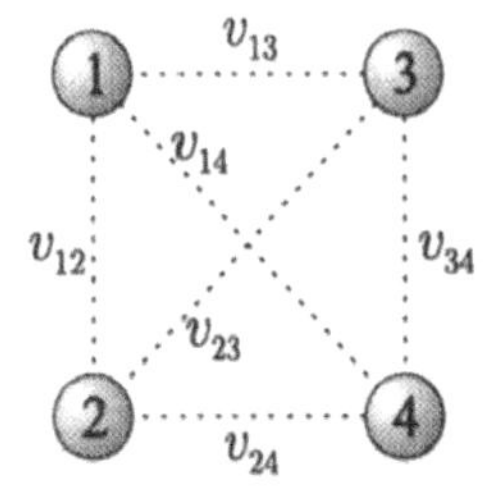

Figure 1. Four-body system.

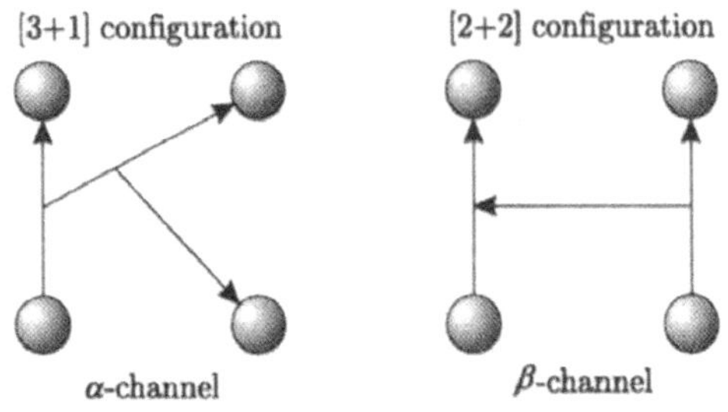

Figure 2. Configurations and channels in four-identical particles.

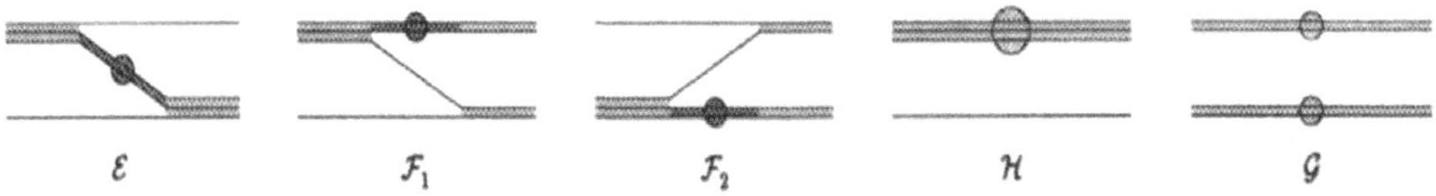

Figure 3. Typical diagrams which appears in the Faddeev-Yakubovsky equation.

interactions (see Fig. 1). The LS eq. is given by

$$T = (v_{12}+v_{13}+v_{14}+v_{23}+v_{24}+v_{34})+(v_{12}+v_{13}+v_{14}+v_{23}+v_{24}+v_{34})\,G_0\,T\,. \quad (1)$$

In order to avoid the well-known δ-function difficulty in the kernel, we must decompose eq.(1) into 18-channel FY equations by Faddeev's technique. The 18-channel system is still complicated, but one can reduce it to two-channel equations (see fig. Fig. 2) in the case of identical particles. One is the [3+1] configuration which is called the α-channel, and the other is the [2+2] configuration, the β-channel.

After the partial wave decomposition and a separable expansion of the sub-amplitudes, the FY equations are written as

$$\begin{pmatrix} \mathcal{M}_{\alpha\alpha} & \mathcal{M}_{\alpha\beta} \\ \mathcal{M}_{\beta\alpha} & \mathcal{M}_{\beta\beta} \end{pmatrix} = \begin{pmatrix} \pm\mathcal{E} & \mathcal{F}_1 \\ 2\mathcal{F}_1^t + 2\mathcal{F}_2^t & 0 \end{pmatrix} + \begin{pmatrix} \pm\mathcal{E} & \mathcal{F}_1 \\ 2\mathcal{F}_1^t + 2\mathcal{F}_2^t & 0 \end{pmatrix} \begin{pmatrix} \mathcal{H} & 0 \\ 0 & \mathcal{G} \end{pmatrix} \begin{pmatrix} \mathcal{M}_{\alpha\alpha} & \mathcal{M}_{\alpha\beta} \\ \mathcal{M}_{\beta\alpha} & \mathcal{M}_{\beta\beta} \end{pmatrix}, \quad (2)$$

where, $\mathcal{M}$'s are the scattering amplitudes, $\mathcal{E}$ the four-body potential for the $\alpha \to \alpha$ transition, and the plus sign corresponds to the boson situation, the minus sign to fermions, $\mathcal{F}_1$ signifies the $\beta \to \alpha$ transition, $\mathcal{F}_1^t$ and $\mathcal{F}_2^t$ the $\alpha \to \beta$ one, subscripts 1 and 2 distinguish the two channels in the [2+2] subsystem. $\mathcal{H}$ and $\mathcal{G}$ are the propagators for the [3+1] and [2+2] subsystems, respectively (see Fig. 3).

For the two-nucleon interactions we employ Yamaguchi-type separable potentials [6] for the $^1S_0$, $^3S_1$-$^3D_1$, $^3P_0$, $^1P_1$, $^3P_1$ and $^3P_2$ states, in which we adopt one parameter set from Ref. [7], where the D-state probability in $^3S_1$-$^3D_1$ state (P_d) is 7.0%, and other parameter sets from Ref. [8] for P_d= 5.5% and 4.0%.

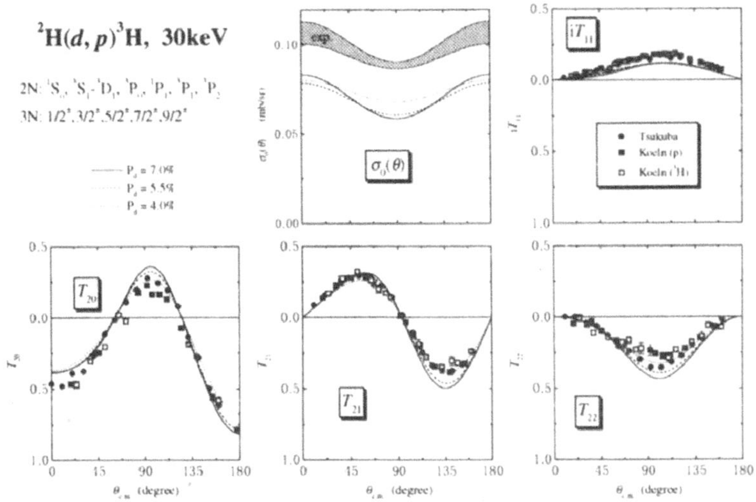

Figure 4. Comparison of unpolarized cross section σ_0, vector analyzing power iT_{11}, and tensor analyzing powers T_{20}, T_{21}, and T_{22} between the theory and the experiment for $^2\text{H}(d,p)^3\text{H}$ reactions at $E_d = 30\text{keV}$. The cross section data [2] are shown by the shaded area between the two thin solid lines. In the vector and tensor analyzing powers, the solid circles are for the 30keV Tsukuba data [3] and the open and solid squares for the 28keV Köln data [4]. The solid lines are calculated by $P_d = 7.0\%$, the dashed lines by $P_d = 5.5\%$, and the dotted lines by $P_d = 4.0\%$.

The amplitudes of the [3+1] and [2+2] subsystems are treated by the Hilbert-Schmidt expansion method. Our four-body calculation is converged at rank-10 for the [2+2] subsystem. In the [3+1] subsystem, there is about a 4% difference between adopting the rank-24 and the rank-16. We take rank-24 in the [3+1] subsystem and rank-10 in the [2+2] subsystem.

The convergence of the four-body amplitudes with respect to the spin and parity of the [3+1] subsystem is achieved at $7/2^{\pm}$, which is confirmed by taking up to $9/2^{\pm}$. In the [2+2] subsystem, all of the allowed states are included. The convergence of the total spin and parity of the four-body system is sufficient for $4^{\pm}$. The FY equations are solved by the Continuum Fraction Method [9] without the Coulomb force. However, Coulomb corrections are taken into account by including a Coulomb penetration factor in the cross section and by introducing Coulomb phase shifts in the scattering amplitude.

The calculated differential cross section and analyzing powers are illustrated in Fig. 4, where the solid lines show the results for $P_d = 7.0\%$, the dashed lines those for $P_d = 5.5\%$, and the dotted lines those for $P_d = 4.0\%$. The anisotropy of the calculated differential cross section agrees well with the data [2] but the magnitude of the calculated cross section is smaller than the measured one

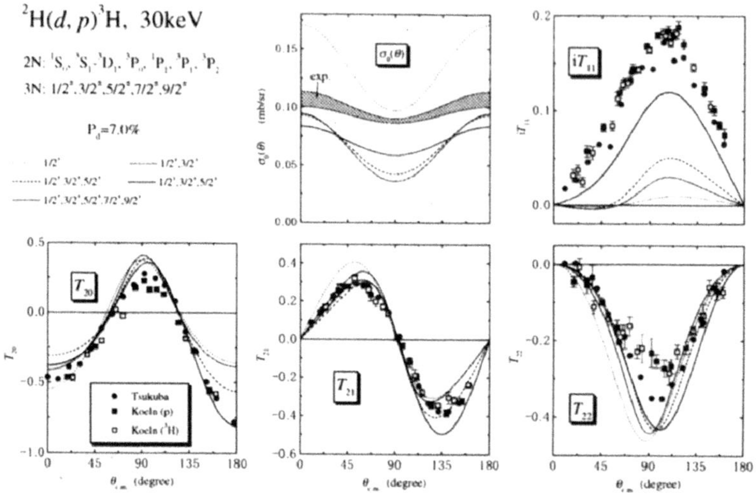

Figure 5. Convergence due to the spin and parity in the [3+1] subsystem. Captions of the experimental data are the same as those in Fig. 4. The dotted lines include only the $1/2^+$ state, the dash-dotted lines include state up to the $3/2^+$, the dashed lines up to the $5/2^+$ state, the dash-two-dotted lines up to the $5/2^\pm$ state, and the solid lines up to the $9/2^\pm$ state. The solid lines and the dash-two-dotted ones are almost overlapped.

by about 30%. One reason for this discrepancy may come from an inadequate treatment of the Coulomb force, in which the incident and target deuterons are treated as point particles where the Coulomb penetration factor is constructed without considering the charge distribution of the deuteron. Another reason may be the Yamaguchi potential. The calculated analyzing powers agree well with the data [3, 4].

Fig. 5 shows the convergence due to the spin and parity in the [3+1] subsystem. The dotted lines describe the calculation which includes only one state $j^\pi = 1/2^+$, where the calculation reproduces most of the tensor analyzing power data. However, the calculated differential cross section is larger than the data by about factor 1.5, the anisotropy of the cross section is enhanced, and the vector analyzing power is almost zero. By including states up to $3/2^+$, the cross section is decreased by about 50% from that of the dotted line, and iT_{11} is increased a little. The change induced by $3/2^-$ and $5/2^+$ states is small. The additional $5/2^-$ state improves the anisotropy of the calculated cross section, though the magnitude of the cross section is still smaller than the measured data. The calculated iT_{11} is increased and comes close to the experimental data.

Fig. 6 shows the effect of including P-waves in the 2N subsystem, where the dashed lines include only $^1S_0$ and $^3S_1$-$^3D_1$ states, and the solid lines include $^1S_0$,

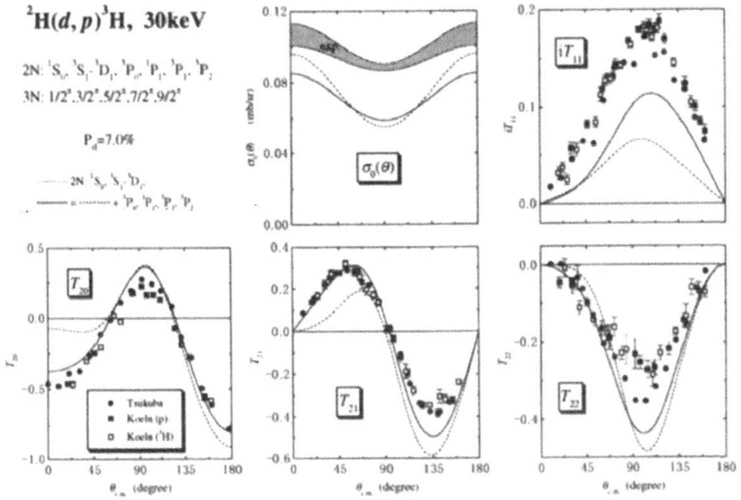

Figure 6. Effects of including P-waves in the 2N subsystem. The dashed lines include the $^1S_0$ and $^3S_1$-$^3D_1$ states, and the solid ones the $^1S_0$, $^3S_1$-$^3D_1$, $^3P_0$, $^1P_1$, $^3P_1$ and $^3P_2$ states.

$^3S_1$-$^3D_1$, $^3P_0$, $^1P_1$, $^3P_1$ and $^3P_2$ states. It is clear that the P-wave component, for which LS forces are effective, plays an important roll not only in the vector analyzing powers but also in the tensor ones. The convergence with respect to the 2N states is not confirmed at present, but it will be examined by using realistic potentials in the future.

Our prediction of the suppression ratio for the Yamaguchi potential is shown in Fig. 7 as a function of Θ, where Θ is the angle between the incident momentum and the polarization axis. The ratio we obtain is 1.15 at $\Theta = 0°$ and 1.03 at $\Theta = 90°$, and is generally larger than 1 for any Θ. This means the total cross section is enhanced by polarization of the incident and target deuterons.

We obtained observables in the low energy $^2\mathrm{H}(d,p)^3\mathrm{H}$ reaction by solving the four-body Faddeev-Yakubovsky equation using Yamaguchi-type two-body potentials. The experimental differential cross section data for an unpolarized beam and the analyzing powers for a polarized one are consistently reproduced by the calculation, though some discrepancies between the measurement and the calculation still remain. We also predict the suppression ratio for the simultaneous polarizations of the beam and the target. Calculations with realistic potentials are now in progress.

References

1. O.A. Yakubovsky: Sov. J. Nucl. Phys. **5**, 937 (1967)

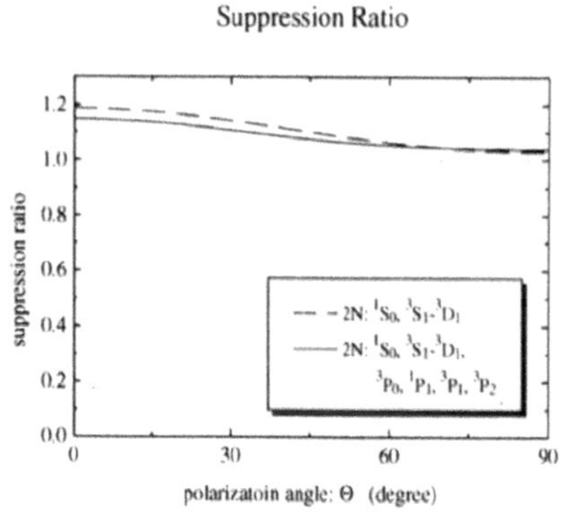

Figure 7. Prediction of the suppression ratio σ_pol/σ_0. Θ indicates the angle between the incident momentum and the polarization axis. The definitions of the lines are the same as those in Fig. 6.

2. R.E. Brown and N. Jarmie: Phys. Rev. **C41**, 1391 (1990)

3. Y. Tagishi, N. Nakamoto, K. Katoh, J. Togawa, T. Hisamune, T. Yoshida, Y. Aoki: Phys. Rev. **C46**, R1155 (1992)

4. B. Becker, R. Randermann, B. Polke, S. Lemaître, R. Reckenfelderbäumer, P. Niessen, G. Rauprich, L. Sydow, H. Paetz gen. Schieck: Few-Body Sys. **13**, 19 (1992)

5. E. Uzu, H. Kameyama, S. Oryu, M. Tanifuji: Few-Body Sys. **22**, 65 (1997); E. Uzu, S. Oryu, M. Tanifuji: Prog. Theor. Phys. **90**, 937 (1993); E. Uzu, S. Oryu, M. Tanifuji: *AIP Conference Proceedings 334 on Few-Body Ploblems in Physics*, Williamsburg, Virginia, 1994, p.447; E. Uzu, S. Oryu, M. Tanifuji: Few-Body Systems Suppl. **99**, 97 (1995); S. Oryu, E. Uzu, T. Hino, S.A. Sofianos: *Proceedings of Advances in Nuclear Physics and Related Areas*, Greece, 1997 (to be published)

6. Y. Yamaguchi: Phys. Rev. **95**, 1628 (1954); Y. Yamaguchi and Y. Yamaguchi: Phys. Rev. **95**, 1635 (1954)

7. P. Doleschall: Nucl. Phys. **A201**, 264 (1973)

8. A.C. Phyllips: Nucl. Phys. **A107**, 209 (1968)

9. J. Horáček and T. Sasakawa: Phys. Rev. **A28**, 2151 (1983); ibid., **A30**, 2274 (1984); ibid., **C32**, 70 (1985)

10. M. Tanifuji: Phys. Lett. **B289**, 233 (1992)

Few-Body Systems Suppl. 12, 497–500 (2000)

Few-Body Systems

$^3\overrightarrow{\text{He}}\ (\vec{d},p)^4$He Reaction at Intermediate Energies

T. Uesaka[1], H. Sakai[1,2], H. Okamura[3], A. Tamii[2], T. Ohnishi[2], Y. Satou[1], N. Sakamoto[1], T. Wakasa[4], K. Itoh[1], K. Sekiguchi[2], K. Yako[2], S. Sakoda[2], K. Suda[3]

[1] The Institute of Physical and Chemical Research (RIKEN), Saitama 351-0198, Japan

[2] Department of Physics, University of Tokyo, Hongo, Bunkyo, Tokyo 113-0033, Japan

[3] Department of Physics, Saitama University, Saitama 338-8570, Japan

[4] Research Center for Nuclear Physics, Osaka University, Osaka 567-0047, Japan

Abstract. We have measured the cross sections and the polarization observables for the ^{3}He$(d,p)^4$He reaction at E_d =140, 200, and 270 MeV. This work aims to investigate the high momentum component of deuteron D-state wave function and the mechanism for the one-nucleon transfer reaction at intermediate energies. The measurement was carried out by using the spectrograph SMART at RIKEN Accelerator Research Facility. A cryogenic ^{3}He gas target with a density of 6.6×10^{20} cm^{-3} was bombarded by a polarized deuteron beam. The data are compared with the calculation based on the one-nucleon exchange approximation.

1 Introduction

The high momentum component of deuteron is of fundamental interest because it is directly connected to the interaction between two bound nucleons in a specific quantum state $J^\pi = 1^+, T = 0$. The ^{3}He$(d,p)^4$He reaction at intermediate energies is considered to provide a unique way to study the high momentum component of deuteron D-state wave function [1]. A prior study was made at $\theta_{\text{lab}} = 4°$ by using a 270 MeV polarized deuteron beam and a polarized ^{3}He target at RIKEN Accelerator Research Facility. The results of this experiment show polarization observables in significant disagreement with the one-nucleon exchange (ONE) approximation calculations. Since the large disagreement is far beyond the uncertainty of the D-state wave function calculated with realistic nucleon-nucleon potential models, it is necessary to reconsider the relevance of reaction mechanism.

For the purpose of a detailed investigation regarding the reaction mechanism, we have measured the angular distributions of differential cross section, vector and tensor analyzing powers for the ^{3}He$(d,p)^4$He reaction at $E_d = 140$, 200, and 270 MeV.

2 Experimental Procedure

The experiments was carried out by using the spectrograph SMART [2].

A cryogenic-gas-target was used in the experiment. The schematic view of the target is shown in Fig. 1. ^{3}He gas was enclosed in a copper container which was contacted to a refrigerator. The target dimensions were 13 mm (20 mm) wide, 15 mm (20 mm) high and 20 mm (10 mm) thick in the case of 270 MeV (140 and 200 MeV) measurement. Entrance and exit windows were 6 μm-thick Havar foils. Temperature of the target was measured by using a diode thermosensor and found to be about 11 K throughout the experiment. Thus, the density of the ^{3}He gas reached 6.6×10^{20} cm^{-3} (6.7 mg/cm^2).

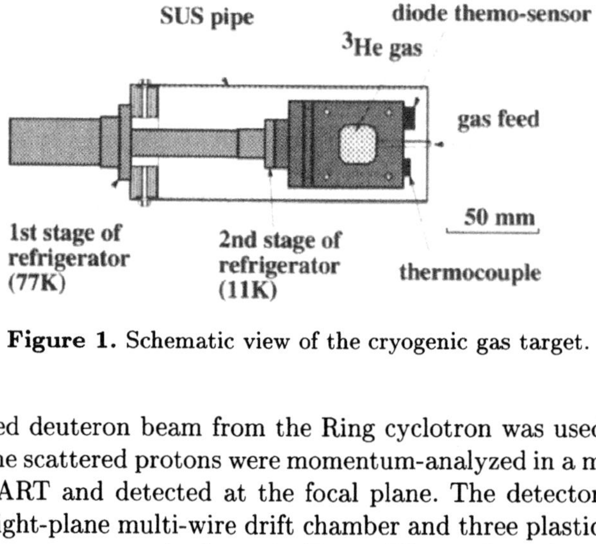

Figure 1. Schematic view of the cryogenic gas target.

A polarized deuteron beam from the Ring cyclotron was used to bombard the target. The scattered protons were momentum-analyzed in a magnetic spectrograph SMART and detected at the focal plane. The detector system consisted of an eight-plane multi-wire drift chamber and three plastic scintillators. Obtained energy spectrum is shown in Fig. 2. The peak at E_x=0 MeV corresponds to events for the ^{3}He$(d,p)^4$He reaction. Other peaks at higher excitation energies are from Havar foils.

3 Results

Very preliminary results of the analyzing powers are shown in Fig.3, together with the ONE calculations (solid lines). Note that the vector analyzing power A_y is identical to zero in the ONE approximation. The tensor analyzing powers A_{yy} and A_{xx} depend strongly on the deuteron wave function, in the framework

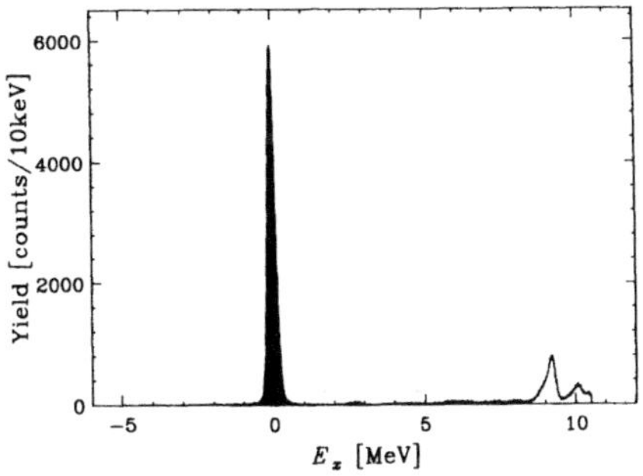

Figure 2. Excitation energy spectrum for the ^{3}He$(d,p)^4$He reaction at E_d = 270 MeV.

of the ONE approximation, and are expressed as,

$$A_{yy} = \frac{-\sqrt{2}u(k_{pn})w(k_{pn}) + \frac{1}{2}w^2(k_{pn})}{u^2(k_{pn}) + w^2(k_{pn})}, \tag{1}$$

$$A_{xx} = \frac{-\sqrt{2}u(k_{pn})w(k_{pn}) + \frac{1}{2}w^2(k_{pn})}{u^2(k_{pn}) + w^2(k_{pn})}(1 - 3\sin^2\theta_{pn}), \tag{2}$$

where u and w are the S- and D-state wave functions of deuteron. The internal momentum k_{pn} is calculated in the infinite momentum frame [1]. θ_{pn} is the angle between the beam axis and the internal momentum.

As shown in Eq.(1), A_{yy} is independent from the scattering angle explicitly. However, since the internal momentum k_{pn} increases as the scattering angle, A_{yy} depends on the scattering angle through the angular dependence of k_{pn}. The modest angular dependence of A_{yy} shown in Fig. 3 is well described by the ONE calculation. In addition, rather drastic change in A_{xx} at forward angles is similar to that expected from the $(1 - 3\sin^2\theta_{pn})$ term in Eq.(2). Thus, the ONE approximation can reproduce the angular dependence of tensor analyzing powers qualitatively.

However, it is clearly seen in Fig. 3 that the magnitudes of A_{yy} and A_{xx} are significantly smaller than the prediction of the ONE approximation at forward angles. The data also show that A_{xx} takes its minimum value at around $\theta_{cm} = 20°$ which corresponds to $\theta_{pn} \sim \pi/4$, though ONE approximation predict that it should be at $\theta_{pn} = \pi/2$ (see Eq.(2)). This minimum point of A_{xx} cannot be reproduced, within the framework of the ONE approximation, even if any kinematical corrections is introduced in the calculation of the internal angle θ_{pn} . These feaures of the tensor analyzing powers, together with the large negative value of the vector analyzing power at θ_{cm} = 15–20°, indicate that the ONE approximation is insufficient to describe the ^{3}He$(d,p)^4$He reaction at this energy region. Mechanisms beyond the ONE, for example, multiple scattering [3], pion rescattering [4], delta excitation in the intermediate channel [5], may be necessary to cover the discrepancy between the experiment and calculation.

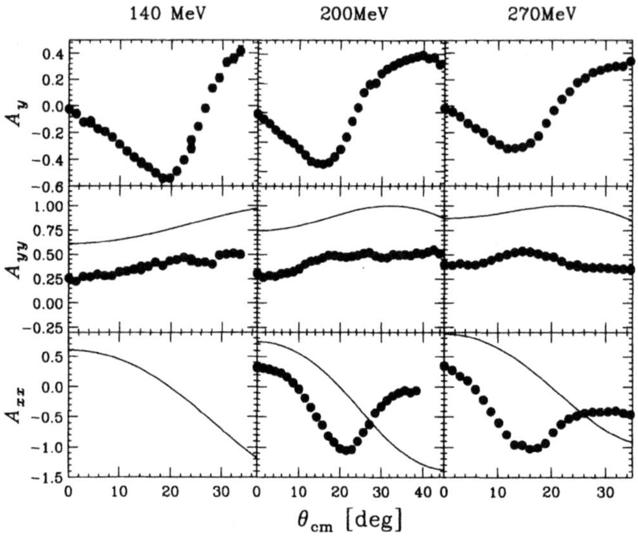

Figure 3. The vector and tensor analyzing powers for the ^{3}He$(d,p)^4$He reaction at E_d=140, 200, and 270 MeV. Solid lines represent the ONE calculations.

4 Summary

We have measured the cross sections and the polarization observables for the ^{3}He$(d,p)^4$He reaction at E_d = 140, 200, and 270 MeV. The results clearly show that the ONE approximation is insufficient to describe the reaction at these energies.

5 Acknowledgement

The authors would like to express their gratitude to the staff of RARF. One of the authors (T.U.) would like to acknowledge the financial support of the Special Researcher's Basic Science Program of RIKEN (the Institute of Physical and Chemical Research).

References

1. T. Uesaka et al.: Phys. Lett. **B467**, 199 (1999)
2. T. Ichihara et al.: Nucl. Phys. **A569**, 287c (1994)
3. A. Tékou: J. of Phys. **G7**, 1439 (1981)
4. J. Källne and P.C. Gugelot: Phys. Rev. **C20**, 1085 (1979)
5. J. Arvieux et al.: Nucl. Phys. **A431**, 613 (1984)

Few-Body Systems Suppl. 12, 501–504 (2000)

Few-Body Systems

Analysis of the $^3\vec{\mathrm{He}}(\vec{\mathrm{d}},\mathrm{p})^4\mathrm{He}$ Reaction at E_{lab} =270 MeV

S. Gojuki[1], H. Kamada[2], E. Uzu[1], S. Oryu[1]

[1] Department of Physics, Faculty of Science and Technology, Science University of Tokyo, Noda, Chiba, 278-8510, Japan
[2] Institut für Theoretische Physik II, Ruhr-Universität Bochum, D-44780 Bochum, Germany

Abstract. Reaction $^3\mathrm{He}(\mathrm{d},\mathrm{p})^4\mathrm{He}$ with a polarized ^{3}He target and a polarized deuterium beam at $E_{\mathrm{lab}} = 270$MeV would be an ambitious experiment in which a constraint on the deuteron D-state probability could be measured directly. Moreover, the reaction is interesting as an example of how to treat the five-body problem from Faddeev theoretical point of view of a multi-channel three-cluster problem. Our approach involves coupled-channel Faddeev equations for the ^{3}He-p-n and d-d-p systems. We obtain the scattering amplitudes "X_{hpn}" and "X_{ddp}" by solving the three-cluster Faddeev equations which is an improvement over our former Born approximation. The calculated X_{hpn} and X_{ddp} are combined by a proper normalization method. We investigate the polarized and unpolarized differential cross sections and the polarization correlation coefficient $C_{//}$. We find that the amplitude X_{hpn} contributes mainly to the forward region of the differential cross section while X_{ddp} to the backward region. $C_{//}$ is not improved.

1 Introduction

As a problem of five nucleon system, the $^3\mathrm{He}(\mathrm{d},\mathrm{p})^4\mathrm{He}$ reaction is a well-known reaction as one of the important processes in primordial nucleosynthesis at low energy. Furthermore, at a higher energy the experiment by using a polarized ^{3}He target with a polarized deuterium beam at $E_{\mathrm{lab}} = 270$MeV by Uesaka et al. in RIKEN (1996)[0] has produced interesting results. It would be interesting to analyze the reaction process in the three-body scattering point of view.

One of the purposes for this experiment[0] was the direct measurement of the D-state probability on the deuteron. In this reaction process, the polarization correlation coefficient $C_{//}$ can be defined by using the tensor analyzing powers,

A_{yy} and A_{xx}, and the spin correlation coefficients, $C_{y,y}$ and $C_{x,x}$, *i.e.*,

$$C_{//}(\theta) = 1 + \frac{1}{4}(A_{yy} + A_{xx}) + \frac{3}{4}(C_{y,y} + C_{x,x}). \tag{1}$$

If one adopt the plane wave Born approximation (PWBA) for the scattering angle at zero degree, the polarization correlation coefficient $C_{//}$ can be written by

$$C_{//}(0^\circ) = \frac{9}{4}\frac{w^2(k)}{u^2(k) + w^2(k)}, \tag{2}$$

where k is the momentum, and $u(k)$ and $w(k)$ are the S-state, the D-state wave functions of the deuteron, respectively. While the D-state probability of the deuteron is defined by

$$P_D = \frac{\int_0^\infty w^2(r)dr}{\int_0^\infty (u^2(r) + w^2(r))dr}, \tag{3}$$

where $u(r)$ and $w(r)$ are the S-state and the D-state wave functions of the deuteron in the configuration space, respectively. Therefore, it is found that the quantity of $C_{//}(0^\circ)$ is closely related to the D-state probability of the deuteron. Furthermore, one could imagine from the spin algebra that the $^3\vec{\mathrm{He}}(\vec{\mathrm{d}},\mathrm{p})^4\mathrm{He}$ reaction could be suppressed in the absence of the D-state of the deuteron.

Therefore the analysis of $^3\vec{\mathrm{He}}(\vec{\mathrm{d}},\mathrm{p})^4\mathrm{He}$ reaction is very interesting not only in verifying the prediction of the spin algebra, but also in establishing the validity of five nucleon calculations.

2 Calculation

Although the $^3\mathrm{He}(\mathrm{d},\mathrm{p})^4\mathrm{He}$ reaction is a five nucleon problem, the problem could be rearranged by several three cluster systems, ^{3}He-p-n, ^{3}H-p-p and d-d-p in the context of the multi-channel three-cluster Faddeev theory. We adopt the Faddeev equations for two three-cluster systems, ^{3}He-p-n and d-d-p without distinguishing the neutron and the proton.

In the Faddeev formalism, we choose rank-1 separable forms not only for the N-N potential but also for d-N, d-d and ^{3}He-N effective potentials *i.e.*, $V_i = g_i\lambda_i g_i (i = \mathrm{dN}, \mathrm{dd}, \mathrm{hN})$ with form factors g_{dN}, g_{dd} and g_{hN}, respectively. We introduce the form-factors by the following relations,

$$g_{\mathrm{dN}}(q) = \sum_\alpha \int dp \Psi_{\mathrm{d}}(p,\alpha)(E - \frac{p^2}{m} - \frac{3q^2}{4m})\Psi_{^3\mathrm{He}}(p,q,\alpha), \tag{4}$$

$$g_{\mathrm{dd}}(q) = \sum_\alpha \int dp dp' \Psi_{\mathrm{d}}(p,\alpha)\Psi_{\mathrm{d}}(p',\alpha)(E - \frac{p^2}{m} - \frac{p'^2}{m} - \frac{q^2}{2m})\Psi_\alpha(p,p',q,\alpha), \tag{5}$$

$$g_{\mathrm{hN}}(r) = \sum_\alpha \int dp dq \Psi_{\mathrm{h}}(p,q,\alpha)(E - \frac{p^2}{m} - \frac{3q^2}{4m} - \frac{2r^2}{3m})\Psi_\alpha(p,q,r,\alpha), \tag{6}$$

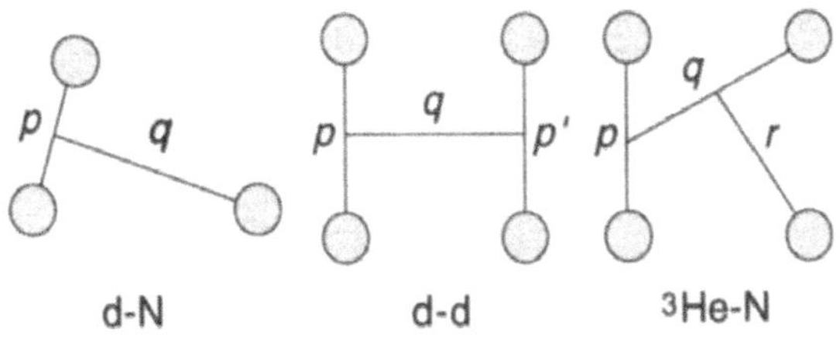

Figure 1. The Jacobi coordinates in the momentum space.

where $\Psi_{\rm d}$, $\Psi_{\rm h}$ and Ψ_α are the wave functions for the deuteron, ^{3}He and ^{4}He, respectively. These three wave functions are obtained with the AV14 N-N potential by using the two-body Lippmann-Schwinger, three-body Faddeev, and the four-body Faddeev-Yakubovsky equations, respectively. The kinematical conditions are expressed by means of the Jacobi-momentum coordinates in Fig.1. Two types of three-cluster Faddeev equations for ^{3}He-p-n and d-d-p systems are written by a set of Amado-Loveace equations in the following forms,

$$X^{\rm hpn}_{\alpha\beta} = Z^{\rm hpn}_{\alpha\beta} + \sum_\gamma Z^{\rm hpn}_{\alpha\gamma}\tau^{\rm hpn}_\gamma X^{\rm hpn}_{\gamma\beta} \tag{7}$$

and

$$X^{\rm ddp}_{\alpha'\beta'} = Z^{\rm ddp}_{\alpha'\beta'} + \sum_{\gamma'} Z^{\rm ddp}_{\alpha'\gamma'}\tau^{\rm ddp}_{\gamma'} X^{\rm ddp}_{\gamma'\beta'}, \tag{8}$$

where $\alpha = (np) - h$, $\beta = (hn) - p$, $\alpha' = (dp) - d$, $\beta' = (dd) - p'$, etc. The total scattering amplitude X which consists of two partial amplitudes $X^{\rm hpn}$ and $X^{\rm ddp}$ systems is represented by using the usual antisymmetrization scheme for the protons and the neutrons,

$$X = \frac{1}{2}X^{\rm hpn}_{\alpha\beta} + \frac{1}{6}X^{\rm ddp}_{\alpha'\beta'}. \tag{9}$$

3 Result and Discussions

Our calculated results for the polarized and unpolarized cross sections shown in Fig.2 and $C_{//}$ in Fig.3.

As can be seen from Figures, the effects of the ^{3}He-p-n three-body system dominate over the d-d-p one at the forward angle. The same tendency also appears in $C_{//}$. The present calculations are improvements over the previous calculations since in which the ^{3}He-p-n and d-d-p systems are solved exactly by the three-cluster Faddeev equations. Here the form factors are systematically introduced based on the AV14 N-N potential. The present model calculations are much superior to the former one in a sense which was calculated with the Born approximation for ^{3}He-p-n system, in which the seRGM and the PEST-type effective potentials for ^{3}He-N interaction and the 13 different N-N potentials were investigated. Nevertheless, the new results could not always represent the experimental data. Especially, the calculated results for $C_{//}$ are

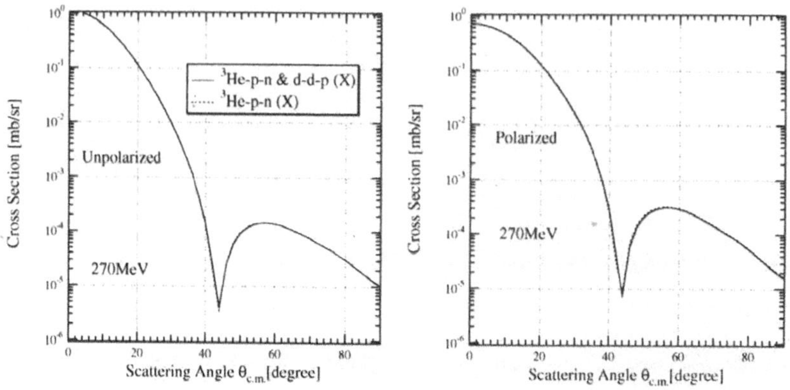

Figure 2. The unpolarized and polarized differential cross sections.

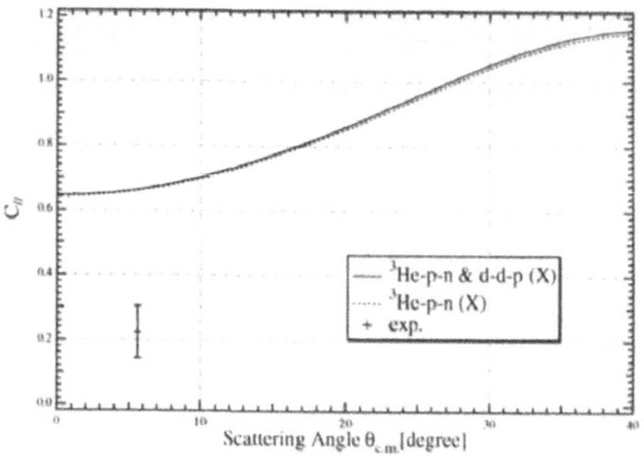

Figure 3. The polarization correlation coefficient $(C_{//})$.

not much improved over the former one. A possible reason for the disagreement may be due to the limited degrees of freedom of the system in which the effective potentials could represent only the ground states. The higher excited states of ^{3}He and ^{4}He in the p-d, the n-^{3}He and the d-d subsystems could appear in the intermediate states. Another possible reason may be due to the channel coupling scheme between ^{3}He-p-n and d-d-p systems which is closely related to the original nature fo five-body problem.

Acknowledgement. This work was supported by a grant of the FRCCS projects in the Science University of Tokyo.

References

1. T. Uesaka: Doctoral Thesis of University of Tokyo 1996
2. S. Oryu et al.: Few-Body Systems Suppl. **10**, 371 (1999)

Few-Body Systems Suppl. 12, 505–512 (2000)

Few-Body Systems

Concluding Remarks

B.F. Gibson*

Theoretical Division, Los Alamos National Laboratory, Los Alamos, New Mexico 87545, USA

Abstract. This represents the author's personal perspective regarding the ideas presented at this first Asia Pacific Conference on Few-Body Problems in Physics. It is not a summary talk in the traditional sense, but is a set of remarks emphasizing physics *not* speakers. Many of the talks covered results of those other than the speaker, and the ideas discussed will hopefully outlive those bringing the ideas to life.

1 Introduction

The organizers produced a program of great interest to the audience. In particular, a number of non traditional few-body physics topics were showcased. Examples include the antiworld, metallic clusters, nuclear astrophysics, and atomic systems.

From the world of antiparticles we heard about the intriguingly long lifetime of a $\bar{p}$ in liquid He, some 3μs compared to 10^{-12} seconds in matter. Relativistic corrections were suggested to play a role in the understanding. The goal is an eventual Lamb shift measurement leading to an uncertainty in $m_{\bar{p}}$ which is smaller than the current uncertainty in the proton mass.

From the study of metallic clusters we heard about the surprising shell structure, analogous to that observed in atoms and in nuclei. One particular question of interest was whether the structure was molecular or solid. Magic number implications will apparently have much to say about the resolution.

In the area of astrophysics, explosive nucleosynthesis was the hot topic, in particular questions related to nuclear cosmochromometry and the age of the universe. Specific rare isotope beam measurements of importance were ^{13}N$(p,\gamma)^{14}$O for the hot CNO cycle and ^{7}Be$(p,\gamma)^{8}$B for the solar neutrino problem. Crucial few-body reactions were suggested to be ^{4}He$(d,\gamma)^{6}$Li and n$(p,\gamma)^{2}$H.

* *Additional Institutions:* Forschungszentrum Juelich, University of Houston, Institute for Nuclear Theory (Seattle), Flinders University, University of Melbourne, and Tokoku University

The complex coordinate rotation method, a powerful calculational technique from atomic physics, was summarized with results reported for the $e^-e^+e^-$ floppy molecule and for resonances in e^--He scattering. Complex scaling methods applied to calculate the properties of unstable nuclei were also explored. It was noted that an analytic potential is required, that a Woods-Saxon form is not useful. Moreover, strong repulsion presents some difficulty.

Areas of emphasis in more traditional few-body hadronic physics fell into four categories: QCD in hadron structure and interactions, few-nucleon scattering and reactions, hypernuclear systems, and rare isotope nuclei. Observations regarding each of these follow in separate sections.

2 QCD in Nuclear Physics

Hadron structure was the key question addressed. This included quark models of baryons, mesons, and dibaryons. Particular emphasis was given to the chiral quark approach. An unanswered question appeared to be how the $q\bar{q}$ pairs exchanged among the quarks in this model are to be related to the physical pion. Physical pions are thought to be complex Fock space entities involving significant $(q\bar{q})^n$ components.

Other interesting questions were also addressed. Chiral symmetry above the critical temperature T_C remains a taunting concept. Also intriguing is how to properly include in the non perturbative QCD models those lower-energy features which are clearly absent in the perturbative QCD limit. Sum rule methods were offered as one constraint. The πN Σ term continues to attract contradictory explanations. One obvious question that still begs for an answer is "where are the gluons in the models?"

In the area of QCD-based models of the nucleon-nucleon (NN) interaction there appeared an interesting conflict. In the quark-model realm, it was suggested that a comparison with standard meson-exchange models may well reveal distinguishing differences at short range. It is at short range where one might anticipate nucleon substructure to play a role. Alternatively, effective field theory (EFT) asserts authoritatively that low-energy properties are not affected by short range (the quark-gluon region) in any measurable manner. Certainly, the low-energy aspects of the NN force do appear to determine the physically observable properties of nuclei.

In the important area of NN interaction models, several interesting points were made. Precision partial-wave analyses (PWAs) define the quality of the data representation to be expected from any NN model. Recent potential models (*e.g.*, Nijmegen, Argonne, Bonn) are effectively equivalent to an amplitude analysis in that they provide descriptions of the physical observables (the scattering data) with $\chi^2/\mathrm{df} \simeq 1$. Alternative approaches (quark, EFT, ...) would appear to require a comparable number of parameters to produce an equivalent fit – that is, more than an "eye ball" fit to the experimental data. It was actually the quantitative improvement of the one-boson-exchange (OBE) models over the then contemporary phenomenological NN models which promoted the OBE hypothesis. What features of the new alternative models distinguish them from the OBE hypothesis? The longer range of the chiral perturbation theory $\pi\pi$ form of the $\sigma \cdot \ell$ term compared to that of the OBE hypothesis may be significant. The quark model in the strangeness -1 sector predicts repulsion at short range in all partial waves, rather than attraction in some as a pure OBE hypothesis suggests.

3 Few-Nucleon Dynamics

3.1 The 2 1/2 Body Problem

There are several "three-body" problems in which one of the three bodies is weakly interacting, where one does not need the power of a Faddeev-like formalism to solve the continuum problem with adequate precision. Three of these were explored at this meeting: 1) proton-proton bremsstrahlung, 2) radiative pion capture by deuterium leading to an $nn\gamma$ final state, and 3) radiative kaon capture by deuterium leading to a $\Lambda n\gamma$ final state.

3.1.1 $pp\gamma$

Precision measurements of this inelastic scattering process provide important model tests. The primary goal is the same as found in $np \to d\gamma$ or in $d(e,e')np$ reaction studies. One explores the fundamental question of what is the important physics in coupling the electromagnetic interaction with the strong interaction in hadronic systems.

3.1.2 $\pi^- d \to nn\gamma$

Pion absorption is utilized as a tool to measure a_{nn} in the final state where the interaction of the photon with the two neutrons can be neglected. The ultimate goal is a comparison of the zero-energy nn interaction with the zero-energy pp interaction, to explore charge symmetry breaking in the NN force.

3.1.3 $K^- d \to \Lambda n\gamma$

One uses the strange K^- to convert a nucleon into a Λ and the three-body $\Lambda n\gamma$ final-state to explore the low-energy Λn force. A pure two-body Λn interaction (in which there is no three-body final-state interaction) is not easily achieved by other means. The goal is separation of the $^1S_0$ and $^3S_1$ amplitudes, so that a comparison with Λp data is possible to determine the charge symmetry breaking in the strangeness -1 sector.

3.2 $N \geq 3$ Dynamics

3.2.1 The Standard Model

The non relativistic hamiltonian approach[1], in which one assumes

1) nuclei consist only of nucleons – other degrees of freedom are suppressed or treated perturbatively,

2) nucleons move slowly within the nucleus – non relativistic dynamics prevail,

3) nucleons interact primarily via pairwise forces,

provides an enormous simplification and has been very successful. The dominant properties of the NN force (for lab energies of less than 350 MeV or below the pion production threshold) are:

- strong spin/isospin dependence,
- strong tensor force,
- important short-range repulsion.

Any realistic model *must* include these aspects of the force.

Extensions of the model include meson exchange currents (MEC), three-body forces (3BF), Δ degrees of freedom, relativistic effects, and quark-gluon substructure. Note that many MEC effects are of order $(v/c)^2$; that is, they are of relativistic order. Care is needed to ensure a consistent treatment of both.

3.2.2 Important Historical Highlights

For the benefit of the students at the meeting, let us review some of the historical highlights:

- Thomas' ^{3}H variational calculation [2, 3] of the mid 1930s demonstrated that the NN force must be of finite range – a zero range force leads to infinite binding of the three-body system.
- Tjon produced in the 1970s the initial calculation [4] of the triton from a "realistic" potential and investigated the resulting electromagnetic form factors.
- Alt demonstrated in a separable potential AGS calculation [5] that the neutron-deuteron quartet scattering length ($^4a_{nd}$) depends only upon the low-energy $^3S_1$ interaction (the spin-triplet scattering length or correspondingly the deuteron binding energy).
- Amado and co-workers demonstrated [6] the importance of analyticity and unitarity in the scattering problem, that the angular shape of the differential cross section is determined by the last rescattering but that the cross section normalization is fixed by the solution of the full integral equation.
- Finally, low energy triton observables "scale" with the binding energy of the system [1] – rms radius, Coulomb energy, D/S asymptotic normalization ratio, $^2a_{nd}$. [If one includes a 3BF to ensure a fit to $B(^3H)$, then one obtains experimental values for the other observables.]

3.3 Three-Body Forces and NN Off-Shell Amplitudes

Because of issues raised at this conference, brief comments about three-body forces, NN off-shell amplitudes, and the deuteron D-state are in order:

- Potentials are not observables; that is, they are not measurable.
- NN off-shell amplitudes are not observables; they are not measurable.
- P_D, the deuteron D-state, is not an observable; it cannot be measured.

Paraphrasing the content of Ref.[7]: If there exists a local potential hamiltonian ($\sim V_{NN} + V_{NNN}$), then there exists a second hamiltonian ($\sim \tilde{V}_{NN}$ and for which $\tilde{V}_{NNN} \equiv 0$) with the same bound states and on-shell scattering amplitudes (phases), **but** $\tilde{V}_{NN}$ is nonlocal **and** $V_{NN} \neq \tilde{V}_{NN}$ off shell. Nonetheless, particular V_{NN} models imply specific P_D and off-shell amplitudes. Data from Nd scattering, $NN\gamma$, etc. provide useful constraints on these model properties.

3.4 Prominent Issues at APFB99

A number of experimental and theoretical issues were raised at the conference:

• Nd scattering at 270 MeV: The data are beautiful. The spin observables provide a strong filter to suppress the dominant spin independent amplitudes, to permit one to examine finer details. The importance of $NN - N\Delta$ coupling and possible three-body force effects was emphasized.

• $nd \rightarrow nnp$ (coincidence studies) and A_y in $\vec{p}d$ scattering: A three-nucleon-force signal continues to be the motivation for many experiments.

• A=4/5 continuum investigations: Core issues in the A=4 system were [31]$\otimes$[22] coupling, the n-^{3}H resonance structure below 10 MeV, and the question of whether the A_y problem in nd scattering has a connection with the nt cross section.

• Electromagnetic interactions of the type $^3\vec{\mathrm{He}}(\vec{e},e'X)$, T_{20} from $^2\vec{\mathrm{H}}(e,e)$, and the measurements of G_E^n and G_M^n: The momentum transfer involved in T_{20} raises the question of the importance of relativistic effects. The use of ^{3}He as a neutron target requires theoretical modeling.

• Chiral symmetry in the πNN system: The imposition of chiral symmetry in model calculations is recognized as an important problem.

4 Hypernuclear Physics

The physics of hypernuclei differs significantly from that of conventional, non strange nuclei. Of the many novel aspects of flavor nuclear physics some of the more interesting strangeness -1 features include:

• The ΛN interaction has no direct one pion exchange component, but it does involve significant ΣN coupling.

• The hypertriton $^3_\Lambda$H is barely bound, in contrast to the conventional triton; ΛN-ΣN coupling in the hyperon-nucleon interaction plays a prominent role.

• The $^4_\Lambda$H, $^4_\Lambda$He isodoublet exhibits sizeable charge symmetry breaking and a large spin-flip E_γ between ground and excited states; again, ΛN-ΣN coupling appears to play an important role.

The strangeness degree of freedom adds a new dimension to our evolving picture of nuclear physics. We should investigate whether the intuition carefully developed in the non strange sector (of conventional nuclei) extrapolates to explain the observations in the strangeness -1 sector of Λ hypernuclei **or** whether these complex models are merely sophisticated interpolation tools valid primarily within the two-dimensional (neutron-proton) isospin space where they were developed.

There were a number of hypernuclear topics highlighted at this meeting, some involving theoretical calculations while others involved new experimental measurements:

• ${}^{n}_{\Lambda}$H isotopes clearly exhibit the role that the Λ can play in enhancing nuclear core binding and converting resonant levels into particle-stable states.

• The beautiful γ de-excitation data for ${}^{7}_{\Lambda}$Li and ${}^{9}_{\Lambda}$Be (from KEK and BNL) provide new information about the ΛN $\vec{\sigma} \cdot \vec{\sigma}$ force, confirm that the ΛN $\vec{\sigma} \cdot \ell$ force must be small (in contrast to that of the NN interaction), and illustrate through B(E2) values that the Λ compresses the nuclear core of "halo" nuclei.

• Weak mesonic decays are first cousin to neutron β decay. Rates and momentum distributions test hypernuclear model wave functions. A measurement of the π^0 decay of ${}^{4}_{\Lambda}$H holds the key to our testing the validity of the $\Delta I = 1/2$ rule in this realm.

• Strangeness -2 data are eagerly anticipated. The existence of such hypernuclei is pertinent to the **H** dibaryon question. They provide our window to investigate the $\Lambda\Lambda$ interaction. The role of $\Lambda\Lambda$-ΞN mixing in $\Lambda\Lambda$ hypernuclei compared to that of ΛN-ΣN mixing in Λ hypernuclei is an important issue.

• The question of narrow states in Σ hypernuclei remains of interest.

• Exotic phenomena continue to attract experimental effort: **H** dibaryon searches, stranglet searches, and searches for hypernuclei with strangeness less than -2.

Hypernuclear physics is clearly a vibrant enterprise in Japan. The experimental and theoretical efforts by nuclear physicists from our host country lead the world in this field.

5 Rare Isotopes and Unstable Nuclei

Rare isotopes with a large neutron or proton excess take us far from the line of stability and play a significant role in nuclear astrophysics. Characteristics of such halo nuclei were reported to include 1) a large interaction cross section due to the extended density, 2) a correspondingly narrow momentum distribution and soft E1 transition strength, and 3) a departure from a simple $r_0 A^{1/3}$ rule that describes stable nuclei.

For few-body physicists these properties would be expected to follow directly from the small separation energy – nothing more. Viewed in that way, the highlighted characteristics do not appear "exotic". The hypertriton ${}^{3}_{\Lambda}$H qualifies as one of the world's largest halo systems: the separation energy is $B_\Lambda = 130$ keV; the distance separating the Λ from the deuteron core is some 6 times larger than the radius of the deuteron core. The complete description of ${}^{3}_{\Lambda}$H comes from solving the same Faddeev equations that one solves to obtain the properties of the normal triton ^{3}H; no exotic modification of the theoretical model is required. The entire difference lies in the properties of the ΛN interaction, $V_{\Lambda N}$ being much weaker than V_{NN}, which leads directly to the small Λ separation energy and corresponding extended radial distribution for the Λ.

A few other observations seem to be in order:

• The n-p force, not the n-n (nor the p-p force), binds nuclei. (If the n-p spin-triplet force were as weak as the n-n, then the triton would be bound by only about 1 MeV rather than 8.5 MeV.) Deviations of shell structure for N$\gg$Z nuclei from that observed near the line of stability should come as no surprise; they should be anticipated.

• The mean field shell model is a phenomenological approximation, as is the radius expression $r = r_0 A^{1/3}$, developed to describe experimental observation along the line of stability (N$\simeq$Z). Deviations far from the line of stability are to be expected, when the weaker n-n force plays a more significant role.

• A Borromean system, in which there exist no pairwise bound states among the three constituents, does not guarantee a halo nucleus. ^{9}Be (α-α-n) is a normal nucleus and ^{6}He (α-n-n) is not a halo system, although ^{11}Li (^{9}Li-n-n) is. A halo nucleus requires only a weak valence particle interaction with the core of the nucleus. ${}^3_\Lambda$H is the quintessential halo nucleus, and it is not Borromean in nature.

Despite these caveats, the investigation of shell structure far from the line of stability contains truly interesting physics. The existence and character of neutron magic numbers as a function of Z and the relationship to the neutron drip line is a fertile area of research. Study of the properties of the He isotopes represented the beginning of this field of study.

6 Summary

It is unfortunately the case in many areas of nuclear physics research that the overlap between the domain of calculations that are easily performed by theorists and the domain of measurements that are easily made by experimentalists is limited. Nonetheless, nuclear physicists are working diligently to expand the overlap between experiment and theory and our fundamental understanding of the underlying physics. We look forward to further progress in unraveling the mysteries of the nucleus by March 2000 when the 16th International few-body conference is scheduled for Taiwan.

In order to give you something to think about during the intervening months, let me leave you with a question to ponder. Everyone has heard of the classic atomic physics question: The sky is blue; what is Avagadro's number? Let me pose for you a classic few-body nuclear physics question: Why should Bethe have known in 1932 that $r^2_{chg}(^3\mathrm{He}) > r^2_{chg}(^3\mathrm{H})$, even before they were measured by Hofstadter in the 1960s? *Hint*: The answer has nothing to do with the Coulomb repulsion between the protons in ^{3}He.

Acknowledgement. The work of the author is performed under the auspices of the U.S. Department of Energy. Travel support was provided by the U.S. National Science Foundation through a Division of International Programs U.S.-Japan Cooperative Research Grant administered through the University of Houston.

References

1. B.F. Gibson: Nucl. Phys. **A543**, 1c (1992)

2. L.W. Thomas: Phys. Rev. **47**, 903 (1935)

3. H.A. Bethe and R.F. Bacher: Rev. Mod. Phys. **8**, 82 (1936)

4. J.A. Tjon et al.: Phys. Rev. Lett. **25**, 540 (1970)

5. E.O. Alt: *Proceedings of the VIIth International Conference on Few-Body Problems in Nuclear and Particle Physics*, edited by A.N. Mitra, I. Slaus, V.S. Bashin, V.K. Gupta, p.76. Amsterdam: North-Holland 1976

6. R.D. Amado: Phys. Rev. **C11**, 719 (1975); **C12**, 1354 (1975); S.K. Adhikari and R.D. Amado: Phys. Rev. **D9**, 1467 (1974)

7. W.N. Polyzou and W. Glöckle: Few-Body Systems **9**, 97 (1993)

List of Participants

Afnan, Iraj
Physics Department
Flinders University
G.P.O. Box 2100
Adelaide SA 5001
Australia
Iraj.Afnan@Flinders.edu.au

Akaishi, Yoshinori
Theory Division
KEK-Tanashi
Tanashi, Tokyo 188-8501
Japan
akaishi@tanashi.kek.jp

Akiyama, Satoru
Department of Physics
Science University of Tokyo
2641 Yamazaki
Noda, Chiba 278-8510
Japan
akiyama@ph.noda.sut.ac.jp

Akiyoshi, Hiromichi
RIBF
RIKEN
2-1 Hirosawa
Wako, Saitama 351-0198
Japan
akiyoshi@postman.riken.go.jp

Aoki, Shozo
Department of Nuclear Engineering
Kyushu University
6-10-1 Hakozaki
Higashi, Fukuoka 812-8581
Japan
aoki@nucl.kyushu-u.ac.jp

Arai, Koji
Department of Physics
Niigata University
Ikarashi 2 - 8050
Niigata 950-2181
Japan
arai@nt.sc.niigata-u.ac.jp

Arima, Masaki
Department of Physics
Osaka City University
3-3-138 Sugimoto
Sumiyoshi, Osaka 558-8585
Japan
arima@ocunp1.hep.osaka-cu.ac.jp

Arisaka, Isamu
General Eeducation
Chiba Institute of Technology
Shibazono 2-1-1
Narashino, Chiba 275-0023
Japan
arisaka@cc.it-chiba.ac.jp

Bentz, Wolfgang
Department of Physics
University of Tokyo
7-3-1 Hongo
Bunkyo, Tokyo 113-0033
Japan
bentz@phys.s.u-tokyo.ac.jp

Blankleider, Boris
School of Chemistry, Physics and Earth Sciences
Flinders University
Bedford Park
SA 5042
Australia
boris.blankleider@flinders.edu.au

Carbonell, Jaume
Institut des Sciences Nucleaires
53, Av. des Martyrs
F-38026 Grenoble
France
carbonel@isn.in2p3.fr

Chandler, Colston
Department of Physics & Astronomy
University of New Mexico
NM 87131-1156
USA
chandler@unm.edu

Chen, Jian-ping
Physics Division
Thomas Jefferson National Accelerator Facility
MS 12H, 12000 Jefferson Avenue
Newport News, VA 23606
USA
jpchen@jlab.org

Chmielewski, Karsten
Institut für Theoretische
Universitaet Hannover
Appelstrasse 2
D-30167 Hannover
Germany
chmiele@itp.uni-hannover.de

Fachruddin, Imam
Institut für Theoretische Physik II
Ruhr-Universitaet Bochum
Universitaetsstrasse 150
D-44780 Bochum
Germany
imamf@hadron.tp2.ruhr-uni-bochum.de

Fearing, Harold W.
TRIUMF
4004 Wesbrook Mall
Vancouver, BC V6T 2A3
Canada
fearing@triumf.ca

Fonseca, Antonio C.
Centro Fisica Nuclear
Universidade Lisboa
Av Prof. Gama Pinto 2
1649-003 Lisboa
Portugal
fonseca@cfnul3a.cii.fc.ul.pt

Fujita, Tadashi
Department of Physics
Kyoto University
Kitashirakawa Oiwake-cho
Sakyo, Kyoto 606-8502
Japan
fujita@ruby.scphys.kyoto-u.ac.jp

Fujiwara, Yoshikazu
Department of Physics
Kyoto University
Kitashirakawa Oiwake-cho
Sakyo, Kyoto 606-8502
Japan
fujiwara@ruby.scphys.kyoto-u.ac.jp

Fukuda, Tomokazu
KEK
1-1 Oho
Tsukuba, Ibaraki 305-0801
Japan
fukuda@jhfpc1.kek.jp

Futami, Yasuhiko
Department of Physics
Science University of Tokyo
2641 Yamazaki
Noda, Chiba 278-8510
Japan
futami@ph.noda.sut.ac.jp

Gao, Haiyan
Physics Department
Massachusetts Institute of Technology
26-413, MIT
77 Massachusetts Avenue
Cambridge, MA 02139
USA
haiyan@mitlns.mit.edu

Gibbs, William
Department of Physics
New Mexico State University
Las Cruces, NM 88006
USA
gibbs@nmsu.edu

Gibson, Benjamin
Theory Division
Los Alamos National Laboratory
T-5, MS-B283
Los Alamos, NM 87545
USA
gibson@t5.lanl.gov

Glöckle, Walter
Institut für Theoretische Physik II
Ruhr-Universitaet Bochum
Universitaetsstrasse 150
D-44780 Bochum
Germany
walterg@hadron.tp2.ruhr-uni-bochum.de

Glozman, Leonid
Institute for Theoretical Physics
University of Graz
Universitatsplatz 5
A-8010 Graz
Austria
lyg@physik.kfunigraz.ac.at

Gojuki, Shuichi
Department of Physics
Science University of Tokyo
2641 Yamazaki
Noda, Chiba 278-8510
Japan
j6299704@ed.noda.sut.ac.jp

Hamahata, Yoshiyuki
Department of Physics
Kyushu University
6-10-1 Hakozaki
Higashi, Fukuoka 812-8581
Japan
hama2scp@mbox.nc.kyushu-u.ac.jp

Hamamoto, Nobukuni
Graduate School of Arts and Sciences
University of Tokyo
3-8-1 Komaba
Meguro, Tokyo 153-8902
Japan
hamamoto@ntl.c.u-tokyo.ac.jp

Harada, Kazuo
Faculty of Food Culture
Kurashiki-Sakuyo University
3515 Tamashima-Nagao
Kurashiki 710-0292
Japan
harada@ksu.ac.jp

Harada, Toru
Department of Social
Sapporo Gakuin University
11-Bunkyodai
Ebetsu 069-8555
Japan
harada@earth.sgu.ac.jp

Hashimoto, Masa-aki
Department of Physics
Kyushu University
4-2-1 Ropponmatsu
Chuo, Fukuoka 810-8560
Japan
hashi@gemini.rc.kyushu-u.ac.jp

Hatano, Michio
Department of Physics
University of Tokyo
7-3-1 Hongo
Bunkyo, Tokyo 113-0033
Japan
hatano@tkynx0.phys.s.u-tokyo.ac.jp

Hayano, Ryugo
Department of Physics
University of Tokyo
7-3-1 Hongo
Bunkyo, Tokyo 113-0033
Japan
hayano@phys.s.u-tokyo.ac.jp

Hiroshige, Noboru
Department of Economics
Hannan University
5-4-33 Amami-higasi Matsubara
Osaka 580-0850
Japan
hiro@hannan-u.ac.jp

Hiyama, Emiko
Muon Siceince Laboratory
RIKEN
2-1 Hirosawa
Wako, Saitama 351-0198
Japan
hiyama@postman.riken.go.jp

Ho, Yew Kam
Institute of Atomic and Molecular Sciences
Academia Sinica
P.O.Box 23-166
Taipei, Taiwan 106
Taiwan
ykho@atom.iams.sinica.edu.tw

Horiuchi, Hisashi
Department of Physics
Kyoto University
Kitashirakawa Oiwake-cho
Sakyo, Kyoto 606-8502
Japan
horiuchi@ruby.scphys.kyoto-u.ac.jp

Howell, Calvin
Department of Physics
TUNL/Duke University
Duke University, Box 90308
Durham, NC 27708-0308
USA
howell@tunl.duke.edu

Hungerford, Ed V.
Department of Physics
University of Houston
Houston, TX 77204
USA
hunger@UH.EDU

Ikeda, Kiyomi
RIKEN
2-1 Hirosawa
Wako, Saitama 351-0198
Japan
k-ikeda@postman.riken.go.jp

Ikeda, Nobuo
Department of Physics
Kyushu University
6-10-1 Hakozaki
Higashi, Fukuoka 812-8581
Japan
ikeda@nucl.phys.kyushu-u.ac.jp

Iseri, Yasunori
Department of Physics
Chiba-Keizai College
Todoroki-cho 4-3-30
Inage, Chiba 263-0021
Japan
iseri@chiba-kc.ac.jp

Ishii, Chikara
Department of Physics
Science University of Tokyo
1-3 Kagurazaka
Shinjuku, Tokyo 162-8601
Japan
ishii@rs.kagu.sut.ac.jp

Ishikawa, Souichi
Department of Physics
Hosei University
2-17-1 Fujimi
Chiyoda, Tokyo 102-8160
Japan
ishikawa@i.hosei.ac.jp

Itagaki, Naoyuki
RIKEN
2-1 Hirosawa
Wako, Saitama 351-0198
Japan
itagaki@postman.riken.go.jp

Ito, Makoto
Department of Physics
Osaka City University
3-3-138 Sugimoto
Sumiyoshi, Osaka 558-8585
Japan
itom@ocunp.hep.osaka-cu.ac.jp

Kajino, Toshiki
NAOJ
2-21-1 Ohsawa
Mitaka, Tokyo 181-8588
Japan
kajino@ferio.mtk.nao.ac.jp

Kalantar-Nayestanaki, Nasser
Kernfysisch Versneller Instituut (KVI)
Zernikelaan 25
9747 AA Groningen
The Netherlands
nasser@kvi.nl

Kamada, Hiroyuki
Institut für Theoretische Physik II
Ruhr-Universitaet Bochum
NB6/127, Universitaetsstrasse 150
D-44780 Bochum
Germany
kamada@tp2lin48.tp2.ruhr-uni-bochum.de

Kamimura, Hiroshi
Science University of Tokyo
1-3 Kagurazaka
Shinjuku, Tokyo 162-8601
Japan
kamimura@klabsun.yy.kagu.sut.ac.jp

Kamimura, Masayasu
Department of Physics
Kyushu University
6-10-1 Hakozaki
Higashi, Fukuoka 812-8581
Japan
kami2scp@mbox.nc.kyushu-u.ac.jp

Kaneko, Tsuneo
Graduate School of Science and Technology
Niigata University
Ikarashi 2 - 8050
Niigata 950-2181
Japan
kaneko@nt.sc.niigata-u.ac.jp

Katayama, Toshiyuki
Department of Computatinal
Hokusei Gakuen University
Ohyachi W.2-3-1
Atsubetsu, Sapporo 004-8631
Japan
katayama@hokusei.ac.jp

Kato, Hiromitsu
Department of Physics
University of Tokyo
7-3-1 Hongo
Bunkyo, Tokyo 113-0033
Japan
hkato@nucl.phys.s.u-tokyo.ac.jp

Kato, Kiyoshi
Division of Physics
Graduate School of Science
Hokkaido University
Kita-10, Nishi-8
Kita, Sapporo 060-0810
Japan
kato@nucl.sci.hokudai.ac.jp

Kawai, Mitsuji
Department of Physics
Kyushu University
6-10-1 Hakozaki
Higashi, Fukuoka 812-8581
Japan
kawa2scp@mbox.nc.kyushu-u.ac.jp

Khin Swe Myint
Department of Physics
Mandalay University
Mandalay
Union of Myanmar

Kim, Yeong E.
Department of Physics
Purdue University
West Lafayette, IN 47907
USA
yekim@physics.purdue.edu

Kino, Yasushi
Department of Chemistry
Tohoku University
Aramaki-aza-Aoba
Aoba, Sendai 980-8578
Japan
kino@mail.cc.tohoku.ac.jp

Koike, Yasuro
Department of Physics
Hosei University
2-17-1 Fujimi
Chiyoda, Tokyo 102-8160
Japan
koike@i.hosei.ac.jp

Korenman, Grigori
Institute of Nuclear
Moscow State University
Institute of Nuclear Physics
Moscow State
Russia
korenman@anna19.npi.msu.su

Korobov, Vladimir
Laboratory of Nuclear Problems
Joint Institute for Nuclear Research
141980, Dubna
Russia
korobov@nu.jinr.ru

Krein, Gastao
Instituto de Fisica Teorica
Universidade Estadual
145 01405-900
Sao Paulo, SP
Brazil
gkrein@ift.unesp.br

Kubodera, Kuniharu
Department of Physics & Astronomy
University of South Carolina
Columbia, South Carolina 29208
USA
kubodera@sc.edu

Ladygin, Vladimir
Laboratory of High Energies
Joint Institute for Nuclear Researches
141980, Dubna
Moscow region
Russia
ladygin@sunhe.jinr.ru

Lodhi, M.A.K.
Department of Physics
Texas Tech University
MS 1051
Lubbock, Texas 79409
USA
b5mak@ttacs.ttu.edu

Maeda, Ichiro
Department of Physics
Osaka City University
3-3-138 Sugimoto
Sumiyoshi, Osaka 558-8585
Japan
maeda@ocunp.hep.osaka-cu.ac.jp

Maeda, Yukie
Department of Science
University of Tokyo
7-3-1 Hongo
Bunkyo, Tokyo 113-0033
Japan
yukie@tkynx0.phys.s.u-tokyo.ac.jp

Maruyama, Masahiro
Department of Physics
Tohoku University
Aramaki-aza-Aoba
Aoba, Sendai 980-8578
Japan
maruyama@nucl.phys.tohoku.ac.jp

Masui, Hiroshi
Meme media laboratory
Hokkaido University
Kita-13, Nishi-8
Kita, Sapporo 060-0810
Japan
masui@nucl.sci.hokudai.ac.jp

Matsuda, Masanori
Faculty of Integrated Arts and Sciences
Division of Materials Science
Hiroshima University
1-7-1 Kagamiyama
Higashi-Hiroshima 739-8521
Japan
masa@ipc.hiroshima-u.ac.jp

Miyagawa, Kazuya
Department of Applied Physics
Okayama University of Science
1-1 Ridai-cho
Okayama 700-0005
Japan
miyagawa@dap.ous.ac.jp

Miyazawa, Hironari
Research Institute for Integrated Science
Kanagawa University
2946 Tsuchiya
Hiratsuka 259-1293
Japan
miyazawa@info.kanagawa-u.ac.jp

Mizoi, Yutaka
Radiation Laboratory
RIKEN
2-1 Hirosawa
Wako, Saitama 351-0198
Japan
mizoi@rikaxp.riken.go.jp

Motoba, Toshio
Laboratory of Physics
Osaka Electro-Communication University
Hatsu-cho
Neyagawa, Osaka 572-8530
Japan
motoba@isc.osakac.ac.jp

Motobayashi, Tohru
Department of Physics
Rikkyo University
3 Nishi-Ikebukuro
Toshima, Tokyo 171-8501
Japan
motobaya@rikkyo.ac.jp

Mukhamedzhanov, Akram
Cyclotron Institute
Texas A&M University
College Station
Texas, 77843
USA
akram@comp.tamu.edu

Myo, Takayuki
Division of Physics
Hokkaido University
Kita-10, Nishi-8
Kita, Sapporo 060-0810
Japan
myo@nucl.sci.hokudai.ac.jp

Nagae, Tomofumi
KEK
1-1 Oho
Tsukuba, Ibaraki 305-0801
Japan
nagae@mail.kek.jp

Nagata, Junichi
Venture Business Laboratory
Hiroshima University
2-313 Kagamiyama
Higashi-Hiroshima 739-8527
Japan
nagata@vbl.hiroshima-u.ac.jp

Nakagawa, Kimiko
College of Science and Technology
Nihon University
7-24-1 Narashino-Dai
Funabashi, Chiba 274-0063
Japan
nakagawa@phys.ge.cst.nihon-u.ac.jp

Nakai, Kozi
Department of Physics
Science University of Tokyo
2641 Yamazaki
Noda, Chiba 278-8510
Japan
nakai@post.kek.jp

Nakamura, Hiroyuki
Department of Integrated Arts
Kitakyushu National College
5-20-1 Sii
Kokura-minami, Kitakyushu 802-0985
Japan
nakamura@kct.ac.jp

Nakano, Jo
Department of Science
Physics Division
University of Tokyo
3-2-1 Midori-cho
Tanashi, Tokyo 188-8501
Japan
jnakano@cns.s.u-tokyo.ac.jp

Narumi, Hajime
Department of Physics
Faculty of Science
Hiroshima University
Higashi-Hiroshima 739-8526
Japan

Nemoto, Shino
Department of Physics
Science University of Tokyo
2641 Yamazaki
Noda, Chiba 278-8510
Japan
nemoto@ph.noda.sut.ac.jp

Nemura, Hidekatsu
Graduate School of Science
Niigata University
Ikarashi 2 - 8050
Niigata 950-2181
Japan
nemura@nt.sc.niigata-u.ac.jp

Nishikawa, Tetsuji
Science University of Tokyo
1-3 Kagurazaka
Shinjuku, Tokyo 162-8601
Japan

Nomachi, Masaharu
Research Center for Nuclear Physics
Osaka University
10-1 Mihogaoka
Ibaraki, Osaka 567-0047
Japan
nomachi@rcnp.osaka-u.ac.jp

Ohnishi, Akira
Division of Physics
Graduate School of Science
Hokkaido University
Kita-10, Nishi-8
Kita, Sapporo 060-0810
Japan
ohnishi@nucl.sci.hokudai.ac.jp

Ohnishi, Tetsuya
Department of Physics
University of Tokyo
7-3-1 Hongo
Bunkyo, Tokyo 113-0033
Japan
oonishi@nucl.phys.s.u-tokyo.ac.jp

Oka, Makoto
Department of Physics
Tokyo Institute of Technology
2-12-1 Oh-Okayama
Meguro, Tokyo 152-8551
Japan
oka@th.phys.titech.ac.jp

Opper, Allena
Department of Physics and Astronomy
Ohio University
Clippinger Labs
Athens, OH 45701
USA
opper@helios.phy.ohiou.edu

Oryu, Shinsho
Department of Physics
Science University of Tokyo
2641 Yamazaki
Noda, Chiba 278-8510
Japan
oryu@ph.noda.sut.ac.jp

Phillips, Daniel
Department of Physics
University of Washington
Box 351560
WA 98195
USA
phillips@phys.washington.edu

Ponomarev, Leonid
Theory Division
RPC Kurchatov Institute
123182 Moscow, Rogova 2-9
MUCATEX
Russia
leonid@mucatex.msk.ru

Rijken, Thomas
Institute for Theoretical Physics
University of Nijmegen
Faculteit Natuurwetenschappen
Toernooiveld, Po Box 9010, Nijmegen
The Netherlands
rijken@sci.kun.nl

Sadovoy, Alexander A.
TH
Russian Federal Nuclear Center (RFNC) - VNIIEF
607200, Sarov
Nizhny Novgorod Region
Russia
ludmila@mucatex.msk.ru

Sagara, Kenshi
Department of Physics
Kyushu University
6-10-1 Hakozaki
Higashi, Fukuoka 812-8581
Japan
sagara@nucl.phys.kyushu-u.ac.jp

Saghai, Bijan
Service de Physique Nucleaire
Commissariat a l'Energie Atomique
Centre d'Etudes de Saclay
F-91191 Gif-sur-Yvette cedex
France
bsaghai@cea.fr

Saha, Arun
Thomas Jefferson National Accelerator Facility
12000 Jefferson Avenue
USA
saha@jlab.org

Sakai, Hideyuki
Department of Physics
University of Tokyo
7-3-1 Hongo
Bunkyo, Tokyo 113-0033
Japan
sakai@tkyvax.phys.s.u-tokyo.ac.jp

Sakai, Tsutomu
Research Center for Nuclear Physics
Osaka University
10-1 Mihogaoka
Ibaraki, Osaka 567-0047
Japan
tsakai@rcnp.osaka-u.ac.jp

Sakoda, Seitaro
Department of Physics
University of Tokyo
7-3-1 Hongo
Bunkyo, Tokyo 113-0033
Japan
sakoda@nucl.phys.s.u-tokyo.ac.jp

Salmè, Giovanni
Istituto Nazionale di Fisica Nucleare
Sezione di Roma 1
P.le A. Moro 2
00185 Rome
Italy
salmeg@roma1.infn.it

Sasakawa, Tatuya
4-21-2, 3-301 Matsushiro
Tsukuba 305-0035
Japan

Sauer, Peter U.
Institut für Theoretische
Universitaet Hannover
Appelstrasse 2
D-30167 Hannover
Germany
sauer@itp.uni-hannover.de

Sawado, Nobuyuki
Department of Physics
Faculty of Science and Technology
Science University of Tokyo
2641 Yamazaki
Noda, Chiba 278-8510
Japan
a6293701@rs.noda.sut.ac.jp

Schuck, Peter
CNRS
Institut des Sciences Nucleaires
3 rue Michel Ange Paris cedex
53, av des Martyrs, F-38026 Grenoble
France
schuck@isnhp4.in2p3.fr

Sekiguchi, Kimiko
Department of Physics School
University of Tokyo
7-3-1 Hongo
Bunkyo, Tokyo 113-0033
Japan
kimiko@nucl.phys.s.u-tokyo.ac.jp

Sekimoto, Michiko
Physics Division III
Institute of Particle and Nuclear Studies (IPNS)
High Energy Accelerator Research Organization (KEK)
1-1 Oho
Tsukuba, Ibaraki 305-0801
Japan
sekimoto@post.kek.jp

Shinmura, Shoji
Department of Basic Science
Gifu University
Yanagido 1-1
Gifu 501-1193
Japan
shinmura@info.gifu-u.ac.jp

Suzuki, Katsuhiko
Research Center for Nuclear Physics
Osaka University
10-1 Mihogaoka
Ibaraki, Osaka 567-0047
Japan
ksuzuki@rcnp.osaka-u.ac.jp

Suzuki, Ryusuke
Division of Physics
Hokkaido University
Kita-10, Nishi-8
Kita, Sapporo 060-0810
Japan
ryusuke@nucl.sci.hokudai.ac.jp

Suzuki, Yasuyuki
Department of Physics
Niigata University
Ikarashi 2 - 8050
Niigata 950-2181
Japan
suzuki@nt.sc.niigata-u.ac.jp

Tachikawa, Yoshiaki
Department of Physics
Kyushu University
6-10-1 Hakozaki
Higashi, Fukuoka 812-8581
Japan
kisho@kutl.kyushu-u.ac.jp

Takashina, Masaaki
Department of Physics
Osaka City University
3-3-138 Sugimoto
Sumiyoshi, Osaka 558-8585
Japan
takasina@ocunp.hep.osaka-cu.ac.jp

Takatsuka, Ttatsuyuki
Faculty of Humanities and Social Sciences
Iwate University
Morioka 020-8550
Japan
takatuka@iwate-u.ac.jp

Tamagaki, Ryozo
Professor Emeritus, Kyoto University
Maedacho 26-5, Kamitakano
Sakyo, Kyoto 606-0097
Japan
tamagaki@yukawa.kyoto-u.ac.jp

Tamii, Atsushi
Department of Physics
University of Tokyo
7-3-1 Hongo
Bunkyo, Tokyo 113-0033
Japan
tamii@phys.s.u-tokyo.ac.jp

Tamura, Hirokazu
Department of Physics
Tohoku University
Aramaki-aza-Aoba
Aoba, Sendai 980-8578
Japan
tamura@lambda.phys.tohoku.ac.jp

Tamura, Keisuke
Department of Physics
Fukui Medical University
Matsuoka 23-3
Fukui 910-1193
Japan
tamura@fukui-med.ac.jp

Tanifuji, Makoto
Department of Physics
Hosei University
2-17-1 Fujimi
Chiyoda, Tokyo 102-8160
Japan

Tanihata, Isao
RIKEN
2-1 Hirosawa
Wako, Saitama 351-0198
Japan
tanihata@rikvax.riken.go.jp

Tominaga, Kouji
Theory Division
Research Center for Nuclear Physics
Osaka University
10-1 Mihogaoka
Ibaraki, Osaka 567-0047
Japan
tomi@rcnp.osaka-u.ac.jp

Tornow, Werner
Department of Physics
TUNL/Duke University
Duke University, Box 90308
Durham, NC 27708-0308
USA
tornow@tunl.duke.edu

Ueda, Tamotsu
Faculty of Science
Ehime University
Bunkyo 2-5
Matsuyama, Ehime 790-8577
Japan
uedatam@ehimegw.dpc.ehime-u.ac.jp

Uesaka, Tomohiro
RIKEN
2-1 Hirosawa
Wako, Saitama 351-0198
Japan
uesaka@riken.go.jp

Usukura, Junko
Graduate School of Science
Niigata University
Ikarashi 2 - 8050
Niigata 950-2181
Japan
usukura@nt.sc.niigata-u.ac.jp

Uzu, Eizo
Department of Physics
Faculty of Science and Technology
Science University of Tokyo
2641 Yamazaki
Noda, Chiba 278-8510
Japan
uzu@tanashi.kek.jp

van den Brand, Johannes
Department of Physics and Astronomy
Vrije Universiteit
Room T2.69, de Boelelaan 1081
1081 hv Amsterdam
The Netherlands
jo@nikhef.nl

Wada, Masanobu
Laboratory of Physics
College of Science & Technology
Nihon University
7-24-1 Narashino-Dai
Funabashi, Chiba 274-8501
Japan
wadam@phys.ge.cst.nihon-u.ac.jp

Watanabe, Kazuyuki
Department of Physics
Science University of Tokyo
1-3 Kagurazaka
Shinjuku, Tokyo 162-8601
Japan
kazuyuki@rs.kagu.sut.ac.jp

Watanabe, Tsutomu
Division of Natural Sciences
International Christian University
3-10-2 Ohsawa
Mitaka, Tokyo 181-8585
Japan
tsutomu@icu.ac.jp

Weller, Henry R.
Triangle Universities Nuclear Laboratory
Duke University
TUNL Box 90308, Duke University
Durham, NC 27708-0308
USA
weller@tunl.duke.edu

Witała, Henryk
Institute of Physics
Jagellonian University
Reymonta 4
PL-30059 Cracow
Poland
witala@sigma.if.uj.edu.pl

Wright, Dennis
TRIUMF
4004 Wesbrook Mall
Vancouver, BC V6T 2A3
Canada
wright@triumf.ca

Yabusaki, Nozomi
Division of Physics
Hokkaido University
Kita-10, Nishi-8
Kita, Sapporo 060-0810
Japan
yabusaki@nucl.sci.hokudai.ac.jp

Yagita, Takanori
Department of Physics
Kyushu University
6-10-1 Hakozaki
Higashi, Fukuoka 812-8581
Japan
yagita@kutl.kyushu-u.ac.jp

Yako, Kentaro
Department of Physics
University of Tokyo
7-3-1 Hongo
Bunkyo, Tokyo 113-0033
Japan
yakou@nucl.phys.s.u-tokyo.ac.jp

Yamada, Taiichi
Laboratory of Physics
Kanto Gakuin University
Mutuura 4834
Kanazawa, Yokohama 236-8501
Japan
yamada@kanto-gakuin.ac.jp

Yamaguchi, Masahiro
Department of Physics
Ehime University
Bunkyo 2-5
Matsuyama, Ehime 790-8577
Japan
yamagu@qws1.sci.ehime-u.ac.jp

Yamaguchi, Yoshio
2-22-11 Hikawadai
Higashikurume 203-0004
Japan

Yamamoto, Yasuo
Physics
Tsuru University
3-8-1 Tahara
Tsuru 402-8555
Japan
yamamoto@tsuru.ac.jp

Yamamura, Hisahiko
Department of Applied Physics
Okayama University of Science
1-1 Ridai-cho
Okayama 700-0005
Japan
yamamu-s@dap.ous.ac.jp

Yang, Shin Nan
Department of Physics
National Taiwan University
Taipei 10617
Taiwan
snyang@phys.ntu.edu.tw

Yasuda, Keisuke
Research Center for Nuclear Physics
Osaka University
10-1 Mihogaoka
Ibaraki, Osaka 567-0047
Japan
yasuda@rcnp.osaka-u.ac.jp

Yoshida, Satoshi
Department of Physics
Hosei University
2-17-1 Fujimi
Chiyoda, Tokyo 102-8160
Japan
s_yoshi@i.hosei.ac.jp

Yoshino, Hiro
Faculty of Health Sciences
Hiroshima International University
Gakuendai
Kurose, Kamo 724-0695
Japan
h-yosino@hs.hirokoku-u.ac.jp

Yoshino, Yutaka
Division of Materials Science
Hiroshima University
1-7-1 Kagamiyama
Higashi-Hiroshima 739-8521
Japan
yutaka@minerva.ias.hiroshima-u.ac.jp

Zhang, Zong-Ye
Institute of High Energy Physics
Academia Sinica
P.O.Box 918
Beijing, 100039
China
zhangzy@alpha02.ihep.ac.cn

Zhou, Zilu
Laboratory for Nuclear Science
Massachusetts Institute of Technology
MIT-LNS, room 26-452
77 Massachusetts Avenue
Cambridge, MA 02139
USA
zzhou@mitlns.mit.edu

List of Authors

Springer-Verlag
and the Environment

WE AT SPRINGER-VERLAG FIRMLY BELIEVE THAT AN international science publisher has a special obligation to the environment, and our corporate policies consistently reflect this conviction.

WE ALSO EXPECT OUR BUSINESS PARTNERS – PRINTERS, paper mills, packaging manufacturers, etc. – to commit themselves to using environmentally friendly materials and production processes.

THE PAPER IN THIS BOOK IS MADE FROM NO-CHLORINE pulp and is acid free, in conformance with international standards for paper permanency.

GPSR Compliance

The European Union's (EU) General Product Safety Regulation (GPSR) is a set of rules that requires consumer products to be safe and our obligations to ensure this.

If you have any concerns about our products, you can contact us on ProductSafety@springernature.com

In case Publisher is established outside the EU, the EU authorized representative is:

Springer Nature Customer Service Center GmbH
Europaplatz 3
69115 Heidelberg, Germany

Batch number: 10732477

Printed by Printforce, the Netherlands